AF676662

Advanced Chemical Sensors for Gas Detection

Advanced Chemical Sensors for Gas Detection

Guest Editor

Shuai Chen

MDPI

Basel • Beijing • Wuhan • Barcelona • Belgrade • Novi Sad • Cluj • Manchester

Guest Editor
Shuai Chen
School of Pharmacy
Jiangxi Science & Technology
Normal University
Nanchang
China

Editorial Office
MDPI AG
Grosspeteranlage 5
4052 Basel, Switzerland

This is a reprint of the Special Issue, published open access by the journal *Chemosensors* (ISSN 2227-9040), freely accessible at: https://www.mdpi.com/journal/chemosensors/special_issues/88970HEN61.

For citation purposes, cite each article independently as indicated on the article page online and as indicated below:

Lastname, A.A.; Lastname, B.B. Article Title. *Journal Name* **Year**, *Volume Number*, Page Range.

ISBN 978-3-7258-5473-8 (Hbk)
ISBN 978-3-7258-5474-5 (PDF)
https://doi.org/10.3390/books978-3-7258-5474-5

© 2025 by the authors. Articles in this book are Open Access and distributed under the Creative Commons Attribution (CC BY) license. The book as a whole is distributed by MDPI under the terms and conditions of the Creative Commons Attribution-NonCommercial-NoDerivs (CC BY-NC-ND) license (https://creativecommons.org/licenses/by-nc-nd/4.0/).

Contents

Contents

About the Editor

Shuai Chen

Shuai Chen has been an associate professor since 2015 at Jiangxi Science & Technology Normal University (China). He earned his bachelor's degree in 2007 from Qufu Normal University (China), his master's degree in 2012 from Jiangxi Science & Technology Normal University, his PhD degree in materials physics and chemistry in 2015 from the University of Chinese Academy of Sciences, and he was a postdoctoral fellow at the University of Utah during 2018–2020. He has served as the chief editor of the Natural Science Edition of the *Journal of Jiangxi Science & Technology Normal University* since 2015. His current research emphasizes the design and fabrication of new materials and devices for bionic electronics and functional coatings, undergraduate and graduate training, and editing and journal development. He has published over 100 peer-reviewed scientific papers and chapters in academic books, and has obtained more then 10 invention patents. He has participated in many national research programs as a coordinator or scientific member, and has guided over 30 outstanding students in obtaining their master's and doctoral degrees.

Preface

This Special Issue, "Advanced Chemical Sensors for Gas Detection", includes two review papers and fifteen full articles that fall within *Chemosensors'* scope.

Fifteen research articles investigate the following applications of semiconductor materials or device techniques for chemosensors towards gaseous environment monitoring:

- $ZnSb_2O_6$ pellets for gas sensing of CO and C_3H_8;
- Modeling frequency–amplitude characteristics of a tunable SAW oscillator;
- $SnS_2/LaFeO_3$ composite for enhanced triethylamine detection;
- Smartphone technology for monitoring of CO and NO_2;
- Functionalized graphene used in an E-nose;
- 2D TiO_2@MoS_2 heterojunction nanosheets for toluene detection;
- P(V3D3-*co*-TFE) copolymer covered TiO_2 sensor for hydrogen detection;
- SnO_2-based CMOS-integrated gas sensor;
- Adsorption/combustion-type microsensors for VOCs detection;
- CRDS technology-based integrated breath sensor for acetone real-time detection;
- Nanosized MnO_2-doped rGO for gas NO_2 detection;
- Intensity-variation RI sensor for multi-variant alcohol detection;
- SnO_2-based sensors with a Pt interdigital electrodes gap for CO_2 detection;
- p-Type Cr_2O_3 and Ti-doped Cr_2O_3 thin films for isobutylene and ammonia detection;
- Low-drift NO_2 sensor based on polyaniline/black phosphorus composites.

Two review articles summarize the progress of chemosensors in gas monitoring as follows:

- Layered double hydroxide-based materials for gas and electrochemical sensing;
- Screening and diagnosis technologies towards pneumoconiosis, from imaging analysis to E-noses.

Shuai Chen
Guest Editor

Preface

This Special Issue, "Advanced Chemical Sensors for Gas Detection", includes two review papers and fifteen full articles that fall within *Chemosensors'* scope.

Fifteen research articles investigate the following applications of semiconductor materials or device techniques for chemosensors towards gaseous environment monitoring:

- $ZnSb_2O_6$ pellets for gas sensing of CO and C_3H_8;
- Modeling frequency–amplitude characteristics of a tunable SAW oscillator;
- $SnS_2/LaFeO_3$ composite for enhanced triethylamine detection;
- Smartphone technology for monitoring of CO and NO_2;
- Functionalized graphene used in an E-nose;
- 2D TiO_2@MoS_2 heterojunction nanosheets for toluene detection;
- P(V3D3-*co*-TFE) copolymer covered TiO_2 sensor for hydrogen detection;
- SnO_2-based CMOS-integrated gas sensor;
- Adsorption/combustion-type microsensors for VOCs detection;
- CRDS technology-based integrated breath sensor for acetone real-time detection;
- Nanosized MnO_2-doped rGO for gas NO_2 detection;
- Intensity-variation RI sensor for multi-variant alcohol detection;
- SnO_2-based sensors with a Pt interdigital electrodes gap for CO_2 detection;
- p-Type Cr_2O_3 and Ti-doped Cr_2O_3 thin films for isobutylene and ammonia detection;
- Low-drift NO_2 sensor based on polyaniline/black phosphorus composites.

Two review articles summarize the progress of chemosensors in gas monitoring as follows:

- Layered double hydroxide-based materials for gas and electrochemical sensing;
- Screening and diagnosis technologies towards pneumoconiosis, from imaging analysis to E-noses.

Shuai Chen
Guest Editor

chemosensors

MDPI

Editorial

Advanced Chemosensors for Gas Detection

Shuai Chen [1,2,3]

1 Jiangxi Provincial Engineering Research Center for Waterborne Coatings, School of Chemistry and Chemical Engineering, Jiangxi Science & Technology Normal University, Nanchang 330013, China; shuaichen@jxstnu.edu.cn
2 Jiangxi Provincial Key Laboratory of Flexible Electronics, Flexible Electronics Innovation Institute, Jiangxi Science & Technology Normal University, Nanchang 330013, China
3 School of Pharmacy, Jiangxi Science & Technology Normal University, Nanchang 330013, China

1. Introduction

The exploration of gas sensing technologies lies at the forefront of modern scientific and technological advancements [1–5]. Chemosensors are powerful and indispensable tools that enable the detection and quantification of chemical substances in diverse gas environments, serving as tools for environmental monitoring, industrial safety, healthcare, food sanitation supervision, and more [6–8]. The demand for advanced chemosensors capable of detecting trace gases with high sensitivity, selectivity, and real-time responsiveness is ongoing. In recent years, a large number of advanced materials and techniques have been developed in this field [9–14].

2. The Special Issue

This Special Issue comprises fifteen high-quality original research papers and two comprehensive review papers, all focused on the latest advances and innovative applications of chemosensors for gas detection. In brief, it covers novel gas-sensitive materials (inorganic, organic, hybrid, composite, etc.), together with their synthesis and characterization, advanced sensor processing technology, special architectural or theoretical design, multi-dimensional performance evaluation and optimization, and their applications or even promising prospects for the future.

On the one hand, the design of inorganic semiconductor composite material systems, especially employing metal-based oxides and heterostructure architectures [15–18], holds significant value and abundant use for the detection of gaseous analytes by chemosensors. Optimized designs based on composition, form, morphology, etc., could enhance sensor performance (such as increasing sensitivity, improving selectivity, accelerating response speed, and enhancing stability), broaden the application scope to accommodate various gaseous analytes, or meet different environmental conditions. For example, Xiaofeng Wang et al. [19] prepared a n-p type SnS_2 nanosheet/$LaFeO_3$ nanoparticle composite using a hydrothermal method together with the sol–gel technique. The optimized chemosensor can be operated at 140 °C and demonstrated strong stability, selectivity, and long-term durability for triethylamine (TEA) vapor detection. Hui Xu et al. [20] constructed a 2D core–shell TiO_2@MoS_2 composite with n-n heterostructures. This chemosensor can monitor trace toluene at 240 °C, exhibiting a superior response (Ra/Rg = 9.8 to 10 ppm), rapid response/recovery kinetics (9 s/16 s), a low detection limit (50 ppb), and excellent selectivity against interfering gases and moisture. Larissa Egger et al. [21] reported a complementary metal oxide semiconductor (CMOS)-integrated SnO_2 thin film gas chemosensor functionalized with mono-, bi-, and trimetallic (Ag-, Pd-, and Ru-) nanoparticles. It can detect carbon monoxide (CO) and a specific hydrocarbon mixture in a concentration range of 5–50 ppm. Moreover, the use of CMOS chips gives such low-power and integrated sensors a chance to be applied in cell phones, watches, etc., which can be used for air quality monitoring at different temperatures and humidity levels. Takeo Hyodo et al. [22] introduced an adsorption/combustion-type microsensor using 5 wt% Pt-loaded aluminas (α-Al_2O_3 and γ-Al_2O_3) with catalytic effect and thermal conductivity to determine the dynamic response of ethanol and toluene vapor in air. Xavier Vilanova and Eduard Llobet et al. [23] studied a selective chemosensor based on reduced graphene oxide (GO)@MnO_2 deposited on different substrates. The sensor deposited on Kapton showed the highest response of 6.6% towards 1 ppm of NO_2 under dry conditions at room temperature, while the sensor on silicon showed the highest response of 18.5% towards 50 ppm of NH_3 under 50% relative humidity (RH) at room temperature. Cristian Eugen Simion et al. [24] realized the sensitive CO_2 response under in-field-like conditions by using SnO_2-deposited Al_2O_3 substrates with Pt-based interdigital electrodes with gaps of 100 μm and 30 μm. Chris Blackman et al. [25] prepared a Ti-doped p-type Cr_2O_3 thin film sensor via an aerosol-assisted chemical vapor deposition method. It can monitor the flammable isobutylene (C_4H_8) and toxic NH_3, showing a reversible response and good sensitivity, which may be explained by the variations in the oxygen vacancy concentration. Héctor Guillén-Bonilla, et al. [26] emphasized the bifunctional use (gas sensing and photocatalytic dye degradation) of $ZnSb_2O_6$ pellets. Khursheed Ahmad and Tae Hwan Oh et al. [27] provided a special review on layered double hydroxide (LDH)-based gas and electrochemical sensors that are capable of monitoring toxic and hazardous gases/compounds due to their layered structure, larger surface area, decent conductivity, and excellent electrochemical properties.

On the other hand, the organic–inorganic composite materials combine the characteristics of organic and inorganic chemosensing materials, enabling more effective capture of gaseous analytes and generating stronger signal responses, thereby enhancing their response sensitivity [28]. Organic materials also can enhance their ability to recognize target gases through molecular design, and the combination of the organic and inorganic feature can reduce signal interference from other gases, achieving selective detection of specific analytes. Moreover, the interaction between the organic and inorganic components in the composite material can promote charge transfer and substance transport, accelerating its response speed to gaseous analyte and improving the real-time monitoring capability. By reasonably designing the composition and structure of the composite material, it can be optimized for different gaseous analytes, enabling the detection of multiple gases and meeting the requirements of different application scenarios and environmental conditions. For example, Thomas Strunskus, Franz Faupel, and Stefan Schröder et al. [29] demonstrated the optimized selective sensing performance and long-term stability (exhibiting high reliability even for more than 427 days) of TiO_2-coated ultra-thin copolymer films of poly(trivinyltrimethylcyclotrisiloxane-*co*-tetrafluoroethylene) (P(V3D3-*co*-TFE)), even in a high humid environment and under an optimized operating temperature of 300 to 350 °C. Sadam Hussain and Mujahid Mehdi et al. [30] detected various alcohol species (methanol, ethanol, propanol, butanol, and pentanol) with a twisted polymer optical fiber (POF) sensor. Such a refractive index sensor represents a significant leap forward in optical sensing technology. Yunbo Shi et al. [31] reported polyaniline (PANI)/black phosphorus (BP), which can sensitively monitor NO_2 within 2–60 ppm at room temperature.

In addition to the research on material systems, researchers are also constantly improving the design, manufacturing, and application technologies of chemosensors for gas detection. For instance, Domenico Suriano et al. [32] integrated a miniaturized chemosensor with a smartphone for mobile monitoring of the frequently changing atmospheric environment, including CO and typical indicators of air quality. The maximum response value for the indoor CO and NO_2 concentrations was 12.3 ppm and 64 ppb, respectively, while for mobile measurements, the maximum concentrations were 8.3 ppm and 38 ppb, respectively. Cristian Viespe et al. [33] proposed a model to assess the frequency–amplitude characteristics of surface acoustic wave (SAW) oscillator, aiming to increase the robustness and interpretability of the sensing behavior.

Moreover, breath analysis technology, typically electronic noses (e-noses), has advantages for obtaining key information of gaseous analytes through non-invasive, real-time, and convenient methods, with great significance in multiple aspects, including medical health, environmental monitoring, public security, and other fields [34–37]. Compared to traditional gas analysis methods, e-noses are more efficient and accurate when using sensor array technology and pattern recognition algorithms [38]. For example, Bingqiang Cao and Chenyu Jiang et al. [39] investigated an integrated breath gas detection system based on cavity ring-down spectroscopy (CRDS). The ring-down time, detection sensitivity, and stability of this system for breath acetone were 1.068 μs, 1 ppb, and 0.13%, respectively. Sonia Freddi et al. [40] fabricated graphene-based sensors functionalized with unconventional in-house synthesized zinc and copper octyl-pyrazinoporphyrazines and commercially available zinc phthalocyanine, which demonstrated excellent performance for detecting NH_3, benzene, and H_2S as a single sensor. After being assembled into an e-nose, it showed remarkable discrimination capability of single gases and their mixtures. Wufan Xuan and Shuai Chen et al. [41] summarized the progress on screening and diagnosis technologies for pneumoconiosis, known as one of the most serious global occupational diseases, emphasizing the application prospect of chemosensing strategies like e-noses compared to conventional and wide-used imaging analysis ways.

In summary, these of chemosensors can provide highly sensitive and selective response to diverse gaseous chemicals via optical, electrical, electrochemical signals, etc. With the advancement of material science, processing technology, miniaturized or integrated devices, Internet of Things (IoT), emerging artificial intelligence (AI) and machine learning (ML) concepts, etc., gas chemosensors will usher in a future of growth. This Special Issue may promote increased interest or research in related fields, bringing together emerging sensing material, device innovation, and commercial applications.

Acknowledgments: The success of this Special Issue would not have been possible without the dedicated contributions of the authors, the expertise of reviewers, and the support of the editorial team at *Chemosensors*. I extend my gratitude to all participants for their invaluable efforts in advancing the frontiers of gas chemosensing research. I also thank the funding support from the Academic Development Project of TongXin Funds (No. 2024161817).

Conflicts of Interest: There are no conflicts of interest.

References

1. Feng, S.B.; Farha, F.; Li, Q.J.; Wan, Y.L.; Xu, Y.; Zhang, T.; Ning, H.S. Review on smart gas sensing technology. *Sensors* **2019**, *19*, 3760. [CrossRef]
2. Alrammouz, R.; Podlecki, J.; Abboud, P.; Sorli, B.; Habchi, R. A review on flexible gas sensors: From materials to devices. *Sens. Actuators A Phys.* **2018**, *284*, 209–231. [CrossRef]
3. Qiu, P.L.; Liu, H.J.; Hu, C.Q.; Liu, J.Q.; Fu, C.; Qin, Y.X. Advances in memristive gas sensors: A review. *Talanta* **2025**, *293*, 128058. [CrossRef]
4. Chowdhury, A.Z.; Oehlschlaeger, M.A. Artificial intelligence in gas sensing: A review. *ACS Sens.* **2025**, *10*, 1538–1563. [CrossRef]
5. Chitaranjan, M. Recent advances in medical gas sensing with artificial intelligence-enabled technology. *Med. Gas Res.* **2025**, *15*, 318–326.

6. Humayun, M.; Bououdina, M.; Usman, M.; Khan, A.; Luo, W.; Wang, C.D. Designing state-of-the-art gas sensors: From fundamentals to applications. *Chem. Rec.* **2024**, *24*, e202300350. [CrossRef] [PubMed]
7. Kou, Y.J.; Hua, L.; Chen, W.J.; Xu, X.X.; Song, L.Z.; Yu, S.J.; Lu, Z.Q. Material design and application progress of flexible chemiresistive gas sensors. *J. Mater. Chem. A* **2024**, *12*, 21583–21604. [CrossRef]
8. Kanan, S.; Obeideen, K.; Moyet, M.; Abed, H.; Khan, D.; Shabnam, A.; El-Sayed, Y.; Arooj, M.; Mohamed, A.A. Recent advances on metal oxide based sensors for environmental gas pollutants detection. *Crit. Rev. Anal. Chem.* **2025**, *55*, 911–944. [CrossRef]
9. Ahmad, K.; Oh, T.H. Recent progress in MOFs and MOF-derived materials for gas sensing applications. *Chemosensors* **2025**, *13*, 100. [CrossRef]
10. Prakash, C.; Yadav, A.K.; Sharma, M.; Singh, V.K.; Dixit, A. Recent developments on 2D-materials for gas sensing application. *J. Phys. Condens. Matter* **2025**, *37*, 193004. [CrossRef] [PubMed]
11. Yang, J.F.; Sun, R.J.; Bao, X.; Liu, J.J.; Ng, J.W.; Tang, B.J.; Liu, Z. Enhancing selectivity of two-dimensional materials-based gas sensors. *Adv. Funct. Mater.* **2025**, *35*, 2420393. [CrossRef]
12. Harun-Or-Rashid, M.; Mirzaei, S.; Nasiri, N. Nanomaterial innovations and machine learning in gas sensing technologies for real-time health diagnostics. *ACS Sens.* **2025**, *10*, 1620–1640. [CrossRef] [PubMed]
13. Yu, S.W.; Jia, X.H.; Zhang, J.X.; Yang, W.C.; Song, H.J. Recent advances in different materials for moisture resistance of metal oxide-based gas sensors: A review. *Chem. Eng. J.* **2025**, *505*, 159639. [CrossRef]
14. Vaishag, P.V.; Noh, J.S. A comparative review of graphene and MXene-based composites towards gas sensing. *Molecules* **2024**, *29*, 4558. [CrossRef] [PubMed]
15. Meng, F.J.; Xin, R.F.; Li, S.X. Metal oxide heterostructures for improving gas sensing properties: A review. *Materials* **2023**, *16*, 263. [CrossRef]
16. Liu, Y.N.; Xiao, S.; Du, K. Chemiresistive gas sensors based on hollow heterojunction: A review. *Adv. Mater. Interfaces* **2021**, *8*, 2002122. [CrossRef]
17. Zhao, J.H.; Wang, H.Q.; Cai, Y.H.; Zhao, J.J.; Gao, Z.D.; Song, Y.Y. The challenges and opportunities for TiO_2 nanostructures in gas sensing. *ACS Sens.* **2024**, *9*, 1644–1655. [CrossRef]
18. Panigrahi, P.K.; Chandu, B.; Puvvada, N. Recent advances in nanostructured materials for application as gas sensors. *ACS Omega* **2024**, *9*, 3092–3122. [CrossRef]
19. Wu, H.; Wang, X.B.; Chen, Y.X.; Wang, X.F. Enhanced triethylamine-sensing characteristics of $SnS_2/LaFeO_3$ composite. *Chemosensors* **2025**, *13*, 228. [CrossRef]
20. Wang, D.H.; Hu, J.W.; Xu, H.; Wang, D.; Li, G.S. Construction of 2D TiO_2@MoS_2 heterojunction nanosheets for efficient toluene gas detection. *Chemosensors* **2025**, *13*, 154. [CrossRef]
21. Egger, L.; Sosada-Ludwikowska, F.; Steinhauer, S.; Singh, V.; Grammatikopoulos, P.; Köck, A. SnO_2-based CMOS-integrated gas sensor optimized by mono-, bi-, and trimetallic nanoparticles. *Chemosensors* **2025**, *13*, 59. [CrossRef]
22. Hyodo, T.; Matsuura, Y.; Inao, G.; Sasahara, T.; Shimizu, Y.; Ueda, T. Effects of base materials (α-alumina and/or γ-alumina) on volatile organic compounds (VOCs)-sensing properties of adsorption/combustion-type microsensors. *Chemosensors* **2025**, *13*, 9. [CrossRef]
23. Alouani, M.A.; Casanova-Chafer, J.; de Bernardi-Martín, S.; García-Gómez, A.; Salehnia, F.; Santos-Ceballos, J.C.; Santos-Betancourt, A.; Vilanova, X.; Llobet, E. The effect of doping rGO with nanosized MnO_2 on its gas sensing properties. *Chemosensors* **2024**, *12*, 256. [CrossRef]
24. Stanoiu, A.; Iacoban, A.C.; Mihalcea, C.G.; Dinu, I.V.; Florea, O.G.; Vlaicu, I.D.; Simion, C.E. CO_2 interaction mechanism of SnO_2-based sensors with respect to the Pt interdigital electrodes gap. *Chemosensors* **2024**, *12*, 238. [CrossRef]
25. Zhou, P.; Tsang, J.-H.; Blackman, C.; Shen, Y.; Liang, J.; Covington, J.A.; Saffell, J.; Danesh, E. A novel mechanism based on oxygen vacancies to describe isobutylene and ammonia sensing of p-type Cr_2O_3 and Ti-doped Cr_2O_3 thin films. *Chemosensors* **2024**, *12*, 218. [CrossRef]
26. Jacob, M.-B.; Héctor, G.-B.; Lucia, I.J.-A.; Alex, G.-B.; Verónica-María, R.-B.; Jorge, A.R.-O.; José, T.G.-B.; María de la, L.O.-A. Synthesis, characterization, and evaluation of photocatalytic and gas sensing properties of $ZnSb_2O_6$ pellets. *Chemosensors* **2025**, *13*, 329.
27. Raza, W.; Ahmad, K.; Oh, T.H. Progress in layered double hydroxide-based materials for gas and electrochemical sensing applications. *Chemosensors* **2025**, *13*, 115. [CrossRef]
28. Dong, R.Q.; Yang, M.N.; Zuo, Y.X.; Liang, L.S.; Xing, H.K.; Duan, X.M.; Chen, S. Conducting polymers-based gas sensors: Principles, materials, and applications. *Sensors* **2025**, *25*, 2724. [CrossRef] [PubMed]
29. Brinza, M.; Schwäke, L.; Zimoch, L.; Strunskus, T.; Pauporté, T.; Viana, B.; Ameri, T.; Adelung, R.; Faupel, F.; Schröder, S.; et al. Influence of P(V3D3-*co*-TFE) copolymer coverage on hydrogen detection performance of a TiO_2 sensor at different relative humidity for industrial and biomedical applications. *Chemosensors* **2025**, *13*, 150. [CrossRef]

30. Ghaffar, A.; Mehdi, R.; Mehdi, I.; Das, B.; Kumar, V.; Hussain, S.; Sher, G.; Memon, K.A.; Ali, S.; Mehdi, M.; et al. An intensity-variation RI sensor for multi-variant alcohol detection with twisted structure using polymer optical fiber. *Chemosensors* **2024**, *12*, 252. [CrossRef]
31. Tang, B.L.; Shi, Y.B.; Liu, J.J.; Zheng, C.D.; Zhao, K.; Zhang, J.H.; Feng, Q.H. Low-drift NO_2 sensor based on polyaniline/black phosphorus composites at room temperature. *Chemosensors* **2024**, *12*, 181. [CrossRef]
32. Suriano, D.; Abulude, F.O.; Penza, M. The use of low-cost gas sensors for air quality monitoring with smartphone technology: A preliminary study. *Chemosensors* **2025**, *13*, 189. [CrossRef]
33. Nicolae, I.; Viespe, C. Modeling the frequency–amplitude characteristics of a tunable SAW oscillator. *Chemosensors* **2025**, *13*, 240. [CrossRef]
34. Verma, G.; Gupta, A. Next-generation chemiresistive wearable breath sensors for non-invasive healthcare monitoring: Advances in composite and hybrid materials. *Small* **2025**, *21*, 2411495. [CrossRef]
35. Alenezy, E.K.; Kandjani, A.E.; Shaibani, M.; Trinchi, A.; Bhargava, S.K.; Ippolito, S.J.; Sabri, Y. Human breath analysis; clinical application and measurement: An overview. *Biosens. Bioelectron.* **2025**, *278*, 117094. [CrossRef] [PubMed]
36. Heng, W.Z.; Yin, S.K.; Chen, Y.L.; Gao, W. Exhaled breath analysis: From laboratory test to wearable sensing. *IEEE Rev. Biomed. Eng.* **2024**, *18*, 50–73. [CrossRef]
37. Yuan, R.; Yang, Y.X.; Zou, B.S.; Zhang, Y.B. MXene-enabled gas sensors for wearable breath monitoring. *Chem. Eng. J.* **2025**, *510*, 161414. [CrossRef]
38. Chen, H.; Huo, D.X.; Zhang, J.L. Gas recognition in e-nose system: A review. *IEEE Trans. Biomed. Circuits Syst.* **2022**, *16*, 169–184. [CrossRef]
39. Sun, J.; Shi, D.X.; Wang, L.; Yu, X.L.; Song, B.H.; Li, W.X.; Zhu, J.K.; Yang, Y.; Cao, B.Q.; Jiang, C.Y. CRDS technology-based integrated breath gas detection system for breath acetone real-time accurate detection application. *Chemosensors* **2024**, *12*, 261. [CrossRef]
40. Freddi, S.; Vaghi, L.; Penoni, A.; Scapinello, L.; Sangaletti, L. A minimal electronic nose based on graphene functionalized with metalated pyrazinoporphyrazines and phthalocyanines for ammonia, benzene, and hydrogen sulfide discrimination. *Chemosensors* **2025**, *13*, 165. [CrossRef]
41. Zhang, Y.Q.; Xuan, W.F.; Chen, S.; Yang, M.N.; Xing, H.K. The screening and diagnosis technologies towards pneumoconiosis: From imaging analysis to e-noses. *Chemosensors* **2025**, *13*, 102. [CrossRef]

Disclaimer/Publisher's Note: The statements, opinions and data contained in all publications are solely those of the individual author(s) and contributor(s) and not of MDPI and/or the editor(s). MDPI and/or the editor(s) disclaim responsibility for any injury to people or property resulting from any ideas, methods, instructions or products referred to in the content.

chemosensors

MDPI

Article

Low-Drift NO_2 Sensor Based on Polyaniline/Black Phosphorus Composites at Room Temperature

Bolun Tang [1,2], Yunbo Shi [1,2,*], Jijiang Liu [3], Canda Zheng [1,2], Kuo Zhao [1,2], Jianhua Zhang [1,2] and Qiaohua Feng [1,2]

1 School of Measurement and Control Technology and Communication Engineering, Harbin University of Science and Technology, Harbin 150080, China; tangbolun@sina.com (B.T.); zhengcanda@outlook.com (C.Z.); iridescen7@163.com (K.Z.); zhangjianhua0919@163.com (J.Z.); fengqiaohua80@126.com (Q.F.)

2 China Higher Educational Key Laboratory for Measuring & Control Technology and Instrumentations of Heilongjiang Province, Harbin 150080, China

3 The 49th Research Institute of China Electronics Technology Group, Harbin 150001, China; liujijianga@163.com

* Correspondence: shiyunbo@hrbust.edu.cn

Abstract: In this paper, a room-temperature NO_2 sensor based on a polyaniline (PANI)/black phosphorus (BP) composite material was proposed to solve the power consumption problem of traditional metal-oxide sensors operating at high temperatures. PANI was synthesized by chemical oxidative polymerization, whereas BP was synthesized by low-pressure mineralization. The PANI/BP composite materials were prepared via ultrasonic exfoliation and mixing. Various characterization techniques, including scanning electron microscope (SEM), X-ray diffraction (XRD), Fourier-transform infrared spectroscopy (FTIR), and X-ray photoelectron spectroscopy (XPS), confirmed the successful preparation of the PANI/BP composites and their excellent structural properties. The sensor demonstrated outstanding gas sensitivity in the NO_2 concentration range of 2–60 ppm. In particular, the sensor showed a response exceeding 2200% at 60 ppm NO_2 concentration when using a 1:1 mass ratio of PANI to BP in the composite material.

Keywords: gas sensors; polyaniline; black phosphorus; room temperature; nitrogen dioxide sensor

1. Introduction

With the rapid industrialization and urbanization of modern society, air pollution has become increasingly serious [1]. Nitrogen dioxide (NO_2) is a major air pollutant that significantly affects human health and the environment. NO_2 is commonly produced by anthropogenic processes such as vehicle exhausts and fossil fuel combustion. Long-term exposure to high NO_2 increases the risk of asthma and other respiratory diseases [2]. In addition, NO_2 is a key factor in the formation of acid rain and photochemical smog. These environmental problems can damage crops, forests, and aquatic ecosystems [3]. Therefore, the detection of NO_2 in air quality warnings is urgently required. Currently, the detection of NO_2 is mainly based on metal-oxide semiconductor (MOS) gas sensors. Although MOS gas sensors have the advantages of high responses and low detection limits [4], they are usually required to operate at temperatures above 300 °C. The following is a list of some common MOS resistive gas sensors operating at 100–400 °C power: 0.3–0.8 W of ceramic tube sensors with Cr–Ni resistor wires [5], 0.5–1.2 W of micro-electromechanical system (MEMS) ceramic chips [6], and 0.03–0.5 W of MEMS silicon chips [7]. With the development of Internet of Things technology, gas sensors, as an important part of the sensing layer, suffer from high power consumption, which is not conducive to their application in battery-driven, low-power end devices [8]. The problem of the high power consumption of gas sensors can be solved effectively by developing gas sensors that operate at room temperature [9].

Conductive polymers (CPs) such as polythiophene, polypyrrole, and PANI have been shown to be effective gas-sensitive materials [10]. PANI, in particular, not only has the advantages of high response and low detection limit but can also realize room-temperature detection [11]. However, PANI has disadvantages such as poor selectivity, repeatability, and zero drift [12]. Researchers have used different materials to modify PANI and improve its gas sensitivity. For example, RK Sonker et al. prepared α-Fe_2O_3-PANI thin films on glass substrates by spin-coating, achieving a high response to 20 ppm NO_2 [13]. Khalifa et al. prepared a flexible high-sensitivity gas sensor for PVDF/PANI/g-C_3N_4 nanocomposites using electrospinning technology, which showed a high response to NO_2 (~92% at 108 ppm) [14]. Drift is a serious problem in the operation of gas sensors. In sensors working at room temperature, the material is at a relatively low temperature, the gas adsorption and desorption process lacks external energy, and there is slower molecular thermal motion. Xia Zhao et al. proposed a drift compensation supervised learning algorithm based on a multi-classifier ensemble, which integrates drift compensation into the classification process with the motivation of improving the classification performance of drift compensation [15]. Hang Liu et al. proposed a new two-dimensional classifier integration strategy to solve the drift problem of metal-oxide gas sensors [16]. Tao Liu et al. proposed a dynamic method called AL-ACR, which uses active learning technology to evenly collect instances from different categories online to effectively deal with the gas sensor drift problem and thus improve the performance of the electronic nose system [17]. Although algorithms can improve the sensor's drift issue to some extent, material optimization, compared to algorithmic methods, can fundamentally reduce drift and provide more robust long-term performance.

Recently, BP has attracted considerable attention from the academic community. BP is the most stable allotrope of phosphorus with a unique folded hexagonal layered structure. Each layer of BP has two crystal orientations, "armchair" (x-direction) and "sawtooth" (y-direction) [18], and can be mechanically exfoliated in a similar way to 2D materials such as graphene [19]. Researchers have used BP in applications such as light absorbers, modulators, detectors, field-effect transistors, memory devices, sensors, and energy devices [20]. Modification and doping are mostly used to enhance the performance of BP in various fields [21]. In the field of gas sensing, Kou et al. studied the adsorption of different gas molecules onto a phosphene monolayer using first-principles calculations. They predicted that the superior sensing performance of phosphene would be comparable to that of other two-dimensional materials such as graphene and MoS_2 [22]. Abbas et al. used field-effect transistors based on multilayer BP for NO_2 detection, and the chemical sensing showed higher conductivity [23]. Valt et al. prepared Ni-modified BP thin films to improve the stability and functionality of BP for gas sensing [24]. Sajedi–Moghaddam et al. reported a PANI/BP hybrid material for supercapacitor applications [25].

In summary, to solve the problem of the high power consumption of gas sensors, PANI, a conductive polymer operating at room temperature, can be used as a gas-sensitive material and compounded with BP to overcome the disadvantages of polymer zero drift. The preparation of monolayer BP is extremely difficult, and it is easy to oxidize, whereas the performance of thick-film BP is poor. The incorporation of PANI enhanced the gas sensitivity of the thick-film BP. In this study, PANI was prepared by chemical oxidative polymerization. BP crystals were prepared using a low-pressure mineralization method. The bulk BP crystals were then stripped using ultrasound and mixed with PANI powder. The PANI/BP powder was mixed with terpinol to prepare a slurry, which was dripped onto an interdigital electrode to prepare the gas sensor. The structure and morphology of PANI/BP were characterized using field-emission scanning electron microscopy (SEM), X-ray diffractometry (XRD), Fourier-transform infrared spectroscopy (FTIR), and X-ray photoelectron spectroscopy (XPS). The performance of the sensor was tested for exposure to the range of 10–60 ppm NO_2, and the gas-sensing mechanism of the sensor was discussed. The results showed that the gas-sensitizing properties of the PANI/BP composites were superior to those of thick BP films and pure PANI. The addition of BP substantially improved the zero drift of PANI during the continuous measurement of NO_2.

2. Materials and Methods

2.1. Synthesis of PANI

Aniline (analytically pure) was distilled and purified at 200 °C and stored at a low temperature for later use; other chemicals were used as received without further purification. Firstly, 5.02 g of ammonium persulfate (APS, $(NH_4)_2S_2O_8$) was weighed and dissolved in 15 mL of deionized water to configure a solution. Then, 2 mL of purified aniline was slowly dripped into 50 mL of 1 M HCl under an ice bath (0 °C) and with continuous stirring. The APS solution was slowly added to a mixture of aniline and HCl and stirred continuously for 10 h in an ice bath. After the dark green sample was obtained by filtration, it was alternately washed three times with deionized water and anhydrous ethanol. The sample was dried under a vacuum at 80 °C for 24 h to obtain the PANI sample. The preparation process is shown in Figure 1a. Aniline, ammonium persulfate and HCl were supplied by Sinopharm Chemical Reagent Co., Ltd. (Beijing, China).

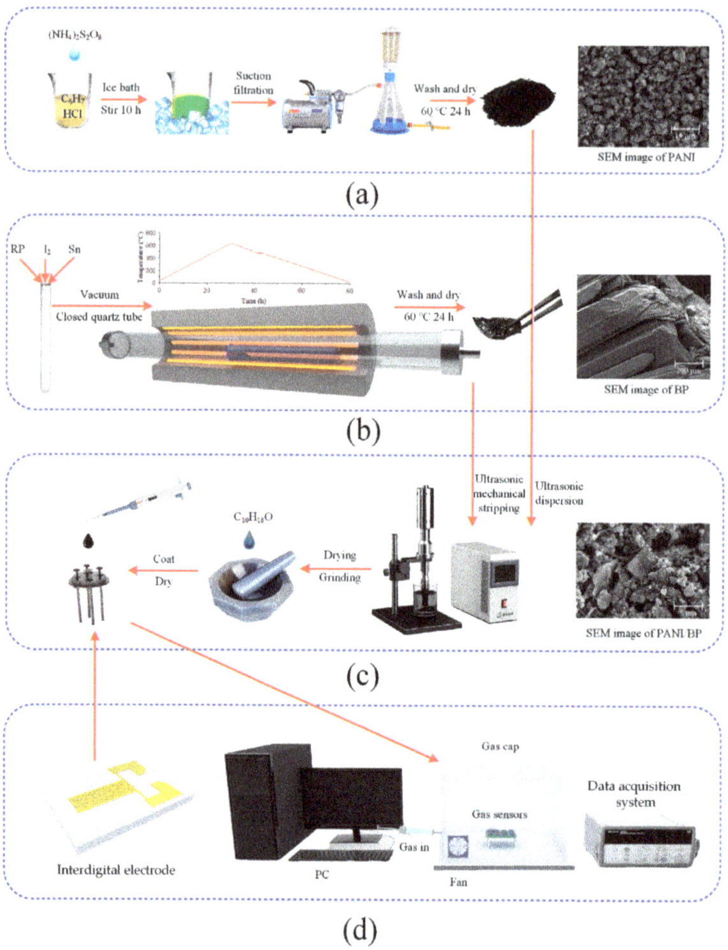

Figure 1. The synthesis processes and SEM images of (**a**) PANI, (**b**) BP, and (**c**) PANI/BP; a (**d**) schematic diagram of the interdigital electrode and test environment.

2.2. Synthesis of BP

BP can be synthesized by high-pressure bismuth melting and mineralization. Considering the advantages and disadvantages of the various methods, low-pressure mineralization was used in this study. Dry red phosphorus (500 mg; BASF Chemical Co., Ltd., Ludwigshafen, Germany), tin powder (100 mg; Sinopharm Chemical Reagent Co., Ltd, Beijing, China), and iodine (10 mg; Fuchen Chemical Reagent Co., Ltd, Tianjin, China) were placed in quartz glass test tubes. The quartz tube was then evacuated while the quartz tube was closed using a high-temperature hydroxide flame gun. Finally, the quartz tube was placed in a tube furnace, such that the raw material was at the highest temperature point at the center of the furnace. The heating process was divided into four stages: (1) 30 h to increase the temperature to 650 °C, (2) a holding time of 2 h, (3) 10 h to reduce the temperature to 460 °C, and (4) cooling to room temperature after a holding time of 2 h. The preparation process is shown in Figure 1b.

2.3. Preparation of Composite Materials and Sensors

The prepared BP samples exhibited black crystal and flake shapes after grinding and crushing. BP and PANI powders were prepared as two samples with mass ratios of 2:1, 1:1, and 1:2, respectively. After simple grinding, BP was placed in deionized water and broken into a few layers of phosphorene (10 W, 30 min) using an ultrasonic cell crusher. For each 10 s of ultrasonic treatment, cooling was suspended for 10 s to prevent the water temperature from becoming too high. PANI was then added and sonicated for 3–5 min to homogenize the PANI/BP mixture. The synthetic process is illustrated in Figure 1c. Finally, the PANI/BP mixed solution was dried at 60 °C, and a suitable amount of terpineol was added. After grinding, it was dropped onto an interdigital electrode (Xinyun Nanotechnology Co., Ltd, Suzhou, China). The interdigital electrode, depicted in Figure 1d, had dimensions of 2 mm × 4 mm and is made of alumina ceramic. The surface electrode was coated with gold (Au). The gap between adjacent electrode fingers is 50 μm.

2.4. Characterization

The micromorphologies of the samples were observed using field-emission scanning electron microscopy (ZEISS Gemini SEM 300, Carl Zeiss AG, Oberkochen, Germany) at an accelerating voltage of 3–15 kV. The crystal structures of the samples were analyzed using Cu Kα radiation from an X-ray diffractometer (SmartLab, Rigaku Corporation, Tokyo, Japan) in the scanning range of 20–70°. Fourier-transform infrared spectroscopy (Nexus 470, Nicolet, Madison, WI, USA) was used to analyze the functional groups and molecular structures of the samples in the range of 400–2000 cm^{-1}. An X-ray spectrometer (ESCALAB 250 Xi, Thermo Fisher Scientific, Waltham, MA, USA) was used to analyze the chemical compositions and oxidation states of the samples. A thermogravimetric analyzer (ZCT-B, Jingyi Gaoke, Beijing, China) was used to analyze the thermal stabilities and compositions of the samples.

2.5. Test Environment

The test environment consisted of a 1 L acrylic enclosed gas chamber, a data acquisition system (34901A, Agilent Technologies Inc., Santa Clara, CA, USA), and a computer terminal. The resistance of the sensor was altered by the gas to be measured. The resistance value was collected using a data acquisition card and uploaded to a computer in real time. A static measurement method was used for testing. A microsyringe was used to measure the volume of pure gas, which was injected into the gas chamber for gas distribution. For example, to test a 10 ppm gas, 10 μL of pure gas was measured and injected into the gas chamber. The test environment is shown in Figure 1d. During the test, the response value ($S\%$) was calculated using Equation (1):

$$S\% = \frac{\Delta R}{R_{gas}} \times 100\% \tag{1}$$

where ΔR is the difference between the resistance of the sensor in air and that of the measured gas, and R_{gas} is the resistance of the sensor to gas. The response–recovery time of the sensor was defined as the time required for a 90% change in the response value.

3. Results and Discussion

3.1. Characterization of Material

Figure 2a shows an SEM image of the PANI, which has diameter of approximately 5–10 μm and is uniformly accumulated to form a loose and porous granular structure. Compared to other polymers with dense surfaces without gaps, PANI has a larger specific surface area. Thus, there are more sites at which gas adsorption can be measured. Figure 2b shows the SEM image of the PANI/BP composites. Owing to ultrasonic crushing, the PANI and BP sheets were evenly mixed. The particle size of PANI in the composite material is approximately 3–5 μm, which is smaller than that of pure PANI. BP treated with an ultrasonic cell crusher appears as a sheet with a length and width of approximately 5–20 μm and a thickness of several nanometers. The BP sheets and PANI particles were tightly interwoven. The BP sheets acted as skeletons to support the larger pores, and the PANI nanoparticles acted as binders, which together created more sites for gas adsorption. Figure 2c, show the total distribution spectra of the composite materials. Figure 2d–g show the energy-dispersive X-ray spectroscopy (EDS) mappings of the composite materials, showing the distributions of C, Cl, N, and P, respectively. X-rays in EDS penetrate the surface of the composite material by approximately 1–5 μm. Limited by the ultrathin lamellar structure of the BP material, it can only show a rough distribution of P. The distributions of C and N in the composite were consistent with the distribution of the particles in the SEM images. The Cl originated from hydrochloric acid during the preparation process. Its distribution was approximately the same as that of the PANI particles.

Figure 3a shows the XRD patterns of the PANI/BP, pure PANI, and BP crystals and the BP standard PDF cards (00-057-0507). The main peaks of the composite correspond to the (020), (111), and (060) planes of the BP crystal. In addition, ultrasonic mechanical stripping expose more BP crystal faces, including (021), (040), (042), and (151). The positions of these peaks are one-to-one, corresponding to the peaks in the standard card (PDF#00-057-0507). This indicates that BP is perfectly stripped and does not chemically react with PANI to form other unknown crystals. Pure PANI has a broad peak between 2θ = 15 and 35°, with peaks at 2θ = 25° and 29°, which correspond to the only two peaks, 2θ = 18.57° and 25.80, in the PANI standard card (PDF#00-047-2481). The polymer is amorphous and does not exhibit a crystalline structure. Compared to the XRD pattern of pure PANI, the PANI in the PANI/BP composite is well preserved. The main functional groups in the prepared materials are determined using FTIR spectroscopy, as shown in Figure 3b. The main characteristic peaks of PANI are consistent with those reported previously. The peak at 3477 cm^{-1} is attributed to the stretching vibration of the N–H band. The peaks at 1382 cm^{-1} and 1347 cm^{-1} are related to the C-N stretching mode of the benzene ring unit. The FTIR spectrum of the PANI/BP composite is consistent with that of pure PANI; however, there is still a slight deviation. The differences at 1646 cm^{-1} and 1581 cm^{-1} indicate that the addition of BP affects some chemical bonds in PANI.

The PANI/BP composites are further characterized using XPS. Figure 4a shows the C 1s spectrum. The main peak at 285 eV is further deconvoluted into three peaks at 284.53, 285.86, and 286.98 eV. These correspond to the C-H, C=N, and C=N+ bonds, respectively. Figure 4b shows the N 1s spectrum. The main peak at 399.6 eV is deconvoluted into three peaks at 399.13, 400.03, and 401.48 eV. Among them, 399.13 eV corresponds to quinoid imine (=N−), 400.03 eV corresponds to protonated amine (−N+), and 401.48 eV corresponds to protonated imine (=N+). Figure 4c shows the P 2p spectrum. The peaks at 130.28 eV and 131.18 eV correspond to P 2p3/2 and P 2p1/2, respectively (spin–orbit splitting of ~0.86 eV). The broad peak at 134.33 eV originates from residual phosphorus oxides, usually in the form of metaphosphate (-PO3). This indicates that BP is unavoidably oxidized by exposure to oxygen in air during stripping in water. Figure 4d shows the

TG-DTA image of PANI. Changes in the TG curve before 140 °C usually result from evaporation of water. At 300–600 °C, PANI is decomposed by thermal decomposition, resulting in obvious weightlessness. DTA indicates that the thermal decomposition of PANI is accompanied by heat absorption. Figure 4d shows the TG-DTA image of PANI/BP. The thermal decomposition temperature of the composite material is significantly increased from 300 °C to 400 °C in comparison with pure PANI.

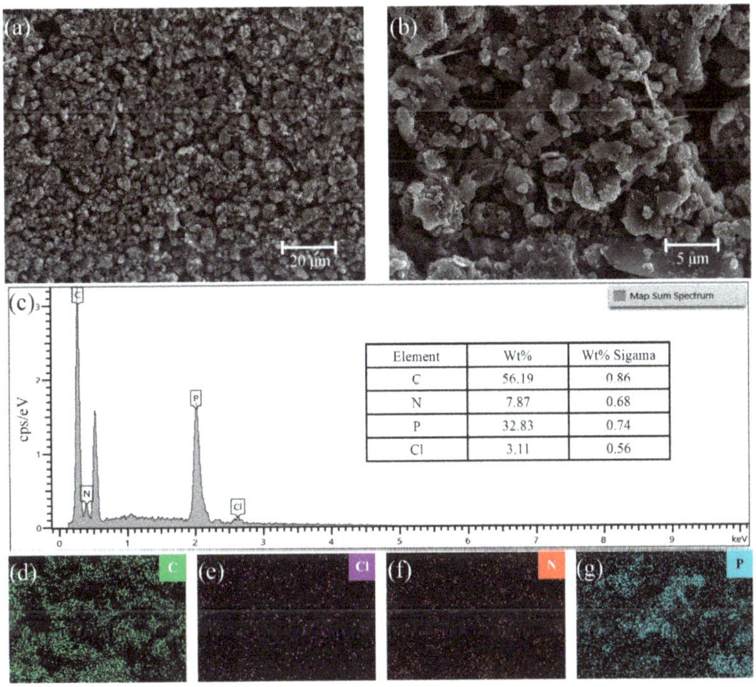

Figure 2. SEM images of (**a**) PANI and (**b**) PANI/BP, and (**c**) EDS spectrum of PANI/BP composite; EDS elemental mapping profiles of (**d**) C, (**e**) Cl, (**f**) N, and (**g**) P in PANI/BP composite.

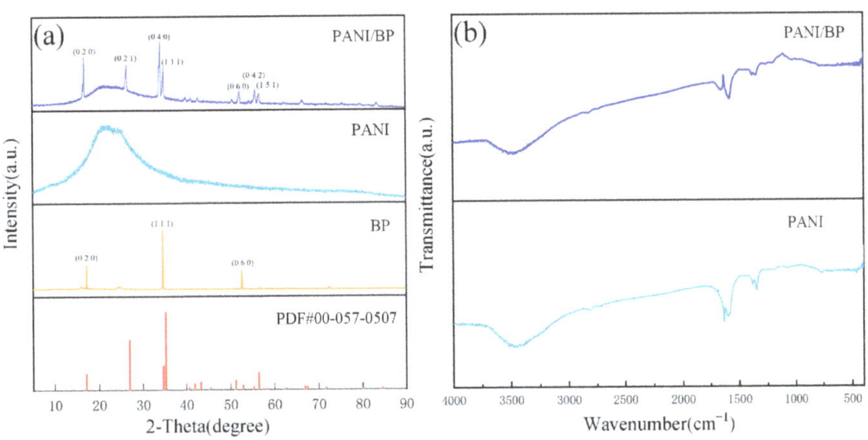

Figure 3. (**a**) XRD patterns of PANI/BP, pure PANI, BP, and the standard card of BP; (**b**) FTIR spectra of PANI/BP and PANI.

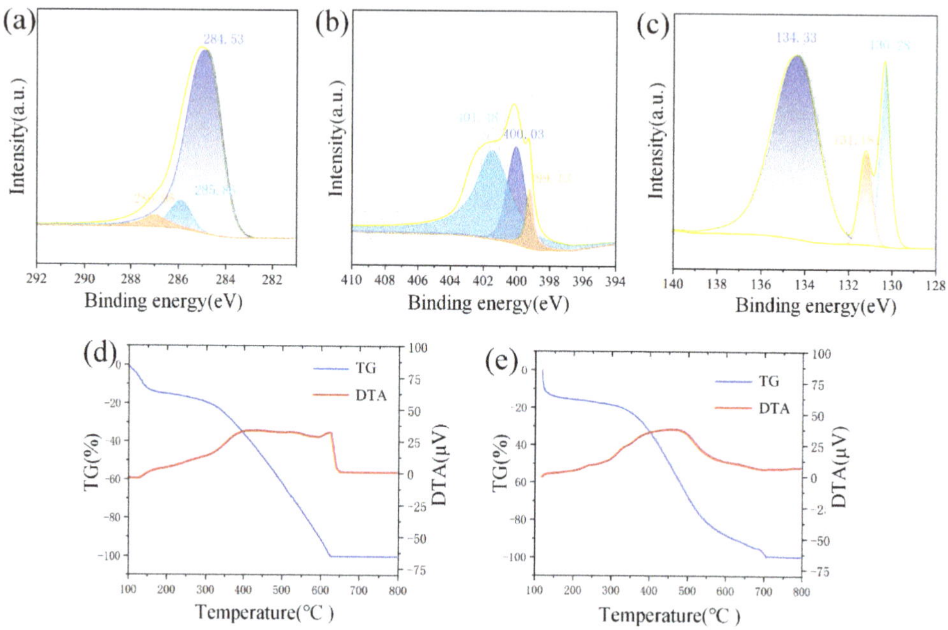

Figure 4. (**a**) C 1s, (**b**) N 1s, and (**c**) P 2p XPS spectra of PANI/BP; TG-TDA curves of (**d**) PANI and (**e**) PANI/BP.

3.2. Gas-Sensitive Performance Testing

To investigate their gas-sensitizing properties, BP, PANI/BP-2:1, PANI/BP-1:1, PANI/BP-1:2, and pure PANI were tested at NO_2 gas concentrations ranging from 2 to 60 ppm. Figure 5a shows the responses of the five sensors at 10, 20, 30, 40, 50, and 60 ppm NO_2 concentrations, respectively. The responses of all five sensors gradually increased with increasing gas concentration. The blue line represents the change in the response of the pure PANI sensor with time. Although pure PANI exhibited the highest response, a significant zero drift was observed, similar to the results of previous studies. Severe drift caused the sensor resistance to increase and eventually exceed the upper limit of the data acquisition system. The response of the pure BP sensors was unsatisfactory, and the response of the PANI/BP composite material was the best. With PANI and BP at a mass ratio of 1:1, the sensor exhibited more than 2200% response at 60 ppm NO_2, which was better than that at a mass ratio of 1:2. When the mass ratio of PANI to BP was 2:1, a significant zero-point drift occurred, similar to that observed in pure PANI. In addition, the PANI/BP sensor improved the severe zero drift of the pure PANI sensor and the response of the pure BP sensor. Figure 5b shows the responses of the four sensors to 2–8 ppm of NO_2. The response trend was similar to that observed at high concentrations. PANI/BP achieved even higher responses at low concentrations than pure PANI. Figure 5c shows a line graph of the response of the four sensors as a function of the NO_2 gas concentration. A binomial fitting was used to fit the PANI/BP (1:1) response curve at $y = 0.13449x + 0.00405x^2 - 0.57552$ with a coefficient of determination $R^2 = 0.998$. This indicates that the sensor had good calibratability. Figure 5d shows the change in resistance over time for five different sensors exposed to 30 ppm NO_2. The initial resistance of the BP sensor was the lowest, while the PANI sensor had the highest resistance. As the doping concentration of BP increased, the resistance gradually decreased. It can be observed that the resistance decreased as the gas is introduced, and after introducing fresh air into the chamber at 400 s, the sensor's resistance gradually recovered.

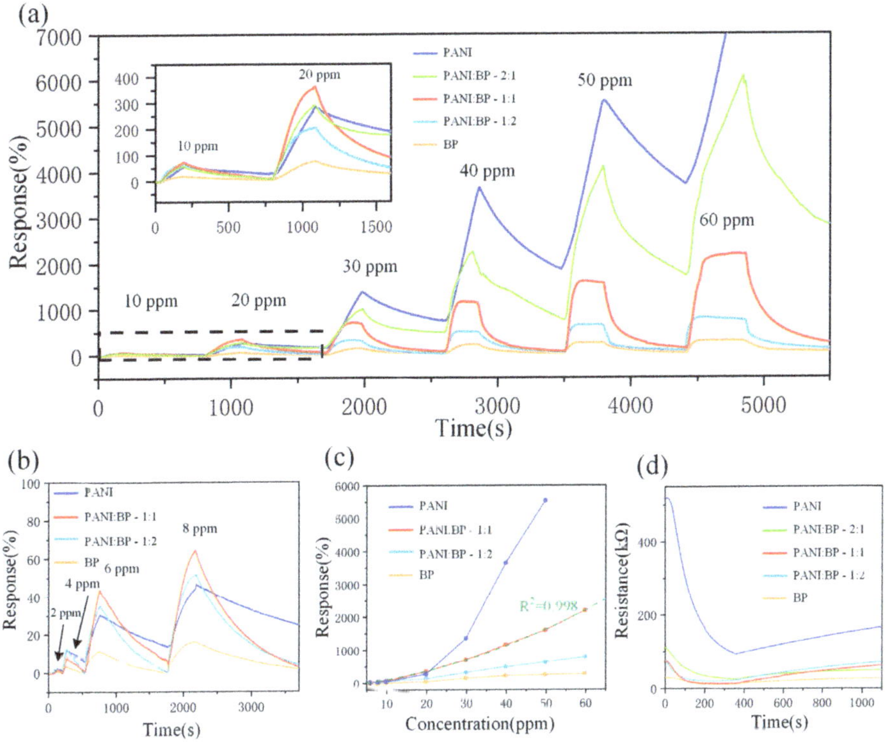

Figure 5. (**a**) Responses of PANI, PANI/BP-2:1, PANI/BP-1:1, PANI/BP-1:2, and BP gas sensors at 10–60 ppm; (**b**) responses of PANI, PANI/BP-1:1 PANI/BP-1:2, and BP gas sensors at 2–8 ppm concentrations of NO_2 over time; (**c**) responses of four sensors at different NO_2 concentrations, (**d**) resistance values of five sensors as a function of time in 30 ppm NO_2.

Performance metrics for evaluating gas sensors include sensor response and recovery time, as well as sensor repeatability, selectivity, and long-term stability. In view of the conclusions reached in the different concentration tests, the PANI/BP (1:1) sensor performed the best and was used in all further tests. Figure 6a shows the response and recovery times of the PANI/BP sensor for 60 ppm NO_2. The response and recovery times of the sensor were 98 and 406 s, respectively. To highlight the good recovery and low zero drift of PANI/BP in the repeatability test, pure PANI was selected for comparison. As shown in Figure 6b, the PANI/BP and pure PANI sensors were tested five times in NO_2 gas with a concentration of 20 ppm. A significant drift of pure PANI was observed. Both the sensors started with a response value of 0. The pure PANI sensor accumulated a total drift of 1032% at the end of the last test, whereas the PANI/BP sensor produced a drift of only 56%. In a long-term stability test lasting up to 4 weeks, we tested five sensors in the same manner, and the sensors were placed at room temperature when not working. As shown in Figure 6c, the response of the sensor started to decrease in the second week but leveled off in weeks 3–4. The polymers were exposed to air for long periods of time. Therefore, it was difficult to maintain long-term stability compared to MOS. This was a problem for the polymer materials themselves, and there is no optimal solution yet. Six gases, NO_2, NH_3, CH_4, H_2S, CO, and H_2, were used to study the selectivity of the sensors. The sensor exhibited good response to both NO_2 and NH_3, which is consistent with the results of many previous polymer-based gas sensor studies. The other four gases did not exhibit better sensitivities.

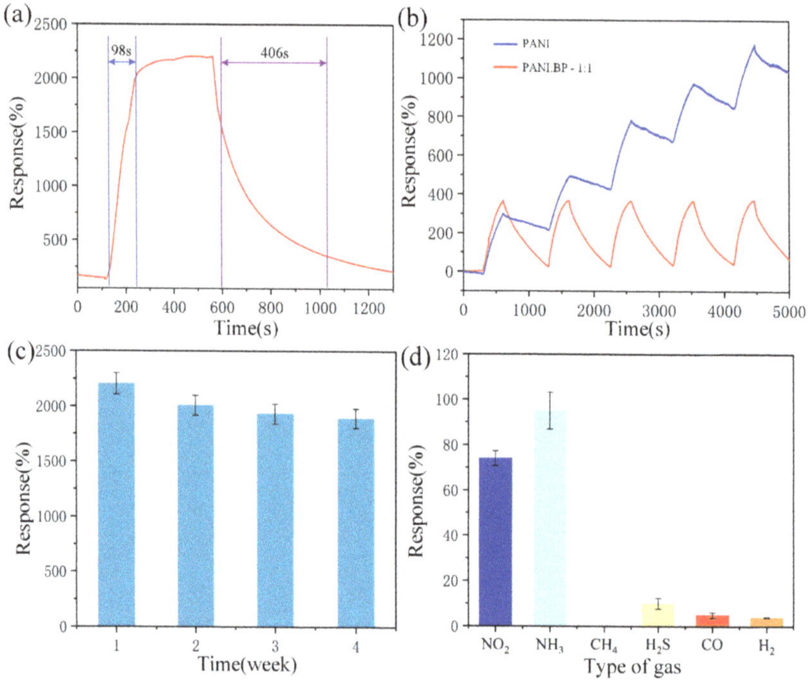

Figure 6. (**a**) Response and recovery time of PANI/BP sensor at 60 ppm concentration NO_2; (**b**) repeated line measurements of PANI/BP and pure PANI sensors at 20 ppm concentration of NO_2; (**c**) response of PANI/BP sensor to 60 ppm concentration of NO_2 over 4 weeks; (**d**) response of PANI/BP sensors to 10 ppm of different target gases.

Table 1 summarizes the sensor performance from other studies, as well as the data from this work. The use of MOS modifications to improve the gas-sensitive properties of PANI has been considered in previous studies on room-temperature sensing. However, MOS materials struggle to perform in low-temperature applications. Other studies have used reduced graphene oxide (rGO) and carbon nanotubes (CNT) combined with PANI, which can improve the gas-sensitive properties of PANI owing to their interesting structures. The PANI/BP composites developed in this study exhibited excellent responses to NO_2 at room temperature.

Table 1. A summary of the latest research on the performance of NO_2 gas sensors based on PANI-sensitive materials.

Materials	Sensing Range	Response	Res/Rec Time	Temperature	Reference
PANI/Ni@ZnO	5–100 ppm	75% (100 ppm)	82/399 s	RT	[26]
SnO_2/PANI	5–50 ppm	810% (37 ppm)	17/25 s	140 °C	[27]
PANI/g-C_3N_4	8–108 ppm	92% (108 ppm)		RT	[14]
PANI/SAN	5–100 ppm	72% (80 ppm)	1/1900 s	30 °C	[28]
PANI/CNT	10–50 ppm	65.9% (50 ppm)	5.2/3.2 s	RT	[29]
PANI/rGO	15–35 ppm	21.56% (30 ppm)	12/14 s	RT	[30]
PANI/BP	2–60 ppm	2204% (60 ppm)	98/406 s	RT	This work

3.3. Gas-Sensitive Mechanism

To elaborate on the gas sensitization mechanism of the PANI/BP composites, it is first necessary to explain the gas sensitization mechanisms of PANI and BP for NO_2 gas and then discuss the sensitization mechanism of the composites. The molecular structure of the PANI/BP-adsorbed NO_2 gas is shown schematically in Figure 7a. PANI is a conductive polymer whose molecular model was proposed by the Nobel Laureate Alan G. MacDiarmid. The PANI chain consisted of two structural units: a reduced [–B–NH–B–NH–] repeat unit and an oxidized [–B–N=Q=N–] repeat unit, where B and Q denote the C_6H_4 rings in the benzenoid and quinonoid forms, respectively. PANI exhibits conductivity in acidic or doped emerald green salt forms and insulation in undoped or emerald green alkaline forms. After treatment with protonic acid or other oxidizing agents, electrons in the PANI chain are transferred to the oxidizing agent, which increases the concentration of holes in PANI to form p-type semiconductors. Pure PANI reacts with NO_2 gas adsorption to generate a nitrogen dioxide anion (NO_2^-), which reduces the resistance value of PANI. The specific reaction is as follows:

$$NO_{2(gas)} + e^- \rightarrow NO^-_{2(ads)} \tag{2}$$

$$NO^-_{2(ads)} + O^-_{ads} \rightarrow 2O^{2-}_{ads} + NO_2 \tag{3}$$

$$NO_{2(gas)} + e^- \rightarrow NO^-_{2(ads)} \tag{4}$$

Researchers have different views on the gas-sensing mechanism of BP for NO_2. From the perspective of morphology, the surfaces and sides of the BP thin films are nanoscale films and stepped stacked nanoscale films, respectively. This special surface structure increases the surface-to-volume ratio of the BP microstrips. Physically, semiconductors with larger bandgaps have a weaker ability to adsorb gas molecules because of their lower carrier concentrations, whereas semiconductors with smaller bandgaps have a harder time producing a change in conductivity because of their higher carrier concentrations. The rich nanoscale films on the surface of the BP microstrip and the lateral step-stacked nanoscale films are not only good for increasing the effective gas-sensing region, but the thickness of these nanoscale films is also expected to be in the optimal bandgap range and carrier concentration, which leads to high sensitivity. On the other hand, Shumao Cui et al. further quantitatively correlated binding energy with sensitivity by establishing a statistical thermodynamic model to evaluate gas adsorption density [31]. For crystalline materials, gas molecules adsorbed from their free phase are considered to be lattice gases, and the interaction between the gas molecules and the solid surface (i.e., the crystalline material) is described by the Morse potential. This reveals that the adsorption density of NO_2 on phosphene (2.8×10^{12} cm^{-2}) is much higher than that of graphene (2.0×10^{10} cm^{-2}).

The enhancement of the gas sensitivity of the PANI/BP composites can be divided into two aspects. First, the response of the composites is enhanced compared to pure BP, which is mainly due to the fact that the addition of PANI allows the original thick film of BP to be better spaced apart, so that the BP exists in the form of more and thinner tiny flakes. The unique two-dimensional layered structure of BP makes the conductivity outside the surface much smaller than that inside the surface, and the charge transfer by gas adsorption in the case of thick films tends not to be uniformly distributed throughout the material but rather aggregates in the top surface region. Thus, the addition of PANI enhances the response to BP. Second, the involvement of BP significantly improves the zero drift of the PANI. The electrical conductivity of PANI depends primarily on its degrees of protonation and oxidation. NO_2 is a strong oxidizing gas; when in contact with the PANI chain, it oxidizes PANI, which is often difficult to recover. This causes the zero drift of the PANI. However, the addition of BP resulted in the formation of a p/p-type Schottky junction at the composite interface. The Fermi level of BP is 4.6 eV, and the band gap of BP is approximately 1.7 eV [32], depending on its thickness. Typically, the Fermi energy level of the unprimed PANI is 4.4 eV with a band gap of 2.8 eV, as shown in Figure 7b,c. When the two materials are in contact with each other, the holes in the BP tend to transfer to the PANI surface, and the Fermi level reaches a new equilibrium. The NO_2 molecule reacts with the chemisorbed oxygen anion, releasing electrons further back to the surface of the material, reducing the hole concentration, and thus affecting the resistance of the material.

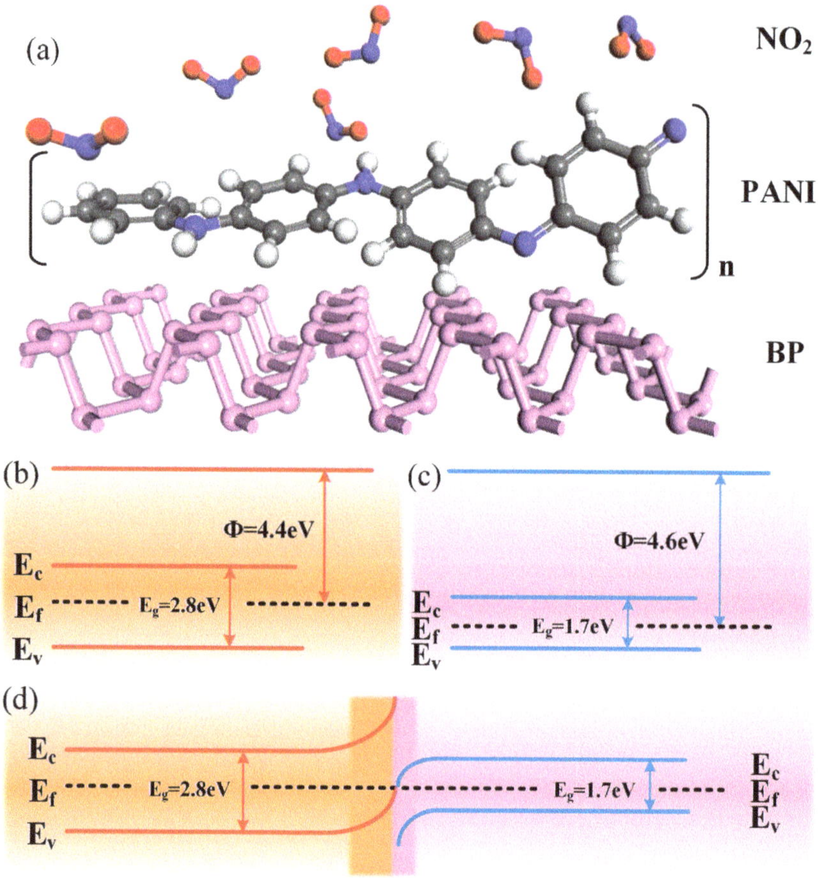

Figure 7. (**a**) Molecular schematic of NO_2 gas adsorption by PANI/BP; band diagrams of (**b**) PANI, (**c**) BP, and (**d**) PANI/BP.

4. Conclusions

This study demonstrated that the PANI/BP composite materials exhibited excellent NO_2 gas-sensing properties at room temperature. When the mass ratio of PANI to BP was 1:1, the sensor showed responses ranging from 1.8% to 2204% in the concentration range of 2–60 ppm NO_2. PANI/BP composites were formed by ultrasonic exfoliation and mixing BP prepared by low-pressure mineralization with PANI synthesized via chemical oxidative polymerization. The composite material significantly mitigated the zero-drift issue of pure PANI sensors and enhanced the response of pure BP sensors. This achievement provides an effective solution for room-temperature NO_2 sensors and highlights the tremendous potential of BP in the gas-sensing field. In the future, the selectivity of PANI/BP gas-sensitive materials still needs to be improved. As a polymer, PANI can be fabricated into flexible film coatings, whereas the bandgap of BP can be modulated under external stress-induced deformation. A flexible device prepared by combining the two is more likely to have more interesting characteristics, and it awaits further exploration. Both PANI and BP are nontoxic and biocompatible, suggesting broad prospects in the field of biosensing.

Author Contributions: Conceptualization, Y.S. and J.L.; methodology, B.T. and Q.F.; software, C.Z. and K.Z.; investigation, Q.F. and J.Z.; writing—original draft preparation, B.T.; writing—review and editing, B.T. and J.Z. All authors have read and agreed to the published version of the manuscript.

Funding: This research was funded by a grant from the National Natural Science Foundation of China, grant number 62271176, and the China National Basic Enhancement Program, 2022-JCJQ-JJ-0438.

Institutional Review Board Statement: Not applicable.

Informed Consent Statement: Not applicable.

Data Availability Statement: The original contributions presented in the study are included in the article, further inquiries can be directed to the corresponding author.

Conflicts of Interest: Author Jijiang Liu was employed by the 49th Research Institute of China Electronics Technology Group. The remaining authors declare that the research was conducted in the absence of any commercial or financial relationships that could be construed as a potential conflict of interest.

References

1. Gamon, L.F.; Wille, U. Oxidative Damage of Biomolecules by the Environmental Pollutants NO_2 and NO_3. *Acc. Chem. Res.* **2016**, *49*, 2136–2145. [CrossRef]
2. Chauhan, A.J.; Inskip, H.M.; Linaker, C.H.; Smith, S.; Schreiber, J.; Johnston, S.L.; Holgate, S.T. Personal exposure to nitrogen dioxide (NO_2) and the severity of virus-induced asthma in children. *Lancet* **2003**, *361*, 1939–1944. [CrossRef]
3. Chapman, J.; Truong, V.K.; Elbourne, A.; Gangadoo, S.; Cheeseman, S.; Rajapaksha, P.; Latham, K.; Crawford, R.J.; Cozzolino, D. Combining Chemometrics and Sensors: Toward New Applications in Monitoring and Environmental Analysis. *Chem. Rev.* **2020**, *120*, 6048–6069. [CrossRef]
4. Zhu, L.Y.; Ou, L.X.; Mao, L.W.; Wu, X.Y.; Liu, Y.P.; Lu, H.L. Advances in Noble Metal-Decorated Metal Oxide Nanomaterials for Chemiresistive Gas Sensors: Overview. *Nanomicro Lett.* **2023**, *15*, 89. [CrossRef] [PubMed]
5. Burgues, J.; Marco, S. Low Power Operation of Temperature-Modulated Metal Oxide Semiconductor Gas Sensors. *Sensors* **2018**, *18*, 339. [CrossRef] [PubMed]
6. Tang, B.; Shi, Y.; Li, J.; Tang, J.; Feng, Q. Design, Simulation, and Fabrication of Multilayer Al_2O_3 Ceramic Micro-Hotplates for High Temperature Gas Sensors. *Sensors* **2022**, *22*, 6778. [CrossRef] [PubMed]
7. Henriquez, D.; Kang, M.; Cho, I.; Choi, J.; Park, J.; Gul, O.; Ahn, J.; Lee, D.S.; Park, I. Low-Power, Multi-Transduction Nanosensor Array for Accurate Sensing of Flammable and Toxic Gases. *Small Methods* **2023**, *7*, e2201352. [CrossRef]
8. Wu, H.; Bu, X.; Deng, M.; Chen, G.; Zhang, G.; Li, X.; Wang, X.; Liu, W. A Gas Sensing Channel Composited with Pristine and Oxygen Plasma-Treated Graphene. *Sensors* **2019**, *19*, 625. [CrossRef]
9. Han, J.K.; Kang, M.; Jeong, J.; Cho, I.; Yu, J.M.; Yoon, K.J.; Park, I.; Choi, Y.K. Artificial Olfactory Neuron for an In-Sensor Neuromorphic Nose. *Adv. Sci.* **2022**, *9*, e2106017. [CrossRef]
10. Farea, M.A.; Mohammed, H.Y.; Shirsat, S.M.; Sayyad, P.W.; Ingle, N.N.; Al-Gahouari, T.; Mahadik, M.M.; Bodkhe, G.A.; Shirsat, M.D. Hazardous gases sensors based on conducting polymer composites: Review. *Chem. Phys. Lett.* **2021**, *776*, 138703. [CrossRef]
11. Wen, J.Y.; Wang, S.; Feng, J.Y.; Ma, J.X.; Zhang, H.; Wu, P.; Li, G.; Wu, Z.H.; Meng, F.Z.; Li, L.Q.; et al. Recent progress in polyaniline-based chemiresistive flexible gas sensors: Design, nanostructures, and composite materials. *J. Mater. Chem. A* **2024**, *12*, 6190–6210. [CrossRef]
12. Sharma, S.; Hussain, S.; Singh, S.; Islam, S.S. MWCNT-conducting polymer composite based ammonia gas sensors: A new approach for complete recovery process. *Sens. Actuators B Chem.* **2014**, *194*, 213–219. [CrossRef]
13. Sonker, R.K.; Yadav, B.C. Development of Fe_2O_3-PANI nanocomposite thin film based sensor for NO_2 detection. *J. Taiwan Inst. Chem. Eng.* **2017**, *77*, 276–281. [CrossRef]
14. Khalifa, M.; Anandhan, S. Highly sensitive and wearable NO_2 gas sensor based on PVDF nanofabric containing embedded polyaniline/g-C_3N_4 nanosheet composites. *Nanotechnology* **2021**, *32*, 485504. [CrossRef] [PubMed]
15. Zhao, X.; Li, P.F.; Xiao, K.T.; Meng, X.N.; Han, L.; Yu, C.C. Sensor Drift Compensation Based on the Improved LSTM and SVM Multi-Class Ensemble Learning Models. *Sensors* **2019**, *19*, 3844. [CrossRef] [PubMed]
16. Liu, H.; Chu, R.Z.; Tang, Z.A. Metal Oxide Gas Sensor Drift Compensation Using a Two-Dimensional Classifier Ensemble. *Sensors* **2015**, *15*, 10180–10193. [CrossRef] [PubMed]
17. Liu, T.; Li, D.Q.; Chen, J.J.; Chen, Y.B.; Yang, T.; Cao, J.H. Gas-Sensor Drift Counteraction with Adaptive Active Learning for an Electronic Nose. *Sensors* **2018**, *18*, 4028. [CrossRef]
18. Kim, H.; Uddin, S.Z.; Lien, D.H.; Yeh, M.; Azar, N.S.; Balendhran, S.; Kim, T.; Gupta, N.; Rho, Y.; Grigoropoulos, C.P.; et al. Actively variable-spectrum optoelectronics with black phosphorus. *Nature* **2021**, *596*, 232. [CrossRef]
19. Dhanabalan, S.C.; Ponraj, J.S.; Guo, Z.N.; Li, S.J.; Bao, Q.L.; Zhang, H. Emerging Trends in Phosphorene Fabrication towards Next Generation Devices. *Adv. Sci.* **2017**, *4*, 1600305. [CrossRef]

20. Zhou, Y.; Zhang, M.X.; Guo, Z.N.; Miao, L.L.; Han, S.T.; Wang, Z.Y.; Zhang, X.W.; Zhang, H.; Peng, Z.C. Recent advances in black phosphorus-based photonics, electronics, sensors and energy devices. *Mater. Horiz.* **2017**, *4*, 997–1019. [CrossRef]
21. Hu, Z.H.; Niu, T.C.; Guo, R.; Zhang, J.L.; Lai, M.; He, J.; Wang, L.; Chen, W. Two-dimensional black phosphorus: Its fabrication, functionalization and applications. *Nanoscale* **2018**, *10*, 21575–21603. [CrossRef] [PubMed]
22. Kou, L.Z.; Frauenheim, T.; Chen, C.F. Phosphorene as a Superior Gas Sensor: Selective Adsorption and Distinct I-V Response. *J. Phys. Chem. Lett.* **2014**, *5*, 2675–2681. [CrossRef]
23. Abbas, A.N.; Liu, B.L.; Chen, L.; Ma, Y.Q.; Cong, S.; Aroonyadet, N.; Köpf, M.; Nilges, T.; Zhou, C.W. Black Phosphorus Gas Sensors. *Acs Nano* **2015**, *9*, 5618–5624. [CrossRef] [PubMed]
24. Valt, M.; Caporali, M.; Fabbri, B.; Gaiardo, A.; Krik, S.; Iacob, E.; Vanzetti, L.; Malagù, C.; Banchelli, M.; D'Andrea, C.; et al. Air Stable Nickel-Decorated Black Phosphorus and Its Room-Temperature Chemiresistive Gas Sensor Capabilities. *Acs Appl. Mater. Interfaces* **2021**, *13*, 44711–44722. [CrossRef]
25. Sajedi-Moghaddam, A.; Mayorga-Martinez, C.C.; Sofer, Z.; Bousa, D.; Saievar-Iranizad, E.; Pumera, M. Black Phosphorus Nanoflakes/Polyaniline Hybrid Material for High-Performance Pseudocapacitors. *J. Phys. Chem. C* **2017**, *121*, 20532–20538. [CrossRef]
26. Jain, S.; Karmakar, N.; Shah, A.; Shimpi, N.G. Development of Ni doped ZnO/polyaniline nanocomposites as high response room temperature NO_2 sensor. *Mater. Sci. Eng. B-Adv. Funct. Solid-State Mater.* **2019**, *247*, 114381. [CrossRef]
27. Xu, H.Y.; Ju, D.X.; Li, W.R.; Gong, H.B.; Zhang, J.; Wang, J.Q.; Cao, B.Q. Low-working-temperature, fast-response-speed NO_2 sensor with nanoporous-SnO_2/polyaniline double-layered film. *Sens. Actuators B Chem.* **2016**, *224*, 654–660. [CrossRef]
28. Reddy, N.R.; Anandhan, S. Polyaniline/poly(styrene-*co*-acrylonitrile) blend nanofibers exhibit enhanced ammonia and nitrogen dioxide sensing characteristics. *J. Mater. Sci. Mater. Electron.* **2016**, *27*, 13329–13337. [CrossRef]
29. Zhang, W.Y.; Cao, S.; Wu, Z.F.; Zhang, M.; Cao, Y.L.; Guo, J.X.; Zhong, F.R.; Duan, H.M.; Jia, D.Z. High-Performance Gas Sensor of Polyaniline/Carbon Nanotube Composites Promoted by Interface Engineering. *Sensors* **2020**, *20*, 149. [CrossRef]
30. Kailasa, S.; Reddy, M.S.B.; Rani, B.G.; Maseed, H.; Rao, K.V. Twisted Polyaniline Nanobelts @ rGO for Room Temperature NO_2 Sensing. *Mater. Lett.* **2019**, *257*, 126687. [CrossRef]
31. Cui, S.M.; Pu, H.H.; Wells, S.A.; Wen, Z.H.; Mao, S.; Chang, J.B.; Hersam, M.C.; Chen, J.H. Ultrahigh sensitivity and layer-dependent sensing performance of phosphorene-based gas sensors. *Nat. Commun.* **2015**, *6*, 8632. [CrossRef] [PubMed]
32. Lei, W.Y.; Zhang, T.T.; Liu, P.; Rodriguez, J.A.; Liu, G.; Liu, M.H. Bandgap- and Local Field-Dependent Photoactivity of Ag/Black Phosphorus Nanohybrids. *Acs Catal.* **2016**, *6*, 8009–8020. [CrossRef]

Disclaimer/Publisher's Note: The statements, opinions and data contained in all publications are solely those of the individual author(s) and contributor(s) and not of MDPI and/or the editor(s). MDPI and/or the editor(s) disclaim responsibility for any injury to people or property resulting from any ideas, methods, instructions or products referred to in the content.

chemosensors

MDPI

Article

A Novel Mechanism Based on Oxygen Vacancies to Describe Isobutylene and Ammonia Sensing of p-Type Cr_2O_3 and Ti-Doped Cr_2O_3 Thin Films

Pengfei Zhou [1,2,3,†], Jone-Him Tsang [2,†], Chris Blackman [2,*], Yanbai Shen [3], Jinsheng Liang [1], James A. Covington [4], John Saffell [5] and Ehsan Danesh [6]

1 Key Laboratory of Special Functional Materials for Ecological Environment and Information, Hebei University of Technology, Ministry of Education, Tianjin 300130, China; zhoupengfei@hebut.edu.cn (P.Z.); liangjinsheng@hebut.edu.cn (J.L.)
2 Department of Chemistry, University College London, London WC1H 0AJ, UK; jone-him.tsang.16@alumni.ucl.ac.uk
3 School of Resources and Civil Engineering, Northeastern University, Shenyang 110819, China; shenyanbai@mail.neu.edu.cn
4 School of Engineering, University of Warwick, Coventry CV4 7AL, UK; j.a.covington@warwick.ac.uk
5 NosmoTech Ltd., Cambridge CB3 0AZ, UK; john@nosmotech.com
6 Advanced Sensing Technologies Ltd., London N12 8FA, UK; e.danesh@ieee.org
* Correspondence: c.blackman@ucl.ac.uk
† These authors contributed equally to this work.

Abstract: Gas sensors based on metal oxide semiconductors (MOS) have been widely used for the detection and monitoring of flammable and toxic gases. In this paper, p-type Cr_2O_3 and Ti-doped Cr_2O_3 (CTO) thin films were synthesized using an aerosol-assisted chemical vapor deposition (AACVD) method. Detailed analysis of the thin films deposited, including structural information, their elemental composition, oxidation state, and morphology, was investigated using XRD, Raman analysis, SEM, and XPS. All the gas sensors based on pristine Cr_2O_3 and CTO exhibited a reversible response and good sensitivity to isobutylene (C_4H_8) and ammonia (NH_3) gases. Doping Ti into the Cr_2O_3 lattice improves the response of the CTO-based sensors to C_4H_8 and NH_3. We describe a novel mechanism for the gas sensitivity of p-type metal oxides based on variations in the oxygen vacancy concentration.

Keywords: Cr_2O_3; Ti-doped; AACVD; thin films; gas sensor; mechanism

1. Introduction

Volatile organic compounds (VOCs), including ammonia and hydrocarbons, are ubiquitous, and their detection has been employed for the monitoring of food and beverages, agricultural produce, pharmaceuticals, and health [1]. Certain VOCs are highly toxic and/or carcinogenic and may cause both short- and long-term health issues (e.g., allergies or cancer) as well as impact our ecosystem [2]. Even though the human nose serves as a highly sensitive sensing system, it still fails when low-concentration or odorless toxic gases—which pose a serious threat to human health—need to be detected [3,4], so the development of a low-cost, simple-to-operate VOC sensor is of great current interest.

Gas sensors based on metal oxide semiconductors (MOSs) are commercially available and have been widely used in the detection and monitoring of flammable and toxic gases [5]. MOS-based sensors are cheap, reliable, consume relatively low power, and have good sensitivity, due to a wide conductivity range and a robust structure, making them an ideal gas-sensing material [6]. Compared to the commercially successful n-type SnO_2 and WO_3 sensors, p-type MOS-based sensors, such as Cr_2O_3, CuO, and NiO, have received relatively little attention [7]. As a typical p-type semiconductor, Cr_2O_3 has been used in gas sensing [8], catalysis [9], and electrochemical devices [10]. It is reported that doping titanium into Cr_2O_3 is an effective method to enhance the gas-sensing performance of pristine Cr_2O_3 [11]. Ti-doped Cr_2O_3 (CTO) MOSs are also good p-type materials for use in gas sensors, due to their tolerance towards humidity, good baseline stability, and reasonable sensitivity. CTO-based sensors are especially good for the detection of trace quantities of reducing gases in air (such as H_2S, CO, or ethanol vapor), with operating temperatures ranging from 300 to 500 °C [12,13]. Shaw [14] reported the deposition of $Cr_{2-x}Ti_xO_3$ thin films on sensor substrates using a variety of techniques—including screen printing, atmospheric pressure chemical vapor deposition (APCVD), and flame fusion methods—and investigated their gas response to CO and ethanol. Du [15] used electrostatic spray-assisted vapor deposition (ESAVD) at 650 °C to prepare $Cr_{2-x}Ti_xO_3$ films for the detection of NH_3. The results showed that the sensitivity of $Cr_{1.8}Ti_{0.2}O_3$ films to 500 ppm NH_3 at 500 °C was around 1.45. To further investigate the effect of different Ti-doping levels on $Cr_{2-x}Ti_xO_3$, Du [16] carried out the synthesis and characterization of $Cr_{2-x}Ti_xO_3$ with different nominal compositions using the ESAVD technique. Of all the CTO sensors—prepared with varying compositions—$Cr_{1.7}Ti_{0.3}O_3$ exhibited the highest gas sensitivity. This sensor exhibited a gas response of 2.90 when exposed to 500 ppm NH_3 at 200 °C and also revealed good sensitivity to ethanol vapor at 400 °C. Conde-Gallardo [17] found that the conductivity of $Cr_{2-x}Ti_xO_3$ films deposited by aerosol-assisted CVD (AACVD), using copper acetylacetonate and titanium butoxide precursors dissolved in isopropanol, was dependent on the degree of Ti-substitution (conductivity decreased with increasing Ti), although they did not measure the gas-sensing properties.

AACVD has been widely implemented to prepare homogenous and uniform thin films and can effectively generate high-purity MOSs [18]. It has a key advantage over alternative methods of sensor material synthesis in that it combines both material synthesis and device integration in a single step. The AACVD process has attracted attention over other CVD processes due to its characteristics of easy operation, low cost, higher deposition rates, and flexibility in tuning thin-film micro/nanostructures [19].

This work focused on the synthesis of Cr_2O_3 and Ti-doped Cr_2O_3 thin films via the AACVD method and demonstrated the relative sensitivity of the materials to the gases ammonia (NH_3) and C_4H_8. NH_3 is a colorless gas with corrosive properties and a strong pungent odor. Severe damage is caused to the throat, lungs, eyes, and skin, even at low NH_3 concentrations [20]. Moreover, it is reported that exposure to an NH_3 atmosphere exceeding 50 ppm for a long time leads to serious pathological changes in organs such as the liver and kidneys [21] and also results in serious health issues, including respiratory distress, eye irritation, and skin problems [22]. NH_3 is also a metabolite in breath exhaled from the human body, which can be used as a marker of end-stage renal disease (ESRD) for a non-destructive diagnosis in clinical medicine (average 4.88 ppm; range 0.82–14.7 ppm) [23]. As far as we know, this report is also the first time that CTO sensors have been used to detect C_4H_8, a model alkene VOC. The chemical composition and morphological structure of the Cr_2O_3 and CTO thin films were characterized using XRD, Raman analysis, SEM, and XPS. To understand how the conductivity of CTO may change upon exposure to these gases, we present the first description of a gas-sensing mechanism for p-type MOSs, based on varying concentrations of oxygen vacancies in the sensing material.

2. Experimental Section

2.1. Synthesis of Cr_2O_3 and CTO Thin Films

The deposition of Cr_2O_3 and CTO thin films was achieved via the AACVD technique in a homemade reactor equipped with a water-cooling system in the proximity of the inlet to minimize the pre-reaction of precursors. A schematic of the AACVD technique is shown in Figure 1.

To prepare the Cr_2O_3 films, alumina gas-sensor platforms (with interdigitated screen-printed gold electrodes) were first cleaned with acetone and deionized water before use and were then set onto a base and covered with a mask. These were then placed into the reactor and heated to 340 °C. A solution of chromium hexacarbonyl (0.1 g, 0.454 mmol) in methanol (40.0 mL) was prepared and aerosols of the solution were formed using an ultrasonic humidifier operated at 2 MHz, with the aerosols then transported to the reactor by a N_2 gas carrier at a flow rate of 1000 standard cubic centimeters per minute (sccm). After ~10 cm^3 of the precursor solution had been transported, the substrate and mask were rotated 90° and the deposition process continued; this was repeated four times to ensure homogeneous coverage of the thin films across the sensors. Once the precursor solution was exhausted, the heating temperature and flow rate were reduced to 100 °C and 300 sccm, respectively. After deposition, the dark green thin films obtained were annealed for 24 h at variable temperatures and, respectively, named as Cr_2O_3-1, Cr_2O_3-2, and Cr_2O_3-3, corresponding to the annealing temperatures of 500, 600, and 700 °C.

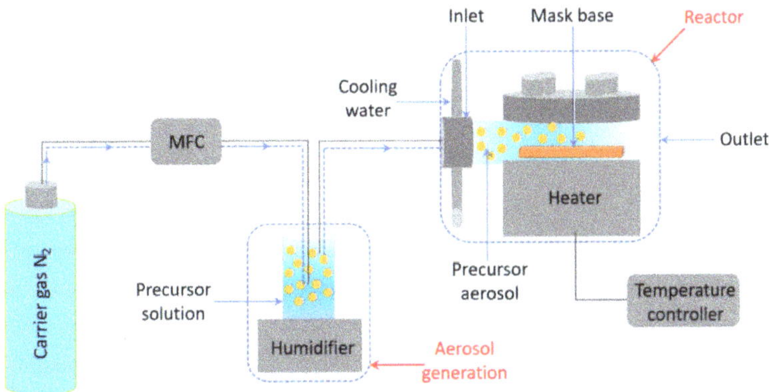

Figure 1. Schematic diagram of an aerosol-assisted CVD.

For the synthesis of CTO thin films, a solution of 0.22 g (0.775 mmol) titanium di-isopropoxide bis-(acetylacetonate) (TDBAA) and 15.0 mL methanol was prepared, then 3.0 cm^3 of this TDBAA–methanol solution was added to the chromium solution (as previously prepared) and stirred to obtain a precursor solution with a Cr:Ti molar ratio of $Cr_{1.5}Ti_{0.5}$. Aerosols of this solution were again generated via a humidifier and transported by a N_2 gas carrier, but this time at a flow rate of 1500 sccm (to provide more homogeneous Ti-incorporation). For the CTO-based sensors, the annealing temperature was maintained at 600 °C and the annealing time varied; CTO-1 is the sample annealed for 6 h, while CTO-2 is the sample annealed for 30 h.

2.2. *Characterization of the Samples*

X-ray diffraction (XRD) patterns were analyzed using a Bruker, LinxEye D8 X-ray diffractometer in reflection mode, using Cu Kα radiation (λ = 1.5406 Å) and operated at 50.0 kV and 1.0 mA. Scans were performed using a detection angle ranging from 20.0° to 70.0°. Raman spectroscopy analysis was carried out using a Renishaw 1000 spectrometer equipped with a 532 nm laser. The Raman system was calibrated using a silicon reference. All films were placed in the spectrometer using an X-Y stage and analyzed in the ranges of 200 to 800 cm^{-1}, with a laser power of 10%, an exposure time of 45 s, and an accumulation of three times per sample. Scanning electron microscopy (SEM) was conducted using a Jeol 6310F microscope. All sample images were collected using a secondary electron detector. The samples were coated with a thin layer of sputtered gold and connected to the metal stage with copper tape. Energy dispersive analysis of the X-rays was conducted using the same instrument to determine the sample composition (for the samples coated with carbon, not gold). X-ray photoelectron spectroscopy (XPS) measurements were performed using a Thermo Kα spectrometer with monochromatic Al Kα radiation, a dual-beam charge compensation system, and a constant pass energy of 50 eV. The binding energies were calibrated with respect to the C 1s peak at 284.6 eV.

2.3. Sensor Fabrication and Gas-Sensing Measurements

A photograph of the MEMS platform is shown in Figure 2a. Once the AACVD-deposited substrates were made, platinum wires were used to connect the alumina substrates to a pin stage, as shown in Figure 2b. Four platinum wires were used for the electrical connections; two were welded to the gold sensor trackpads on the top side of the sensor platform, and the other two were connected to the platinum heater trackpads on the bottom side. Gas-sensing measurements of the as-prepared Cr_2O_3 and CTO sensors were tested at Alphasense Ltd., UK. The working temperature and resistance of the sensors were measured and controlled by a Sensor Management System (SMS), which is designed to work with up to eight metal oxide gas sensors. It has circuits to accurately control the sensor heater temperature based on a constant-resistance setup with a digital control. The variance in the concentration of the tested gas was obtained by altering the flow rate of each gas using a mass flow controller (MFC, UFC 1100, Brooks), which was controlled by a computer program written in LabVIEW (National Instrument 2016). In a typical test cycle, the enclosed system containing the sensors was first purged with 50% humid air for 30 min; afterwards, a flow of concentrated gas analyte was passed through for 30 min, and then a flow of just humid air passed for 30 min again to clean the analyte.

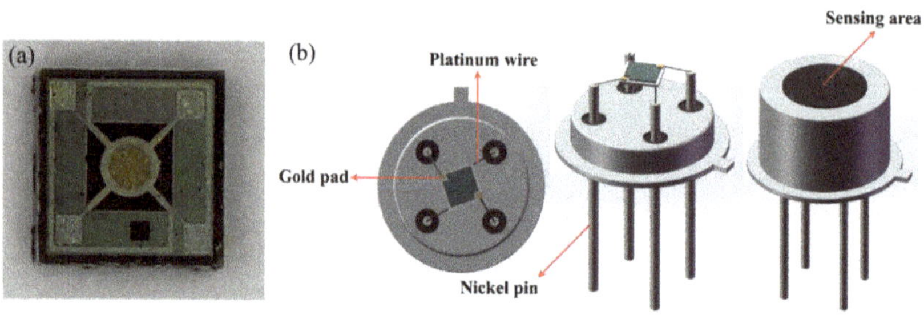

Figure 2. (**a**) Photograph of sensor platform, and (**b**) schematic diagram of a sensor connected to nickel pins via platinum wires.

3. Results and Discussion

3.1. Material Characterization

EDX analysis gave the composition of the samples as Cr1.95Ti0.05 (97.5% Cr, 2.5% Ti on a metals basis), indicating decreased incorporation of Ti into the film relative to the concentration in the precursor solution; this is commonly observed in multicomponent films deposited by AACVD and occurs due to differences in the thermal decomposition behaviour of the individual precursors. The XRD patterns of Cr_2O_3 and CTO-based sensors are shown in Figure 3, as shown in Figure 3a, the characteristic peaks at 2θ = 24.5°, 33.6°, 36.2°, 41.2°, 50.2°, 54.9°, and 57.1° can be assigned to the (012), (104), (110), (113), (024), (116), and (211) planes of hexagonal phase eskolaite (JCPDS # 38-1479) [24]. The XRD patterns of corundum indicate that all of the diffraction peaks match well with the hexagonal phase Al_2O_3 (JCPDS # 46-1212), which is expected as the substrates that are made from alumina. Notably, the sharp diffraction peaks of gold (JCPDS # 04-0784) were also detected that come from the ink used for the sensing electrodes. No obvious 2θ peak shift was observed when Cr_2O_3 thin films were annealed at different temperatures (500, 600, and 700 °C) on the surface of alumina sensing platform. Figure 3b exhibits the XRD patterns of CTO thin films and the results indexed to the crystal structure of eskolaite (Cr_2O_3), corundum (Al_2O_3), and gold. The CTO patterns are in close alignment with the deposited Cr_2O_3 thin films and the result shows that no measurable change in the crystal structure is observed and peaks of TiO_2 are not present, which indicates that the full incorporation of Ti dopant into the lattice of Cr_2O_3 and no phase separation occurred within the limit of detection of XRD. Furthermore, no other impurity peaks were detected in all of the XRD patterns, demonstrating the high purity of the samples.

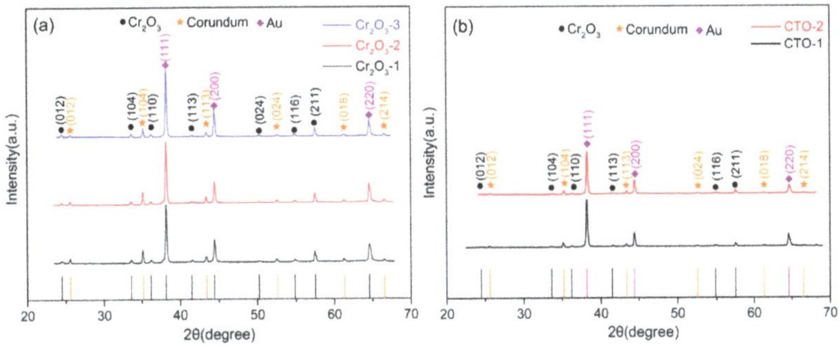

Figure 3. XRD patterns of (**a**) Cr_2O_3 thin film-based sensors exposed to different annealing temperatures and (**b**) CTO thin film-based sensors subjected to different annealing times.

The average crystal sizes of the Cr_2O_3 and CTO thin films were calculated using Scherrer's formula:

$$D = k\lambda / \beta \cos\theta. \quad (1)$$

Herein, D is the average crystal size, and all angles are in radians. For spherical crystallites, in most cases k is a constant of about 0.9. λ represents the X-ray wavelength (1.5406Å), and β is the full width at half maximum (FWHM) of the peaks of the Cr_2O_3 and CTO thin films. For the Cr_2O_3 thin films, the calculated values of the crystallite sizes are 26, 31, and 30 nm, corresponding to the different annealing temperatures of 500, 600, and 700 °C. From the results, it can be seen that the crystallite sizes may become a little bigger with the increase in annealing temperatures, but the difference is likely to be insignificant within the precision of the Scherrer estimate. In the CTO thin films, the crystallite sizes of CTO-1 and CTO-2 are 25 and 36 nm, respectively, demonstrating that the incorporation of Ti into Cr_2O_3 crystal causes no obvious influence on its crystallite size (CTO-1), but extending the annealing time from 6 h to 30 h appears to contribute to a growth in crystallite size (CTO-2). Table 1 summarizes the preparation conditions and Scherrer sizes of the different sensors.

Table 1. Preparation conditions and Scherrer sizes of the different sensors.

Samples	Annealing Temperatures (°C)	Annealing Times (h)	Scherrer Size (nm)
Cr_2O_3-1	500	24	26
Cr_2O_3-2	600	24	31
Cr_2O_3-3	700	24	30
CTO-1	600	6	25
CTO-2	600	30	36

The crystal structures were also investigated using Raman spectroscopy. Figure 4 displays the Raman spectra of the Cr_2O_3 and CTO thin films. In Cr_2O_3, there are a total of four Raman modes observed (three E_g modes and one A_{1g} mode). Pristine Cr_2O_3 thin films exhibited three E_g vibration modes at 303, 351, and 612 cm^{-1} and one intense A_{1g} mode at 553 cm^{-1}. The absence of characteristic TiO_2 Raman peaks for the CTO sample correlates well with the results from XRD characterization—namely, there was no phase separation on the film. The broad peak located at 721 cm^{-1} has been seen previously and has been ascribed to the existence of local vibration [25].

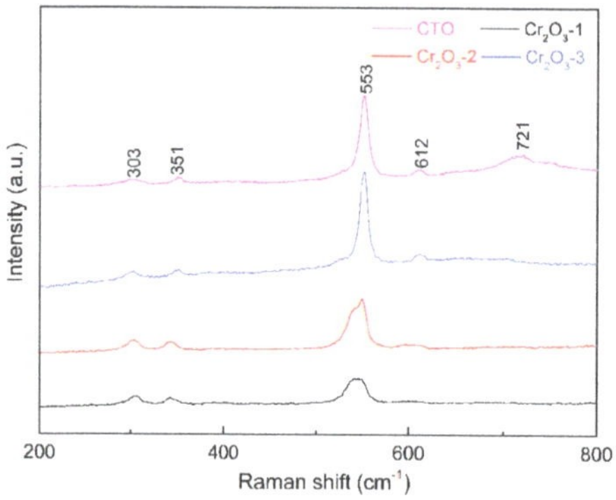

Figure 4. Raman spectra of the Cr_2O_3 and CTO thin films.

The morphologies of the as-prepared CTO thin films were characterized using SEM. As seen from the high-resolution picture (Figure 5a), the CTO thin films consisted of many uniform and well-dispersed sub-micron particles, which are spherical-like, with a diameter of 300–500 nm. It can be speculated that the particles observed in SEM are agglomerates of many smaller crystallites, due to the disparity between the particle size (SEM) and the crystallite size (Scherrer). As shown in the low-magnification SEM image (Figure 5b), the CTO particles are self-assembled to form a lamellar structure. SEM images of the $Cr_{1.95}T_{0.05}O_3$ films previously deposited (at higher temperatures) by APCVD exhibited dense and spheroidal platelets sized 1–4 μm [14], whilst the $Cr_{1.8}Ti_{0.2}O_3$ films deposited by the ESAVD technique had a more ordered porous structure, due to the introduction of polymeric additives [15].

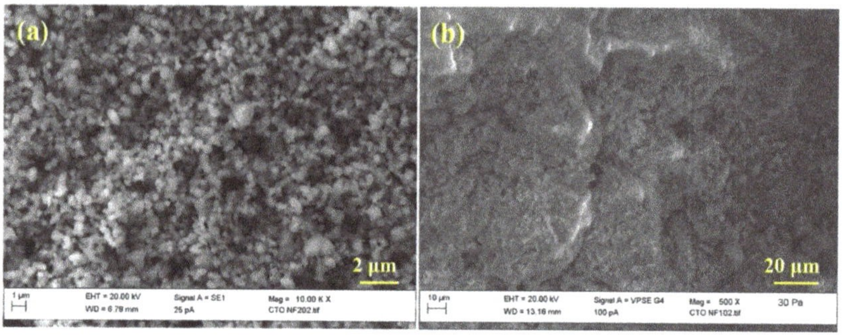

Figure 5. (**a**) High-resolution and (**b**) low-resolution SEM images of the CTO thin film.

The surface chemical composition of the as-synthesized CTO was analyzed by XPS and the results are shown in Figure 6. The XPS surface analysis was conducted using C 1s as calibration at 284.6 eV. It can be observed that the sharp peaks in the survey spectrum (Figure 6a) can be attributed to the elements of C, Ti, O, and Cr, demonstrating the successful doping of Ti into the Cr_2O_3 thin film. The high-resolution spectrum of Ti^{4+} 2p (shown in Figure 6b) is split into two peaks—of Ti^{4+} $2p_{1/2}$ and Ti^{4+} $2p_{3/2}$—centered at 464.5 and 458.6 eV. This is in contradiction with the work of Conde-Gallardo et al. [17], who found that their $Cr_{2-x}Ti_xO_3$ films deposited by AACVD featured Ti^{3+}—although substitution by Ti^{4+} would be consistent with the decrease in conductivity they observed with increasing Ti incorporation, due to electronic compensation through the formation of electrons (which are expected to recombine with the native p-type carriers in Cr_2O_3). The Cr 2p spectrum of CTO in Figure 6c could be divided into two peaks of Cr^{3+} $2p_{1/2}$ and Cr^{3+} $2p_{3/2}$, located at 586.6 and 576.7 eV, respectively. Figure 6d shows that the XPS spectrum of

O 1s is deconvoluted into three peaks, corresponding to Ti–O–Ti (534.2 eV), surface oxygen species (-CO_x, -OH) (532.3 eV), and Cr–O–Cr (530.4 eV), respectively [26].

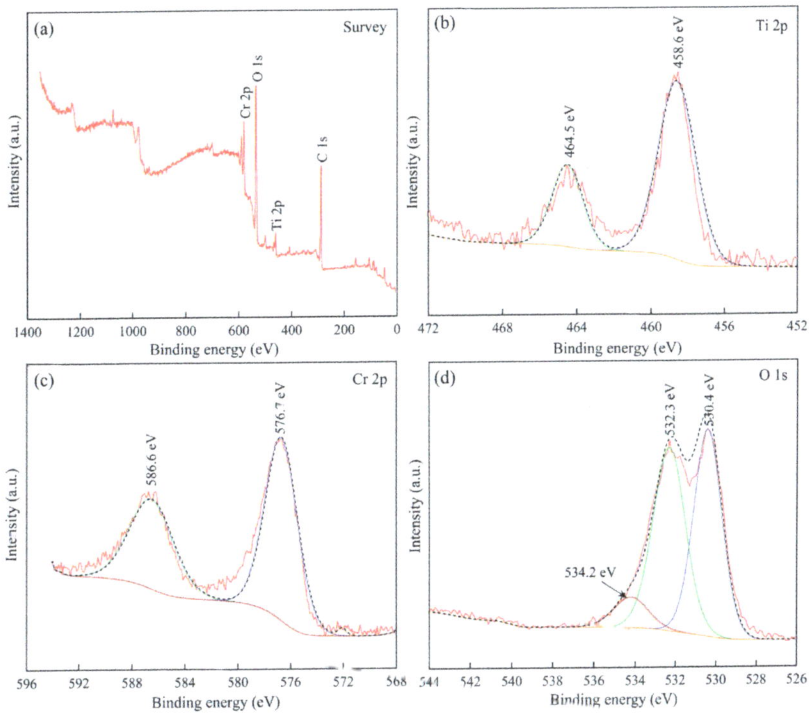

Figure 6. XPS spectra of the CTO film: (**a**) survey scan, (**b**) Ti 2p, (**c**) Cr 2p, and (**d**) O 1s.

3.2. Gas Response

Figure 7a,b show the dynamic responses of the Cr_2O_3 and CTO sensors towards C_4H_8 gas, respectively. The test conditions were set at two different concentrations of C_4H_8 gas at a relative humidity of 50% and operating temperatures of 400 and 450 °C. Each temperature was tested with three pulses of C_4H_8 gas (20-5-20 ppm). Figure 8a,b show the sensors response (resistance in gas over resistance in air, Rg/Ra) towards the different concentrations of C_4H_8 at different temperatures. The dynamic response results (Figure 7) show that the response values remained stable after the continuous test with different gas concentrations at different working temperatures. Each pulse towards the analyte was 30 min, then followed by 30 min of humid air. It can be observed that resistance values are higher for CTO as compared to resistance values of Cr_2O_3, as expected for substitution of Cr^{3+} with Ti^{4+}, again suggesting that Ti was fully incorporated into the host lattice rather than as a separate phase. The CTO sensors also exhibited a typical p-type response upon exposure to C_4H_8 gas. All the CTO sensors prepared by AACVD show a higher response than that of the Cr_2O_3 sensors (Figure 8), although the CTO-1 sensor annealed for 6 h displayed a lower response compared to the one annealed for 30 h (CTO-2).

All the sensors were then exposed to NH_3, at the same relative humidity and operating temperature as the C_4H_8 gas but using different concentrations (75-25-75 ppm), and the results are shown in Figure 9. Obviously, the resistance values of the sensors also all increase upon exposure to NH_3, again showing the typical p-type property associated with Cr_2O_3/CTO. As previously, the Cr_2O_3-2 film annealed at 600 °C shows the largest response of the three Cr_2O_3-based sensors (Figure 8a). After doping Ti into the Cr_2O_3 thin films, the as-prepared CTO sensors exhibited a higher response to NH_3 compared to the

Cr_2O_3-based sensors (Figure 8b). The highest response for Cr_2O_3 at 400 °C and 75 ppm NH_3 is about 1.09, whereas the response for CTO under the same conditions is about 1.45. It is noteworthy that the annealing time influences the response of the CTO sensors to C_4H_8 and NH_3 gases.

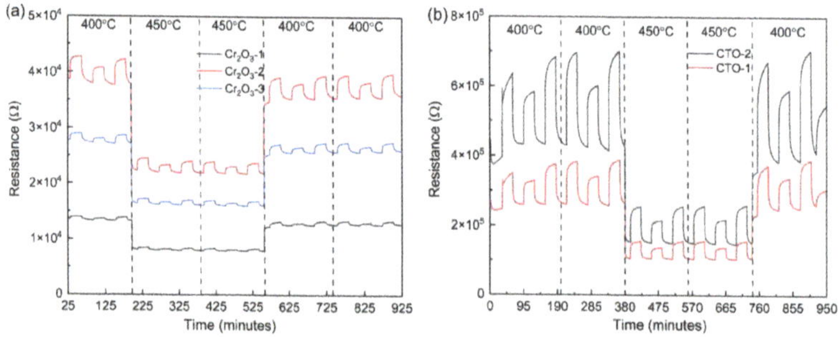

Figure 7. Change in resistance of the (**a**) Cr_2O_3-based sensors and (**b**) CTO-based sensors to C_4H_8 gas.

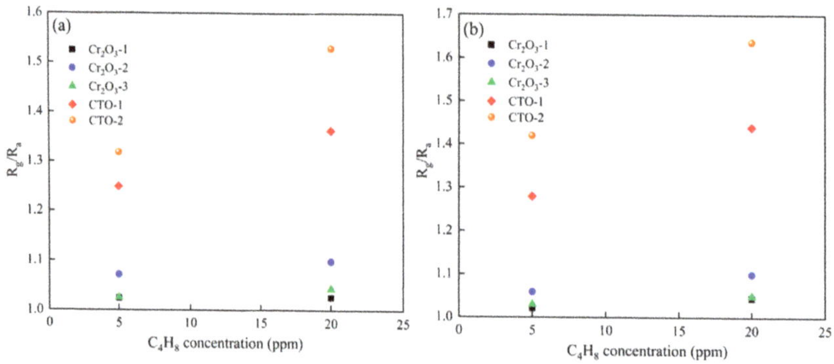

Figure 8. R_g/R_a against C_4H_8 concentration for each Cr_2O_3 and CTO sensor at different temperatures: (**a**) 400 °C; (**b**) 450 °C.

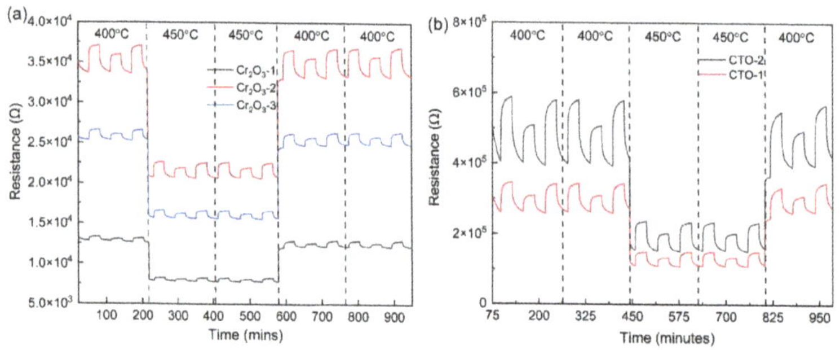

Figure 9. Change in resistance of the (**a**) Cr_2O_3-based sensors and (**b**) CTO-based sensors to NH_3.

3.3. Sensing Mechanism

In general, electron depletion theory is the most widely accepted sensing mechanism for the majority of MOSs [27]. However, an alternative self-consistent description of electronic change mediated by oxygen vacancies, rather than ionosorbed oxygen species, has been postulated for n-type MOS [28–30]. A similar mechanism for p-type materials is currently missing, despite the commonly employed electron depletion ex-planations for p-type materials requiring the surface oxygen acceptor species to be at vastly different energy to those expected in n-type materials, which must be close to the conduction band maximum in order for their adsorption and desorption to be thermodynamically accessible, for the descriptions to work (in fact below the valence band maximum), despite there being no rationalization for such a difference to exist. However, in direct contrast to n-type materials, oxygen vacancies found in p-type ma-terials are typically 'deep' in the bandgap (further from the conduction band edge than in n-type). Consequently, they will not be ionized at normal sensor operating temper-atures (Equation (2), the electrons arising from formation of the oxygen vacancy are not ionized to free electrons but rather are 'trapped' at the vacancy). Due to the pres-ence of the (relatively) numerous (majority carrier) holes in p-type materials it is ex-pected to be thermodynamically favorable for the electrons trapped at oxygen vacan-cies (within the bandgap) to recombine with holes (in the valence band). This means formation of oxygen vacancies would reduce the concentration of primary carriers in a p-type material (Equation (3)), hence leading to a decrease in conductivity (increase in resistance) with increasing oxygen vacancy concentration (a similar explanation ex-plains the observed increase in Cr_2O_3 sensor resistance with Ti(IV)-doping (CTO); to maintain electroneutrality for every two Ti(IV) substituted onto a Cr(III) site an oxygen vacancy must be formed).

$$2O_O \rightarrow O_2 + 2V_O'' \tag{2}$$

$$2V_O'' + 4h^{\cdot} \rightarrow 2V_O^x. \tag{3}$$

Therefore, under exposure to reducing conditions, the hole concentration is expected to decrease [31] and hence the resistance is expected to increase, at least at the surface of the material. As shown in Figure 10, exposure of Cr_2O_3 or CTO to NH_3 or C_4H_8 would cause the surface lattice oxygen concentration (O_O) to decrease as the lattice oxygen is consumed by oxidizing the gas species (R), i.e., the oxidation of the analyte gas causes the oxygen vacancy concentration in the sensing material to increase (V_O), which annihilates holes (4) and causes the resistance to rise, as observed in our tests.

$$2O_O + 4h^{\cdot} + 2R \rightarrow 2RO + 2V_O^x \tag{4}$$

Exposure to pure air would allow the surface to re-oxidize, reducing the concentration of oxygen vacancies and hence decreasing the (surface) resistance.

Assuming that the increases in resistance observed upon introduction of the analyte gas are due to their reaction with lattice oxygen, rather than some form of direct donor (analyte)/acceptor (sensor) behavior—which seems reasonable given that the strength of any donor interaction at the elevated temperatures used here (400–450 °C) must be questionable—the relative sensitivity towards these different reducing gases would then relate either to the different thermodynamics (for a steady-state 'saturated' response) or kinetics (for a transient response) for the oxidation of the gas species by lattice oxygen, i.e., the reactivity of a given analyte species towards oxidation. Whilst we do not have a direct comparison available in our test data, a comparison of the change in resistance of our CTO sensors towards 20 ppm C_4H_8 and 25 ppm NH_3 at 450 °C (closest to steady-state response) suggests that it is thermodynamically more favorable for C_4H_8 to be oxidized by CTO (either due to the ease of oxidation or due to a more complete oxidation) at an elevated temperature (higher resistance change observed for C_4H_8 at a similar concentration).

In addition, the difference in the sensing performance between Cr_2O_3 and CTO (e.g., Figure 8) may then be attributed to the difference in the relative change in oxygen vacancy concentration between the two materials (the lower initial hole concentration in CTO, displayed schematically in Figure 10, with each Ti inducing the loss of one hole, meaning a greater relative change is measured)—although without additional analysis techniques, it cannot be discounted that the presence of Ti has a direct (e.g., catalytic) effect on the reaction.

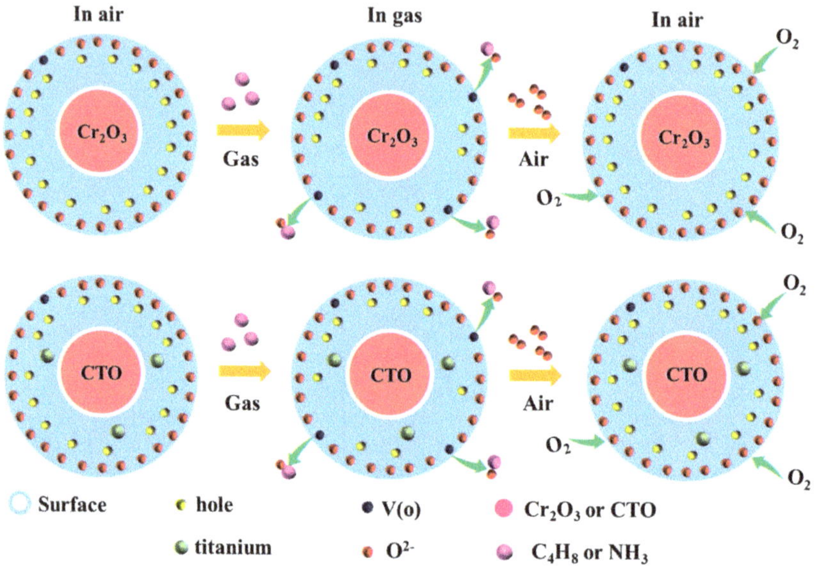

Figure 10. Schematic illustration of the possible sensing mechanism for Cr_2O_3 and CTO thin films when exposed to C_4H_8 or NH_3.

4. Conclusions

In this work, Cr_2O_3 and CTO thin films, from [Cr(CO)$_6$] and TBDAA precursors in a methanol solution, were deposited on the surface of an alumina platform using AACVD. Except for the patterns coming from the platform, XRD patterns for the CTO sensors only displayed the crystalline phase of eskolaite (Cr_2O_3), which shows no indication of phase separation. XPS analysis confirmed the presence of Ti^{4+} in the CTO thin films. Doping with Ti^{4+} increased the resistance, characteristic of an n-type dopant. All the as-prepared sensors based on Cr_2O_3 and CTO thin films exhibited good sensitivity and reproducibility in response to isobutylene and ammonia gases at a humidity of 50% RH. We have postulated a new mechanism for gas sensitivity in p-type metal oxides based on a varying oxygen vacancy concentration.

Author Contributions: Conceptualization, P.Z. and J.-H.T.; methodology, C.B.; validation, J.A.C., J.S. and E.D.; formal analysis, J.L.; investigation, J.-H.T.; resources, C.B.; data curation, P.Z.; writing—original draft preparation, P.Z.; writing—review and editing, P.Z. and C.B.; visualization, Y.S.; supervision, C.B.; project administration, C.B.; funding acquisition, C.B. All authors have read and agreed to the published version of the manuscript.

Funding: This research was funded by the Engineering and Physical Sciences Research Council (grant number EP/R512400/1), the China Scholarship Council (grant number 202006080083), and the Natural Science Foundation of Hebei Province (grant number E2023202041).

Institutional Review Board Statement: Not applicable.

Informed Consent Statement: Not applicable.

Data Availability Statement: The data presented in this study are available on request from the corresponding author.

Acknowledgments: The authors thank the Engineering and Physical Sciences Research Council (EP/R512400/1), the China Scholarship Council (CSC, No. 202006080083), and the Natural Science Foundation of Hebei Province (No. E2023202041) for providing financial support. The authors would also like to thank Alphasense Ltd. for packaging the sensors.

Conflicts of Interest: Author John Saffell employed by the company Nosmotech Ltd.; Author Ehsan Danesh employed by the company Advanced Sensing Technologies Ltd. The remaining authors declare that the research was conducted in the absence of any commercial or financial relationships that could be construed as a potential conflict of interest.

References

1. Ayyala, S.K.; Covington, J.A. Nickel-Oxide Based Thick-Film Gas Sensors for Volatile Organic Compound Detection. *Chemosensors* **2021**, *9*, 247. [CrossRef]
2. Agbroko, S.O.; Covington, J. A novel, low-cost, portable PID sensor for the detection of volatile organic compounds. *Sens. Actuators B* **2018**, *275*, 10–15. [CrossRef]
3. Ding, C.; Ma, Y.; Lai, X.; Yang, Q.; Xue, P.; Hu, F.; Geng, W. Ordered Large-Pore Mesoporous Cr_2O_3 with Ultrathin Framework for Formaldehyde Sensing. ACS Appl. *Mater. Interfaces* **2017**, *9*, 18170–18177. [CrossRef] [PubMed]
4. Jin, H.; Huang, Y.; Jian, J. Plate-like Cr_2O_3 for highly selective sensing of nitric oxide. *Sens. Actuators B* **2015**, *206*, 107–110. [CrossRef]
5. Zhang, M.; Sui, N.; Wang, R.; Zhang, T. The effect of shell thickness on gas sensing properties of core-shell fibers. *Sens. Actuators B* **2021**, *332*, 129456. [CrossRef]
6. Yamazoe, N. Toward innovations of gas sensor technology. *Sens. Actuators B Chem.* **2005**, *108*, 2–14. [CrossRef]
7. Moumen, A.; Kumarage, G.C.W.; Comini, E. P-Type Metal Oxide Semiconductor Thin Films: Synthesis and Chemical Sensor Applications. *Sensors* **2022**, *22*, 1359. [CrossRef]
8. Song, B.-Y.; Liu, L.-H.; Yang, M.; Zhang, X.-F.; Deng, Z.-P.; Xu, Y.-M.; Huo, L.-H.; Gao, S. Biotemplate-directed synthesis of Cr_2O_3 mesoporous monotubes for enhanced sensing to trace H_2S gas. *Sens. Actuators B* **2022**, *369*, 132294. [CrossRef]
9. Liu, B.; Han, W.; Li, X.; Li, L.; Tang, H.; Lu, C.; Li, Y.; Li, X. Quasi metal organic framework with highly concentrated Cr_2O_3 molecular clusters as the efficient catalyst for dehydrofluorination of 1,1,1,3,3-pentafluoropropane. *Appl. Catal. B* **2019**, *257*, 117939. [CrossRef]
10. Cheng, J.; Li, Y.; Zhong, J.; Lu, Z.; Wang, G.; Sun, M.; Jiang, Y.; Zou, P.; Wang, X.; Zhao, Q.; et al. Molecularly imprinted electrochemical sensor based on biomass carbon decorated with MOF-derived Cr_2O_3 and silver nanoparticles for selective and sensitive detection of nitrofurazone. *Chem. Eng. J.* **2020**, *398*, 125664. [CrossRef]
11. Atkinson, A.; Levy, M.R.; Roche, S.; Rudkin, R.A. Defect properties of Ti-doped Cr_2O_3. *Solid State Ion.* **2006**, *177*, 1767–1770. [CrossRef]
12. Henshaw, G.S.; Dawson, D.H.; Williams, D.E. Selectivity and composition dependence of response of gas-sensitive resistors. Part 2.—Hydrogen sulfide response of $Cr_{2-x}Ti_xO_{3+y}$. *J. Mater. Chem.* **1995**, *5*, 1791–1800. [CrossRef]
13. Blacklocks, A.N.; Atkinson, A.; Packer, R.J.; Savin, S.L.P.; Chadwick, A.V. An XAS study of the defect structure of Ti-doped α-Cr_2O_3. *Solid State Ion.* **2006**, *177*, 2939–2944. [CrossRef]
14. Shaw, G.A.; Pratt, K.F.E.; Parkin, I.P.; Williams, D.E. Gas sensing properties of thin film (≤3 μm) $Cr_{2-x}Ti_xO_3$ (CTO) prepared by atmospheric pressure chemical vapour deposition (APCVD), compared with that prepared by thick film screen-printing. *Sens. Actuators B* **2005**, *104*, 151–162. [CrossRef]
15. Du, J.; Wu, Y.; Choy, K.-L. Controlled synthesis of gas sensing $Cr_{2-x}Ti_xO_3$ films by electrostatic spray assisted vapour deposition and their structural characterisation. *Thin Solid Films* **2006**, *497*, 42–47. [CrossRef]
16. Du, J.; Wu, Y.; Choy, K.-L.; Shipway, P.H. Structure, properties and gas sensing behavior of $Cr_{2-x}Ti_xO_3$ films fabricated by electrostatic spray assisted vapour deposition. *Thin Solid Films* **2010**, *519*, 1293–1299. [CrossRef]
17. Conde-Gallardo, A.; Cruz-Orea, A.; Zelaya-Angel, O.; Bartolo-Pèrez, P. Electrical and optical properties of $Cr_{2-x}Ti_xO_3$ thin films. *J. Phys. D Appl. Phys.* **2008**, *41*, 205407. [CrossRef]
18. Vallejos, S.; Di Maggio, F.; Shujah, T.; Blackman, C. Chemical Vapour Deposition of Gas Sensitive Metal Oxides. *Chemosensors* **2016**, *4*, 4. [CrossRef]
19. Ehsan, M.A.; Shah, S.S.; Basha, S.I.; Hakeem, A.S.; Aziz, M.A. Recent Advances in Processing and Applications of Heterobimetallic Oxide Thin Films by Aerosol-Assisted Chemical Vapor Deposition. *Chem. Rec.* **2022**, *22*, e202100278. [CrossRef]
20. Srirattanapibul, S.; Nakarungsee, P.; Issro, C.; Tang, I.M.; Thongmee, S. Enhanced room temperature NH_3 sensing of rGO/Co_3O_4 nanocomposites. *Mater. Chem. Phys.* **2021**, *272*, 125033. [CrossRef]
21. Dey, A. Semiconductor metal oxide gas sensors: A review. *Mater. Sci. Eng. B* **2018**, *229*, 206–217. [CrossRef]
22. Natarajamani, G.S.; Kannan, V.P.; Madanagurusamy, S. Synergistically enhanced NH_3 gas sensing of graphene oxide-decorated Nano-ZnO thin films. *Mater. Chem. Phys.* **2024**, *316*, 129036. [CrossRef]
23. Li, D.; Han, D.; Chen, Y.; Hong, Y.; Duan, Q.; Wang, H.; He, X.; Zhao, L.; Wang, W.; Sang, S. GaN/rGO nanocomposite gas sensor for enhanced NH_3 sensing performances at room temperature. *Sens. Actuators B* **2024**, *403*, 135209. [CrossRef]
24. Song, B.-Y.; Zhang, X.-F.; Huang, J.; Cheng, X.-L.; Deng, Z.-P.; Xu, Y.-M.; Huo, L.-H.; Gao, S. Porous Cr_2O_3 Architecture Assembled by Nano-Sized Cylinders/Ellipsoids for Enhanced Sensing to Trace H_2S Gas. *ACS Appl. Mater. Interfaces* **2022**, *14*, 22302–22312. [CrossRef]
25. Dutta, S.; Pandey, A.; Leeladhar; Jain, K.K. Growth and characterization of ultrathin TiO_2-Cr_2O_3 nanocomposite films. *J. Alloys Compd.* **2017**, *696*, 376–381. [CrossRef]

26. Baraskar, P.; Chouhan, R.; Agrawal, A.; Choudhary, R.J.; Sen, P. Weak ferromagnetism at room temperature in Ti incorporated Cr_2O_3 thin film. *Phys. B* **2019**, *571*, 36–40. [CrossRef]
27. Williams, D.E. Semiconducting oxides as gas-sensitive resistors. *Sens. Actuators B Chem.* **1999**, *57*, 1–16. [CrossRef]
28. Blackman, C. Do We Need "Ionosorbed" Oxygen Species? (Or, "A Surface Conductivity Model of Gas Sensitivity in Metal Oxides Based on Variable Surface Oxygen Vacancy Concentration"). *ACS Sens.* **2021**, *6*, 3509–3516. [CrossRef]
29. Kucharski, S.; Ferrer, P.; Venturini, F.; Held, G.; Walton, A.S.; Byrne, C.; Covington, J.A.; Ayyala, S.K.; Beale, A.M.; Blackman, C. Direct in situ spectroscopic evidence of the crucial role played by surface oxygen vacancies in the O_2-sensing mechanism of SnO_2. *Chem. Sci.* **2022**, *13*, 6089–6097. [CrossRef]
30. Cox, D.F.; Fryberger, T.B.; Semancik, S. Oxygen vacancies and defect electronic states on the SnO_2 (110)-1 $\times$ 1 surface. *Phys. Rev. B* **1988**, *38*, 2072–2083. [CrossRef]
31. Gunkel, F.; Christensen, D.V.; Chen, Y.Z.; Pryds, N. Oxygen vacancies: The (in)visible friend of oxide electronics. *Appl. Phys. Lett.* **2020**, *116*, 120505. [CrossRef]

Disclaimer/Publisher's Note: The statements, opinions and data contained in all publications are solely those of the individual author(s) and contributor(s) and not of MDPI and/or the editor(s). MDPI and/or the editor(s) disclaim responsibility for any injury to people or property resulting from any ideas, methods, instructions or products referred to in the content.

chemosensors

MDPI

Article

CO_2 Interaction Mechanism of SnO_2-Based Sensors with Respect to the Pt Interdigital Electrodes Gap

Adelina Stanoiu [1], Alexandra Corina Iacoban [1], Catalina Gabriela Mihalcea [1,2], Ion Viorel Dinu [1], Ovidiu Gabriel Florea [1], Ioana Dorina Vlaicu [1] and Cristian Eugen Simion [1,*]

[1] National Institute of Materials Physics, Atomistilor 405A, 077125 Magurele, Romania; adelina.stanoiu@infim.ro (A.S.); alexandra.iacoban@infim.ro (A.C.I.); catalina.mihalcea@infim.ro (C.G.M.); vdinu@infim.ro (I.V.D.); ovidiu.florea@infim.ro (O.G.F.); ioana.vlaicu@infim.ro (I.D.V.)

[2] Faculty of Physics, University of Bucharest, Atomistilor 405, 077125 Magurele, Romania

* Correspondence: simion@infim.ro; Tel.: +40-21-2418134

Abstract: The tuning sensitivity towards CO_2 detection under in-field-like conditions was investigated using SnO_2-sensitive material deposited onto Al_2O_3 substrates provided with platinum electrodes with interdigital gaps of 100 μm and 30 μm. X-ray diffraction, low-magnification and high-resolution transmission electron microscopy, and electrical and contact potential difference investigations were employed to understand the sensing mechanism involved in CO_2 detection. The morpho-structural analysis revealed that the SnO_2 nanoparticles exhibit well-defined facets along the (110) and (101) crystallographic planes. Complex phenomenological investigations showed that moisture significantly affects the gas sensing performance. The experimental results corroborated the literature evidence, highlighting the importance of Pt within the interdigital electrodes subsequently reflected in the increase in the CO_2 sensing performance with the decrease in the interdigital gap. The catalytic efficiency is explained by the distribution of platinum at the gas-Pt-SnO_2 three-phase boundary, which is critical for enhancing the sensor performance.

Keywords: SnO_2 nanoparticles; CO_2 sensitivity; platinum electrodes; variable interdigital gap; catalytic activity of platinum electrodes

1. Introduction

The origin of metal oxide semiconductor (MOS)-based gas sensors may be traced back to the early 1970s, when Taguchi introduced the first gas sensing device utilising tin dioxide (SnO_2) [1]. In addition to its use as a transparent coating element for electrode material in solar cells or displays, SnO_2 is primarily involved in gas sensing and catalysis. The SnO_2-based sensors have been utilised in various industrial sectors, including automotive, chemical, environmental detection and control, food, medicine, military, and safety. The initial requirement involves monitoring the emissions from the vehicle, which mostly consist of carbon in the form of small particles, unburned hydrocarbons, carbon dioxide, carbon monoxide, nitrogen oxides, sulphur oxides, water vapour, and various other low-level chemicals. The need to monitor the levels of different gases has led to extensive research and development efforts to create sensors based on SnO_2 that are widely applicable in various fields [2]. Moreover, there are plenty of studies related to doped or un-doped SnO_2 specific to the aforementioned target sectors. When it comes to the detection of carbon dioxide (CO_2), MOS-based gas sensors are almost insensitive under normal working in-field conditions (e.g., the presence of oxygen and relative humidity—RH).

Carbon dioxide is an inert gas that absorbs short-wavelength light, contributing significantly to the greenhouse effect. It is commonly found in the atmosphere at concentrations between 300 and 400 parts per million (ppm). As the concentration of CO_2 in the environment rises, the human body will progressively encounter discomfort, difficulty breathing, impaired sensory function, dizziness, shock, and perhaps death [3]. Monitoring the concentration of CO_2 in respiration can provide valuable insights into respiratory status and circulation disorders [4]. It can also help in the timely detection of respiratory dysfunction and other abnormal situations during general anaesthesia, making it highly valuable in both the operating room and the intensive care department [5]. From an industrial perspective, the extensive utilisation of fossil fuels leads to an increase in the concentration of CO_2 in the atmosphere. Hence, the monitoring of CO_2 is interconnected with the well-being, health, and safety of individuals, as well as the government's oversight of carbon emissions and precise decision-making regarding energy structure regulation [6].

Currently, CO_2 detection is performed with optical sensors [7], resistive sensors [8], potentiometric sensors [9], amperometric sensors [10], capacitive sensors [11], surface acoustic wave sensors [12], and quartz crystal microbalance sensors [13], among others. Chemical gas sensors, which rely on the changes in the electrical resistance (conductance) of MOS under variable surrounding conditions, are the simplest, cheapest, and most user-friendly alternative [14]. In this respect, binary and ternary MOSs with high porosity, a large specific surface area, and defect structures have attracted the attention of researchers. For instance, ternary metal oxides exhibit Schottky, Frenkel, and interstitial atom defects, due to variations in crystal structure, cation species, and valence states. In addition, the presence of doped cations occupying the A/B sites will result in the formation of certain non-intrinsic defects, which in turn contribute to the absorption of the target gas. The primary ternary metal oxide materials utilised for CO_2 sensing are the perovskite structure (ABO_3) and spinel structure (AB_2O_4) oxides. These include barium titanate ($BaTiO_3$), lanthanum ferrite ($LaFeO_3$), nickel ferrite ($NiFe_2O_4$), and magnesium ferrite ($MgFe_2O_4$). Typically, they are manufactured to create integrated structures or heterostructures by combining them with metal oxides to improve their gas-sensing abilities [15,16]. In the work of Chavali et al., metal oxide nanoparticles are described by their gas-sensing performance concerning hierarchical structure, shape, and size. Not only the gas-sensing applications are discussed, but also their complex applications in nanotechnologies [17]. Unfortunately, the atmospheric relative humidity (RH) significantly affects the absorption and release of carbon dioxide. The impact of RH on the sensing mechanism was investigated using density functional theory (DFT). Under conditions of low humidity (RH < 35%), the interaction between carbon dioxide (CO_2) and tin dioxide (SnO_2) with pre-adsorbed oxygen ions (O^{2-} and O^-) from the (110) crystal surface was minimal. At a higher RH, CO_2 molecules were adsorbed onto the (110) crystal surface of SnO_2. During this process, CO_2 reacts with OH^- ions (from the dissociative adsorption of water) and leads to the formation of carbonate. Additionally, an exchange of electrons between CO_2 and SnO_2 occurs. The results suggest that the pre-adsorbed OH^- on the (110) crystal facet enhances the CO_2 sensing properties [18]. This indicates that certain crystal surfaces have a strong ability to adsorb or react with various atomic, ionic, and molecular groups. Xiong et al. investigated the sensing properties of SnO_2 films doped with La, Gd, and Lu, as well as pure SnO_2 films, for the detection of CO_2 in an environment devoid of oxygen [19]. La significantly enhanced the performance of CO_2 sensing, with the highest performance observed when the La doping content reached 4%. The rate at which carbonates develop and the amount of oxygen vacancies have a significant impact on the features of gas sensing. Consequently, the researchers analysed the XPS spectra of the O 1 s and C 1 s samples. They then determined the ratio of peak areas for residual carbon (organic carbon pollution) and carbonates, as well as the ratio of peak areas for oxygen (O^- and O^{2-}). The optimal doping ratio was determined to be 8 at. % when the concentrations of O_2 in the carrier gas reached 21% (in-field oxygen level), and it should increase to 16 at.% in the presence of free oxygen.

In the present study, we aimed to address a realistic scenario of the gas-surface interaction mechanisms involved in CO_2 detection under in-field-like conditions characterised by the presence of oxygen and variable relative humidity. Therefore, we performed DC electrical investigations and simultaneous electrical resistance measurements assisted by the contact potential differences on planar sensors, which were obtained by screen-printed SnO_2 thick layers on substrates with different interdigital gaps of 100 and 30 μm, respectively.

2. Materials and Methods

2.1. Powders Synthesis

As previously reported [20], a sodium hydroxide solution was added dropwise over a solution of tin (IV) chloride containing CTAB (hexadecyltrimethylammonium bromide). The mixture was stirred for half an hour at room temperature, was afterward sealed in a Hydrothermal Synthesis Autoclave Reactor with a PTFE Lined Vessel of 100 mL, and was left to react at 160 °C for 18 h. SnO_2 powder was isolated by centrifugation, washed repeatedly, and air-dried at 120 °C. To complete the synthesis process, the dried powder was thermally treated at 550 °C. The chemicals were used as purchased, without any further purification: tin (IV) chloride ($SnCl_4$, 99.999%, Acros Organics, Geel, Belgium), sodium hydroxide (NaOH, >98%, Honeywell Fluka, Seelze, Germany), and hexadecyltrimethylammonium bromide (CTAB, 99%, Acros Organics, Geel, Belgium). A final thermal treatment at 550 °C has completed the preparation processes.

2.2. Structural and Morphological Investigations

Structural investigations were performed using powder X-ray diffraction with a D8 ADVANCE diffractometer (BRUKER-57AXS GmbH, Karlsruhe, Germany) with Ni-filtered Cu radiation (λ = 1.54184 Å) at room temperature in Bragg–Brentano geometry in the range of 2θ from 20° to 95°. The XRD data were analysed using the Rietveld refinement with MAUD version 2.99 software to determine the lattice parameters and average crystallite size.

The morphology of the SnO_2 sample was investigated using a JEOL JEM-ARM200F transmission electron microscope (JEOL LTT., Japan, Tokyo), operated at 200 kV. To prepare the TEM sample, a small quantity of the powder was mixed with ethanol, then it was ultra-sonicated for 10 min, and a small droplet from this mix was drop-cast on a TEM copper grid provided with a carbon membrane.

2.3. Layer Deposition and Sensing Investigations

The layer deposition process is the subject of a national patent application, OSIM/Nr. A 00110/18 March 2024. In brief, the as-prepared sensitive powders were combined with 1,2 Propanediol and grounded to produce a paste with a medium viscosity. Subsequently, the paste was applied onto commercial alumina substrates using the screen-printing technique to form thick layers. The sensors obtained were gradually dehydrated and thermally treated at 500 °C in air. This procedure enables the thorough elimination of the organic solvent, guaranteeing the porosity of the layer and enhancing its adhesion to the substrate. The alumina substrates are made using planar technology and are provided with platinum electrodes and a heater on the backside. The interdigital electrodes on the substrate have gaps measuring 100 μm and 30 μm, respectively. The obtained samples were labelled "SnO_2 100 μm" and "SnO_2 30 μm". The electrical power through the heater is adjusted to control the temperature of the sensor layer, which in turn modulates the chemical interaction between the MOS layer and the test gases. The substrates were passivated with glass on both sides to avoid parasitic catalytic effects, thus highlighting the catalytic processes that are taking place solely on the sensitive layer or at the three-phase boundary (e.g., gas–electrodes–SnO_2).

A Gas Mixing System (GMS) was used to simulate the in-field atmosphere (Figure 1). The GMS is managed by specialised software and comprises eleven gas channels fitted with mass flow controllers, solenoid valves, and high-purity gas bottles. The system functions in a dynamic state, maintaining a total flow rate of 200 mL/min. A relative humidity (RH) is accomplished using a distinct pathway involving a gas-washing bottle containing moisture. The real-time acquisition of the electrical resistance changes in sensors positioned inside the sensor chamber was performed using a dedicated electrometer Keithley 6517A (Tektronix, OHIO, Solon, OH, USA) operated under constant 3V bias, while the contact potential differences (CPD) were recorded using a McAllister 6500 Kelvin Probe (McAllister Technical Services, CALIFORNIA, Berkeley, CA, USA).

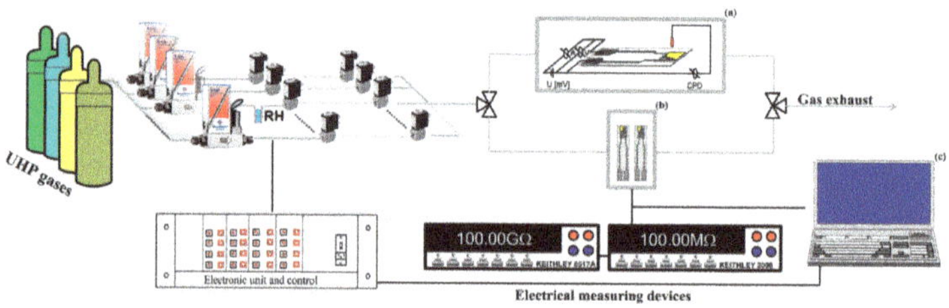

Figure 1. Gas Mixing System provided with a Kelvin probe involved in DC and contact potential difference measurements (**a**), a sensor chamber (**b**), and a PC dedicated to evaluation and control (**c**).

3. Results and Discussions

3.1. XRD Characterisation

The powder X-ray diffraction (XRD) of the SnO_2 sample prepared by the hydrothermal method at a temperature of 160 °C presents the characteristic diffraction lines of the SnO_2 (Figure 2). It exhibits a tetragonal structure with a symmetry space group of $P4_2/mnm$, according to CIF no. 1526637 from the COD database.

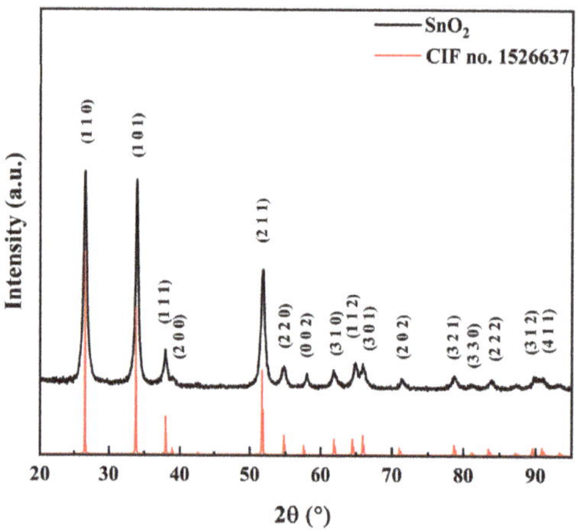

Figure 2. XRD pattern of SnO_2.

Through Rietveld refinement, using the MAUD version 2.99 software, the lattice parameters and the average crystallite size of 21.47 ± 0.19 nm were determined. The crystallographic parameters for the lattice are slightly different than the reference data.

3.2. Analytical TEM Characterisation

The low-magnification TEM images were used to characterise the sample's morphology. Figure 3a shows that the sample consists of nanoparticles with the following different dimensions: smaller quasi-spherical nanoparticles (having ~9 nm) and larger elongated nanoparticles (having ~35 nm).

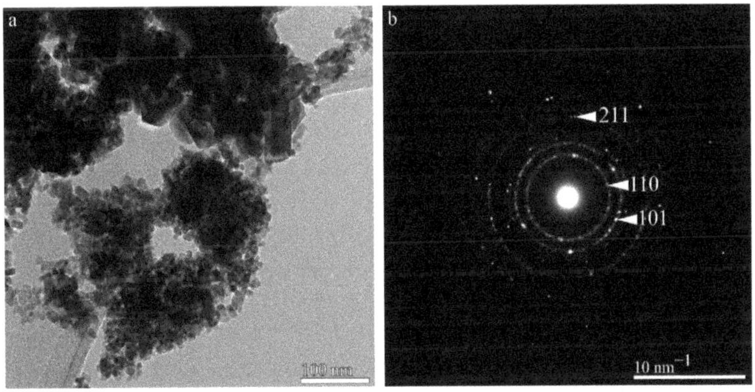

Figure 3. TEM image of SnO_2 (**a**) and the corresponding SAED pattern, revealing the tetragonal structure of SnO_2 (**b**).

The structure of the sample was determined using selected area electron diffraction (SAED). The SAED pattern shown in Figure 3b is typical for a polycrystalline material, and it reveals the tetragonal structure of SnO_2, with the space group $P4_2/mnm$. The white arrows are pointing towards crystallographic planes (110), (101), and (211), as indexed according to cif no. 2104754 from COD database.

The high-resolution TEM (HRTEM) investigations revealed the presence of (110) and (101) crystallographic facets (Figure 4). Kuncser et al. [20], using VESTA modelling, showed that (110) and (101) facets mostly have oxygen terminations.

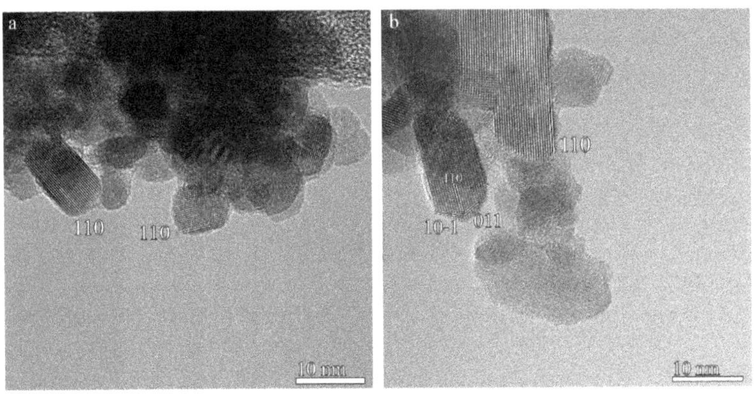

Figure 4. HRTEM images showing faceted nanoparticles having different dimensions.

Figure 4a shows two nanoparticles having the edges along (110) crystallographic planes. The double white lines in Figure 4b indicate the (110), (10-1), and (011) crystallographic planes of a nanoparticle oriented along the [1–11] axis.

The average dimension was determined by measuring ~400 nanoparticles using low-magnification TEM images similar to the TEM image shown in Figure 3a. The size distribution is shown in Figure 5. After fitting the as-obtained histogram with a log-normal function, we obtained a mean value of 15.4 nm and a standard deviation of 6.9 nm. Within the error limits, the mean value of the particle size obtained from TEM is in accordance with the value of the crystallite size obtained from Rietveld refinement.

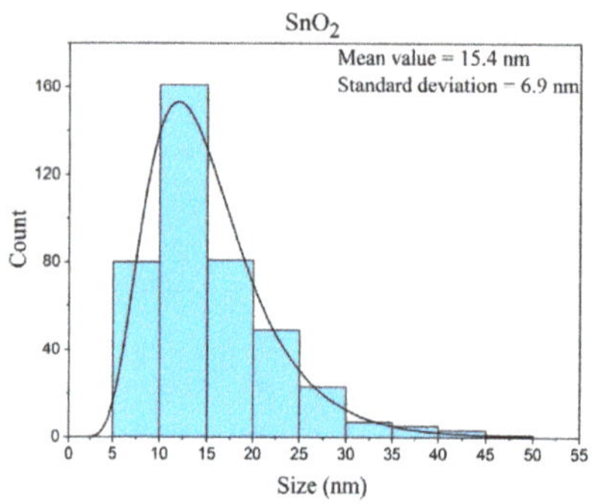

Figure 5. SnO_2 nanoparticles size distribution histogram fitted using a log-normal function.

In conclusion, the morpho-structural TEM investigations have shown that the sample consists of nanoparticles with different dimensions, having well-defined facets along (110) and (101) crystallographic planes [21–23].

3.3. Sensing Characterisation

The response of MOS sensors depends on the operating temperature of the gas-sensitive material, which determines the adsorption of atmospheric oxygen in its molecular or ionic forms. The response capacity decreases at higher working temperatures due to the different rates of the adsorption/desorption processes [24]. Consequently, we conducted an initial experiment to observe the changes in the sensor signal as the operating temperature varied in the range of 100–400 °C for sensors based on SnO_2 thick layers deposited onto interdigital Pt electrodes with alternative gaps of 100 and 30 μm. The sensor signal was determined by calculating the ratio of the electrical resistance measured under a reference atmosphere (e.g., a synthetic air 5.0 background with 50% relative humidity) to the electrical resistance measured while exposed to 3000 ppm of CO_2 under the same RH conditions. The highest level of sensitivity was achieved at an operating temperature of 350 °C (Figure 6a).

When the sensors operate at moderate temperatures, oxygen physisorption on the MOS surface is generally accepted. This process does not entail any transfer of electrical charge from the sensitive material to the adsorbed molecular oxygen. When the temperature rises, oxygen chemisorption occurs at the surface of the SnO_2 grains. Consequently, electron-depleted layers are formed near the surface of the grains due to the electron transfer from the sensitive material to the adsorbed O^- oxygen species. In terms of energy, the depleted layers act as potential barriers, impeding the flow of electrons between the boundaries of the grains [25].

Considering the interdigital gap as $d \approx N \times g$ (where N is the number of grains and g is the average grain size), the larger d is, the more grains we have and thus the more intergranular barriers. From the electrical point of view, this situation is associated with electrical resistance in dry air conditions, explaining the higher resistance in the case of SnO_2 deposited on the substrate with an interdigital gap of 100 μm (Figure 6b).

In humid conditions, the electrical resistance decreases due to the known reducing effect of RH [26].

$$2Sn + O^- + H_2O \rightarrow 2(Sn^{\delta+} - OH^{\delta-}) + e^- \quad (1)$$

A similar behaviour is related to the presence of CO_2, although it is known as a rather inert gas [27]. Herein, the resistance of SnO_2 decreases under CO_2 exposure, suggesting interactions with previously chemisorbed O^- (Equation (2)) or OH^- (Equation (3)) species, leading to the release of electrons in CB.

$$Sn + O^- + CO_2 \rightarrow Sn - CO_3{}^- + V_O{}^+ + e^- \quad (2)$$

$$(Sn^{\delta+} - OH^{\delta-}) + CO_2 \rightarrow Sn + HCO_3{}^- + V_O{}^{2+} + 2e^- \quad (3)$$

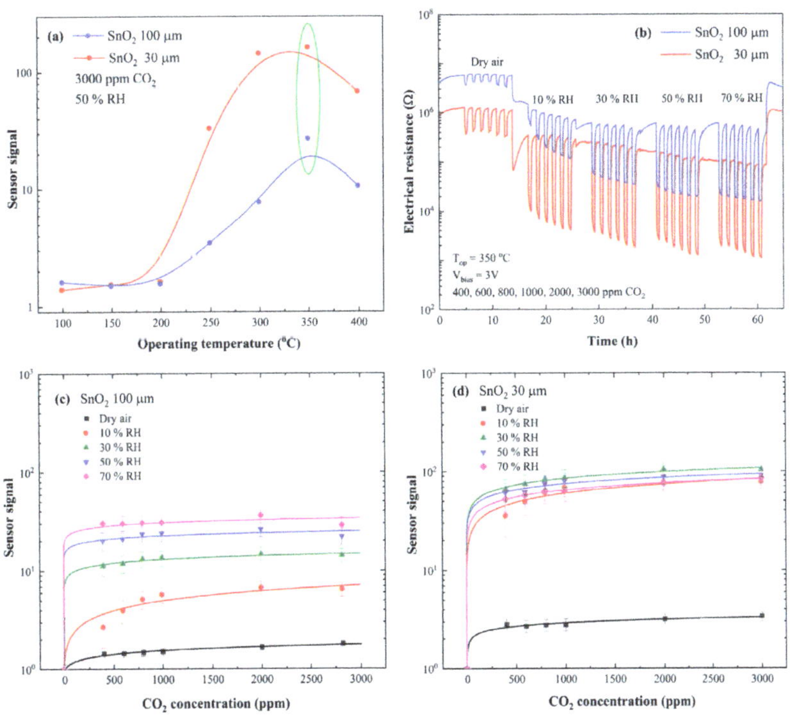

Figure 6. The dependence of the sensor signal on the operating temperature (**a**), the behaviour of the electrical resistance in the atmosphere with variable RH and CO_2 concentrations for an operating temperature of 350 °C (**b**), and the sensor signal to CO_2 for SnO_2 100 μm versus SnO_2 30 μm (**c**,**d**).

Translating the electrical resistance changes in the sensor signals for both sensitive structures (Figure 6c,d), it can be observed that the smaller interdigital gap (30 μm), the higher the sensor signal is. Moreover, one can see that both sensors exhibit a special relation with water, as long as the CO_2 sensitivity increases in the presence of RH compared to dry air. Complex phenomenological investigations (e.g., simultaneous electrical resistance and contact potential differences) provide the perspective of an explanation regarding this behaviour (Figure 7).

The CPD measurements provide insights into the changes in work function ($\Delta\Phi$), allowing us to discriminate between physisorption (associated with changes in the electronic affinity, $\Delta\chi$) and ionosorption (associated with changes in the surface potential barrier, $q\Delta V_s$). Considering the electrochemical potential (μ) of the SnO_2 constant, we can write

$$q \times \Delta CPD = \Delta\Phi = \Delta\chi + q\Delta V_s = \Delta\chi - k_B T ln R_{air}/R_{gas} \quad (4)$$

where q is the electron charge, ΔCPD is the contact potential difference measured with the Kelvin Probe, k_B is the Boltzmann constant, T is the operating temperature, R_{air} is the electrical resistance measured under humid synthetic air, and R_{gas} is the electrical resistance measured under CO_2 exposure.

As shown in Figure 7a,b, $\Delta\Phi$ does not vary because it is the same SnO_2 material deposited onto substrates with different interdigital gaps. On the other hand, the greater the water effect ($\Delta\chi$), the greater the potential barrier variation ($q\Delta V_s$).

At this point, it is important to understand whether the differences induced by the interdigital gap are determined by their electrical contribution to the total resistance of the sensors or by the catalytic activity of the platinum electrodes.

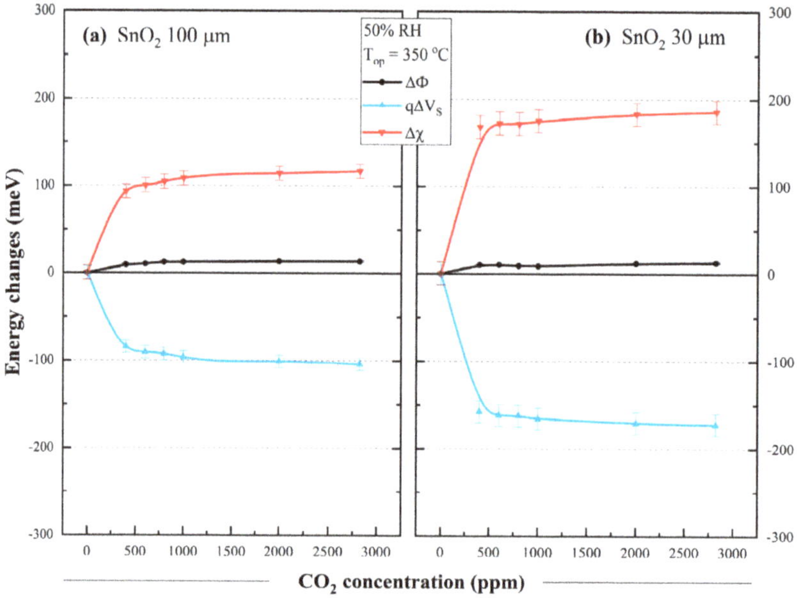

Figure 7. CO_2 influence over the potential changes: SnO_2 100 μm (**a**) and SnO_2 30 μm (**b**) for T_{op} = 350 °C and 50% RH.

The first hypothesis is supported by the geometrical approach of a SnO_2 polycrystalline sensitive material, considering incompletely depleted grains, arranged between Pt electrodes with interdigital gaps (L) of 100 and 30 μm (Figure 8a). The so-called surface state model [28] was used to represent the surface and bulk energetic dependencies with respect to the percolation path phase of the electric current (Figure 8b).

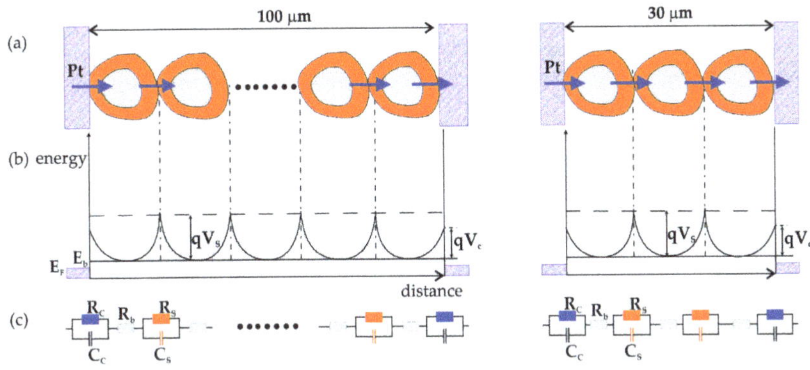

Figure 8. Schematic representation of the sensitive structure based on SnO_2 deposited on a substrate with Pt electrodes with different interdigital gaps L (**a**), energy band bending (**b**), and the equivalent electrical circuit (**c**).

For gas sensors consisting of a thick polycrystalline-sensitive MOS layer, the conduction mechanism is dominated by surface phenomena. Exposure to an in-field-like atmosphere and operating temperature favours oxygen and water vapour chemisorption, determining the covering of the surface with the negative charge. This causes the appearance of a depletion layer near the surface due to the Coulomb repulsion towards the electrons in the bulk. Thus, the conduction is limited by the Schottky barriers (qVs) formed at the grain boundaries. In terms of the electrical equivalent circuit (EEC), the surface is associated with a resistance Rs in parallel with a capacity Cs (Figure 8c). These are controlled by the charging/discharging of the surface traps which are directly modulated by the surface chemical gas interactions [29].

The resistance Rb corresponds to the electrons in the bulk and is weakly influenced [30]. The SnO_2-Pt electrode contact is associated with the RcCc parallel circuit.

The electrons will flow from the semiconductor towards the metal because the SnO_2 work function (4.53 eV) is lower than the Pt work function (5.64 eV) [31,32]. Accordingly, the energy bands bend up due to the depletion of electrons in the semiconductor, and the electrons flowing towards the Pt electrode must cross over the potential barrier qVc, as in the case of the grain-to-grain Schottky element. But, taking into account that the Pt resistance is much lower than the SnO_2 surface resistance ($Rc \ll Rs$), its contribution can be neglected, whether the interdigital gap L is 100 or 30 μm.

By close observation and through the aforementioned reasons, the electrical influence of the interdigital electrodes on a thick film gas sensor can be considered negligible.

The second hypothesis is supported by the known spill-over effect of Pt [33,34]. In brief, when a reducing gas is adsorbed on Pt, its dissociation is activated, followed by migration (spill-over) on the MOS surface where it interacts with oxygen, increasing the conductivity in the case of an n-type semiconductor [35].

In addition to the effects induced by impregnation/doping with noble metals, the role of Pt electrodes must be considered in terms of the geometric and catalytic effects. Several authors [36,37] have pointed out the significant role of the geometry influence of the electrodes placed at the bottom region of the sensitive MOS layer. When a certain gas molecule hits the MOS surface, it diffuses through the porous film and induces an electrical effect at the bottom electrodes (so-called "three-phase boundary regions"). Starting from the empiric demonstration of the differences in sensitivity induced by the variation in the electrode gaps, Hoefer et al. [38,39] propose a rigorous work aiming to establish precisely the role of interdigital geometry on the sensing performance. As such, a sensor array with a highly asymmetric electrode configuration could develop a low-cost sensor system with improved selectivity (e.g., the selective detection of NO_2 versus CO by varying the interdigital electrode gaps).

On the other hand, the Pt electrode shows increased activity in the direct conversion of CO_2, with implications on the electrical properties of the SnO_2 sensitive layer [40]. Besides direct conversion, the spillover effect activates the gas adsorption on the Pt, which subsequently leads to the diffusion or migration of the adsorbed species on the neighbouring gains [41]. By spanning the literature, one can find that the chemical activation of CO_2 as a quite inert gas is still a matter of debate, especially with MOS-based gas sensors. The complex work of Wang et al. [42] stated that Pd and Pt represent the most favourable metals for CO_2 conversion. The role of Pt as a single-atom catalyst has attracted extensive attention due to the superior catalytic performance towards CO_2 conversion via a reverse water gas shift reaction (RWGS). Thus, the work of He et al. [43] presented a simple route to load 6.4 wt %Pt on silicon carbide with superior catalytic properties. Mondoza-Núñez et al. [44] investigated the effect of Pt addition on CO_2 methanation catalysed by ZrO_2-supported Rh. Quite similar results have been obtained by Ramos Gonzalez et al. [40] on pristine Pt-catalysed SnO_2 hybrid nanostructures when operated at low temperatures (100 °C), envisaging a gas sensing mechanism based on the adsorption of OH^- ions onto the SnO_2 surface. The sensing mechanism was attributed to the electron transfer from the porous silicon-based band to the conduction band of the SnO_2. In the case of ZnO nano-ribbons [45], the role of Pt was used alongside Pd, Fe, Ag, and Au as simple adatom catalysts towards sensing H_2, H_2S, and CO_2.

With the aforementioned review of the literature and considering the technological details of our samples (Figure 9), it is beyond doubt that the noble parts of the sensor (e.g., at the three-phase boundary between gas–Pt interdigital electrodes and SnO_2) have a great share in the catalytic activity of the CO_2 gas.

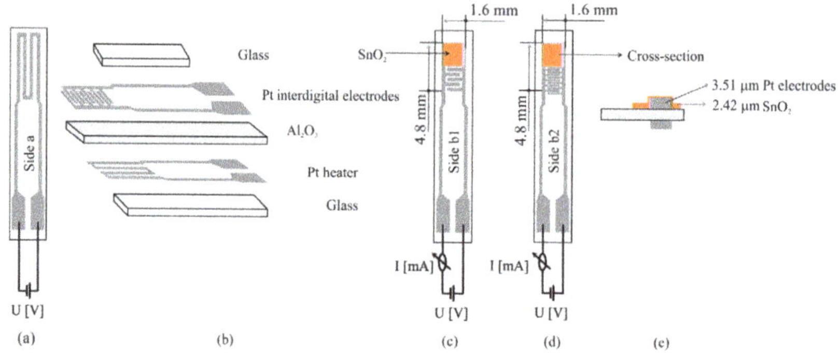

Figure 9. Planar substrate sensor overview. Side a represents the Pt heater meander (**a**) and sensor components parts (**b**); Side b1 represents the interdigital electrodes with a 100 µm gap (**c**); Side b2 represents the interdigital electrodes with a 30 µm gap (**d**) and a cross-section of the sensor (**e**).

Specifically, the thickness of the SnO_2 layer (2.42 µm) and the Pt electrodes (3.51 µm) was determined with the Ambios XP-100 Profilometer, and their ratio suggests that the dominating amount of Pt controls the overall sensing effect.

It is worth mentioning that besides the Pt interdigital electrodes (e.g., where solely catalytic effects take place), other Pt regions are covered with glass to avoid parasitic CO_2 catalytic effects.

In this context, we strongly believe that the sensitivity to CO_2 increases with the proportion of Pt in the sensitive structure. This explains the higher sensor signal when SnO_2 is deposited over electrodes with an interdigital distance of 30 µm, compared to 100 µm.

These results are consistent with the results from the literature which highlights the role of Pt in CO_2 reactivity; although, in the field of chemical sensors, the findings are less true for MOS-sensitive materials.

4. Conclusions

SnO_2 powder was obtained by hydrothermal growth at a temperature of 160 °C. A final thermal treatment at 550 °C has completed the synthesis processes. X-ray diffraction has shown a tetragonal structure, and the Rietveld refinement revealed a crystallite size of ~21.5 ± 0.19 nm. The morpho-structural TEM/HRTEM investigations highlight nanoparticles with different dimensions, having a mean value of 15.4 nm for a standard deviation of 6.9 nm and well-defined facets along (110) and (101) crystallographic planes. The SnO_2 powder was transformed into a paste by mixing it with 1,2 Propanediol, and then it was screen-printed as a thick layer onto commercial Al_2O_3 substrates having Pt electrodes with alternative interdigital gaps of 100 µm and 30 µm. The obtained sensors labelled "SnO_2 100 µm" and "SnO_2 30 µm" were thermally treated at 500 °C in air to eliminate the organic solvent and ensure SnO_2 adhesion to the substrate. The electrical power through the substrate backside heater controlled the temperature of the SnO_2 layer, favouring the chemical interaction with gases. A Gas Mixing System was used to simulate the in-field-like test atmosphere (synthetic air having different relative humidity and CO_2 concentration). The contact potential differences (CPDs) induced by the chemical interactions on the SnO_2 surface were recorded using a McAllister 6500 Kelvin Probe. Thus, it was possible to differentiate between water physisorption associated with changes in the electronic affinity and oxygen ionosorption associated with changes in the surface potential barrier and, consequently, with the electrical resistance. The differences induced by the interdigital gap over the sensor signal to CO_2 were analysed in terms of the electrical equivalent circuit and catalytic activity of the platinum electrodes. The thickness ratio of 2.42 µm/3.51 µm between the SnO_2 layer and the Pt electrodes allowed us to conclude that the amount of Pt specific for a 30 µm interdigital gap determines the increase in the sensing response to CO_2 in accordance with the catalytic effect reported in the literature.

Author Contributions: Methodology, writing—review and editing, funding acquisition, A.S.; investigation, writing—original draft, A.C.I.; investigation, writing—original draft, C.G.M.; formal analysis, writing—review and editing, I.V.D.; investigation, validation, O.G.F.; methodology, writing—original draft, I.D.V.; conceptualization, writing—original draft, writing—review and editing, C.E.S. All authors have read and agreed to the published version of the manuscript.

Funding: This research was funded by the Ministry of Research, Innovation and Digitization CNCS-UEFISCDI through the project PN-III-P4-PCE-2021-0384 within PNCDI III and the Core Program within the National Research Development and Innovation Plan 2022–2027, carried out with the support of the Ministry of Research, Innovation and Digitization, project no. PC1-PN23080101. The contribution of Catalina G. Mihalcea (C.G.M.) to this work is part of the PhD "Nanostructured materials for gas sensing: correlations between functional, electronic and microstructural properties" supported by CERIC-ERIC.

Institutional Review Board Statement: Not applicable.

Informed Consent Statement: Not applicable.

Data Availability Statement: Data are available on request from the authors.

Conflicts of Interest: The authors declare no conflicts of interest. The funders had no role in the design of the study; in the collection, analyses, or interpretation of data; in the writing of the manuscript, or in the decision to publish the results.

References

1. Taguchi, N. Gas Detecting Device. U.S. Patent US3695848A, 3 October 1972.
2. Kong, Y.; Li, Y.; Cui, X.; Su, L.; Ma, D.; Lai, T.; Yao, L.; Xiao, X.; Wang, Y. SnO_2 nanostructured materials used as gas sensors for the detection of hazardous and flammable gases: A review. *Nano Mater. Sci.* **2022**, *4*, 339–350. [CrossRef]
3. Zheng, Z.; Wang, Y.-S.; Wang, M.; Zhao, G.-H.; Hao, G.-P.; Lu, A.-H. Anomalous enhancement of humid CO_2 capture by local surface bound water in polar carbon nanopores. *Nat. Commun.* **2024**, *15*, 8919. [CrossRef]
4. Ramanathan, S.; Malarvili, M.B.; Gopinath, S.C.B. Assessing respiratory complications by carbon dioxide sensing platforms: Advancements in infrared radiation technology and IoT integration. *Arab. J. Chem.* **2023**, *16*, 104478. [CrossRef]
5. Fleming, L.; Gibson, D.; Song, S.; Li, C.; Reid, S. Reducing N_2O induced cross-talk in a NDIR CO_2 gas sensor for breath analysis using multilayer thin film optical interference coatings. *Surf. Coat. Technol.* **2018**, *336*, 9–16. [CrossRef]
6. Zubair, M.; Chen, S.; Ma, Y.; Hu, X. A Systematic Review on Carbon Dioxide (CO_2) Emission Measurement Methods under PRISMA Guidelines: Transportation Sustainability and Development Programs. *Sustainability* **2023**, *15*, 4817. [CrossRef]
7. Almawgani, A.H.M.; Fathy, H.M.; Elsayed, H.A.; Ali, Y.A.A.; Mehaney, A. A promising ultra-sensitive CO_2 sensor at varying concentrations and temperatures based on Fano resonance phenomenon in different 1D photonic crystal designs. *Sci. Rep.* **2023**, *13*, 15028. [CrossRef]
8. Maier, C.; Egger, L.; Köck, A.; Reichmann, K. A Review of Gas Sensors for CO_2 Based on Copper Oxides Their Derivatives. *Sensors* **2024**, *24*, 5469. [CrossRef]
9. Molino, D.; Ferraro, G.; Lettieri, S.; Zaccagnini, P.; Etzi, M.; Astorino, C.; Nardo, E.D.; Bartoli, M.; Lamberti, A.; Pirri, C.F.; et al. Enhanced CO_2 Detection Using Potentiometric Sensors Based on PIM-1/DBU Imidazolate Membranes. *Adv. Sustain. Syst.* **2024**, 2400415. [CrossRef]
10. Steininger, F.; Revsbech, N.P.; Koren, K. Total Dissolved Inorganic Carbon Sensor Based on Amperometric CO_2 Microsensor and Local Acidification. *ACS Sens.* **2021**, *6*, 2529–2533. [CrossRef]
11. Héraly, F.; Sikdar, A.; Chang, J.; Yuan, J. Capacitive CO_2 sensor made of aminated cellulose nanofibrils: Development and optimization. *New J. Chem.* **2024**, *48*, 6064–6070. [CrossRef]
12. Wang, C.; Ding, Y.; Li, M.; Li, H.; Xu, S.; Li, C.; Qian, L.; Yang, B. Surface acoustic wave sensor based on Au/TiO_2/PEDOT with dual response to carbon dioxide and humidity. *Anal. Chim. Acta* **2022**, *1190*, 339264. [CrossRef]
13. Park, K.; Koh, M.; Yoon, C.; Kim, H.; Kim, H. The behavior of quartz crystal microbalance in high pressure CO_2. *J. Supercrit. Fluids* **2004**, *29*, 203–212. [CrossRef]
14. Domènech-Gil, G.; Samà, J.; Fàberga, C.; Gràcia, I.; Cané, C.; Barth, S.; Romano-Rodríguez, A. Highly sensitive SnO_2 nanowire network gas sensors. *Sens. Actuators B Chem.* **2023**, *383*, 133545. [CrossRef]
15. Casanova-Chafer, J.; Garcia-Aboal, R.; Llobet, E.; Atienzar, P. Enhanced CO_2 sensing by oxygen plasma treated perovskite-graphene nanocomposites. *ACS Sens.* **2024**, *9*, 830–839. [CrossRef] [PubMed]
16. Jiao, A.; Zhang, Y.; Yang, L.; Zhao, X.; Wu, C.; Chen, T.; Zhan, R.; Huang, Z.; Lin, H. Enhanced CO_2 response of $La_{1\text{-}x}FeO_{3\text{-}\delta}$ perovskites with A-site deficiency synthesized by flame spray pyrolysis. *Ceram. Int.* **2023**, *49*, 591–599. [CrossRef]
17. Chavali, M.S.; Nikolova, M.P. Metal oxide nanoparticles and their application in nanotechnology. *SN Appl. Sci.* **2019**, *1*, 607. [CrossRef]
18. Batzill, M. Surface Science Studies of Gas Sensing Materials: SnO_2. *Sensors* **2006**, *6*, 1345–1366. [CrossRef]
19. Xiong, Y.; Xue, Q.; Ling, C.; Lu, W.; Ding, D.; Zhu, L.; Li, X. Effective CO_2 detection based on LaOCl-doped SnO_2 nanofibers: Insight into the role of oxygen in carrier gas. *Sens. Actuators B Chem.* **2017**, *241*, 725–734. [CrossRef]
20. Kuncser, A.C.; Vlaicu, I.D.; Dinu, I.V.; Simion, C.E.; Iacoban, A.C.; Florea, O.G.; Stanoiu, A. The impact of the synthesis temperature on SnO_2 morphology and sensitivity to CO_2 under in-field conditions. *Mater. Lett.* **2022**, *325*, 132855. [CrossRef]
21. Sopiha, K.V.; Malyi, O.I.; Persson, C.; Wu, P. Chemistry of Oxygen Ionosorption on SnO_2 Surfaces. *ACS Appl. Mater. Interfaces* **2021**, *13*, 33664–33676. [CrossRef]
22. Masuda, Y. Recent advances in SnO_2 nanostructure based gas sensors. *Sens. Actuators B Chem.* **2022**, *364*, 131876. [CrossRef]
23. Masuda, Y. Facet controlled growth mechanism of SnO_2 (101) nanosheet assembled film via cold crystallization. *Sci. Rep.* **2021**, *11*, 11304. [CrossRef] [PubMed]
24. Rescalli, A.; Marzorati, D.; Gelosa, S.; Cellesi, F.; Cerveri, P. Temperature Modulation of MOS Sensors for Enhanced Detection of Volatile Organic Compounds. *Chemosensors* **2023**, *11*, 501. [CrossRef]

25. Barsan, N.; Weimar, U. Conduction Model of Metal Oxide Gas Sensors. *J. Electroceram.* **2001**, *7*, 143–167. [CrossRef]
26. Barsan, N.; Koziej, D.; Weimar, U. Metal oxide-based gas sensor research: How to? *Sens. Actuators B Chem.* **2007**, *121*, 18–35. [CrossRef]
27. Staerz, A.; Weimar, U.; Barsan, N. Current state of knowledge on the metal oxide based gas sensing mechanism. *Sens. Actuators B Chem.* **2022**, *358*, 131531. [CrossRef]
28. Morrison, S.R. *The Chemical Physics of Surfaces*; Springer: New York, NY, USA, 1990. [CrossRef]
29. Franke, M.E.; Koplin, T.J.; Simon, U. Metal and Metall Oxide Nanoparticles in Chemiresistors: Does the Nanoscale Matter? *Small* **2006**, *2*, 36–50. [CrossRef]
30. Oprea, A.; Bârsan, N.; Weimar, U. Work function changes in gas sensitive materials: Fundamentals and applications. *Sens. Actuators B Chem.* **2009**, *142*, 470–493. [CrossRef]
31. Islam, M.N.; Hakim, M.O. Electron affinity and work function of polycrystalline SnO_2 thin film. *J. Mater. Sci. Lett.* **1986**, *5*, 63–65. [CrossRef]
32. Hölzl, J.; Schulte, F.K. Work Functions of Metals. In *Solid Surface Physics*; Hölzl, J., Schulte, F.K., Wagner, H., Eds.; Springer: Berlin/Heidelberg, Germany, 1979. [CrossRef]
33. Nulmi, S.; Thangadurai, V. Editors' Choice—Review—Solid-State Electrochemical Carbon Dioxide Sensors: Fundamentals, Materials and Applications. *J. Electrochem. Soc.* **2020**, *167*, 037567. [CrossRef]
34. Yuan, G.; Zhong, Y.; Chen, Y.; Zhuo, Q.; Sun, X. Highly sensitive and fast-response ethanol sensing of porous CO_3O_4 hollow polyhedral via palladium reined spillover effect. *RSC Adv.* **2022**, *12*, 6725–6731. [CrossRef] [PubMed]
35. Goel, N.; Kunal, K.; Kushwaha, A.; Kumar, M. Metal oxide semiconductors for gas sensing. *Eng. Rep.* **2023**, *5*, e12604. [CrossRef]
36. Simion, C.E.; Mihalcea, C.G.; Iacoban, A.C.; Dinu, I.V.; Predoi, D.; Vlaicu, I.D.; Florea, O.G.; Stanoiu, A. Influence of Synthesis Method and Electrode Geometry on GHG-Sensing Properties of 5%Gd-Doped SnO_2. *Chemosensors* **2024**, *12*, 148. [CrossRef]
37. Lee, S.P. Electrodes for Semiconductor Gas Sensors. *Sensors* **2017**, *17*, 683. [CrossRef]
38. Hoefer, U.; Steiner, K.; Wagner, E. Contact and sheet resistance of SnO_2 thin films from transmission-line model measurements. *Sens. Actuators B Chem.* **1995**, *26*, 59–63. [CrossRef]
39. Hoefer, U.; Böttner, H.; Wagner, E.; Kohl, C.D. Highly sensitive NO_2 sensor device featuring a JFET-like transducer mechanism. *Sens. Actuators B Chem.* **1998**, *47*, 213–217. [CrossRef]
40. Gonzalez, D.M.R.; Kumar, Y.; Ramón, J.A.R.; Bogyreddy, N.K.R.; Olive-Méndez, S.F.; Karthik, T.V.K.; Becerra, D.; Pérez-Tijerina, E.; Agarwal, V. CO_2 sensing performance enhanced by Pt-catalyzed SnO_2/porous-silicon hybrid structures. *Sens. Int.* **2022**, *3*, 100165. [CrossRef]
41. Chai, H.; Zheng, Z.; Liu, K.; Xu, J.; Wu, K.; Luo, Y.; Liao, H.; Debliquy, M.; Zhang, C. Stability of Metal Oxide Semiconductor Gas Sensors: A review. *IEEE Sens. J.* **2022**, *22*, 5470–5481. [CrossRef]
42. Wang, Y.; Chen, E.; Tang, J. Insight on Reaction Pathways of Photocatalytic CO_2 Conversion. *ACS Catal.* **2022**, *12*, 7300–7316. [CrossRef]
43. He, Y.; Huang, D. Single-Atom Platinum Catalyst for Efficient CO_2 Conversion via Reverse Water Gas Shift Reaction. *Molecules* **2023**, *28*, 6630. [CrossRef]
44. Mondoza-Núñez, E.M.; Fierro-Conzalez, J.C.; Zepeda, T.A.; Solis-Garcia, A. Effect of platinum addition on the reaction mechanism of the CO_2 methanation catalyzed by ZrO_2-supported Rh. *Mol. Catal.* **2022**, *533*, 112801. [CrossRef]
45. Shaheen, A.; Othman, W.; Ali, M.; Tit, N. Catalyst-induced gas-sensing selectivity in ZnO nanoribbons: Ab-initio investigation at room temperature. *Appl. Surf. Sci.* **2020**, *505*, 144602. [CrossRef]

Disclaimer/Publisher's Note: The statements, opinions and data contained in all publications are solely those of the individual author(s) and contributor(s) and not of MDPI and/or the editor(s). MDPI and/or the editor(s) disclaim responsibility for any injury to people or property resulting from any ideas, methods, instructions or products referred to in the content.

chemosensors

MDPI

Article

An Intensity-Variation RI Sensor for Multi-Variant Alcohol Detection with Twisted Structure Using Polymer Optical Fiber

Abdul Ghaffar [1,†], Rehan Mehdi [2,†], Irfan Mehdi [2], Bhagwan Das [3], Vicky Kumar [4], Sadam Hussain [1,*], Gul Sher [2], Kamran Ali Memon [5], Sikandar Ali [6], Mujahid Mehdi [7,*] and Khurram Karim Qureshi [5]

1 Key Laboratory of Air-Driven Equipment Technology of Zhejiang Province, College of Mechanical Engineering, Quzhou University, Quzhou 324000, China; 92ghaffar@gmail.com
2 Dr. M. A. Kazi Institute of Chemistry, University of Sindh, Jamshoro 76080, Pakistan; rehanmehdi000@gmail.com (R.M.); irfanmehdi512@gmail.com (I.M.); gulsherkhaskheli10@gmail.com (G.S.)
3 Design and Creative Technology Vertical, Torrens University, Ultimo, NSW 2007, Australia; bhagwan.das@torrens.edu.au
4 Civil and Environmental Engineering, Universiti Teknologi Petronas, Seri Iskandar 32610, Malaysia; vicky.kumar@utp.edu.my
5 IRC-CSS, King Fahd University of Petroleum & Minerals, KSA, Dhahran 31261, Saudi Arabia; ali.kamran77@kfupm.edu.sa (K.A.M.); kqureshi@kfupm.edu.sa (K.K.Q.)
6 University of Chinese Academy of Sciences, Beijing 101408, China; alisikandar26@mails.ucas.ac.cn
7 Faculty of Design, Aror University of Art Architecture Design & Heritage Sindh, Sukkur 65200, Pakistan
* Correspondence: sadamhussain@qzc.edu.cn (S.H.); mmehdi.faculty@aror.edu.pk (M.M.); Tel.: +86-155-3690-3568 (S.H.); Fax: +86-570-8015112 (S.H.)
† These authors contributed equally to this work.

Abstract: This research introduces an RI sensor for detecting various alcohol species with a designed twisted polymer optical fiber (POF) sensor. The sensor is developed via a straightforward twisting technique to form an effective coupling mechanism. The sensor works on intensity variation where coupled intensity varies when different types of alcohol are added. The structure relies on the twisting of two fibers, where one fiber is used as the illuminating fiber and the other fiber is used as the receiving fiber. Five different types of alcohol are tested (methanol, ethanol, propanol, butanol, and pentanol) as a substant. The experimental results reveal that the sensor is able to detect all five distinct substants effectively by optical power intensity variation. Moreover, the sensor's sensitivity is analyzed with different factors such as the influence of the bending radius and the coupling length, which reveals that the sensing parameters could be customized depending on specific requirements. The sensor demonstrated consistent responses in repeatability tests, with minimal variation across multiple measurements, highlighting its stability. Additionally, the study explores temperature's influence, revealing a sensitivity shift for every degree Celsius of change. This POF-based alcohol sensor represents a significant leap forward in optical sensing technology.

Keywords: POF-based sensor; alcohol sensor; coupling method; intensity-based sensor

1. Introduction

The need for a precise, rapid, and trustworthy method to detect alcohol levels is crucial across various domains such as the food-processing industry, clinical settings, medical research, and environmental and agricultural evaluations [1–3]. Consequently, continuous monitoring of atmospheric alcohol content is vital for safeguarding workplace health [4–7]. Numerous methods for detecting alcohol have been utilized to examine different substances within complex samples. These techniques include chromatography, immunochromatography, mass spectrometry, nuclear magnetic resonance, polymerase chain reaction, ultraviolet–visible spectroscopy, Fourier-transform infrared spectroscopy (FTIR), standard infrared spectroscopy, surface-enhanced Raman spectroscopy (SERS),

Raman spectroscopy, circular dichroism spectroscopy, spectrofluorimetry, and more, all of which remain highly pertinent [8–18].

Despite their usefulness, these methods have constraints, with some suffering from insufficient precision and accuracy issues. Issues related to ease of movement, delayed reaction times, and complex operational steps emphasize the necessity for fresh analytical strategies that offer swift targeted assessments and clear understanding of outcomes. Exploring the creation of optical fiber sensors as an alternative to established techniques is a potential solution to overcome these difficulties [19,20]. One of the key benefits of optical sensors is their heightened sensitivity and selectivity, enabling exact detection and measurement. These sensors operate in a non-intrusive manner, preserving the original state of the sample and preventing any contamination. They also stand out due to their robust resistance to electromagnetic interference, guaranteeing precise functionality even in adverse conditions. With their long-lasting durability and minimal maintenance needs, optical sensors have solidified their reputation as a dependable option across various sectors [21]. These advantages showcase optical fiber's significant potential for application in both chemical and biological sensing fields.

Optical sensing techniques have long been recognized for their effectiveness in alcohol detection. Fiber optic near-infrared spectroscopy was introduced as a tool for measuring alcohol levels about three decades ago [22]. Since then, significant advancements have been made to enhance sensitivity. A remarkable innovation was the integration of fiber optics with multi-walled carbon nanotubes and Co_3O_4, which achieved heightened sensitivity—35 counts per ppm for ethanol and 51 counts per ppm for ammonia—while also offering exceptional discriminatory capabilities in gaseous substance analysis [23]. Another sensor in alcohol detection technology includes an surface plasmon resonance (SPR)-enabled glass rod sensor design, where the sensor excelled in selectively sensing water and ethanol [24]. Another optical configuration involved an alcohol sensor constructed with lipophilic Reichardt's dyes embedded in polymer membranes, but its functionality is constrained by the influence of pH variations in the solution [25]. Penza et al. combined single-walled carbon nanotubes with silica optical fiber and a quartz crystal microbalance to create a novel alcohol sensor design [26]. Furthermore, an optical microfiber sensor was employed for the detection of alcohol [27]. However, the number of researchers has detected alcohol based on the refractive index change to modulate the optical power intensity.

These sensors work based on the principle that alcohol presence or concentration can change how light travels through a material. Specifically, alcohol alters the material's refractive index. By measuring this change in the refractive index, the sensors can accurately determine the level of alcohol present [28–30]. Morisawa et al. [28] reported a polymer optical fiber-based alcohol sensor in their work, wherein changes in the refractive index were induced by Nonolac resin, which swells upon exposure to alcohol. The SPR optical fiber sensor system is also used for alcohol detection, with an integrated gold layer developed. Notably, this design omitted the need for a rotating stage or spectrometer. Its efficacy in precisely quantifying various alcohol types was then successfully demonstrated through validation tests [29]. A demonstration showcased a tapered fiber optic sensor proficient in identifying aliphatic alcohols, including methanol, ethanol, 1-propanol, and 2-propanol [30]. The sensor operates by detecting alterations in the refractive index within the evanescent field of the fiber's tapered zone, induced by the presence of these alcohols, which in turn modifies the output power. These illustrations demonstrate the advancement in and potential of optical sensing systems in alcohol detection, although with several challenges to overcome.

This research utilizes a POF RI sensor to detect various types of alcohol. The sensor design uses the fiber-twisting technique to obtain intensity variation as the different types of alcohol used for detection purposes. Extensive experimental results were conducted to evaluate the sensor's response and evaluate the sensor's sensitivity performance. This paper's second section discusses the sensor's design and fabrication process, as well as its underlying operational principles and the specifics of the experimental setup. Results and discussions are subsequently presented in the third section.

2. Fabrication and Operation Principle

2.1. Sensor Fabrication

This research employs a commercial POF (1 mm step-index SK-40) manufactured by Mitsubishi Co., Ltd. In the SK-40 step-index fiber, the RI experiences an abrupt shift at the interface of the core and cladding layers. Here, the fiber core portion is responsible for light transmission and is made up of polymethyl methacrylate (PMMA), a prevalent transparent polymer material, and the cladding is made of a fluorinated polymer exhibiting a reduced refractive index. This difference ensures the occurrence of total internal reflection of the light rays. Specifically, the POF has a core diameter of 98 μm, paired with a refractive index of 1.49. The core is an encircled 20 μm-thick cladding layer with a refractive index of 1.41, which establishes the vital optical disparity crucial for efficacious light guidance.

For the sensing probe, two POF fibers are used to create twisting structures, as a single fiber is not able to detect alcohol. The twisting creates a coupling structure where the coupled power is affected by a coupling medium. Creating a twist in two POFs involves intertwining them to form a helical structure, similar to how one might twist together two strands of rope or wire. This process can be beneficial for several reasons in optical fiber applications. The twisting structure keeps the fibers closely bound, preventing gaps from forming between them. In contrast, a parallel fiber structure carries the risk of separation between the fibers. When a twisted pair of fibers is bent or pulled, the twist helps distribute the mechanical forces evenly between the fibers, reducing the likelihood of damage. In certain specialized applications, intentionally twisting fibers can enhance the coupling efficiency of light between fibers.

The process of creating a twist in POFs typically involves manually rotating one fiber around the other while maintaining tension control to ensure even twisting and prevent over-twisting. The number and tightness of the twist (number of twists per unit length) can be adjusted based on the specific application requirements. We kept 1 cm for one twist. In the case of the third configuration (extended length), there were 5 twists and the length was 50 mm. After twisting, the fibers can be secured in position using a sheathing material to maintain the twisted configuration. The light source was coupled to the first fiber, and the intensity was measured from the forward end of the fiber. A visual representation of the completed work is provided in Figure 1. A laser diode operating at 635 nanometers (TLS001-635 Thorlabs) was connected with the illuminating fiber and a pair of photodiodes (S120 Thorlabs) was connected with the forward end of the fibers. In previous research efforts [31–34], the concept of a twisted structure has been employed by various scientists for diverse applications.

2.2. Sensor Principle

The twisting structure depends on three sensing mechanisms. In a few cases, the twisted structure was used for a rotational sensor [31,32], where the sensing mechanism relies on the coupling of macrobend loss. The continuous bending of the fiber to induce light leakage is a means of detecting movement or rotational motion. In another work [34], a force sensor was developed specifically to work on the principle of cladding mode frustrating total internal reflection. This refers to a situation where the total internal reflection of light within the fiber is disturbed by the introduction of a perturbation in term of force, allowing some light to escape. Contrasting with these prior works, our present study introduces a different operational principle. Instead of relying on macrobend loss or cladding mode disruptions, we employ a change in the refractive index as the fundamental mechanism driving our sensor's function. This approach offers unique advantages and sensing capabilities tailored to the detection and measurement of variations in the refractive properties of the medium surrounding or interacting with the optical fiber.

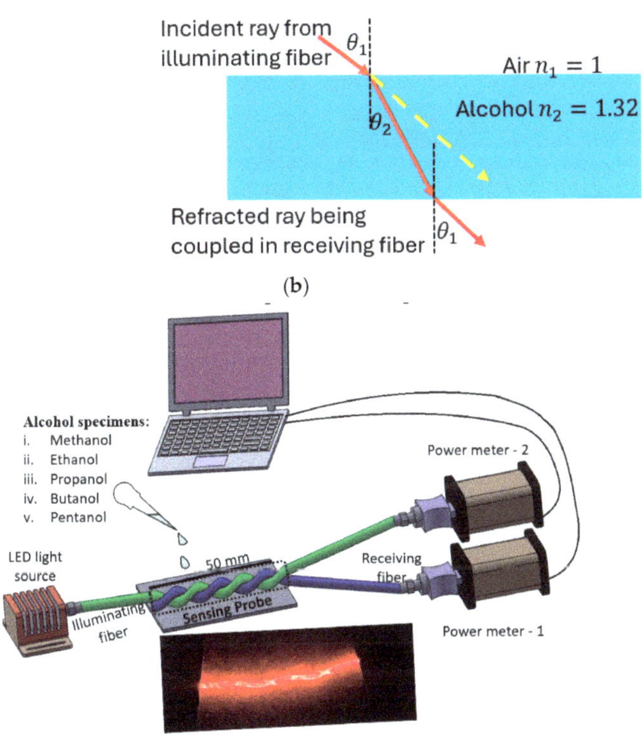

Figure 1. (**a**) Sensing mechanism and (**b**) illustration of sensor fabrication and experimental setup.

Boundary Conditions: In coupled mode theory, solving the differential equations for governing power transfer between fibers requires appropriate boundary conditions specific to the sensor's configuration. For our twisted POF sensor, the boundary conditions are established based on the physical setup of the fibers and the light propagation behavior:

Input Condition: At the start of the illuminated fiber ($z = 0$), the power in the illuminating fiber (P_1) is equal to the input power from the light source. This is defined as $P_1(z = 0) = P_{input}$, where P_{input} is the optical power injected into the illuminated fiber.

Coupled Power Condition: At the receiving fiber's output ($z = L$ where L is the length of the twisted region), the coupled power (P_2) is measured. This is influenced by the coupling coefficient κ and the refractive index of the surrounding medium. The conditions are:

$$P_2(z = L) = \kappa P_1(z = L) \quad (1)$$

In our experiments, the coupling coefficient is directly influenced by the refractive index of the surrounding alcohol medium. For instance, higher refractive indices (e.g., butanol, pentanol) result in increased κ, leading to higher power transfer. Lower refractive indices (e.g., ethanol, water) exhibit reduced κ, leading to lower coupled power. Moreover, the boundary conditions for the twist length (L) and fiber geometry (diameters of the illuminating and receiving fibers) are varied experimentally to optimize coupling efficiency.

Coupled Mode Theory Derivation: When a POF undergoes twisting, its functionality evolves into that of an efficient coupling mechanism. In coupled mode theory, the coupling between two optical fibers is affected by several factors, including changes in the surrounding refractive index. In our case, each alcohol sample has a unique RI. When the refractive index of the medium surrounding the fibers changes, it can alter the coupling efficiency between the fibers by impacting the propagation constant and the overlap of the evanescent fields. This relationship can be described using coupled mode equations. For two coupled fibers, the power transfer between them can be expressed as [35]:

$$\frac{dP_1}{dz} = -\kappa P_2 \;\&\; \frac{dP_2}{dz} = -\kappa P_1 \tag{2}$$

where P_1 and P_2 are the optical powers in fibers 1 and 2, respectively, z is the propagation direction, and κ (kappa) is for the coupling coefficient between the fibers. The coupling coefficient κ depends on the overlap of the mode fields and the surrounding refractive index.

Coupling Coefficient Dependency on Refractive Index: For a refractive index n, the coupling coefficient κ can be approximated as:

$$\kappa \propto e^{-\alpha \frac{n_{fiber} - n_{medium}}{n_{fiber}}} \tag{3}$$

where n_{fiber} is the refractive index of the fiber core; n_{medium} is the refractive index of the surrounding medium, which is alcohol's RI; and α is a constant that depends on the fiber geometry and other parameters.

As the refractive index of the surrounding medium n_{medium} approaches that of the fiber core n_{fiber}, the coupling coefficient κ increases, allowing more power to transfer between the fibers. Conversely, if the surrounding refractive index is significantly different from the core's refractive index, the evanescent fields are less effective in overlapping, reducing the coupling efficiency.

Evanescent Field Penetration Depth and Its Role: The evanescent field associated with cladding modes penetrates the external medium adjacent to the fiber. The penetration depth d_p of this evanescent field is a measure of how far into the external medium this field extends before its intensity drops to a negligible level. This depth is crucial for understanding and designing sensors. The penetration depth d_p can be mathematically expressed as [36]:

$$d_p = \frac{\lambda}{2\pi \sqrt{n_{clad}^2 \sin^2\theta - n_{ext}^2}} \tag{4}$$

where λ is the wavelength of the light in a vacuum, n_{clad} is the refractive index of the cladding, n_{ext} is the refractive index of the external medium, and θ is the incident angle. Thus, a change in the refractive index of the surrounding medium directly affects the amount of power coupled between the fibers. This principle is particularly used in our sensing applications, where changes in the surrounding refractive index can modulate coupling efficiency, enabling the detection of different alcohols.

2.3. Characterization of Alcohol Species

The research encompassed investigations on five distinct alcohol species: methanol, ethanol, propanol, butanol, and pentanol. These diverse alcohol specimens were acquired from Sinopharm Chemical Reagent Co., Ltd., Shanghai, China. The refractive indices of pure alcohol forms were determined utilizing an Abbe refractometer, with all readings taken at ambient temperature (20 °C). The refractive indices of various alcohols are important for understanding their optical properties. The RIs of methanol (methyl alcohol), ethanol (ethyl alcohol), propanol (propyl alcohol), butanol (butyl alcohol), and pentanol (amyl alcohol) are approximately 1.329, 1.361, 1.377, 1.397, and 1.410, respectively as shown in Table 1. To conduct the tests, a precise quantity of the alcohol sample was extracted from the flask using a dropper and applied to the twisted structure's sensing probe. Light propagated from one fiber to another fiber is disrupted with a specific RI medium, as shown in Figure 1b. These observed modifications are fundamentally linked to changes in the refractive index and alcohol-specific absorption traits.

Table 1. The RI of alcohol species.

No	Species	RI
1	Methanol (methyl alcohol)	1.329
2	Ethanol (ethyl alcohol)	1.361
3	Propanol (propyl alcohol)	1.377
4	Butanol (butyl alcohol)	1.397
5	Pentanol (amyl alcohol)	1.410

3. Results and Discussion

Fiber irregularities, including cuts or adjustments similar to those employed in sophisticated technologies such as SPR sensors or D-shaped sensors, can impact the sensor's output [37,38]. Nonetheless, these methods are typically associated with high expenses or fiber damage. In response to non-destructive and cost-effective sensing techniques, the fiber-twisting approach is an alternative option. This approach entails establishing a deliberate coupling zone within the fiber, designed to facilitate enhanced light interaction with the ambient medium, such as water or alcohol, thereby optimizing sensing capabilities. Figure 2 shows the primary sensor detection response to alcohol, where a straight fiber was also tested. Figure 2 displays the intensity (mW) at the forward end of the illumination fiber over time (sec) under different conditions: without coupling, and with coupling in air and alcohol. The coupled power, denoted as normalized optical power (P_2/P_1), was calculated using the outputs, where P_2 represents the final output power and P_1 the initial output power from the same port. The sensor's sensitivity was defined as:

$$S = (\Delta normalized\ optical\ power/\Delta n) \times \% \quad (5)$$

where Δn denotes the change in the RI of the tested alcohol, and Δ *normalized optical power* represents the alteration in the coupling ratio. There is no response to the presence of alcohol (see the red curve of Figure 2a). The red line shows a constant intensity over time, indicating that when the fiber is straight and connected to a light source and power meter without any imperfections or coupling, it is not sensitive to the alcohol application in this setup. The sensor response to detecting alcohol is shown in black in Figure 2. The blue and black lines show variations in coupled power when alcohol is applied to the fiber sensor section. The blue line indicates the response when air is replaced with alcohol in the coupling region, showing a noticeable drop in intensity, which demonstrates the sensor's sensitivity to alcohol.

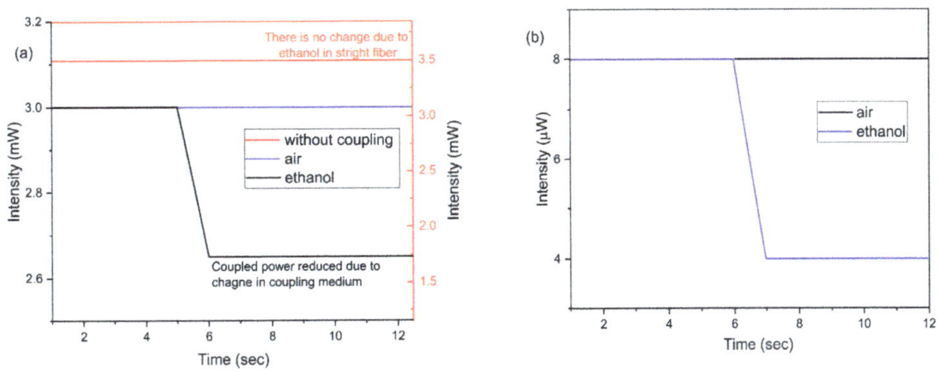

Figure 2. Intensity response of the sensor at the end of both fibers over time: (**a**) illuminating fiber and (**b**) receiving fiber.

Investigation of the spectral characteristics of the twisted structure was conducted and the experimental setup featured a broadband light source (Ocean Optics DH-2000) that spanned wavelengths ranging from 215 to 2000 nanometers, alongside a spectrometer (Ideaoptics NOVA) for spectral analysis. This setup facilitated exact measurements of the optical responses of the fiber system to various media. The experimental protocol involved illumination connected with a broadband light source, and each forward end was connected with a spectrometer. The resultant data are presented in Figure 3. In Figure 3, the vertical axis labeled "Photon Counts" represents the number of photons detected by the photodetector during the measurement process. Photon counts are directly proportional to the intensity of light that is transmitted through the POF and subsequently detected in the receiving fiber. It reflects the decrease in photon counts as light interacts with the alcohol medium surrounding the fiber. In the twisted structure, the transmitting and coupling ends exhibit a significant increase in coupled loss compared to an untwisted fiber when alcohol is added. The coupling power is further reduced in the receiving fiber when alcohol is added.

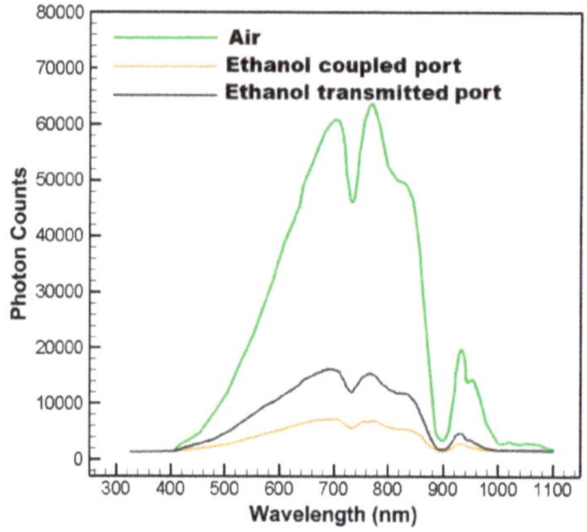

Figure 3. Response of the transmission spectra in the transmitted port and in the coupled port.

The operational effectiveness of this sensor in distinguishing between different types of alcohol is visually demonstrated in Figure 4. The twisting of the fiber structure introduces a considerable improvement in sensing technology. By intentionally introducing twists, a specialized region is created within the fiber where light can couple or interact more dynamically with the substance outside the fiber, such as alcohol. This coupling region functions as a sensing probe, enabling the fiber to perceive alterations in the external environment. When alcohol is added to the coupling region, it causes a significant decrease in the coupled intensity. This decline in light intensity is in response to the presence of alcohol. Consequently, this methodology offers an economically prudent and fiber-preserving strategy for harnessing optical fibers in sensing applications. Further, a series of different detailed assessments and characterizations were executed to analyze the sensor's performance and credibility. These assessments are aimed at analyzing the sensor's sensitivity response, which could be useful for customizing the parameters depending on specific requirements.

Figure 5 illustrates the alcohol-sensing capabilities of the sensor, which is configured with an illuminated fiber and a receiving fiber. Figure 5 highlights the sensor's ability to differentiate between various alcohols based on their refractive indices and the corresponding changes in power output. With the twisted section measuring precisely 8 mm, the graph discloses a clear linear correlation between the coupling ratio, expressed as normalized optical power, and the RI of the external medium. Notably, as the RI ascends beyond 1.329, coinciding with the POF's cladding RI, there is a decline in normalized optical power. On the flip side, in cases where the RI dips beneath 1.410, as observed with pentanol, an increase in normalized optical power is evident. This observed trend stems from the conversion of cladding modes into radiation modes, a phenomenon triggered when the RI of the surrounding alcohol matches the cladding RI, thereby lessening the coupling efficiency. Thus, the sensor excels in detecting RIs that exceed the inherent RI of the fiber cladding. Moreover, the sensor manifests a linear response within the RI spectrum spanning from 1.329 to 1.410.

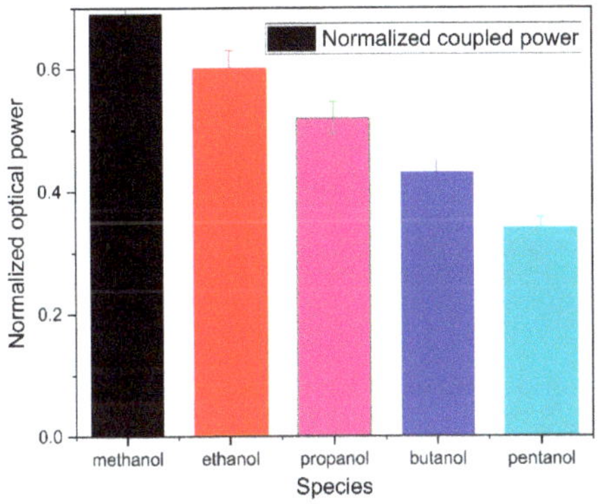

Figure 4. Measured output power of the twisted sensor under various alcohols.

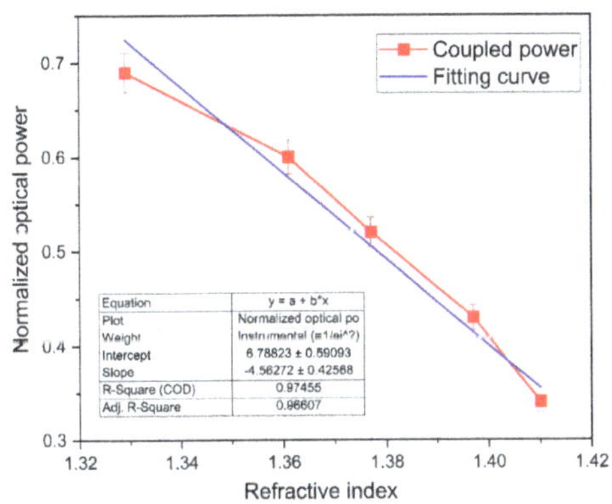

Figure 5. Demonstrated alcohol-sensing capabilities configured with an illuminated fiber and a receiving fiber.

Figure 6 illustrates the results of alcohol-sensing experiments conducted on a series of sensors, all featuring an illuminating fiber with a different bend radius. These sensors incorporated bending with four different radii: 10 mm, 15 mm, 20 mm, and 25 mm. When an optical fiber is bending, some of the guided light is radiated out of the core, leading to bending loss. The amount of light that is lost depends on the radius of the bend; smaller bend radii cause greater bend loss due to the tighter curvature. In a twisted optical fiber sensor, the bending induces more significant radiation loss. Meanwhile more radiated power will be coupled. The light-coupling intensity has a direct impact on the bend radii. Thus, for a smaller bend radius, the optical power loss increases more rapidly as the refractive index of the surrounding medium changes. This is because the evanescent field, which extends into the surrounding medium, is more strongly affected by the refractive index changes in a tightly bent fiber.

As the refractive index increases, more light is coupled out of the core, leading to a larger change in the detected optical power. Therefore, sensors with smaller bend radii (10 mm and 15 mm) exhibit higher sensitivity to refractive index changes, as even slight variations in the surrounding medium's refractive index cause significant changes in the optical power. In contrast, sensors with larger bend radii (20 mm and 25 mm) experience less bend loss and weaker mode coupling. The evanescent field interaction with the surrounding medium is less pronounced, resulting in smaller changes in optical power for the same refractive index variations. As a result, these sensors show reduced sensitivity to refractive index changes because the optical power remains relatively stable, even as the surrounding medium's refractive index changes. Overall, the sensitivity response of the sensors is shown in Table 2.

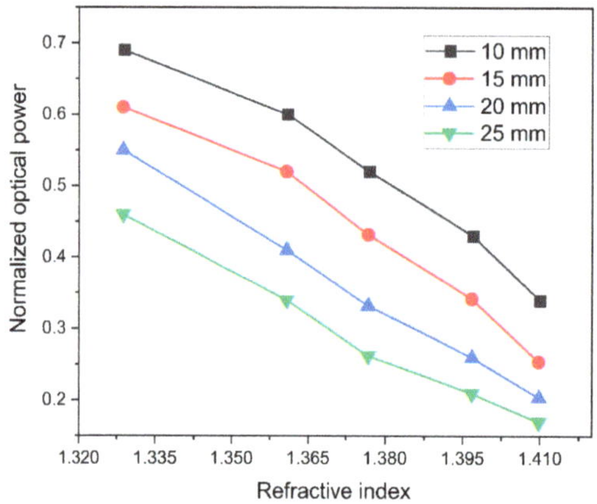

Figure 6. Effect of different bend radii on sensor measurement response.

Table 2. Sensitivity response of the sensors.

No	Species	R^2	Sensing Length	Bend Radius
Without bend	178%/RIU	0.9745	10 mm	-
With bend	416%/RIU	0.9821	-	10~50 mm
Extended length	572%/RIU	0.9891	10~50 mm	-

It was observed that the sensor's sensitivity improves when the smaller bend radius is based on the experimental outcomes, whereas the reverse scenario leads to reduced sensitivity. Because a small bend radius generates a higher radiation mode. Meanwhile, it is worth noting that the sensing length also effects the sensor's performance [32,33]. In the context of optical fiber sensors, the influence of the twisted region's length on sensing performance was also investigated, as shown in Figure 7. With increased twisted region length, the interaction between the light in the core and the surrounding medium (i.e., the alcohol solution) is enhanced, as the twisting causes the light to repeatedly interact with the cladding and the external environment. This results in more significant changes in the optical power in response to variations in the RI of the surrounding medium.

A longer twisted region means that the cumulative effect of the twisting perturbations on the light propagation is greater. Each twist contributes to the modulation of the light path, increasing the sensitivity of the sensor to changes in the external refractive index. As the length of the twisted region increases from 10 mm to 50 mm, the overall effect of the twisting is more pronounced, leading to higher sensitivity. This is because the longer twisted region allows for more substantial light interaction with the surrounding medium, amplifying the sensor's response to refractive index changes. The twisted region effectively increases the interaction length between the guided light and the surrounding medium. A longer interaction length means that even small changes in the refractive index can cause more significant alterations in the light propagation, resulting in higher sensitivity. The sensitivity of the alcohol-sensing performance increases with the length of the twisted region due to enhanced mode coupling, accumulated effects of twisting, and increased interaction length. These factors collectively amplify the sensor's response to changes in the surrounding medium's refractive index, leading to higher sensitivity with longer twisted regions.

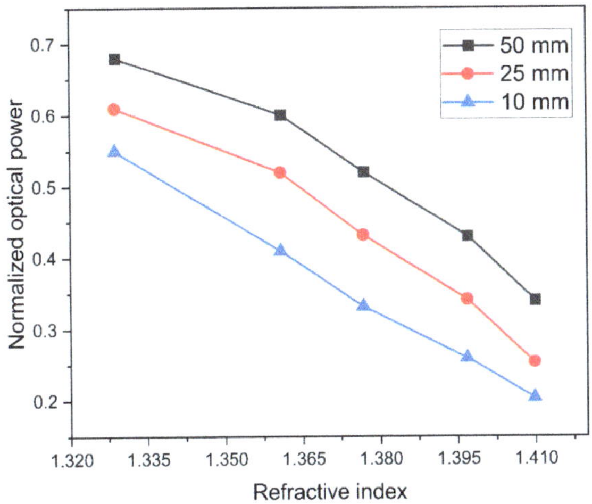

Figure 7. Influence of the twisted region's length on the alcohol sensor's performance.

Figure 8 demonstrates the sensor's repeatability by illustrating that the sensor consistently responds when the same alcohol is tested multiple times. The error bars in this figure represent the variability observed across repeated trials, which underscores the reliability and consistency of the sensor's performance. The normalized optical power is plotted against the refractive index of the alcohol samples for three successive measurements, denoted as the first, second, and third time. The data points for the first (squares), second (circles), and third (triangles) measurements show consistent trends across the different refractive indices. This indicates good repeatability of the sensor, as the results from the three measurements are closely aligned. The overlap of data points from the three measurements for each refractive index value demonstrates the sensor's ability to produce reproducible results. There is minimal deviation between the measurements, which signifies the sensor's reliability in repeated use. As the refractive index increases from 1.329 to 1.410, the normalized optical power decreases, following a consistent pattern across all three measurements. This consistent decrease highlights the sensor's sensitivity and stable performance in detecting changes in refractive index.

Figure 9 illustrates the effect of the temperature on the sensor's sensing capability. For the temperature response, the twisted region was placed on a heating plate, and we slowly changed the temperature from 20 °C to 55 °C. We found that for every 5 °C the temperature changes, the sensor's response changes a little. This small change happens because of two main reasons: the thermo-optic effect and heat, which makes the fiber grow or shrink a bit (thermal expansion). When it is warmer, the fiber's ability to bend light changes, which can change light propagated inside the fiber. Also, as the fiber heats up or cools down, it gets bigger or smaller, causing tiny changes in its shape. Temperature changes also cause physical expansion or contraction of the fiber material due to thermal expansion. In the case of twisted structure, this can lead to minute adjustments in the fiber's geometry, such as its diameter, the tightness of the twist (pitch), or the spacing between fibers. These geometric alterations can disrupt the way light is guided and coupled within the fiber, thereby influencing the sensor's output and contributing to its temperature sensitivity.

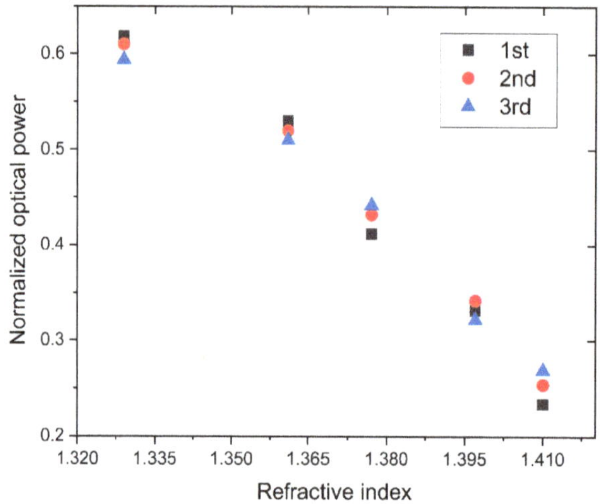

Figure 8. Repeatability sensor response.

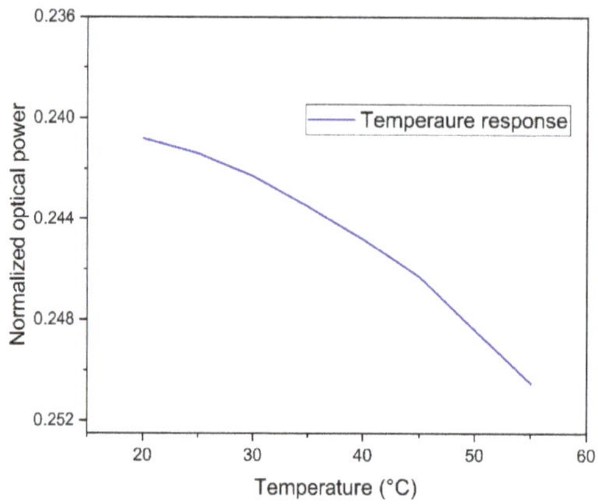

Figure 9. Influence of temperature on the sensor's performance.

A comparison of fiber optic-based alcohol sensors is shown in Table 3. Its non-intrusive design supports repeated use, and its simplicity in manufacturing promises affordability. It also has a negligible response to temperature fluctuations, flexible fiber configurations, and a mostly straightforward reaction pattern, so integrating this sensor into monitoring equipment is effortless. It is also chemically resilient, compatible with a wide array of solvents, and can be adjusted to suit specific needs, making it a formidable asset for meticulous alcohol detection in applications like chemical analyses, quality checks, and healthcare diagnostics. The sensor's working principle is based on intensity modulation, and concentration and different kinds of isomers are not considered in this study. The sensor's structure will be modified to investigate isomers or concentrations of alcohol. Further studies could involve the incorporation of surface modifications or coatings specifically tailored to interact with certain chemical groups present in alcohol but absent in other volatile organic compounds (VOCs). By introducing these modifications, the sensor can interact with alcohols through additional mechanisms such as hydrogen bonding, hydrophobic interactions, or specific binding, rather than just relying on changes in the refractive index. This approach can significantly enhance the specificity of the sensor, allowing it to distinguish between alcohol and other VOCs more effectively.

Table 3. A comparison of fiber optic-based alcohol sensors.

Ref.	Method	Variant	Sensitivity	Fiber
[7]	Finite-element-method	Methanol, ethanol, and propanol,	92%	Photonic crystal fiber
[22]	Spectroscopy	Ethanol	-	Silica fiber
[24]	SPR	Ethanol	-	Silica fiber
[28]	Refractive index	Ethanol, methanol	-	POF
[39]	COMSOL	Ethyl alcohol	44.85%	PCF
[40]	Evanescent field	Alcohol	3.87 dB/%	SMS
[41]	fluorescence intensity	Methanol, ethanol, and propanol	5–90% v/v	-
[42]	Piezoresistive	Ethanol		Poly acrylamide hydrogels
This work	Intensity variation	Methanol ethanol, propanol, butanol, pentanol	572%/RIU	POF

4. Conclusions

This research introduces an RI sensor for detecting various alcohol species using a twisted polymer optical fiber sensor. The sensor is developed via a straightforward twisting and bending principle. The sensor operates on intensity modulation, which changes due to different alcohols altering the coupling medium. The system consists of two fibers: The first fiber launches the light, and the second fiber measures the coupled intensity. The experimental results reveal that the sensor is capable of detecting all five distinct substances. Moreover, the sensor's sensitivity is analyzed with respect to various factors, such as the influence of the sensor bending radius and sensor length. The sensor demonstrates high stability and repeatability, achieving a sensitivity of about 572%/RIU. Additionally, this study explores the influence of temperature, revealing a sensitivity shift for every degree Celsius of change. This POF-based alcohol sensor represents a significant leap forward in optical sensing technology. In future, the sensor's capability to detect additional alcohol types and other volatile organic compounds will be investigated to broaden its applicability in various industries. Long term stability and durability tests will be conducted to assess the sensor's performance over extended periods and in harsh conditions, improving its robustness for industrial applications.

Author Contributions: Conceptualization, S.H.; Data curation, S.A.; Formal analysis, I.M., G.S. and M.M.; Methodology, A.G., V.K. and S.A.; Validation, R.M., K.K.Q. and K.A.M.; Writing—original draft, A.G.; Writing—review & editing, B.D., K.A.M., K.K.Q. and M.M. All authors have read and agreed to the published version of the manuscript.

Funding: This research was supported by Zhejiang Provincial Natural Science Foundation of China under Grant No. TGN23E050001.

Institutional Review Board Statement: Not applicable.

Informed Consent Statement: Not applicable.

Data Availability Statement: The data presented in this study are available on reasonable request from the corresponding authors.

Conflicts of Interest: No, I declare that the authors have no competing interests as defined by Nature Research, or other interests that might be perceived to influence the results and/or discussion reported in this paper.

References

1. Sultana, J.; Islam, M.S.; Ahmed, K.; Dinovitser, A.; Ng, B.W.-H.; Abbott, D. Terahertz detection of alcohol using a photonic crystal fiber sensor. *Appl. Opt.* **2018**, *57*, 2426–2433. [CrossRef] [PubMed]
2. Hamburg, M.A. *Advancing Regulatory Science*; American Association for the Advancement of Science: Washington, DC, USA, 2011; Volume 331, p. 987.
3. Rachana, M.; Charles, I.; Swarnakar, S.; Krishna, S.V.; Kumar, S. Recent advances in photonic crystal fiber-based sensors for biomedical applications. *Opt. Fiber Technol.* **2022**, *74*, 103085. [CrossRef]

4. Triyana, K.; Sembiring, A.; Rianjanu, A.; Hidayat, S.N.; Riowirawan, R.; Julian, T.; Kusumaatmaja, A.; Santoso, I.; Roto, R. Chitosan-based quartz crystal microbalance for alcohol sensing. *Electronics* **2018**, *7*, 181. [CrossRef]
5. Pohanish, R.P. *Sittig's Handbook of Toxic and Hazardous Chemicals and Carcinogens*; William Andrew: Norwich, NY, USA, 2017.
6. Ayad, M.M.; Torad, N.L. Alcohol vapours sensor based on thin polyaniline salt film and quartz crystal microbalance. *Talanta* **2009**, *78*, 1280–1285. [CrossRef] [PubMed]
7. Iqbal, F.; Biswas, S.; Bulbul, A.A.-M.; Rahaman, H.; Hossain, M.B.; Rahaman, M.E.; Awal, M.A. Alcohol sensing and classification using PCF-based sensor. *Sens. Bio-Sens. Res.* **2020**, *30*, 100384. [CrossRef]
8. García-Cañas, V.; Simo, C.; Herrero, M.; Ibáñez, E.; Cifuentes, A. *Present and Future Challenges in Food Analysis: Foodomics*; ACS Publications: Washington, DC, USA, 2012.
9. Martinez, A.W.; Phillips, S.T.; Whitesides, G.M.; Carrilho, E. *Diagnostics for the Developing World: Microfluidic Paper-Based Analytical Devices*; ACS Publications: Washington, DC, USA, 2010.
10. Dzantiev, B.B.; Byzova, N.A.; Urusov, A.E.; Zherdev, A.V. Immunochromatographic methods in food analysis. *TrAC Trends Anal. Chem.* **2014**, *55*, 81–93. [CrossRef]
11. Wei, W.Y.; White, I.M. Inkjet-printed paper-based SERS dipsticks and swabs for trace chemical detection. *Analyst* **2013**, *138*, 1020–1025.
12. Ali, E.M.; Edwards, H.G. The detection of flunitrazepam in beverages using portable Raman spectroscopy. *Drug Test. Anal.* **2017**, *9*, 256–259. [CrossRef]
13. Ayad, M.M.; Salahuddin, N.; Minisy, I.M. Detection of some volatile organic compounds with chitosan-coated quartz crystal microbalance. *Des. Monomers Polym.* **2014**, *17*, 795–802. [CrossRef]
14. Pérez-Ponce, A.; Garrigues, S.; de La Guardia, M. Vapour generation–fourier transform infrared direct determination of ethanol in alcoholic beverages. *Analyst* **1996**, *121*, 923–928. [CrossRef]
15. Allan, J.T.; Geoffrey, H.L.; Easton, E.B. The effect of the gas diffusion layer on the performance of fuel cell catalyst layers in ethanol sensors. *Sens. Actuators B Chem.* **2018**, *254*, 120–132. [CrossRef]
16. Jung, Y.; Kim, J.; Awofeso, O.; Kim, H.; Regnier, F.; Bae, E. Smartphone-based colorimetric analysis for detection of saliva alcohol concentration. *Appl. Opt.* **2015**, *54*, 9183–9189. [CrossRef] [PubMed]
17. Hwang, R.-J.; Beltran, J.; Rogers, C.; Barlow, J.; Razatos, G. Measurement of uncertainty for blood alcohol concentration by headspace gas chromatography. *Can. Soc. Forensic Sci. J.* **2017**, *50*, 114–124. [CrossRef]
18. Ozoemena, K.I.; Musa, S.; Modise, R.; Ipadeola, A.K.; Gaolatlhe, L.; Peteni, S.; Kabongo, G. Fuel cell-based breath-alcohol sensors: Innovation-hungry old electrochemistry. *Curr. Opin. Electrochem.* **2018**, *10*, 82–87. [CrossRef]
19. Zhang, W.; Lang, X.; Liu, X.; Li, G.; Singh, R.; Zhang, B.; Kumar, S. Advances in tapered optical fiber sensor structures: From conventional to novel and emerging. *Biosensors* **2023**, *13*, 644. [CrossRef]
20. An, G.; Jia, P.; Liang, T.; Hong, Y.; Wang, H.; Ghaffar, A.; Xiong, J. Double-sided polished ultra-stable and ultra-sensitive optical fiber sensor. *Plasmonics* **2020**, *15*, 1471–1479. [CrossRef]
21. Hu, Y.; Hou, Y.; Ghaffar, A.; Liu, W. A narrow groove structure based plasmonic refractive index sensor. *IEEE Access* **2020**, *8*, 97289–97295. [CrossRef]
22. Cavinato, A.G.; Mayes, D.M.; Ge, Z.; Callis, J.B. Noninvasive method for monitoring ethanol in fermentation processes using fiber-optic near-infrared spectroscopy. *Anal. Chem.* **1990**, *62*, 1977–1982. [CrossRef]
23. Parthibavarman, M.; Sangeetha, S.; Renganathan, B.; BoopathiRaja, R. High-performance fiber optic gas sensor-based Co_3O_4/mwcnt composite by a novel microwave technique. *J. Iran. Chem. Soc.* **2019**, *16*, 2463–2472. [CrossRef]
24. Mitsushio, M.; Masunaga, T.; Yoshidome, T.; Higo, M. Alcohol selectivity and measurement of ethanol concentrations in liquors using teflon® af2400-coated gold-deposited surface plasmon resonance-based glass rod sensor. *Prog. Org. Coat.* **2016**, *91*, 33–38. [CrossRef]
25. Blum, P.; Mohr, G.J.; Matern, K.; Reichert, J.; Spichiger-Keller, U.E. Optical alcohol sensor using lipophilic Reichardt's dyes in polymer membranes. *Anal. Chim. Acta* **2001**, *432*, 269–275. [CrossRef]
26. Penza, M.; Cassano, G.; Aversa, P.; Antolini, F.; Cusano, A.; Cutolo, A.; Giordano, M.; Nicolais, L. Alcohol detection using carbon nanotubes acoustic and optical sensors. *Appl. Phys. Lett.* **2004**, *85*, 2379–2381. [CrossRef]
27. Chiam, Y.S.; Lim, K.S.; Harun, S.W.; Gan, S.N.; Phang, S.W. Conducting polymer coated optical microfiber sensor for alcohol detection. *Sens. Actuators A-Phys.* **2014**, *205*, 58–62. [CrossRef]
28. Morisawa, M.; Amemiya, Y.; Kohzu, H.; Liang, C.X.; Muto, S. Plastic optical fibre sensor for detecting vapour phase alcohol. *Meas. Sci. Technol.* **2001**, *12*, 877–881. [CrossRef]
29. Mitsushio, M.; Higashi, S.; Higo, M. Analyses of alcohols using a gold-deposited optical-fiber refractive index sensor system. *Bunseki Kagaku* **2003**, *52*, 433–438. [CrossRef]
30. Phang, S.W.; Yang, H.Z.; Harun, S.W.; Arof, H.; Ahmad, H. Simple fiber optic sensor based on tapered fiber for aliphatic alcohol detection. *J. Optoelectron. Adv. Mater.* **2011**, *13*, 604–608.
31. Hussain, S.; Ghaffar, A.; Musavi, S.H.A.; Mehdi, M.; Ahemd, M.; Jianping, Y.; Lei, C.; Mehdi, R.; Mehdi, I. Fiber optic sensor embedded in robotic systems for 3-D orientation assessment using polymer fiber. *Opt. Fiber Technol.* **2023**, *81*, 103559. [CrossRef]
32. Shi, J.; Ghaffar, A.; Li, Y.; Mehdi, I.; Mehdi, R.; Soomro, F.A.; Hussain, S.; Mehdi, M.; Li, Q.; Li, Z. Dynamic rotational sensor using polymer optical fiber for robot movement assessment based on intensity variation. *Polymers* **2022**, *14*, 5167. [CrossRef]

33. Ghaffar, A.; Li, Q.; Mehdi, M.; Hussain, S.; Jia, Y.M.; Onyekwena, C.C.; Xie, Q.; Ali, N. Analysis of Force Sensor Using Polymer Optical Fiber Based on Twisting Structure. *IEEE Sens. J.* **2022**, *22*, 23960–23967. [CrossRef]
34. Ghaffar, A.; Li, Q.; Mehdi, M.; Das, B.; Alvi, I.H.; Xie, Q.; Ma, J. Multiplexing sensors technique for angle and temperature measurement using polymer optical fiber. *Infrared Phys. Technol.* **2023**, *130*, 104585. [CrossRef]
35. Snyder, A.W. Coupled-mode theory for optical fibers. *JOSA* **1972**, *62*, 1267–1277. [CrossRef]
36. Snyder, A.W.; Love, J. *Optical Waveguide Theory*; Springer Science & Business Media: Berlin/Heidelberg, Germany, 2012.
37. Xia, F.; Song, H.; Zhao, Y.; Zhao, W.-M.; Wang, Q.; Wang, X.-Z.; Wang, B.-T.; Dai, Z.-X. Ultra-high sensitivity SPR fiber sensor based on multilayer nanoparticle and Au film coupling enhancement. *Measurement* **2020**, *164*, 108083. [CrossRef]
38. Ying, Y.; Si, G.-y.; Luan, F.-j.; Xu, K.; Qi, Y.-w.; Li, H.-n. Recent research progress of optical fiber sensors based on D-shaped structure. *Opt. Laser Technol.* **2017**, *90*, 149–157. [CrossRef]
39. Podder, E.; Jibon, R.H.; Hossain, M.B.; Bulbul, A.A.-M.; Biswas, S.; Kabir, M.A. Alcohol sensing through photonic crystal fiber at different temperature. *Opt. Photonics J.* **2018**, *8*, 309. [CrossRef]
40. Patrialova, S.N.; Hatta, A.M.; Sekartedjo, S. Alcohol sensor based on U-bent hetero-structured fiber optic. In Proceedings of the Second International Seminar on Photonics, Optics, and Its Applications (ISPhOA 2016), Bali, Indonesia, 24–25 August 2016; pp. 221–225.
41. Petrova, S.; Kostov, Y.; Jeffris, K.; Rao, G. Optical ratiometric sensor for alcohol measurements. *Anal. Lett.* **2007**, *40*, 715–727. [CrossRef]
42. Erfkamp, J.; Guenther, M.; Gerlach, G. Hydrogel-Based Sensors for Ethanol Detection in Alcoholic Beverages. *Sensors* **2019**, *19*, 1199. [CrossRef]

Disclaimer/Publisher's Note: The statements, opinions and data contained in all publications are solely those of the individual author(s) and contributor(s) and not of MDPI and/or the editor(s). MDPI and/or the editor(s) disclaim responsibility for any injury to people or property resulting from any ideas, methods, instructions or products referred to in the content.

chemosensors

MDPI

Article

The Effect of Doping rGO with Nanosized MnO_2 on Its Gas Sensing Properties

Mohamed Ayoub Alouani [1,2,3], Juan Casanova-Chafer [4], Santiago de Bernardi-Martín [5], Alejandra García-Gómez [5], Foad Salehnia [1,2,3], José Carlos Santos-Ceballos [1,2,3], Alejandro Santos-Betancourt [1,2,3], Xavier Vilanova [1,2,3,*] and Eduard Llobet [1,2,3,*]

[1] Department of Electronic Engineering, School of Engineering, Universitat Rovira i Virgili, MINOS, Avda. Països Catalans 26, 43007 Tarragona, Spain; mohamedayoub.alouani@urv.cat (M.A.A.); foad.salehnia@urv.cat (F.S.); josecarlos.santos@urv.cat (J.C.S.-C.); alejandro.santos@urv.cat (A.S.-B.)

[2] IU-RESCAT, Research Institute in Sustainability, Climatic Change and Energy Transition, Universitat Rovira i Virgili, Joanot Martorell 15, 43480 Vila-seca, Spain

[3] TecnATox—Centre for Environmental, Food and Toxicological Technology, Universitat Rovira i Virgili, Avda. Països Catalans 26, 43007 Tarragona, Spain

[4] Research Institute for Materials Science and Engineering, Université de Mons, Avenue Copernic 3, 7000 Mons, Belgium; juan.casanova@umons.ac.be

[5] Gnanomat. C/Faraday 7, 28049 Madrid, Spain; santiago.bernardi@gnanomat.com (S.d.B.-M.); alejandra.garcia@gnanomat.com (A.G.-G.)

* Correspondence: xavier.vilanova@urv.cat (X.V.); eduard.llobet@urv.cat (E.L.)

Abstract: Manganese dioxide (MnO_2) has drawn attention as a sensitiser to be incorporated in graphene-based chemoresistive sensors thanks to its promising properties. In this regard, a rGO@MnO_2 sensing material was prepared and deposited on two different substrates (silicon and Kapton). The effect of the substrate nature on the morphology and sensing behaviour of the rGO@MnO_2 material was thoroughly analysed and reported. These sensors were exposed to different dilutions of NO_2 ranging from 200 ppb to 1000 ppb under dry and humid conditions (25% RH and 70% RH) at room temperature. rGO@MnO_2 deposited on Kapton showed the highest response of 6.6% towards 1 ppm of NO_2 under dry conditions at RT. Other gases or vapours such as NH_3, CO, ethanol, H_2 and benzene were also tested. FESEM, HRTEM, Raman, XRD and ATR-IR were used to characterise the prepared sensors. The experimental results showed that the incorporation of nanosized MnO_2 in the rGO material enhanced its response towards NO_2. Moreover, this material also showed very good responses toward NH_3 both under dry and humid conditions, with the rGO@MnO_2 sensor on silicon showing the highest response of 18.5% towards 50 ppm of NH_3 under 50% RH at RT. Finally, the synthetised layers showed no cross-responsiveness towards other toxic gases.

Keywords: reduced graphene oxide; manganese dioxide (MnO_2); gas sensor; NO_2 detection; NH_3 detection; Kapton; SiO_2/Si

1. Introduction

Technological and economic advances increasingly threaten the environment and air quality, necessitating urgent solutions. The demand for low-cost and effective gas sensors, crucial for detecting toxic agents such as nitrogen dioxide NO_2 [1], ammonia NH_3 [2], and carbon monoxide CO [3] in various fields, has never been more pressing. While traditional methods like infrared spectrophotometry (IRSP) [4], non-dispersive infrared analysis (NDIR) [5], and gas chromatography-mass spectrometry (GC-MS) [6] have been widely used for detecting toxic gases, they are not without significant drawbacks. Their high cost and complexity have underscored the need for a new, more efficient approach. This realisation has led to the development of chemoresistive devices, known for their ease of operation, low production cost, fast response time, and ease of miniaturisation [7].

Since their first use in the 1960s, metal oxides (MO_X) such as ZnO [8,9], SnO_2 [10,11], WO_3 [12,13], and many more [14] have been studied as sensitive films to be used in the fabrication of chemoresistive devices for gas sensing and especially for detecting NO_2. Even though MO_X are effective and sensitive, the need for a pervasive, widespread monitoring of air pollutants has oriented research to the discovery of new, less power-hungry, gas-sensitive materials. This has, for example, led to the emergence of graphene-based chemoresistive sensors. The unique and exceptional properties of graphene that made it a solid gas sensing material candidate are thermal stability, mechanical robustness, high conductivity, high carrier mobility at low to room temperatures, a large surface area of up to 2630 m^2/g for single-layer graphene, low electrical noise [15] and, most importantly, the fact that its electronic properties are easily affected by the adsorption of gas molecules [16]. Graphene, as a chemoresistive layer, is a p-type material. The interaction between a reducing gas such as NH_3 and the already adsorbed oxygen species on the surface of the graphene layer results in a transfer of electrons back to the surface, hence the increase in the surface resistance of the graphene film [17]. Meanwhile, when in contact with an oxidising gas such as NO_2, the transfer of electrons happens from the sensitive layer to the adsorbed NO_2 molecules, leading to a decrease in the surface resistance [18]. Table 1 displays a collection of works found in the literature using graphene-based chemoresistive sensors doped with different MO_X for the gas detection of different analytes:

Table 1. Gas sensors fabricated from graphene-based materials and their response towards different toxic gases.

Material	Target Gas	Gas Concentration (ppm)	Operating Condition	Response (%)	Ref.
Graphene/MoS_2 hybrid aerogel	NO_2	0.5	200 °C	9	[19]
rGO/SnO_2 nanoparticles	NO_2	5	50 °C	3.31	[20]
ZnO nanorods/rGO) mesoporous nanocomposites	NO_2	1	RT	119	[21]
Graphene nanoplatelets/polyaniline nanocomposite	NH_3	1	RT	0.7	[22]
rGO/PtNPs nanocomposite	NH_3	1	RT	2.87	[23]
Polyaniline (PANI)/graphene oxide (GO)/ZnO hybrid LbL film	NH_3	50	65% RH	3	[24]
Graphene oxide	H_2	200	RT	10	[25]
rGO/TiO_2 decorated by Pd/Pt nanoparticles	H_2	500	180 °C	92%	[26]

One of the graphene derivatives is reduced graphene oxide (rGO). It was reported in the literature to be the best and most used graphene-based sensing material for detecting NO_2 [27,28] and NH_3 [29,30] due to its numerous defect sites and functional groups, which facilitate gas adsorption. rGO has been reported to be able to detect chemical warfare agents and explosives at trace levels (ppb) [31]. Moreover, the synthesis of rGO can be achieved via straightforward and inexpensive processes that reduce GO via chemical and thermal routes or even using UV light [32]. Studies have shown that pristine rGO gas sensors exhibit slow response and recovery dynamics. Hence, the hybridisation of rGO with MO_X has been a solution often explored to enhance its sensing properties (e.g., ameliorating response dynamics and extending the number of gases that can be detected). D. Tripathi et al. explored this process, and an enhancement of the sensitivity and selectivity of the rGO material towards ammonia by incorporating WO_3 nanomaterial in the sensing layer was achieved [33]. Another reported work by D. Milad et al. showed that the synthesis of a TiO_2/rGO composite exhibited improved gas sensing properties towards methanol and ethanol [34]. This is achieved via the MO_X nanoparticles supported on rGO behaving as catalysts or as electronic sensitisers, favouring the occurrence of heterojunctions at the MO_X/rGO interface.

Some of the reported MO_X nanoparticles used for the loading of the rGO layers for gas sensing are ZnO [35], SnO_2 [36], and WO_3 [37]. Still, in recent years, MnO_2 has gained popularity due to its low toxicity, low cost, high stability, and ease of fabrication. It was also used in a wide range of applications and fields such as energy storage [38], biomedical field [39], and in developing gas sensors [40]. However, in the gas sensing field, there is a limited amount of reported work discussing the incorporation of the MnO_2 nanomaterial into rGO to achieve a sensitive layer towards different toxic gases. One of the few works reported is of Hui Zhang et al., where they successfully synthesised an rGO-coated Ni foam-supported MnO_2 for the enhanced detection of NO_2 at a concentration of 50 ppm while the sensor was operated at room temperature [41]. Meanwhile, Alexander et al. modified rGO with doping of MnO_2 nanoparticles and tested the rGO/MnO_2 composite as a gas sensor for different gases such as 25 ppm of NO_2, 500 ppm of H_2 and 1000 ppm of CH_4 under dry conditions with heating at 85 °C [42]. Another reported work was of Ghosal et al., where they prepared different hybrids for alcohol vapour detection. One of the hybrids, the rGO/MnO_2 nanoflowers binary composite, showed good responses towards ethanol and methanol vapours in the range of 5–100 ppm while heating at 150 °C [43]. Lastly, Ahmad et al. made a ternary nanocomposite of PANI@rGO@MnO_2 using a multi-step process for NH_3 detection. The tests were made at 100 °C under dry conditions [44]. The fact that very few works have been reported on rGO@MnO_2 gas sensors so far makes exploring its gas-sensing properties interesting and worthwhile.

In this paper, rGO@MnO_2 sensitive layers were successfully synthesised and deposited on different transducing substrates (Kapton and silicon with gold electrodes). The gas sensing performance of the rGO@MnO_2 sensors was studied for different reducing and oxidising species. The effect of ambient moisture on sensor response was evaluated as well. Results are presented and thoroughly discussed. A sensing mechanism for the detection of ammonia and nitrogen dioxide is presented.

2. Materials and Methods

In this section, a detailed explanation of the preparation of the doped reduced graphene oxide with nanosized MnO_2 at 95/5 wt.%, alongside the process of depositing it on the Kapton and silicon substrates via the spray coating technique, is presented. The section also describes the techniques employed in the study of the morphology and composition of the hybrid material and in the test of its gas sensing properties.

2.1. Preparation of Reduced Graphene Oxide Doped with MnO_2

Commercial rGO obtained from LayerOne company (Norway) was doped with nanosized MnO_2 (rGO/MnO_2 95/5 wt.%) using a process based on patented procedures (Patent number ES2678419A1) by GNANOMAT company. Briefly, reduced graphene oxide was dispersed in oxalic acid, in which the starting Mn_3O_4 had been previously dissolved at 50 °C. After homogenisation, nanosized MnO_2 was slowly precipitated on reduced graphene oxide by adding a basic solution (NaOH 5 M) under vigorous agitation. The solid was filtered and dried at 90 °C overnight. Synthesis parameters, such as temperature, stirring speed, addition rate, or MnO_2/rGO proportion, were controlled to obtain the desired crystallinity that provides the material with optimal properties. The manganese oxide phase was checked via XRD; Figure S1 (in Supplementary Materials) shows the XRD diffractogram.

2.2. Substrates Choice, Preparation, and Material Deposition

The substrate material choice was based on several factors. Kapton substrate is considered important in the synthesis of new flexible and wearable sensors since it has a very high flexibility. Ilghar R. et al. used this material as a substrate to fabricate a wearable Kapton graphene biosensor for the detection of toxic gases [45]. Additionally, Kapton has very good thermal stability and is known to be a hydrophobic material. Meanwhile, SiO_2/Si substrates have different advantages to offer. Initially, the substrate starts as a silicon wafer and an insulating layer of SiO_2 is usually obtained via a thermal oxidation process. Such substrates show very high thermal stability and standard processes from the semiconductor industry can be used to pattern interdigitated electrodes or heating resistors.

Ten milligrams of the rGO@MnO_2 was weighed and suspended in a 10 mL ethanol solution via a 30 min sonication. Subsequently, suspended material was deposited by spray coating onto two different substrates (i.e., silicon and Kapton). During the coating process, substrates were heated at 50 °C to promote the evaporation of the solvent and the formation of a homogeneous film. Both substrates were prepared differently. The SiO_2/Si substrate was prepared by thermally growing SiO_2 on top of the silicon wafer and later cut in specific dimensions, as shown in Figure 1a, with a thickness of 0.5 mm. Meanwhile, the Kapton substrate was prepared by sticking the Kapton piece to a small plastic support with dimensions same as the silicon one shown in Figure 1b with a total thickness of 0.58 mm.

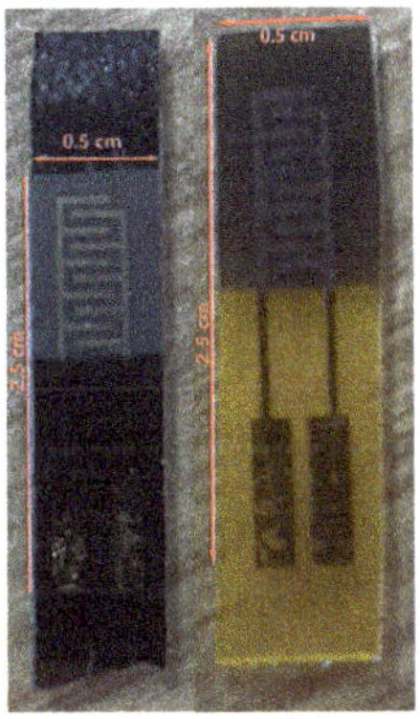

(**a**) (**b**)

Figure 1. Pictures of the prepared sensors on the (**a**) silicon substrate and (**b**) Kapton substrate.

The interdigitated gold electrodes were deposited on the substrates using different processes. For Kapton substrates, 9 nm of gold was sputtered using a shadow mask to form the electrodes. In contrast, a two-step approach was used for the silicon substrates. At first, a laser lithography technique (DWL 66fs, Heidelberg Instruments, Heidelberg, Germany) was used to pattern a photoresist that coated an oxidised silicon wafer in the shape of the electrodes. In the second step, a titanium adhesion layer was sputtered with a thickness of 10 nm, and then a gold layer was sputtered on top with a thickness of 100 nm. A final lift-off process was conducted to obtain the electrodes. The silicon wafer was then diced. Both fabricated substrates have the same configurations regarding the electrodes, which are listed below in Table 2.

Table 2. Configurations of the electrodes of the fabricated sensors.

Parameter	Dimension
Electrodes dimension	15.00 × 2.87 nm
Active area dimension	2.78 × 6.00 mm
Size of fingers	0.216 mm
Number of fingers	5
Gap between fingers	0.3 mm

2.3. Material Characterisation and Gas Sensing Measurements

A set of numerous characterisation techniques was used to study the different morphological and sensing aspects of the synthesised materials. To assess their crystallinity, a Raman spectrometer (Renishaw, plc., Wotton-under-Edge, UK) via a green laser with a wavelength of 514 nm was used. The morphology of the surface and the nanosized material's distribution were studied using a Field Emission Scanning Electron Microscope (FESEM) employing a Carl Zeiss AG-Ultra 55 (ZEISS, Jena, Germany). A JEOL F200 TEM ColdFEG (Tokio, Japan) operated at 200 kV was used for the high-resolution transmission electron microscopy (HRTEM) characterisation. TEM images and electron diffraction patterns were acquired with a Gatan OneView camera, a CMOS-based and optical fibre-coupled detector of 4096 by 4096 pixels. Gatan Digital Micrograph program was used to process the (S)TEM images. STEM images (1024 × 1024 pixels) were recorded from the JEOL bright-field (BF) and high-angle annular dark-field (HAADF) detectors with a camera length of 200 mm. Samples were inserted in a JEOL beryllium double-tilt holder for energy-dispersive X-ray spectroscopy (EDS). STEM-EDS mapping was recorded from an EDS Centurio detector (silicon drift) with an effective area of 100 mm^2 and 133 eV of energy resolution. STEM-EDS maps (512 × 512 pixels) were processed with the JEOL Analysis software to check the shape of the nanosized MnO_2 and its incorporation in the graphene layers. Finally, a Jasco FTIR/IR-6700 infrared spectrometer (Oklahoma City, OK, USA) with a diamond crystal kit was used to conduct the ATR-IR analysis of the interaction between the molecules of the rGO@MnO_2 material. The obtained sensors were placed in a 35 cm^3 volume Teflon chamber. Later, a set of Bronkhorst mass-flow controllers (Ruurlo, The Netherlands) with a 100 mL/min flow were used to pass a continuous stream of dry air (Air Premier, 99.995% purity) and other target analytes with different concentrations through the Teflon chamber. The same set of mass flow controllers was utilised to dilute the target gases from calibrated bottles (NO_2-1 ppm, CO-100 ppm, NH_3-100 ppm, ethanol-20 ppm, and Benzene-10 ppm balanced in dry air). The resistance changes were obtained in real-time via an Agilent HP 34972A multimeter (Santa Clara, CA, USA). The ambient moisture effect on the responses of the sensors was evaluated by using a controller evaporator mixer (CEM) to obtain low humidity levels with a minimum of 25% RH and a maximum of 50% RH at room temperature. For higher values of relative humidity, the flow of the dry air with the corresponding concentration of gas was humidified, passing through a bubbling water system at room temperature. The sensing response is defined as the relative change in resistance and was calculated using the following formula:

$$R\ (\%) = ((\,|\,R_g - R_a\,|\,)/R_a) \times 100$$

R_g and R_a correspond respectively to the resistance values prior to and following gas exposure. Figure 2 shows the schematic representation of the gas detection process and equipment used.

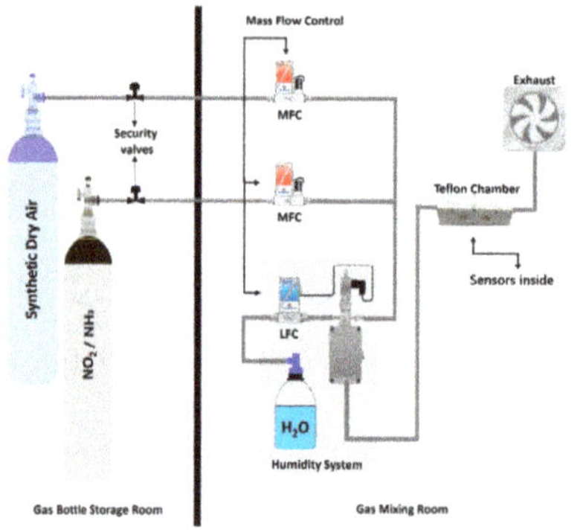

Figure 2. Schematic representation of the gas detection process and used equipment.

3. Results

This section first presents and discusses the results of the characterisation tests made for the prepared sensitive layers (rGO and rGO@MnO_2), which are based on the Raman, FESEM and HRTEM techniques. Secondly, the gas testing results for these sensors towards

NO_2 and NH_3 at room temperature under dry conditions are presented, followed by the results gathered at different humidity levels. Additionally, selectivity tests are reported. The results of the gas sensing tests are compared to those found in the literature. Finally, a sensing mechanism for the detection of NO_2 and NH_3 is introduced.

3.1. Sensitive Layer Characterisation

3.1.1. Raman

The study of the molecular structure of carbon products and the assessment of disorders and defects in the material can be performed using Raman spectroscopy analysis. Two specific peaks always appear when analysing graphene: the G-band and the D-band. The first one, placed at around 1500 cm^{-1}, corresponds to the first order scattering of the E_2g phonons at the Brillouin zone centre and originates from the in-plane vibrations of the sp^2 carbon atoms [46]. Meanwhile, the D-band is observed around 1300 cm^{-1} and represents the formation of j-point photons of A_1g symmetry; it is also associated with double bonds C=C, meaning the more intense the band is, the higher the presence of sp^2 domains is. Furthermore, D-band peak intensity depends highly on the presence of disorders and defects like vacancies and edges in the carbon lattice and grain boundaries [47]. To determine the degree of oxidation of the graphene, a simple calculation of the intensity ratio of both the D and G band peaks is enough, i.e., ID/IG; the higher this ratio is, the lower the oxidation level is [48].

Finally, the second-order bands are observed from 2500 cm^{-1} to 3200 cm^{-1}, containing one always visible peak at around 2700 cm^{-1}, known as the 2D band. They are used generally to determine the layers of the graphene since graphene is susceptible to stacking [49]. The chosen name of this band comes from the fact that it is the overtone of the D band, and two of the same phonons responsible for the D band are involved in the 2D band. Two other bands are sometimes reported when studying graphene Raman spectra, which are the D+G band, which can be seen around 2900 cm^{-1}, and the combined overtone of the D and G bands, the 2G band around 3200 cm^{-1}, which is attributed to the overtone of the G band [50]. In our case, we are working with rGO, which means that the stacking of the layers is random, and since the width of the peaks is relative to the disorder, that can lead to an overlapping of the 2D band peak with the D+G and 2G, resulting in a bump like peak observed in the range of 2600 cm^{-1} to 3100 cm^{-1} [51].

Figure 3a,b shows the Raman spectra of bare rGO and rGO doped with nanosized MnO_2, respectively. For both cases, D bands are located at 1355 cm^{-1} and G bands around 1590 cm^{-1}. For the second-order bands, three visible peaks corresponding to the 2D, D+G, and 2G bands at respectively 2697 cm^{-1}, 2940 cm^{-1}, and 3188 cm^{-1} are present in the Raman spectra of the reduced graphene oxide; meanwhile, in the Raman spectra of the MnO_2 doped reduced graphene oxide we observe a bump-like peak around 2926 cm^{-1}, which is in agreement with the explanation made previously. Moreover, the ID/IG intensity ratio is 0.89 for rGO@MnO_2 and 0.85 for rGO, showing a slightly lower oxidation level in the doped rGO and the presence of a higher number of defects, which can be caused by the doping process. Finally, it was noticed that the Raman signal in the rGO@MnO_2 sample shows a higher intensity than the pristine rGO sample, which can be correlated to the presence of the nanosized MnO_2 since the presence of MO_X usually leads to this increase in the intensity [52].

3.1.2. FESEM

Figure 4a shows the obtained FESEM images of the layers present on the surface of the graphene loaded with MnO_2 on the silicon substrate using a back-scattered electron detector (BSE). A very homogenous layer is observed covering the totality of the surface inspected. MnO_2 cannot be clearly seen, even when using a BSE detector, because of the low concentration of the nanosized material and the small size, but when performing an EDS analysis of the surface, its presence was detected (see Figure S2 in the Supplementary Materials). Figure 4b shows the FESEM image of the graphene doped with MnO_2 deposited on Kapton. Again, a very good coverage of the substrate surface is seen. However, in this case, the surface of the substrate is getting rapidly charged because of the effect of the magnetic field coming from the BSE detector, leading to the formation of very bright areas. Similarly, in the samples on silicon, an EDS analysis (see Figure S3 in the supplementary materials) detected the presence of MnO_2 in the samples deposited over Kapton. Still, the MnO_2 crystals are too small to be seen in the FESEM images, just as for the sensing layer deposited on silicon.

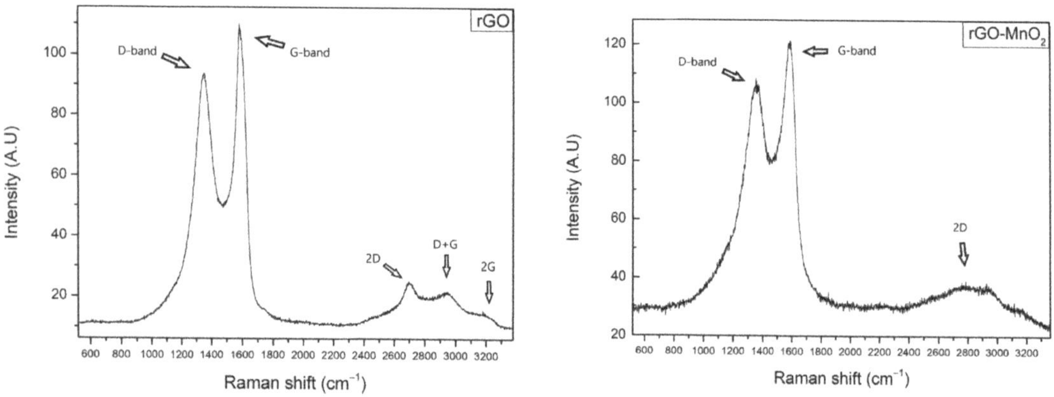

(a) (b)

Figure 3. (**a**) Raman spectra of rGO and (**b**) Raman spectra of rGO@MnO_2.

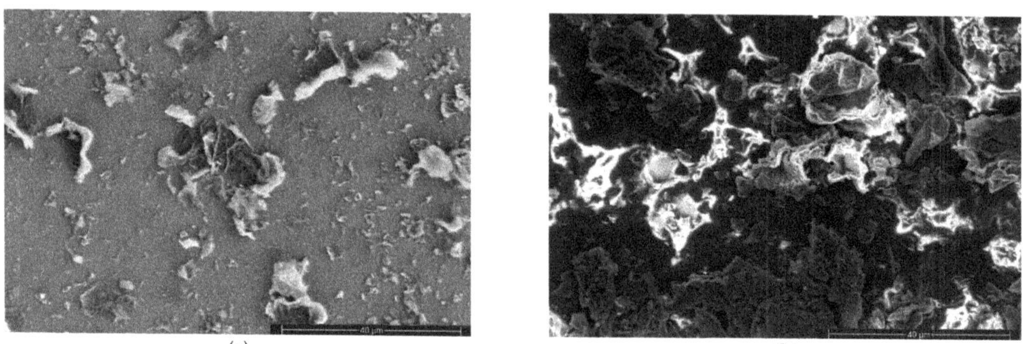

(a) (b)

Figure 4. FESEM images of (**a**) the surface of the graphene doped with MnO_2 deposited on the silicon substrate and (**b**) the surface of the graphene doped with MnO_2 deposited on the Kapton substrate.

3.1.3. HRTEM

An HR-TEM analysis was conducted to examine better the morphology of the nanosized MnO_2 and its incorporation in the graphene layer. Figure 5a shows an HR-TEM image of layers of graphene in addition to an interesting structure on the top right side; zooming in on this structure (Figure 5b), a sponge-like shaped nanosized material was observed, which was attributed to the MnO_2 after performing an EDS analysis. Moreover, when chemically mapping the chosen area's surface, a high concentration of Mn is located in the same position as the sponge-like structure, proving the presence and the likely shape of the nanosized MnO_2 shown in Figure 5c. Meanwhile, Figure 5d,e shows the EDS mapping of the other present elements, which are O and C, respectively. Moreover, Figure 5f shows an overlay image of all the maps of the elements carbon, manganese and oxygen. The overlapping of the elements Mn and O maps with the colours blue and red created a pink zone where the nanosized MnO_2 is concentrated. Finally, Figure 5g presents the EDS map spectrum of the same area, showing the elements present in the mapping appearing as Mn with the highest concentration; the other elements, except for C and O, come from the grid of the HRTEM.

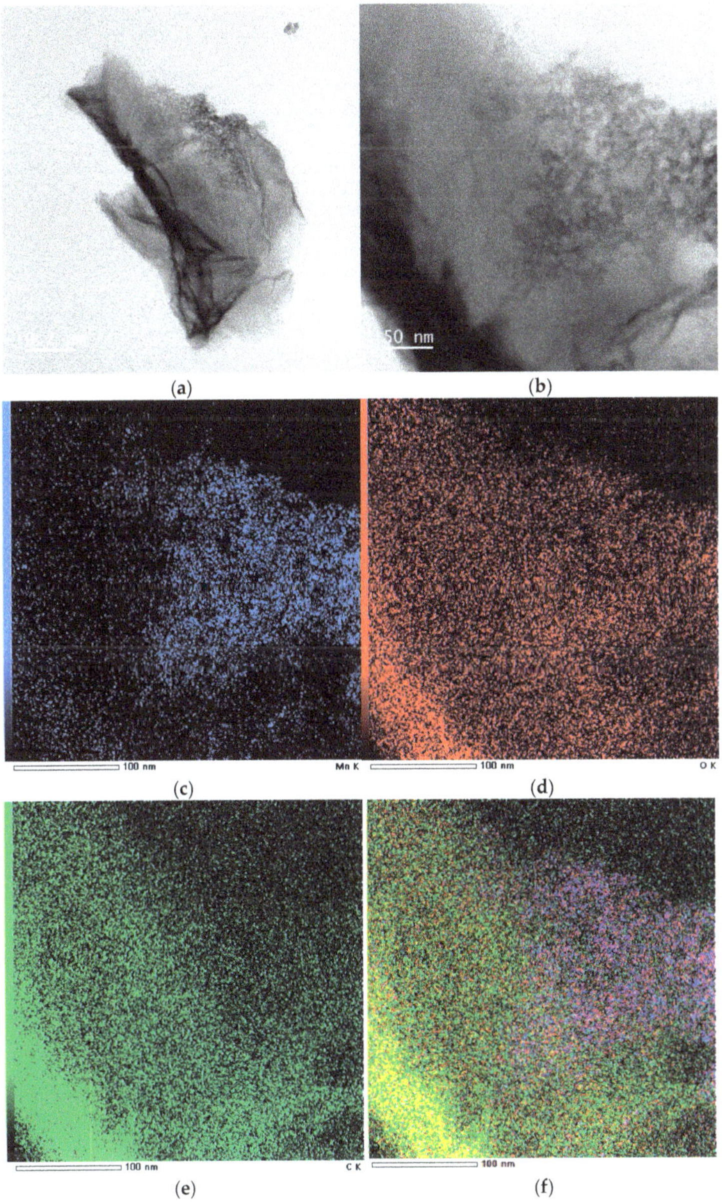

Figure 5. *Cont.*

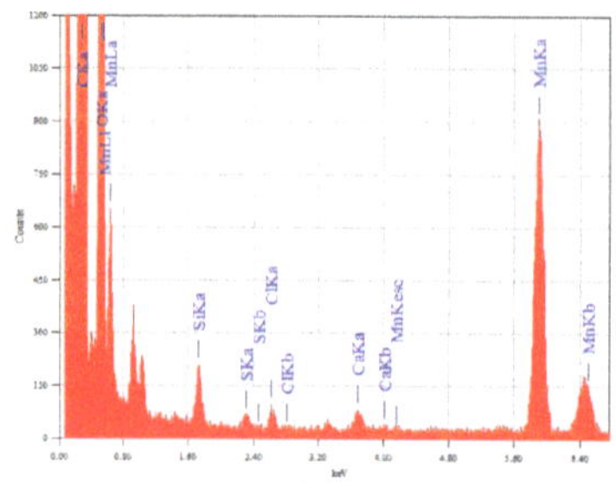

(g)

Figure 5. (**a**) HRTEM image of layered graphene doped with nanosized MnO_2; (**b**) a zoomed HRTEM image of a scale of 50 nm of the same material; (**c**) EDS mapping showing Mn concentration on the area of analysis; (**d**) EDS map of O element in the mapped area; (**e**) EDS map of C element in the same mapped area; (**f**) overlay image of all the maps of the elements C (green), O (red) and Mn (blue); (**g**) EDS map spectrum of the studied area.

3.1.4. ATR-IR

Figure 6 shows the ATR-IR spectrum of both rGO@MnO_2 on silicon and Kapton substrates. Examining Figure 6a,b, we can see rGO specific bands for the stretches of C=C and C–O, a specific band for the Mn–OH vibration proving the successful association of Mn and rGO producing MnO_2@rGO and finally, a specific band for the Mn–O vibration. These bands are present in both materials deposited on silicon and on Kapton substrate. The only difference is the presence of the OH band and also the position of the previously mentioned peaks. For the MnO_2@rGO deposited on silicon, the C=C and C–O stretches are located at around 1144 and 1012 cm^{-1}; meanwhile, for the MnO_2@rGO deposited on Kapton, those peaks are located around 1577 and 1017 cm^{-1} [53]. The Mn–OH stretch is located at 784 cm^{-1} for the sensitive material deposited on silicon and at 870 cm^{-1} for the same material on Kapton. The Mn–O stretch is present for the rGO@MnO_2 at 491 cm^{-1}; meanwhile, the material deposited on Kapton is at 501 cm^{-1}. Moreover, Kapton-specific peaks were clearly seen in the spectrum shown in Figure 6b; these peaks were identified and reported in the work of Evie L. et al. [54].

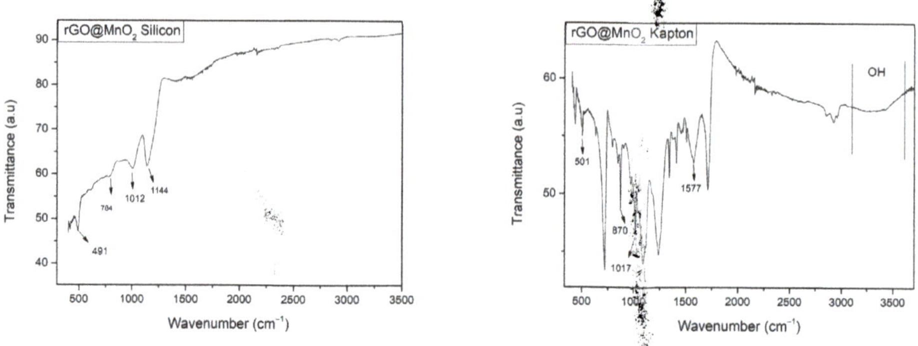

(a) (b)

Figure 6. ATR-IR spectra of (**a**) rGO@MnO_2 on silicon substrate and (**b**) rGO@MnO_2 on Kapton substrate.

Finally, in the ATR-IR spectrum of the material deposited on Kapton, we can observe a bump between 3000 and 3600 cm^{-1}, specific for the OH stretching vibration, which is absent in the Spectra of the same material deposited on silicon. Thus, the substrate generates differences in the film, and Kapton particularly favours hydroxylation, which has an impact on sensing behaviour.

3.2. Gas Sensing Results

A selection of different toxic gases and vapours was used to study the sensing properties of the pristine rGO and rGO@MnO_2 sensors. First, NO_2 was thoroughly studied with different dilutions ranging from 200 ppb up to 1000 ppb under dry air as well as under ambient moisture conditions (close to real conditions). Sensors were always operated at room temperature. Then, NH_3 was also tested as an interferent gas with a concentration of 50 ppm under the same conditions used for NO_2. The registered baseline resistance of both the materials when conducting these tests for rGO@MnO_2 on silicon is near 100 kΩ while the value is near 6 MΩ when deposited on Kapton. For pristine rGO material films, the baseline resistance over the silicon substrate is near 13.2 kΩ and for the Kapton substrate, it is near 28.5 kΩ. This difference in baseline resistance values of the same material on different substrates is directly affected by the sensitive layer thickness; Figures S4 and S5 show the deposited layers' thickness on top of each substrate. Noticeably, the material on top of silicon has a higher thickness and, thus, a lower resistance value. Meanwhile, materials on Kapton have lower thickness and, thus, higher resistance values. Figure 7a shows the response of the different sensors towards different concentrations of NO_2, and it was noticed that the type of substrate used does not affect the response of the sensitive layer towards the analyte. rGO on silicon and Kapton has almost the same response through the studied range, with an average difference of 0.8%. The same behaviour was also seen for the rGO@MnO_2 sensors, where the average difference between the responses was 0.4%. Moreover, the most important aspect to note is that the sensors based on rGO incorporating MnO_2 show a superior response to the pristine ones (a two-fold increase in response). The loading of rGO with MnO_2 is effective at increasing sensitivity towards NO_2. In fact, rGO@MnO_2 on Kapton exhibits a higher sensitivity of 3% ppm^{-1} compared to the 1% ppm^{-1} for the pristine rGO; meanwhile, for the materials deposited on silicon, the sensitivity of the doped material is slightly better than its pristine counterpart with 1.8% ppm^{-1} for rGO@MnO_2 on silicon and 1.5% ppm^{-1} for on rGO silicon. The sensitivity values were evaluated from the slope of the line obtained from the linear regression of the responses of the sensor towards different concentrations of the gas. Figure 7b shows the resistance changes of the rGO@MnO_2 on silicon substrate for 600 ppb of NO_2, Figure 7c shows the resistance changes of the rGO@MnO_2 on Kapton substrate for 600 ppb of NO_2 at 25% RH, and Figures S6–S8 show the resistance changes of the same material under the same atmosphere but towards different concentrations of the target gas (NO_2).

Further studies were conducted where ambient moisture was introduced via two different methods to check its effect on the sensing properties of the sensors. The first method consisted of using a CEM to obtain 25% RH, and the second method consisted of using a bubbling water glass bottle that was installed between the mass flow and the chamber to humidify the air and the gas to reach a maximum RH of 70%. The ambient temperature was kept constant at 25 °C throughout the measurement period. Figure 8a,b shows the calibration curves for the studied sensors at 25 and 70% of relative humidity, respectively. Comparing the results shown in Figure 8a (dry conditions) and 6a (25% RH), one can notice that the response of the MnO_2-doped rGO sensors under humid conditions increases by a factor of 2.5 than when under dry conditions. For example, the rGO@MnO_2 on Kapton sensor response for NO_2 1000 ppb at 25% humidity is 17.6%, while it is 6.4% under dry conditions. Interestingly, the responses of the pristine rGO sensors at 25% RH were enhanced by factors of 3.5 and 4, reaching similar response intensities to those recorded for MnO_2-doped rGO sensors. For example, rGO on silicon and rGO on Kapton responses to NO_2 1000 ppb were 13.8% and 12.7%, respectively, whereas under a dry atmosphere, their responses were 4.1% and 3.2%, respectively. Sensitivity values were calculated following the slope of the linear regression of the response values towards different dilutions of NO_2 and compiled in Table 3.

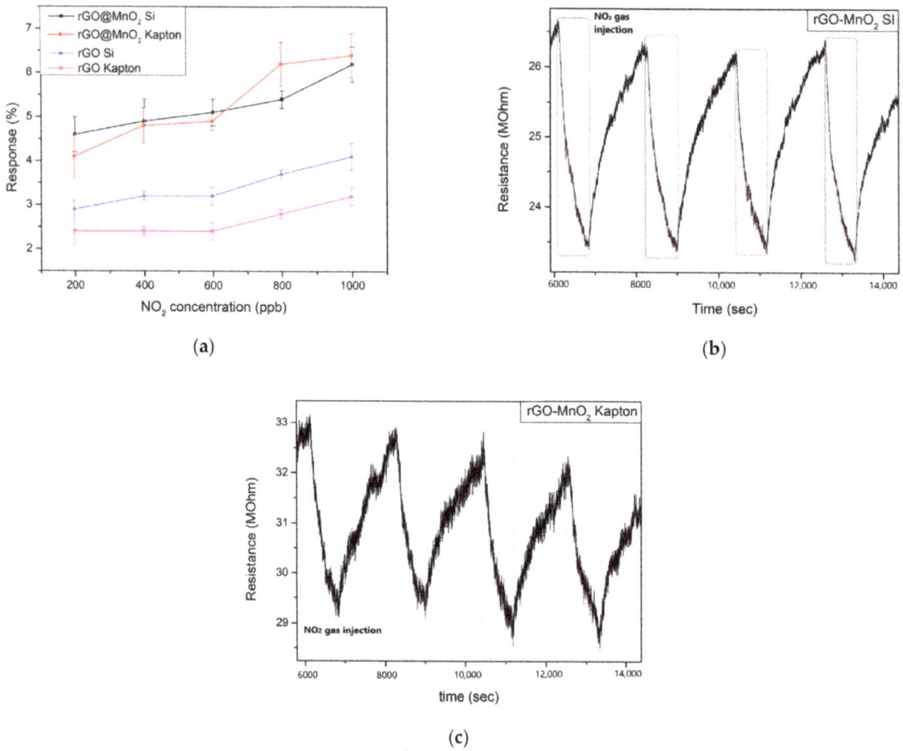

Figure 7. (**a**) Calibration curve of the responses of the fabricated sensors towards different concentrations of NO_2 at room temperature and under dry conditions; (**b**) resistance changes of the rGO@MnO_2 on silicon substrate for 600 ppb of NO_2 at 25% RH; (**c**) resistance changes of the rGO@MnO_2 on Kapton substrate for 600 ppb of NO_2 at 25% RH.

Table 3. Sensitivity values of the different sensors under 25% and 70% relative humidity at room temperature.

Material	Sensitivity (% ppm^{-1})	Relative Humidity (%)
rGO-MnO_2 Silicon	9.8	25
	27.2	70
rGO-MnO_2 Kapton	12.4	25
	16.6	70
rGO Silicon	10.9	25
	13.5	70
rGO Kapton	10.2	25
	14.5	70

Meanwhile, Figure 8b reveals the calibration curves of the sensors under 70% ambient moisture. It is noticed that when increasing the concentration of water vapour, the response of the pristine rGO sensors is much more enhanced than the corresponding doped ones, but the increase in the sensitivity is not so significant. When measuring 1000 ppb of NO_2 and increasing the RH levels from 25 to 70%, the response of the pristine rGO sensors is doubled, while the sensitivity just increased by a factor of around 1.3. In the case of the rGO@MnO_2 sensors, the increase in the response is only in the order of a factor of 1.2, but the increase in the sensitivity is higher than in the previous case, especially for the sensor on the silicon substrate, as can be seen in Table 3.

This behaviour of the pristine rGO layers is expected since the same material was already reported in the literature as a humidity detector, such as in the work of Muhammed et al., where they fabricated rGO and rGO/Fe_2O_3 components for humidity detection and the pristine material showed a high sensitivity towards RH and it increased more with the incorporation of Fe_2O_3 [55]. Zhou et al. also managed to make humidity sensors with the sensitive layer of rGO/SnO_2. Initially, they tested the pristine SnO_2 sensitivity and response towards 75% RH, and they saw these results improve by adding rGO and making rGO/SnO_2 porous film, indicating the fact that rGO is a very sensitive material towards humidity [56]. Although in this work, pristine rGO response towards RH increases with the increase of the humidity level, the doping of rGO with MnO_2 made the response less affected by the RH levels, but the sensitivity increases when the level of humidity increases. Theoretical LOD was calculated following this formula: LOD = 3.3 * (Sy/S), and the values are compiled and shown in Table S3.

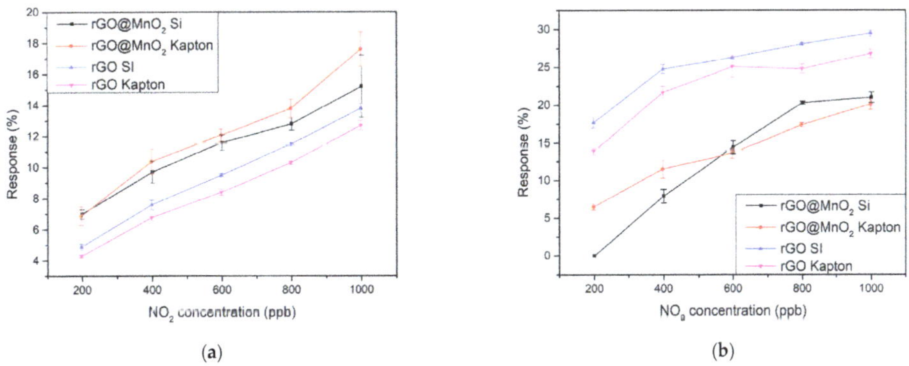

Figure 8. (**a**) calibration curves of the different sensors under 25% relative humidity at room temperature and (**b**) calibration curves under 70% relative humidity at room temperature.

Table 4 compares the results reported here with those of the literature. The sensors we report are more sensitive to NO_2 than those found in the literature. In addition, the concentrations tested in the literature are generally higher than the ones reported here, which indicates that our material is more sensitive in the low ppm concentration range. While most works totally overlook the effect of ambient humidity in the sensing properties, our material is shown to be able to detect NO_2 in a wide range of ambient moisture levels.

Table 4. Comparison of the sensing performance to NO_2 of different materials and rGO-based compounds.

Material	NO_2 Concentration (ppm)	Response (%)	Condition	Sensitivity (%ppm^{-1})	T (°C)	Ref.
Nano-MnO_2/xanthan	7	1.21	Dry	0.17	RT	[57]
δ-MnO_2-Epitaxial Graphene-Silicon Carbide Heterostructures	5	0.27	55% RH	0.14	RT	[58]
Porous MnO_2 /rGO	50	5.9	Dry	0.118	RT	[41]
ZnO/rGO	10	5.1	Dry	0.51	RT	[59]
rGO pomegranate peels	1	3.04	Dry	2.94	100	[60]
Phosphate doped rGO	1	4.5	Dry	4.5	RT	[61]
VO_2/rGO	5	1.63	Dry	0.326	RT	[62]
MnO_2 doped rGO	1	6.2	Dry	9.8	RT	This work
MnO_2 doped rGO	1	21	70% RH	27.2	RT	This work

The selectivity of the different sensors we tested was studied under the same experimental conditions used for NO_2 detection. Different species, namely, CO (50 ppm), NH_3 (50 ppm), H_2 (500 ppm), ethanol (20 ppm) and benzene (1 ppm) under dry conditions for sensors operating at room temperature were measured. Figure 9 shows the responses to these different gases or vapours. As can be seen, none of the sensors showed any response to H_2 and benzene. Noticeably, the inclusion of the nanosized MnO_2 reduced the response towards CO and Ethanol, making it utterly unresponsive to these interfering gases. Therefore, the incorporation of MnO_2 improved the sensor's selectivity. Nevertheless, all the sensors showed very significant responses towards 50 ppm of NH_3.

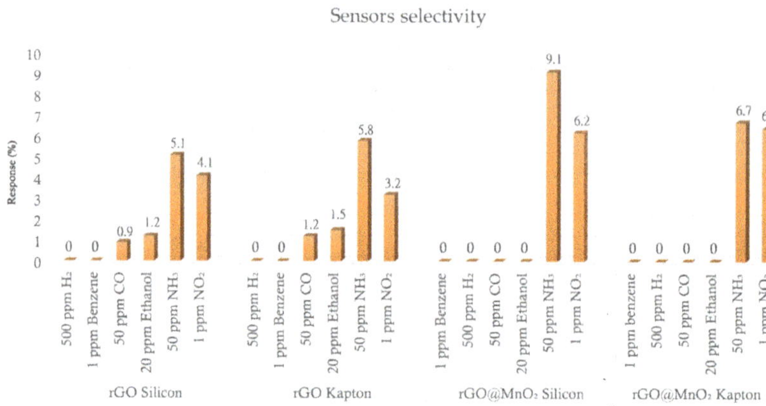

Figure 9. Comparison of the responses of the different sensors towards different gases at dry conditions to study the selectivity of the sensitive layer.

Taking into account the good responses observed for NH_3, the effect of moisture in the sensor responses to this gas was analysed. The sensors were exposed to 50 ppm of NH_3 under different humidity conditions (dry, 25% RH and 50% RH), always at room temperature. Figure 10a shows the responses of the sensors to 50 ppm of ammonia for the three different humidity conditions studied, and Figure 10b shows resistance changes of the sensor pristine rGO on Kapton when exposed to NH_3. As seen in the figure, when exposed to an ammonia analyte, the sensor resistance decreases in contrast to what is expected for a p-type material; this behaviour was explained later in the mechanism part. It was also noticed that the response of the rGO@MnO_2 on the silicon sensor was the highest throughout all the conditions. The response of this sensor reaches a value of 18.5% at 50% RH, which is 4 times higher than the response of the pristine rGO on silicon. rGO@MnO_2 and pristine rGO on Kapton show basically the same behaviour, and the doped one shows a slightly higher response towards NH_3, with values of 6.7% and 5.8% for rGO@MnO_2 and rGO, respectively, under dry conditions, 7.5% and 6% for 25% RH and 8% and 6.7% at 50% RH. Pristine rGO on silicon shows the lowest response values towards ammonia, with a value of 4.6% at 50% RH. In essence, pristine rGO on Kapton and on Silicon substrates shows a linear-like behaviour throughout the different RH levels tested, with very little increase in sensitivity and increasing moisture levels. To have a better understanding of the behaviour of the sensors towards ammonia under humidity, it could be explained as follows: Since we are working in a humid environment, the sensing layers have already adsorbed water molecules on its surface, saturating to an extent the adsorption sites, especially of the pristine rGO layers. Later on, when these layers are exposed to an NH_3 gas flow, another phenomenon happens in the working atmosphere, and it can be attributed to the characteristics of ammonia itself. In fact, both H_2O and NH_3 have a strong tendency to form H bonds. Moreover, the electronegativity of the atoms determines the possibility of forming hydrogen bonds, and since oxygen is more electronegative than nitrogen, the O atom from H_2O rapidly creates a hydrogen bond with NH_3 [33], as shown in Figure 10c. Therefore, when considering the silicon substrate sensors, the response of the pristine rGO sensor remains practically unchanged because of the phenomenon previously explained, which prevents ammonia molecules from getting adsorbed on the surface. Meanwhile, for the rGO@MnO_2 sensor, the significant increase in the response, despite the occurrence of the hydrogen bonding of the ammonia and water molecules, can be explained by the presence of the nanosized MnO_2, which plays a compensatory role by creating more adsorption sites in the layer, meaning more space for the ammonia and water molecules to be adsorbed also it has been previously reported as a good NH_3 adsorbing agent [63], which explains the increase of the response of the MnO_2@rGO sensor. As for the sensors deposited on Kapton, both pristine rGO and rGO@MnO_2 showing similar behaviour can be explained by the fact that the substrate is made of a very strong hydrophobic material. Therefore, water molecules are getting repelled off of the surface, resulting in poor H_2O adsorption and a low dependency on these sensors for ambient moisture.

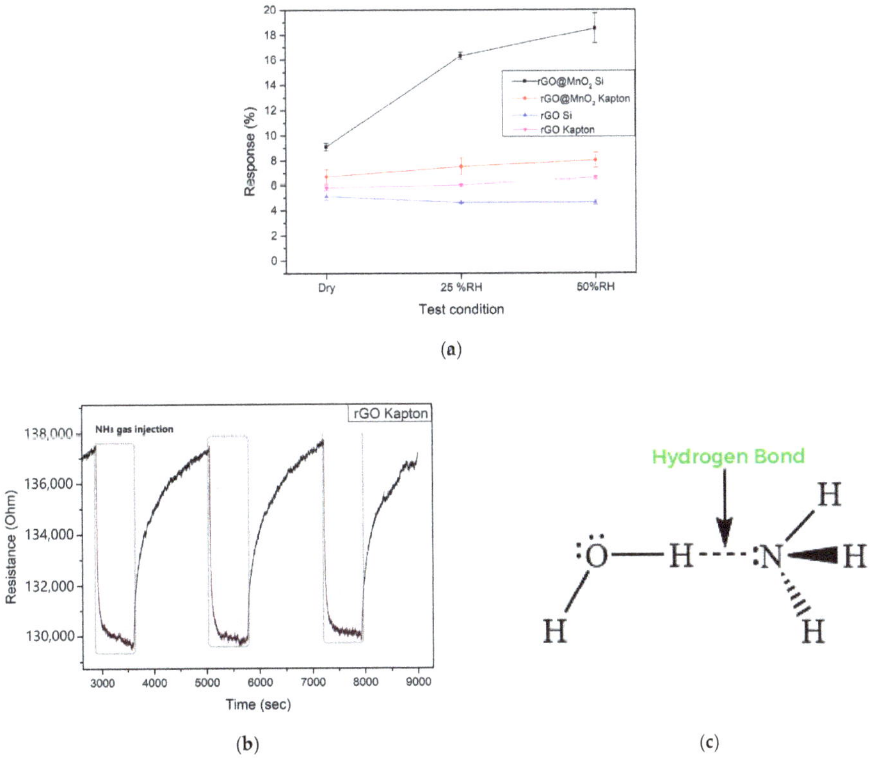

Figure 10. (**a**) Calibration curve of the responses of the fabricated sensors towards different test conditions (Dry, 25% RH and 50% RH); (**b**) resistance changes of the sensor pristine rGO on Kapton when exposed to NH_3 at 25% RH; (**c**) hydrogen bonding of water and ammonia molecules.

To check the position of this work in the literature regarding ammonia detection, a set of data such as response and sensitivity of other materials and sensors analysing NH_3 gas were collected and compiled in Table 5 and put in comparison with our results. Considering the same NH_3 concentration, $NiFe_2O_4$/rGO had a response of 1.17; meanwhile, Pani@MnO_2@rGO had a response of 15.5 while heating up to 100 °C. Both these materials showed lower responses than our work, which was 18.6% at 50% RH. It is true that $FeCo_2O_4$/WO_3/rGO have a slightly higher response of 19.8% at dry conditions, but in this work, NH_3 concentration is 100 ppm, and the working temperature is 200 °C; meanwhile, we are working at RT and half of the NH_3 concentration.

Table 5. Comparison of the sensing performance to NH_3 of different materials and rGO-based compounds.

Material	NH_3 Concentration (ppm)	Response (%)	Condition	T (°C)	Ref.
PANI@MnO_2@rGO	50	15.5	Dry	100	[44]
$NiFe_2O_4$/rGO	50	1.17	Dry	0	[64]
rGO/WO_3	40	8.03	55% RH	35	[65]
$FeCo_2O_4$/WO_3/rGO	100	19.8	Dry	200	[66]
$CoFe_2O_4$/rGO	25	1.06	Dry	RT	[67]
rGO@MnO_2	50	18.6	50% RH	RT	This work

3.3. Sensing Mechanism

Graphene and its derivatives, such as rGO, are p-type materials, which implies that usually, the interaction between rGO and oxidising gases such as NO_2 causes a change in the local carrier concentration and, therefore, a decrease in graphene-based sensor resistance meanwhile when exposed to reducing gas such as NH_3 an increase in the resistance takes place [68]. Meanwhile, MnO_2 is an n-type nanomaterial, and when exposed to an ambient environment, a chemisorption of the oxygen molecules takes place, capturing electrons from it and releasing different oxygen species such as O_2^-, O^{2-} and O^- [69]. Moreover, the incorporation of the MO_X nanomaterial (in our case, nanosized MnO_2) in the rGO results in the formation of a p-n heterojunction, causing the flow of the electrons from the MnO_2 to rGO, implying the formation of a depletion layer on the area of contact of both materials, also increasing the electron concentration in the rGO and the hole concentration in MnO_2 [70]. The exposure of the rGO@MnO_2 to air leads to the adsorption of oxygen on the surface of the p-n heterojunction material and the transfer of electrons from its conduction band to the oxygen, resulting in the formation of O_2^- ions following these equations [71]:

$$O_2 \text{ gas} \rightarrow O_2 \text{ (ads)}$$

$$O_2 \text{ (ads)} + e^- \rightarrow O_2^- \text{ (ads)}$$

When exposed to NO_2, it gets adsorbed on the rGO@MnO_2 surface and reacts with the oxygen ions and electrons from the layer following this equation, causing the decrease of the resistance of the sensor:

$$NO_2\text{(ads)} + O_2^- \text{ (ads)} + 2e^- \rightarrow NO_2^- + 2O^- \text{ (ads)}.$$

As expected, our material showed the exact same behaviour explained previously, where the baseline resistance of the sensors decreased when put in contact with NO_2 gas and recovered again when the gas flow stopped.

Although NH_3 is a strong reducing gas, the baseline resistance should increase when in contact with the gas, but not in our case, where the resistance of our sensors decreased. This behaviour has been reported previously by A. Umar et al. when exposed to NH_3, the interaction between the analyte and the sensitive layer results in the release of electrons back to the conduction band of the nanosized MnO_2, which is believed to be the cause of the decrease of the resistance of the sensor [41]. This behaviour has been observed also for pristine rGO and was reported in the work of X. Xiao et al. [72]. More interestingly, D. Han et al. comprehensively studied the NH_3 sensing mechanisms, revealing that by changing from a pure air atmosphere to an environment with the presence of NH_3, a significant change in the width of the depletion layer can be induced [73]. Thus, it is worth noting that our sensitive film is essentially p-type owing to the major presence of rGO acting as a transducing element. However, the presence of the MnO_2 doping element, which is n-type, creates a p-n heterojunction, resulting in a depletion layer. Therefore, when the NH_3 interact with the MnO_2, this probably leads to a decrease in the width of the depletion layer, resulting in a more efficient transport and mobility of charge carriers. Consequently, the conductivity of the film is improved, resulting in the observed decrease of resistance when the sensor is exposed to NH_3 in comparison to a pure air atmosphere.

Finally, graphene-based sensors are well known to be sensitive towards ambient moisture, enhancing its responses [74]. Since the working temperature is at room temperature, water molecules most likely play the role of a mediated adsorption site for the target gas, thus increasing the sensor's sensitivity and responses towards the chosen analyte [75], which is in accordance with the results we obtained, where the responses of the sensors increased under the ambient moisture conditions.

4. Conclusions

Incorporating nanosized MnO_2 in rGO flakes to form a MnO_2@rGO nanomaterial to be integrated into chemoresistive sensors is a novel approach since a very limited number of papers exist in the literature reporting the use of this material as a sensitive layer for gas detection. Additionally, since the same material was deposited on two different substrates (Kapton and silicon), the effect of the substrate material on the morphological and sensing behaviour of the sensing material was thoroughly studied and reported in this work. Our nanomaterial exhibits better performance and properties than other approaches previously reported in the literature, showing a set of very promising results and high responses towards low concentrations of NO_2 and NH_3. When deposited either onto rigid (silicon) or flexible (polyimide) transducing substrates, the material shows good response properties when operated at room temperature and even in the presence of ambient humidity. Furthermore, sensors show very small cross-sensitivity to other species such as hydrogen, ethanol and benzene vapours or carbon monoxide. All these properties make MnO_2@rGO an excellent candidate nanomaterial for the inexpensive, chemoresistive detection of nitrogen dioxide or ammonia in real-life environments using either the flexible substrate for wearable sensors for real-time gas monitoring or using silicon substrate for the possibility of integrating gas sensors in the semiconductor industry.

Supplementary Materials: The following supporting information can be downloaded at: https://www.mdpi.com/article/10.3390/chemosensors12120256/s1, Figure S1: XRD graph of rGO@MnO_2 powder showing the presence and crystalline phase of manganese oxide; Figure S2: The extracted spectrum from the EDS analysis for rGO@MnO_2 on Silicon; Figure S3: The extracted spectrum from the EDS analysis for rGO@MnO_2 on Kapton; Figure S4: Sensitive layer thickness on top of silicon substrate; Figure S5: Sensitive layer thickness on top of Kapton substrate; Figure S6: Resistance changes of the rGO@MnO_2 sensor when exposed to 200 ppb of NO_2; Figure S7: Resistance changes of the rGO@MnO_2 sensor when exposed to 400 ppb of NO_2; Figure S8: Resistance changes of the rGO@MnO_2 sensor when exposed to 800 ppb of NO_2; Table S1: Characteristics of the elements present in the studied sensitive layer; Table S2: Characteristics of the elements present in the studied sensitive layer; Table S3: Calculated theoretical values of LOD for the different sensitive layers at different conditions.

Author Contributions: Conceptualisation, X.V. and J.C.-C.; methodology, M.A.A., X.V.; Software, A.S.-B.; validation, M.A.A.; formal analysis, M.A.A., X.V., J.C.-C.; investigation, M.A.A.; resources, F.S., J.C.S.-C., A.S.-B., S.d.B.-M. and A.G.-G.; data curation, M.A.A.; writing—original draft preparation, M.A.A.; writing—review and editing, J.C.-C., E.L., X.V., S.d.B.-M., A.G.-G. and F.S.; visualisation, M.A.A.; supervision, J.C.-C. and X.V.; project administration, X.V. and E.L.; funding acquisition, X.V. and E.L. All authors have read and agreed to the published version of the manuscript.

Funding: This research was funded in part by MICINN and FEDER grant no. PDC2022-133967-I00 and AGAUR grant no. 2021 SGR 00147. E.L. is supported by the Catalan Institute for Advanced Studies (ICREA) via the 2023 Edition of the ICREA Academia Award. J.C.-C. is supported by the Marie Skłodowska-Curie grant agreement No. 101066282—GREBOS.

Data Availability Statement: Data used in this paper are available upon demand.

Acknowledgments: The authors want to acknowledge Sergi Plana Ruiz for his help and important discussions about the HR-TEM analysis results, Mariana Stefanova Trifonova for her assistance in the FESEM images preparation and Eric Pedrol Ripoll for his help with the Raman analysis.

Conflicts of Interest: The authors declare no conflicts of interest.

References

1. Zhang, C.; Luo, Y.; Xu, J.; Debliquy, M. Room temperature conductive type metal oxide semiconductor gas sensors for NO_2 detection. *Sens. Actuators A Phys.* **2019**, *289*, 118–133. [CrossRef]
2. Chen, Q.; Wang, S.; Bai, S.; Shen, L.; Peng, J.; Zhang, Y.; Cui, X.; Yu, C.; Zhang, R.; Li, Y.; et al. Sub-ppt NH_3 detection by MoS_2@sulfur nanosheets. *J. Alloys Compd.* **2024**, *985*, 174070. [CrossRef]
3. Naseem, N.; Tariq, F.; Malik, Y.; Zahid, W.A.; El-Fattah, A.A.; Ayub, K.; Iqbal, J. Sensing ability of carbon nitride (C_6N_8) for the detection of carbon monoxide (CO) and carbon dioxide (CO_2). *Sens. Actuators A Phys.* **2024**, *366*, 114947. [CrossRef]
4. Namduri, H.; Nasrazadani, S. Quantitative analysis of iron oxides using Fourier transform infrared spectrophotometry. *Corros. Sci.* **2008**, *50*, 2493–2497. [CrossRef]
5. Shi, J.; Jiang, Y.; Duan, Z.; Li, J.; Yuan, Z.; Tai, H. Designing an optical gas chamber with stepped structure for non-dispersive infrared methane gas sensor. *Sens. Actuators A Phys.* **2024**, *367*, 115052. [CrossRef]
6. Okoffo, E.D.; Thomas, K.V. Quantitative analysis of nanoplastics in environmental and potable waters by pyrolysis-gas chromatography–mass spectrometry. *J. Hazard. Mater.* **2024**, *464*, 133013. [CrossRef]
7. Gaiardo, A.; Novel, D.; Scattolo, E.; Crivellari, M.; Picciotto, A.; Ficorella, F.; Iacob, E.; Bucciarelli, A.; Petti, L.; Lugli, P.; et al. Optimization of a Low-Power Chemoresistive Gas Sensor: Predictive Thermal Modelling and Mechanical Failure Analysis. *Sensors* **2021**, *21*, 783. [CrossRef]
8. Sun, K.; Zhan, G.; Zhang, L.; Wang, Z.; Lin, S. Highly sensitive NO_2 gas sensor based on ZnO nanoarray modulated by oxygen vacancy with Ce doping. *Sens. Actuators B Chem.* **2023**, *379*, 133294. [CrossRef]
9. Krishna, K.G.; Umadevi, G.; Parne, S.; Pothukanuri, N. Zinc Oxide based gas sensors and their derivatives: A critical review. *J. Mater. Chem. C* **2023**, *11*, 3906–3925. [CrossRef]
10. Kong, Y.; Li, Y.; Cui, X.; Su, L.; Ma, D.; Lai, T.; Yao, L.; Xiao, X.; Wang, Y. SnO_2 nanostructured materials used as gas sensors for the detection of hazardous and flammable gases: A review. *Nano Mat. Sci.* **2022**, *4*, 339–350. [CrossRef]
11. Liu, P.; Wang, J.; Jin, H.; Ge, M.; Zhang, F.; Wang, C.; Sun, Y.; Dai, N. SnO_2 mesoporous nanoparticle-based gas sensor for highly sensitive and low concentration formaldehyde detection. *RSC Adv.* **2023**, *13*, 2256–2264. [CrossRef] [PubMed]
12. Staerz, A.; Somacescu, S.; Epifani, M.; Kida, T.; Weimar, U.; Barsan, N. WO_3-Based Gas Sensors: Identifying Inherent Qualities and Understanding the Sensing Mechanism. *ACS Sens.* **2020**, *5*, 1624–1633. [CrossRef] [PubMed]
13. Mazur, M.; Kapuścik, P.; Weichbrodt, W.; Domaradzki, J.; Mazur, P.; Kot, M.; Flege, J.I. WO_3 Thin-Film Optical Gas Sensors Based on Gasochromic Effect towards Low Hydrogen Concentrations. *Materials* **2023**, *16*, 3831. [CrossRef] [PubMed]
14. Morrison, S.R. Semiconductor gas sensors. *Sens. Actuators* **1981**, *2*, 329–341. [CrossRef]
15. Schedin, F.; Geim, A.K.; Morozov, S.V.; Hill, E.W.; Blake, P.; Katsnelson, M.I.; Novoselov, K.S. Detection of individual gas molecules adsorbed on graphene. *Nat. Mater.* **2007**, *6*, 652–655. [CrossRef]
16. Novoselov, K.S.; Geim, A.K.; Morozov, S.V.; Jiang, D.; Zhang, Y.; Dubonos, S.V.; Grigorieva, I.V.; Firsov, A.A. Electric field effect in atomically thin carbon films. *Science* **2004**, *306*, 666–669. [CrossRef]
17. Gautam, M.; Jayatissa, A.H. Ammonia gas sensing behaviour of graphene surface decorated with gold nanoparticles. *Solid-State Electron.* **2012**, *78*, 159–165. [CrossRef]
18. Ko, G.; Kim, H.-Y.; Ahn, J.; Park, Y.-M.; Lee, K.-Y.; Kim, J. Graphene-based nitrogen dioxide gas sensors. *Curr. Appl. Phys.* **2010**, *10*, 1002–1004. [CrossRef]
19. Long, H.; Harley-Trochimczyk, A.; Pham, T.; Tang, Z.; Shi, T.; Zettl, A.; Carraro, C.; Worsley, M.; Maboudian, R. High Surface Area MoS_2/Graphene Hybrid Aerogel for Ultrasensitive NO_2 Detection. *Adv. Funct. Mater.* **2016**, *26*, 5158–5165. [CrossRef]
20. Zhang, H.; Feng, J.; Fei, T.; Liu, S.; Zhang, T. SnO_2 nanoparticles-reduced graphene oxide nanocomposites for NO_2 sensing at low operating temperature. *Sens. Actuators B Chem.* **2014**, *190*, 472–478. [CrossRef]
21. Xia, Y.; Wang, J.; Xu, J.-L.; Li, X.; Xie, D.; Xiang, L.; Komarneni, S. Confined Formation of Ultrathin ZnO Nanorods/Reduced Graphene Oxide Mesoporous Nanocomposites for High-Performance Room-Temperature NO_2 Sensors. *ACS Appl. Mater. Interfaces* **2016**, *8*, 35454–35463. [CrossRef] [PubMed]
22. Wu, Z.; Chen, X.; Zhu, S.; Zhou, Z.; Yao, Y.; Quan, W.; Liu, B. Enhanced sensitivity of ammonia sensor using graphene/polyaniline nanocomposite. *Sens. Actuators B Chem.* **2013**, *178*, 485–493. [CrossRef]
23. Karaduman, I.; Er, E.; Çelikkan, H.; Erk, N.; Acar, S. Room-temperature ammonia gas sensor based on reduced graphene oxide nanocomposites decorated by Ag, Au and Pt nanoparticles. *J. Alloys Compd.* **2017**, *722*, 569–578. [CrossRef]
24. Andre, R.; Shimizu, F.; Miyazaki, C.; Riul Jr, A.; Manzani, D.; Ribeiro, S.; Oliveira, O.; Mattoso, L.; Correa, D. Hybrid layer-by-layer (LbL) films of polyaniline, graphene oxide and zinc oxide to detect ammonia. *Sens. Actuators B Chem.* **2016**, *238*, 795–801. [CrossRef]
25. Wang, J.; Rathi, S.; Singh, B.; Lee, I.; Joh, H.; Kim, G. Alternating Current Dielectrophoresis Optimization of Pt-Decorated Graphene Oxide Nanostructures for Proficient Hydrogen Gas Sensor. *ACS App. Mater. Interfaces* **2015**, *7*, 13768–13775. [CrossRef]

26. Esfandiar, A.; Ghasemi, S.; Irajizad, A.; Akhavan, O.; Gholami, M. The decoration of TiO_2/reduced graphene oxide by Pd and Pt nanoparticles for hydrogen gas sensing. *Int. J. Hydrogen Energy* **2012**, *37*, 15423–15432. [CrossRef]
27. Li, D.; Lu, J.; Zhang, X.; Jin, D.; Jin, H. Engineering of ZnO/rGO towards NO_2 Gas Detection: Ratio Modulated Sensing Type and Heterojunction Determined Response. *Nanomaterials* **2023**, *13*, 917. [CrossRef]
28. Fellah, M.F. The reduced graphene oxide/WO_3: Sensing properties for NO_2 gas detection at room temperature. *Diam. Relat. Mater.* **2021**, *119*, 108593. [CrossRef]
29. Thomas, A.; Jeyaprakash, B.G. Selective detection of ammonia by rGO decorated nanostructured ZnO for poultry and farm field applications. *Synth. Met.* **2022**, *290*, 117140. [CrossRef]
30. Naresh, B.; Sreekanth, T.; Krishna, K.G.; Kumar, K.S.; Suma, C.; Yoo, K.; Kim, J. Enhanced ammonia detection using rGO-wrapped $Zn_3V_2O_8$ nanostructures. *Sens. Actuators A Phys.* **2024**, *379*, 115928. [CrossRef]
31. Meng, F.-L.; Guo, Z.; Huang, X.-J. Graphene-based hybrids for chemiresistive gas sensors. *Trends Anal. Chem.* **2015**, *68*, 37–47. [CrossRef]
32. Majhi, S.M.; Mirzaei, A.; Kim, H.W.; Kim, S.S. Reduced Graphene Oxide (rGO)-Loaded Metal-Oxide Nanofiber Gas Sensors: An Overview. *Sensors* **2021**, *21*, 1352. [CrossRef] [PubMed]
33. Tripathi, D.; Chauhan, P.; Rawat, R. A synergistic approach to enhance sensitivity and selectivity of room temperature operable ammonia gas sensor with humidity assistance using RGO/WO_3 nanocomposite. *Nanotechnology* **2023**, *35*, 065503. [CrossRef] [PubMed]
34. Daneshnazar, M.; Jaleh, B.; Eslamipanah, M.; Varma, R.S. Optical and gas sensing properties of TiO_2/RGO for methanol, ethanol and acetone vapors. *Inorg. Chem. Commun.* **2022**, *145*, 110014. [CrossRef]
35. Abideen, Z.U.; Kim, H.W.; Kim, S.S. An ultra-sensitive hydrogen gas sensor using reduced graphene oxide-loaded ZnO nanofibers. *Chem. Commun.* **2015**, *51*, 15418–15421. [CrossRef] [PubMed]
36. Lee, J.-H.; Katoch, A.; Choi, S.-W.; Kim, J.-H.; Kim, H.W.; Kim, S.S. Extraordinary Improvement of Gas-Sensing Performances in SnO_2 Nanofibers Due to Creation of Local p–n Heterojunctions by Loading Reduced Graphene Oxide Nanosheets. *ACS App. Mater. Interfaces* **2015**, *7*, 3101–3109. [CrossRef]
37. Hingangavkar, G.M.; Kadam, S.A.; Ma, Y.-R.; Bandgar, S.S.; Mulik, R.N.; Patil, V.B. Tailored formation of WO_3-rGO nanohybrids for dependable low temperature NO_2 sensing. *Ceram. Int.* **2023**, *49*, 38866–38876. [CrossRef]
38. Zhang, Q.Z.; Zhang, D.; Miao, Z.C.; Zhang, X.L.; Chou, S.L. Research Progress in MnO-Carbon Based Supercapacitor Electrode Materials. *Small* **2018**, *14*, e1702883. [CrossRef]
39. Wu, M.; Hou, P.; Dong, L.; Cai, L.; Chen, Z.; Zhao, M.; Li, J. Manganese dioxide nanosheets: From preparation to biomedical applications. *Int. J. Nanomed.* **2019**, *14*, 4781–4800. [CrossRef]
40. Malook, K.; Khan, H.; Shah, M.; Haque, I.-U. Highly selective and sensitive response of Polypyrrole–MnO_2 based composites towards ammonia gas. *Polym. Compos.* **2019**, *40*, 1676–1683. [CrossRef]
41. Zhang, H.; Ou, K.; Guan, R.; Cao, Y.; Sun, Y.; Li, X. A Highly Sensitive Room-Temperature NO_2 Gas Sensor based on Porous MnO_2 /rGO Hybrid Composites. *Curr. Nanosci.* **2022**, *19*, 401–409. [CrossRef]
42. Zöpfl, A.; Lemberger, M.-M.; König, M.; Ruhl, G.; Matysik, F.-M.; Hirsch, T. Reduced graphene oxide and graphene composite materials for improved gas sensing at low temperature. *Faraday Discuss* **2014**, *173*, 403–414. [CrossRef] [PubMed]
43. Ghosal, S.; Bhattacharyya, P. Fabrication, Characterization, and Gas Sensing Performance of Pd, RGO, and MnO_2 Nanoflowers-Based Ternary Junction Device. *IEEE Trans. Electron Devices* **2019**, *66*, 3982–3987. [CrossRef]
44. Umar, A.; Akbar, S.; Kumar, R.; Ahmed, F.; Ansari, S.A.; Ibrahim, A.A.; Alhamami, M.A.; Almehbad, N.; Algadi, H.; Almas, T.; et al. Unveiling the potential of PANI@MnO_2@rGO ternary nanocomposite in energy storage and gas sensing. *Chemosphere* **2024**, *349*, 140657. [CrossRef] [PubMed]
45. Rezaei, I.; Haghverdi, A.B.; Soldoozy, A.; Aghaee, T.; Biabanifard, S. Wearable Kapton graphene biosensor for detection of toxic gases. *J. Hazard. Mater. Adv.* **2024**, *15*, 100452. [CrossRef]
46. Yin, F.; Wu, S.; Wang, Y.; Wu, L.; Yuan, P.; Wang, X. Self-assembly of mildly reduced graphene oxide monolayer for enhanced Raman scattering. *J. Solid State Chem.* **2016**, *237*, 57–63. [CrossRef]
47. Hurtado, R.B.; Cortez-Valadez, M.; Aragon-Guajardo, J.R.; Cruz-Rivera, J.J.; Martínez-Suárez, F.; Flores-Acosta, M. One-step synthesis of reduced graphene oxide/gold nanoparticles under ambient conditions. *Arab. J. Chem.* **2020**, *13*, 1633–1640. [CrossRef]
48. Yavuz, S.; Bandaru, P.R. Ag nanowire coated reduced graphene oxide/n-silicon Schottky junction based solar cell. In Proceedings of the IEEE Conference on Technologies for Sustainability (SusTech), Phoenix, AZ, USA, 9–11 October 2016; pp. 265–269. [CrossRef]
49. Hidayah, N.M.S.; Nur, H.; Liu, W.W.; Lai, C.W.; Noriman, N.Z.; Khe, C.-S.; Hashim, U.; Lee, H.C. Comparison on graphite, graphene oxide and reduced graphene oxide: Synthesis and characterization. *AIP Conf. Proc.* **2017**, *1892*, 150002.
50. Ma, B.; Rodriguez, R.D.; Ruban, A.; Pavlov, S.; Sheremet, E. The correlation between electrical conductivity and second-order Raman modes of laser-reduced graphene oxide. *Phys. Chem. Chem. Phys.* **2019**, *21*, 10125–10134. [CrossRef]
51. Sharma, M.; Rani, S.; Pathak, D.K.; Bhatia, R.; Kumar, R.; Sameera, I. Temperature dependent Raman modes of reduced graphene oxide: Effect of anharmonicity, crystallite size and defects. *Carbon* **2021**, *184*, 437–444. [CrossRef]
52. Rajguru, G.M.; Mishra, R.K.; Kharat, P.B.; Khirade, P.P. Structural, microstructural and optical characteristics of rGO-ZnO nanocomposites via hydrothermal approach. *Opt. Mater.* **2024**, *154*, 115720. [CrossRef]

53. Papadopoulou, E.; Morselli, D.; Prato, M.; Barcellona, A.; Athanassiou, A.; Bayer, I.S. An Efficient Pure Polyimide Ammonia, Sensor. *J. Mater. Chem. C* **2016**, *4*, 7790–7797. [CrossRef]
54. Susee, S.K.; Sandhya, S.; Kumar, M.S.; Chidhambararajan, B. Reduced graphene oxide–MnO_2 nanocomposites by hydrothermal method for histamine sensor and photocatalytic activity. *J. Mater. Sci. Mater. Electron.* **2024**, *35*, 1–12. [CrossRef]
55. Muhammad, Y.; Shah, M.; Safi, M.A.; Khattak, S.G.; Aziz, A.; Hassan, H. highly selective and sensitive humidity sensor using reduced graphene oxide-based iron oxide nanocomposites. *Mater. Sci. Eng.* **2024**, *303*, 117324. [CrossRef]
56. Li, Z.; Gardner, D.W.W.; Xia, Y.; Zhao, S.; Pan, A.; Goel, N.; Bart, S.; Liu, C.; Yi, J.; Carraro, C.; et al. Ordered porous RGO/SnO_2 thin films for ultrasensitive humidity detection. *J. Mater. Chem. C Mater.* **2023**, *11*, 9586–9592. [CrossRef]
57. Fahmy, A.; Saeed, A.M.; Dawood, U.; Abdelbary, H.; Altmann, K.; Schönhals, A. Nano-MnO_2/xanthan gum composite films for NO_2 gas sensing. *Mater. Chem. Phys.* **2023**, *296*, 127277. [CrossRef]
58. Pedowitz, M.D.; Kim, S.; Lewis, D.I.; Uppalapati, B.; Khan, D.; Bayram, F.; Daniels, K.M. Fast Selective Sensing of Nitrogen-Based Gases Utilizing δ -MnO_2-Epitaxial Graphene-Silicon Carbide Heterostructures for Room Temperature Gas Sensing. *J. Microelectromech. Syst.* **2020**, *29*, 846–852. [CrossRef]
59. Lu, J.; Li, D.; Chen, X. ZnO/reduced graphene oxide nanocomposite with synergic enhanced gas sensing performance for the effective detection of NO_2 at room temperature. *J. Nanopart.* **2022**, *24*, 265. [CrossRef]
60. Kacem, K.; Ameur, S.; Casanova-Chafer, J. Bio-reduction of graphene oxide using pomegranate peels for NO_2 sensing and photocatalysis applications. *J. Mater. Sci. Mater. Electron.* **2022**, *33*, 16099–16112. [CrossRef]
61. Hasanov, B.E.; Casanova-Chafer, J.; Deokar, G.; Gouveia, J.D.; Nematulloev, S.; Gomes, J.R.; Llobet, E.; Costa, P.M. Amplified sensing of nitrogen dioxide with a phosphate-doped reduced graphene oxide powder. *Carbon* **2024**, *226*, 119207. [CrossRef]
62. Liang, J.; Wu, W.; Lou, Q. Room temperature NO_2 sensing performance enhancement of VO_2(B) composited rGO structure. *J. Mater. Sci. Mater. Electron.* **2022**, *33*, 15473–15482. [CrossRef]
63. Sovizi, M.R.; Mirzakhani, S. highly sensitive detection of ammonia gas by 3D flower-like γ-MnO_2 nanostructure chemiresistor. *J. Taiwan Inst. Chem. Eng.* **2020**, *111*, 293–301. [CrossRef]
64. Ganesan, M.; Ganapathi, B.; Parasuraman, B.; Thangavelu, P. Sensitivity enhancement of ammonia gas sensor based on $NiFe_2O_4$/rGO nanocomposite. *Chem. Phys. Impact* **2024**, *8*, 100616. [CrossRef]
65. Jeevitha, G.; Abhinayaa, R.; Mangalaraj, D.; Ponpandian, N.; Meena, P.; Mounasamy, V.; Madanagurusamy, S. Porous reduced graphene oxide (rGO)/WO_3 nanocomposites for the enhanced detection of NH_3 at room temperature. *Nanoscale Adv.* **2019**, *1*, 1799–1811. [CrossRef]
66. Zhang, H.; Li, Y.; Yuan, Z.; Lei, Y.; Li, X.; Meng, F. Enhanced Ammonia Sensing Performance Based on $FeCo_2O_4$/WO_3/rGO Ternary Nanocomposites. *IEEE Sens. J.* **2023**, *23*, 25698–25707. [CrossRef]
67. Ganesan, M.; Ganapathi, B.; Govindasamy, P.; Parasuraman, B.; Shanmugam, P.; Boddula, R.; Pothu, R.; Thangavelu, P. $CoFe_2O_4$/rGO nanocomposite: Synthesis and enhanced ammonia gas sensing properties at room temperature. *Results Chem. Chem.* **2024**, *7*, 101342. [CrossRef]
68. Alouani, M.A.; Casanova-Cháfer, J.; Güell, F.; Peña-Martín, E.; Ruiz-Martínez-Alcocer, S.; de Bernardi-Martín, S.; García-Gómez, A.; Vilanova, X.; Llobet, E. ZnO-Loaded Graphene for NO_2 Gas Sensing. *Sensors* **2023**, *23*, 6055. [CrossRef]
69. Umar, A.; Ibrahim, A.A.; Kumar, R.; Albargi, H.; Zeng, W.; Alhmami, M.A.M.; Alsaiari, M.A.; Baskoutas, S. Gas sensor device for high-performance ethanol sensing using α-MnO_2 nanoparticles. *Mater. Lett.* **2021**, *286*, 129232. [CrossRef]
70. Chatterjee, S.G.; Chatterjee, S.; Ray, A.K.; Chakraborty, A.K. Graphene–metal oxide nanohybrids for toxic gas sensor: A review. *Sens. Actuators B Chem.* **2015**, *221*, 1170–1181. [CrossRef]
71. Zhang, D.; Liu, A.; Chang, H.; Xia, B. Room-temperature high-performance acetone gas sensor based on hydrothermal synthesized SnO_2-reduced graphene oxide hybrid composite. *RSC Adv.* **2015**, *5*, 3016–3022. [CrossRef]
72. Xiao, X.; Jin, W.; Tang, C.; Qi, X.; Li, R.; Zhang, Y.; Zhang, W.; Yu, X.; Zhu, X.; Ma, Y.; et al. Thermal reduced graphene oxide-based gas sensor for rapid detection of ammonia at room temperature. *J. Mater. Sci.* **2023**, *58*, 11016–11028. [CrossRef]
73. Han, D.; Wang, Y.; Wang, Y.; Duan, Q.; Li, D.; Ge, Y.; He, X.; Zhao, L.; Wang, W.; Sang, S. Machine-learning-assisted n-GaN-Au/PANI gas sensor array for intelligent and ultra-accurate ammonia recognition. *Chem. Eng. J.* **2024**, *495*, 153705. [CrossRef]
74. Lv, C.; Hu, C.; Luo, J.; Liu, S.; Qiao, Y.; Zhang, Z.; Song, J.; Shi, Y.; Cai, J.; Watanabe, A. Recent Advances in Graphene-Based Humidity Sensors. *Nanomaterials* **2019**, *9*, 422. [CrossRef]
75. Casanova-Cháfer, J.; Navarrete, E.; Noirfalise, X.; Umek, P.; Bittencourt, C.; Llobet, E. Gas Sensing with Iridium Oxide Nanoparticle Decorated Carbon Nanotubes. *Sensors* **2019**, *19*, 113. [CrossRef]

Disclaimer/Publisher's Note: The statements, opinions and data contained in all publications are solely those of the individual author(s) and contributor(s) and not of MDPI and/or the editor(s). MDPI and/or the editor(s) disclaim responsibility for any injury to people or property resulting from any ideas, methods, instructions or products referred to in the content.

chemosensors

MDPI

Article

CRDS Technology-Based Integrated Breath Gas Detection System for Breath Acetone Real-Time Accurate Detection Application

Jing Sun [1], Dongxin Shi [2], Le Wang [3], Xiaolin Yu [4], Binghong Song [5], Wangxin Li [2], Jiankun Zhu [5], Yong Yang [5], Bingqiang Cao [1,*] and Chenyu Jiang [2,*]

1 School of Materials Science and Engineering, University of Jinan, Jinan 250022, China; sunj@sibet.ac.cn
2 Suzhou Institute of Biomedical Engineering and Technology, Chinese Academy of Sciences, Suzhou 215163, China; shidx@sibet.ac.cn (D.S.); liwx@sibet.ac.cn (W.L.)
3 Physical Education Department, Shandong Traditional Chinese Medicine University, Jinan 250355, China; 18253121729@163.com
4 Jinan Science and Technology Innovation Promotion Center, Jinan 250014, China; yuxiaolin@jn.shandong.cn
5 Jinan Guoke Medical Technology Development Co., Ltd., Jinan 250001, China; songbh@sibet.ac.cn (B.S.); zhujk@sibet.ac.cn (J.Z.); yangyong@sibet.ac.cn (Y.Y.)
* Correspondence: mse_caobq@ujn.edu.cn (B.C.); jiangcy@sibet.ac.cn (C.J.); Tel.: +86-150-5312-5919 (B.C.); +86-134-0261-7742 (C.J.)

Abstract: The monitoring of acetone in exhaled breath is expected to provide a noninvasive and painless method for dynamic monitoring of summarized physiological metabolic status during obesity treatment. Although the commonly used Mass Spectrometry (MS) technology has high accuracy, the long detection time and large equipment size limit the application of daily bedside detection. As for the real-time and accurate detection of acetone, the gas sensor has become the best choice of gas detection technology, but it is easy to be disturbed by water vapor in breath gas. An integrated breath gas detection system based on cavity ring-down spectroscopy (CRDS) is reported in this paper, which is a laser absorption spectroscopy technique with high-sensitivity detection and absolute quantitative analysis. The system uses a 266 nm single-wavelength ultraviolet laser combined with a breath gas pretreatment unit to effectively remove the influence of water vapor. The ring-down time of this system was 1.068 μs, the detection sensitivity was 1 ppb, and the stability of the system was 0.13%. The detection principle of the integrated breath gas detection system follows Lambert–Beer's law, which is an absolute measurement with very high detection accuracy, and was further validated by Gas Chromatography–Mass Spectrometer (GC-MS) testing. Significant differences in the response of the integrated breath gas detection system to simulated gases containing different concentrations of acetone indicate the potential of the system for the detection of trace amounts of acetone. Meanwhile, the monitoring of acetone during obesity treatment also signifies the feasibility of this system in the dynamic monitoring of physiological indicators, which is not only important for the optimization of the obesity treatment process but also promises to shed further light on the interaction between obesity treatment and physiological metabolism in medicine.

Keywords: obesity treatment; CRDS; breath acetone; breath pretreatment; real-time; accurate detection

1. Introduction

Obesity has become a serious problem threatening human health which is a chronic metabolic disease caused by a variety of factors and may cause many kinds of metabolic disorders such as hypertension, diabetes, cardiovascular disease, and so on. A recent study showed that the global population of obese people was upwards of about 1 billion in 2022 [1]. Controlling weight and maintaining a healthy body condition are the main goals of obesity management. Currently, lifestyle interventions, pharmacotherapy, gene therapy, and weight loss surgery are usually used for obesity management and provide targeted guidance for obese patients and their clinicians in weight management, and the effects are usually evaluated by weight loss [2]. This evaluation method is intuitive and relatively convenient to realize, but it has nuance and limitations for fully and accurately responding to the status of human fat metabolism.

Breath gas detection has been a non-invasive method for disease diagnosis and metabolic status monitoring for the past few years. The changes in the composition and content of trace Volatile Organic Compounds (VOCs) contained in exhaled gas reflect the changes in the internal environment of the human body, which could be applied for the noninvasive detection of many diseases, such as lung cancer, cirrhosis of the liver, breast cancer, and diabetes mellitus [3–8]. Among the VOCs from breath gas, acetone is able to effectively respond to the level of lipolysis and metabolism when the body's glucose supply is insufficient, and it is an important biomarker of human glucose metabolism in respiratory gases. For instance, the amount of acetone produced will increase when the patient's fat metabolism accelerates during obesity treatment. Meanwhile, acetone is highly volatile, and the excess acetone will appear in human exhaled gas at concentrations much higher than the upper level in the range of healthy people (0.5–2.0 ppm) during gas exchange or blood circulation in the human body [9,10]; so, it could provide an important basis for real-time accurate monitoring of human metabolism during obesity treatment. Therefore, accurate real-time detection of acetone content in respiratory gas has become a potential method for monitoring the effectiveness of obesity treatment, which makes up for the inadequacy of the current unitary and limited means of detection.

Gas chromatography (GC), Mass Spectrometry (MS), and electrochemistry (ECM) all can realize the accurate detection of acetone in respiratory gases, but they have drawbacks such as cumbersome sample preconcentration, poor detection selectivity, baseline drift, and time-consuming detection, which lead to these methods not being suitable for real-time and large-scale clinical testing applications [11–16]. On the other hand, laser spectroscopy has the advantages of robustness, short response time, and high detection sensitivity [17–19]. Among the kinds of spectral detection technologies, the cavity ring-down spectroscopy (CRDS) technique has a very high detection sensitivity by increasing the equivalent absorption path through the optical properties of the resonant cavity. However, the sensitivity and stability of CRDS detection is largely influenced by the water vapor content of the ambient gas. Therefore, elucidating how to reduce the water vapor concentration has become the key to accurate detection of respiratory acetone by CRDS technology. Wang et al. [20] successfully eliminated water vapor interference by heating the decay chamber to 40 °C. Hancock [21] et al. achieved water vapor removal by cooling the breathing gas with pretreatment at −20 °C. Jiang et al. [22] used a 0.22 μm filter to deal with the moisture content in the breathing gas. But the methods above are complicated in operation, high in cost, and poor in adsorption performance, and there is an urgent need for a breath acetone detection system based on CRDS technology which is less disturbed by moisture and has high sensitivity and accuracy.

In this study, we designed and constructed an integrated detection system for respiratory acetone with respiratory gas pretreatment based on CRDS technology by utilizing the unique spectral absorption property of the acetone molecule at 266 nm and the water vapor adsorption property of porous activated carbon material. The breath acetone detection accuracy and sensitivity of the designed system were verified through comparative experiments. Further, the breath acetone concentration of obese patients during treatment was tracked and monitored to verify the clinical value of the integrated breath gas detection system in the dynamic monitoring of physiological indexes of obesity. The experiment results showed that the detection system can reduce the interference of water vapor on the accuracy of acetone detection through the breath gas pretreatment unit, the detection accuracy can reach ppb level, and its detection results have good consistency compared with GC-MS. The system has the advantages of high sensitivity, good stability, and fast response speed for real-time clinical testing applications of breath acetone.

2. Materials and Methods

2.1. Basic Principles of CRDS Technology

CRDS is a laser absorption spectroscopy technique with high-sensitivity detection and absolute quantitative analysis, and the basic schematic of CRDS technology is shown in Figure 1.

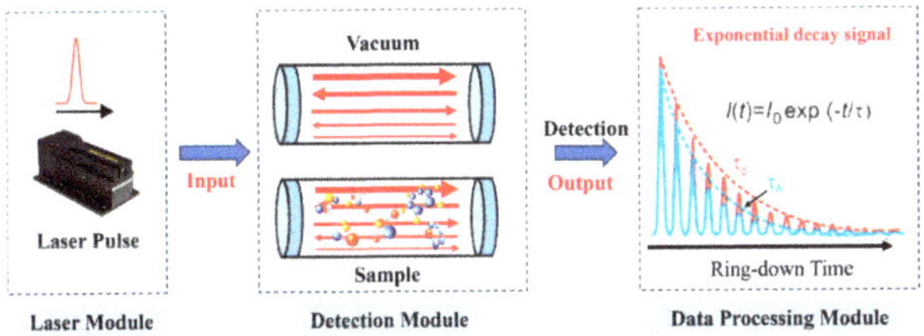

Figure 1. Basic schematic of CRDS technology.

The incident beam is reflected back and forth in a resonant cavity consisting of a highly reflective mirror and a cavity, while the intensity of the beam transmitted after absorption by the sample is detected, according to Lambert–Beer's law [9,23–25]:

$$I(v) = I_0 e^{-\alpha(v)d} \tag{1}$$

where I_0 is the initial intensity of the incident beam, $I(v)$ is the intensity of the outgoing beam, v is the wavelength, e is the natural constant, d is the length of the resonant cavity, and $\alpha(v)$ is the absorption coefficient of the substance to be measured. Therefore, the light intensity in the resonant cavity shows a mono-exponential decay.

It can be calculated from the relationship between $\sigma(v)$ (absorption cross section of the substance to be measured) and $\alpha(v) = n * \sigma(v)$:

$$\tau_0 = \frac{d}{c(1-R)} \tag{2}$$

$$\tau = \frac{d}{c(1-R+\sigma(v)nd)} \tag{3}$$

Therefore, the absorption of the gas sample is as follows:

$$A = \sigma(v)nd = \frac{d}{c}\left(\frac{1}{\tau} - \frac{1}{\tau_0}\right) \tag{4}$$

It follows that the content of the substance to be measured can be obtained by measuring τ_0 and τ.

2.2. Construction of an Integrated Breath Gas Detection System Based on CRDS Technology

The integrated breath gas detection system designed and developed in this paper is based on the basic principle of CRDS mentioned above, and mainly consists of three modules: a laser module, breath acetone detection module, and data processing module, as shown in Figure 2a. Among them, the breath acetone detection module consists of a breath gas pretreatment unit (shown in Figure 2b), a resonance cavity, and a sample injection system. The laser used in the integrated breath gas detection system is an Nd:YAG pumped all-solid-state laser with a center wavelength of 266 nm from Changchun New Industry, which is an electro-optic Q-tuned pulsed laser. The resonant cavity is composed of a cylindrical stainless-steel cavity and a pair of high-reflectance mirrors (Los Gatos Research, U.S.; Purchased in 2016), the length of the stainless-steel cavity is 50 cm, and the cavity is accompanied by three welded standard ports, which are connected to the pressure transducer, vacuum pump, and injection system, respectively. Adjustable flanges are used in this system to fix the high-reflectance mirrors at both ends of the resonant cavity. The detector is a PMT from Hamamatsu, Japan, which has an integrated high-voltage generator module, and the gain adjustment can be realized by adjusting the control voltage. The data processing module mainly adopts the background deduction algorithm for data processing to obtain the concentration of acetone in the measured sample.

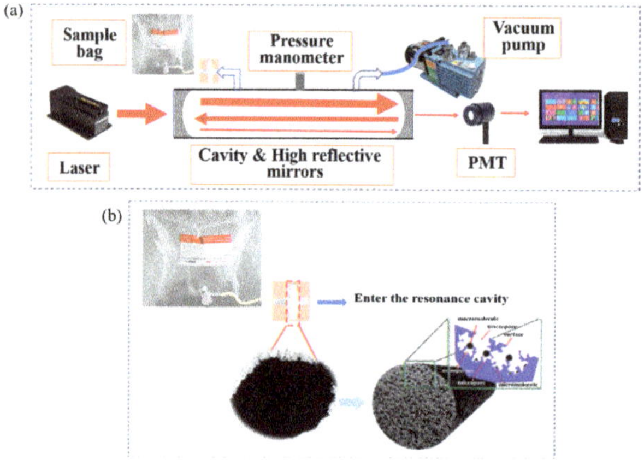

Figure 2. Schematic diagram of (**a**) the structure of the integrated breath gas detection system and (**b**) the breath gas pretreatment unit.

2.3. Preparation of Activated Carbon by Ester Hydrolysis for Breath Gas Pretreatment Unit

First, the graded porous activated carbon material [26,27] prepared in the pre-laboratory stage was soaked in 1 mol/L alkaline solution (KOH) at a ratio of 1:10 for 24 h and then centrifugally washed. Then, the centrifugally washed samples were placed in ethyl acetate solution and stirred and soaked in a stirring heater at 80 °C for 48 h. Finally, the soaked activated carbon was vacuum dried at 100 °C for 24 h to obtain the H-PACs.

The morphologies and microstructures of the samples were examined by scanning electron microscopy (SEM, Quanta 250, FEI, Oxford Instruments, Oxford, UK), transmission electron microscopy (TEM, JEM-2100F, JEOL, Tokyo, Japan), and Fourier Transform Infrared Spectroscopy (FT-IR, iS50, Nicolet Instrument Co., Madison, WI, USA). Samples for FT-IR measurements were obtained over the frequency range of 400–2000 cm^{-1}. Water vapor adsorption experiment data were provided by a micromeritics test.

2.4. Calculation of Acetone Concentration Based on Single-Wavelength Background Deduction Method

According to the basic principle of CRDS, the integrated breath gas detection system designed in this paper adopts the acetone concentration calculation method based on the single-wavelength background deduction method to realize the accurate detection of breath acetone [28,29]. Assuming that the high content of molecules (e.g., N_2, O_2, etc.) in the breath gas is exactly the same in the breath and air as well as the negligible absorption of other VOCs at 266 nm, air is used as the detection background during the specific measurement process, and the decay time constants τ_0, τ_{atm}, and τ_{breath} of vacuum, air, and samples are obtained by alternating the measurements of the vacuum, air, and the sample to be tested, which finally obtains the effective absorption of the air and the sample to be tested:

$$A_{atm} = \sigma(v)nd = \frac{d}{c}\left(\frac{1}{\tau_{atm}} - \frac{1}{\tau_0}\right) \tag{5}$$

$$A_{breath} = \sigma(v)nd = \frac{d}{c}\left(\frac{1}{\tau_{breath}} - \frac{1}{\tau_0}\right) \tag{6}$$

Therefore, the concentration of breath acetone is obtained:

$$\Delta A = A_{breath} - A_{atm} = \sigma(v)nd \quad (7)$$

2.5. Breath Sample Collection Program

A total of 10 obese subjects were recruited for this paper, and breath acetone gas samples were collected every 10 min during obesity treatment. All of the subjects were non-smokers and non-alcoholics and no pregnant or lactating female subjects were recruited. Written informed consent was obtained from all subjects participating in this study, and no cost was subsidized. First, a disposable mouthpiece was attached to the valve of a sampling bag (FEP sampling bag) that had been repeatedly rinsed using high-purity nitrogen. Then, the subject took an appropriate breath, held it in for 3 s, and then exhaled the single breath into the sampling bag. Finally, when the gas volume reached 80% of the sampling bag, the valve of the sampling bag was closed, and the samples were numbered and placed in the holding box.

2.6. Sample Test Methods

Testing of all samples was accomplished on our designed integrated breath gas detection system (shown as Figure 3). First, after the system was warmed up, a baseline test of the system was performed using the ambient gas as a reference. Then, the collected samples were connected to the breath gas pretreatment unit, and after moisture adsorption by activated charcoal, they entered into the resonant cavity for acetone concentration testing, and the sample gases in the resonant cavity were extracted by a vacuum pump after the test was completed. Finally, at the end of each sample test, the resonant cavity is cleaned using high-purity nitrogen gas and repeatedly rinsed at least three times.

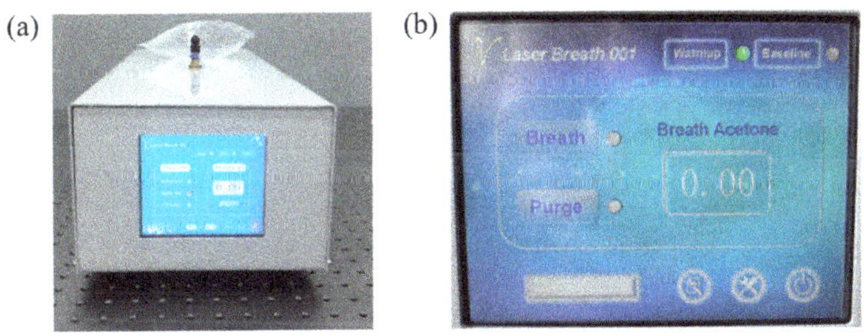

Figure 3. Sample test systems and interfaces: (**a**) Integrated Breath Gas Detection System, and (**b**) Sample test interfaces.

In addition, we tested the breath acetone samples using GC-MS under the following conditions: Analytes absorbed on the SPME fibers were thermally desorbed in a gas chromatography injector at 280 °C. The initial temperature of the column was 5 °C and the column was maintained with carbon dioxide for 3 min. The column is initially heated to 5 °C and maintained with carbon dioxide for 3 min. The temperature was then increased to 250 °C at a rate of 5 °C/min and held for 2 min.

3. Results and Discussion

3.1. Performance of Breath Gas Pretreatment Units

The hydrophilic porous activated carbon material in the breath pretreatment unit is based on the shaped porous activated carbon material prepared in the previous research of our group [26,27], and surface modification was carried out by the ester hydrolysis method to prepare a porous activated carbon material with excellent water vapor adsorption performance (Figure 4). Under alkaline conditions, some oxygen-containing functional groups were formed on the surface of the shaped porous activated carbon by the hydrolysis of ethyl acetate. Its unique pore structure and surface functionalization are more conducive to the adsorption of water molecules. Breath treated with this material can be free from moisture interference in the test, improving the accuracy of the test.

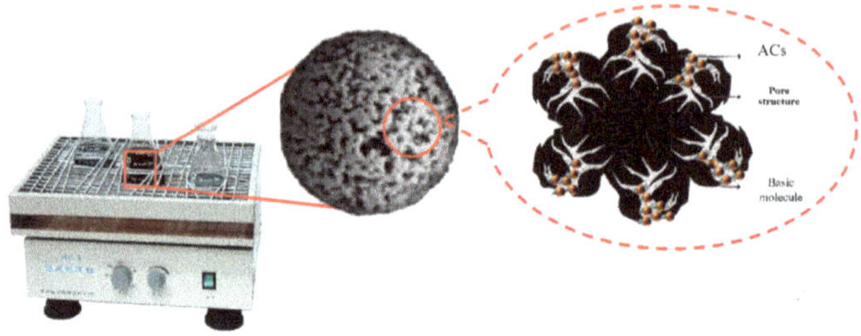

Figure 4. Hydrophilic activated carbon prepared by ester hydrolysis method.

We performed a series of characterizations of the prepared hydrophilic activated carbon materials. Figure 5 shows the SEM and TEM images of H-PACs, respectively. The results showed that the modified activated carbon samples had a good three-dimensional porous structure. Among them, the microporous structure provided more sites for the adsorption of water molecules. From the TEM images, it can be seen that the modified ACs have a unique porous structure as well as a large number of microporous structures, which is conducive to the adsorption of water vapor in the micropores.

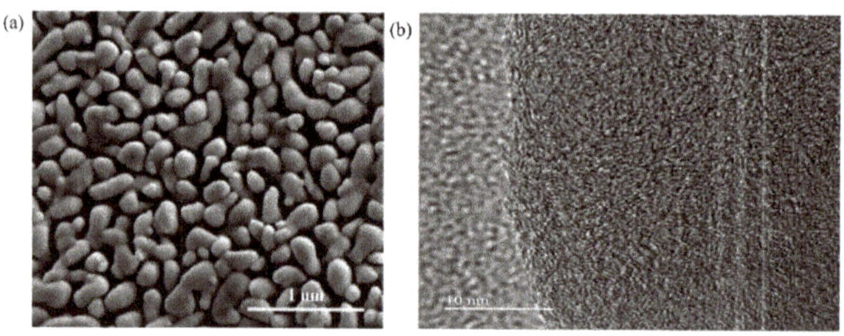

Figure 5. Electron microscope images of H-PACs. (**a**) SEM image; (**b**) TEM image.

Figure 6 compares the N_2 adsorption–desorption isotherms and pore size distribution curves of the H-PACs. According to IUPAC classification, N_2 adsorption isotherms are type I, indicating that they are typical microporous materials [30,31]. As shown in Figure 6a, the type I isotherms of the H-PACs suggest the presence of micropores in the H-PACs, whereas the H4 type hysteresis loops of the H-PACs indicate the presence of both micropores and mesopores. According to the pore size distribution curves of the H-PACs shown in Figure 6b, micropores (0.8 nm~1.5 nm) are dominant in the H-PACs, and there is a unique bimodal pore structure that exists in the H-PACs. Studies have shown that more micropores and smaller mesoporous pores are conducive to water vapor adsorption [32]. Therefore, the H-PACs have the best layered porous structure, which can provide more reaction sites for water vapor adsorption.

Figure 7a shows the FT-IR spectra of the H-PACs. From the figure, it can be seen that the H-PACs prepared by ester hydrolysis modification showed obvious absorption peaks at around 1640, 1410, 1380, and 870 cm^{-1}, which were the vibrational absorption peaks of -C=O and -C-H, respectively [33–35]. This indicates that a large number of functional groups are most likely to be introduced on the surface of H-PACs. Figure 7b shows the adsorption isotherms of the modified sample to water vapor. All of them are V-shaped according to IUPAC classification, which is a typical isotherm for the adsorption of water vapor by activated carbon [36]. From the figure, it can be seen that the adsorption of H-PACs increases with increasing relative pressure at room temperature. This indicates that the functional groups on the surface of the activated carbon have a greater effect on water vapor adsorption at lower relative pressures, which may be due to the fact that the functional groups act as active centers inducing the movement of water molecules and forming hydrogen bonds with them for water molecule adsorption. Long et al. [37] suggested that the hydrophilic groups bind to water molecules through hydrogen bonding. The adsorption of H-PACs becomes larger as the relative pressure increases. This is mainly due to the fact that water molecules continue to enter into the activated carbon and continue to adsorb at the sites of previously adsorbed water molecules and form water molecule clusters, which break off from each other when they grow to a certain size, thus filling in the micropores [38]. This indicates that the hydrophilic activated carbon prepared by the improved ester hydrolysis method has good water vapor adsorption properties, and it can effectively remove the interference of water in the pretreatment stage of breath to ensure the accuracy of breath acetone detection.

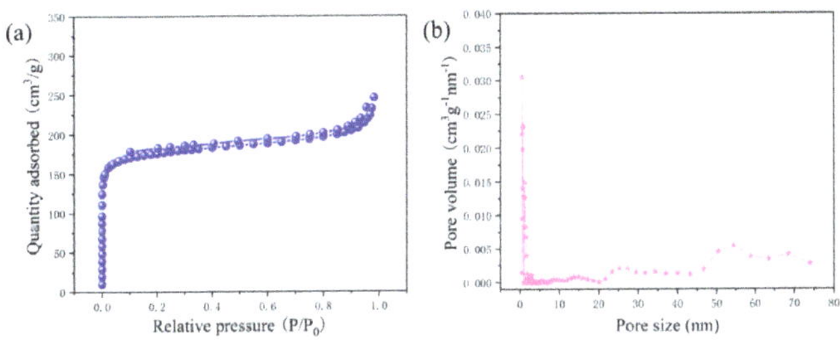

Figure 6. (**a**) N_2 adsorption–desorption isotherms and (**b**) pore size distribution curves of H-PACs.

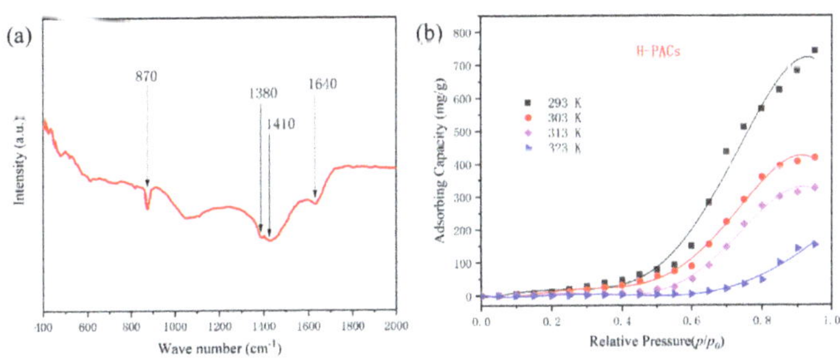

Figure 7. FT-IR (**a**) and isothermal adsorption curve (**b**) of water vapor in H-PACs.

3.2. Performance of the Integrated Breath Gas Detection System

In order to ensure that the absorption of gases within the vacuum decay chamber adheres to the Lambert–Beer law, the ring-down signal under the vacuum of an integrated detection system for breath acetone based on CRDS technology was measured. Figure 8a shows a typical ring-down signal under vacuum acquired by an oscilloscope, with the relative time in the horizontal coordinate and the signal intensity in the vertical coordinate and fitted to the results as shown in Figure 8b. From the figure, it can be seen that the typical ring-down signal shows exponential decay. Further logarithmic operation on the fitted section (0-t) shows that the fitted correlation coefficient is greater than 0.9995, and the fitted residuals and residual sum of squares are 4.25×10^{-4} and 3.16×10^{-4}, respectively, which suggests that the decaying signals obtained by this analytical system are close to the mono-exponential decay conforming to the Lambert–Behr law.

To test the baseline stability of the ring-down time, we continuously measured the vacuum ring-down time constant for 8 h, with each data point being the result of averaging 128 ring-down signals. The typical baseline stability of the system is shown in Figure 9. The results show that the vacuum ring-down time constant t is 1.068 μs with a standard deviation of 1.35 ns. The stability is defined as the ratio of the standard deviation of the ring-down time constant to the mean over a period of time, i.e., the stability is 0.13%, which is within the satisfied error tolerance.

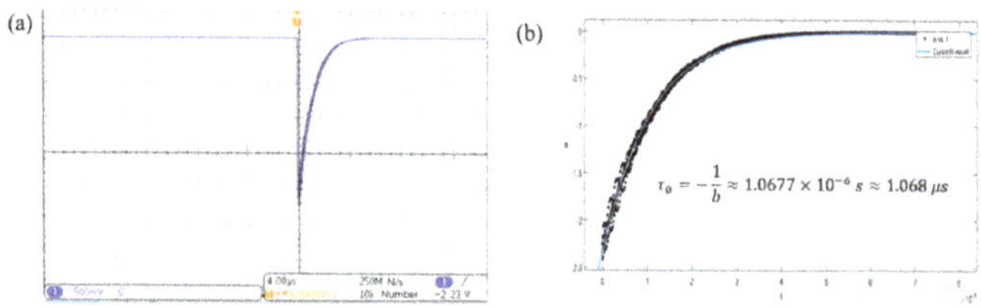

Figure 8. (**a**) Typical ring-down curves; (**b**) single-exponential fitting result curves.

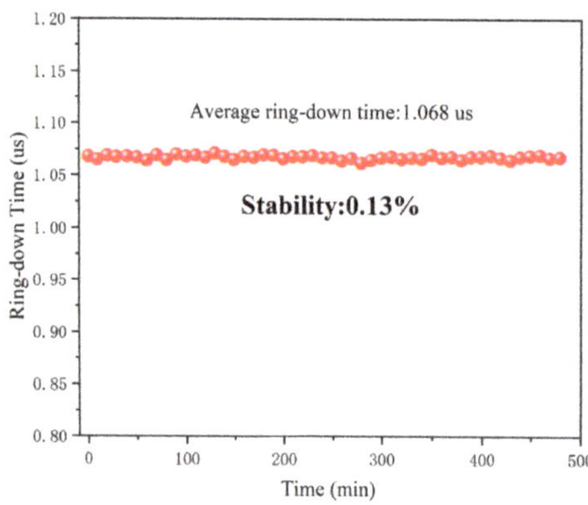

Figure 9. Stability curve of the integrated breath gas detection system.

The accuracy of the measurement by this detection system was further verified by comparing the results with those of GC-MS. The GC-MS test was performed on breath gas samples from healthy individuals in the ion monitoring mode of operation, and the acetone profile obtained is shown in Figure 10a. Figure 10b shows the breath acetone concentrations independently measured using the two methods; the horizontal and vertical coordinates are the acetone concentrations measured using this system and GC-MS, respectively. Based on the test results, the two sets of data were linearly fitted, and the correlation coefficient of the linear fit between the two was $R^2 = 0.99924$, and the linear relationship was y = 0.9853x + 0.08024, which indicates the consistency of the measurement results of the integrated breath gas detection system based on CRDS technology with the results of GC-MS.

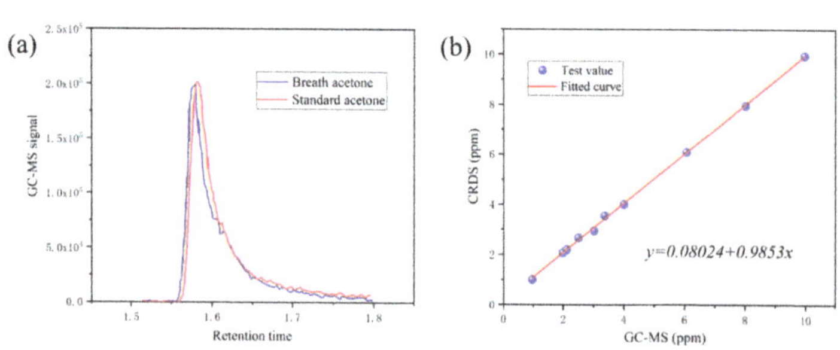

Figure 10. (**a**) GC-MS profiles of breath acetone and (**b**) comparison curve between CRDS-based breath acetone detection system and GC-MS measurement results.

3.3. Application of Integrated Breathing Acetone Detection System

Furthermore, in order to demonstrate the feasibility of this integrated breath gas detection system in the dynamic monitoring of physiological states, the detection system was investigated to monitor the acetone content in the exhaled gas of subjects during obesity treatment (shown as in Figure 11a). Subjects were treated for obesity by massage for 10 min followed by a 5 min rest period, and breath gas samples were collected before, during (10, 15, 25, 30, 40, and 45 min), and in a relaxed state at the end of their massage treatment. The subjects were tested for their response to exhaled gas at different treatment stages, and the corresponding acetone content was calculated based on the ring-down time. As shown in Figure 11b, the acetone content in the breath of the subjects basically remained in a relatively stable state in the calm state before the massage treatment, with its concentration fluctuating between 0.75 and 1.75 ppm, which belongs to the state of the breath acetone level in a normal human body [10].

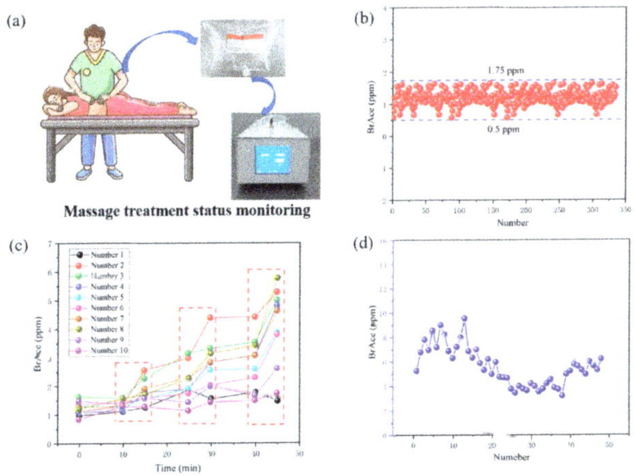

Figure 11. (**a**) Schematic diagram of the response results of the massage treatment process monitoring and detection system. (**b**) The average resting breath acetone concentration profiles for all subjects. (**c**) Curves of changes in breath acetone concentration in subjects during the treatment of obesity with each massage treatment. (**d**) Curves of the breath acetone concentration of subject #2 throughout the course of massage treatment.

The changes in breath acetone during the treatment are shown in Figure 11c. From the figure, it can be seen that after 10 min of massage treatment, the acetone level showed an increasing trend with the increase in massage treatment time, and then, in the relaxed state at the end of the treatment, this gradually declined, which was related to the metabolic process of acetone in the body, and further declined to a level close to that in the resting state before massage. Although subject #1 appeared to have no significant change in breath acetone during a single measurement, his level of breath acetone increased to some degree throughout the course of treatment. During the whole follow-up process of obesity massage therapy, we took subject #2 as an example to analyze the changes in breath acetone concentration during the whole process, and the results were shown in Figure 11d. As can be seen from the figure, subject #2's breath acetone concentration decreased over time as the treatment process progressed, but the lowest concentration at this stage was higher than 4 ppm. In the middle of treatment, the concentration of breath acetone plateaued. As the massage treatment continued, the breath acetone concentration increased again. These results indicate that the developed integrated breath acetone detection system has a promising application in the dynamic real-time monitoring of physiological metabolic states.

4. Conclusions

This study designed and built an integrated detection system for breath acetone detection based on the CRDS technology, which implemented a breath gas pretreatment function by using porous activated carbon material with water vapor adsorption characteristics. This detection system utilized the unique absorption spectrum of acetone molecules at 266 nm and could reduce the interference of water vapor on the accuracy of acetone detection through the breath gas pretreatment unit. The experiment results showed that this novel breath acetone detection system had good consistency compared with GC-MS, so as to realize the accurate detection of breath acetone. Meanwhile, this system had the advantages of high sensitivity, good stability, and a fast response time. Through tracking and monitoring of breath acetone concentration in eight obese patients during the treatment of obesity with a massage method and the changes in body weight indexes, it was proved that the accurate detection of breath acetone by this system could further effectively evaluate the effect of the massage method in the treatment of obesity. The results of the experimental tests showed that the concentration of breath acetone raised with the increase in the treatment time during the treatment of obesity, and the concentration of breath acetone changed significantly after the rest interval of the massage. This may be related to the physiological changes in the body's fat consumption and decomposition as well as the discharging of products from the body after the decomposition. It proved that the integrated breath gas detection system developed in this study had good clinical application prospects for the dynamic real-time monitoring of physiological metabolic states.

Author Contributions: Conceptualization, B.C. and C.J.; methodology, J.S., D.S., X.Y. and L.W.; validation, J.S., D.S., B.S., W.L. and J.Z.; formal analysis, J.Z. and L.W.; investigation, J.Z., W.L. and Y.Y.; resources, B.C. and C.J.; data curation, J.S., D.S., Y.Y. and X.Y.; writing—original draft preparation, J.S. and D.S.; writing—review and editing, B.S. and L.W.; visualization, W.L., B.S. and X.Y.; supervision, J.Z. and Y.Y. All authors have read and agreed to the published version of the manuscript.

Funding: This research received no external funding.

Institutional Review Board Statement: Ethics Committee of the Third Affiliated Hospital of Shandong First Medical University (FY2023037).

Informed Consent Statement: Informed consent was obtained from all subjects involved in this study.

Data Availability Statement: Data are contained within this article.

Acknowledgments: This work is supported by the Taishan Industrial Experts Program (tscx202306125), the Quancheng 5150 Project, and the National Key Research and Development Program (NKRDP): Development of a sterile body fluid culture, monitoring and enrichment system (2022YFC2406204).

Conflicts of Interest: Authors Binghong Song, Jiankun Zhu and Yong Yang were employed by the company Jinan Guoke Medical Technology Development Co., Ltd. The remaining authors declare that the research was conducted in the absence of any commercial or financial relationships that could be construed as a potential conflict of interest.

References

1. Phelps, N.H.; Singleton, R.K.; Zhou, B.; Heap, R.A.; Mishra, A.; Bennett, J.E.; Barbagallo, C.M. Worldwide trends in underweight and obesity from 1990 to 2022: A pooled analysis of 3663 population-representative studies with 222 million children, adolescents, and adults. *Lancet* **2024**, *403*, 1027–1050. [CrossRef] [PubMed]
2. Perdomo, C.M.; Cohen, R.V.; Sumithran, P.; Clément, K.; Frühbeck, G. Contemporary medical, device, and surgical therapies for obesity in adults. *Lancet* **2023**, *401*, 1116–1130. [CrossRef] [PubMed]
3. Romani, A.; Marrone, G.; Celotto, R.; Campo, M.; Vita, C.; Chiaramonte, C.; Carretta, A.; Di Daniele, N.; Noce, A. Utility of SIFT-MS to evaluate volatile organic compounds in nephropathic patients' breath. *Sci. Rep.* **2022**, *12*, 10413. [CrossRef]
4. Tsou, P.-H.; Lin, Z.-L.; Pan, Y.-C.; Yang, H.-C.; Chang, C.-J.; Liang, S.-K.; Wen, Y.-F.; Chang, C.-H.; Chang, L.-Y.; Yu, K.-L.; et al. Exploring Volatile Organic Compounds in Breath for High-Accuracy Prediction of Lung Cancer. *Cancers* **2021**, *13*, 1431. [CrossRef] [PubMed]

5. Sana, S.R.G.L.; Lv, Y.; Chen, G.; Guo, L.; Li, E. Analysis of the volatile organic compounds of epidural analgesia-ameliorated metabolic disorder in pregnant women with gestational diabetes mellitus based on untargeted metabolomics. *Front. Endocrinol.* **2023**, *14*, 1009888. [CrossRef]
6. De Vincentis, A.; Pennazza, G.; Santonico, M.; Vespasiani-Gentilucci, U.; Galati, G.; Gallo, P.; Vernile, C.; Pedone, C.; Antonelli Incalzi, R.; Picardi, A. Breath-print analysis by e-nose for classifying and monitoring chronic liver disease: A proof-of-concept study. *Sci. Rep.* **2016**, *6*, 25337. [CrossRef]
7. Liu, J.; Chen, H.; Li, Y.; Fang, Y.; Guo, Y.; Li, S.; Xu, J.; Jia, Z.; Zou, J.; Liu, G.; et al. A novel non-invasive exhaled breath biopsy for the diagnosis and screening of breast cancer. *J. Hematol. Oncol.* **2023**, *16*, 63. [CrossRef]
8. Nakayama, Y.; Hanada, M.; Koda, H.; Sugimoto, M.; Takada, M.; Toi, M. Reduction of bacterial volatile sulfur compound production by licoricidin and licorisoflavan A from licorice. *J. Breath. Res.* **2022**, *17*, 016006. [CrossRef] [PubMed]
9. Wang, C.; Sahay, P. Breath Analysis Using Laser Spectroscopic Techniques: Breath Biomarkers, Spectral Fingerprints, and Detection Limits. *Sensors* **2009**, *9*, 8230–8262. [CrossRef]
10. Anderson, J.C. Measuring breath acetone for monitoring fat loss: Review. *Obesity* **2015**, *23*, 2327–2334. [CrossRef] [PubMed]
11. Güntner, A.T.; Abegg, S.; Königstein, K.; Gerber, P.A.; Schmidt-Trucksäss, A.; Pratsinis, S.E. Breath Sensors for Health Monitoring. *ACS Sens.* **2019**, *4*, 268–280. [CrossRef]
12. Chapman, E.A.; Baker, J.; Aggarwal, P.; Hughes, D.M.; Nwosu, A.C.; Boyd, M.T.; Mayland, C.R.; Mason, S.; Ellershaw, J.; Probert, C.S.; et al. GC-MS Techniques Investigating Potential Biomarkers of Dying in the Last Weeks with Lung Cancer. *Int. J. Mol. Sci.* **2023**, *24*, 1591. [CrossRef] [PubMed]
13. Du, H.; Sun, R.; Su, J.; Sun, Y.; Xia, K.; Cong, L.; Cui, H. Highly selective acetone detector based on a separation channel and semiconduc-tor gas analyzer. *Meas. Sci. Technol.* **2021**, *32*, 085102. [CrossRef]
14. MG, G.; Bastide, B.H.; Remund, A.L.; Oosthuizen, D.N.; Derron, N.; Gerberb, P.A.; Weber, I.C. Handheld device quantifies breath acetone for real-life metabolic health monitoring. *Anal. Diagn.* **2023**, *2*, 918–928.
15. Zhou, D.; Wang, Q.; Lan, Z.; Chen, Y.; Peng, Z.; Zhang, L.; Liu, Y. Liquid-crystal-based fiber laser ana-lyzer for non-invasive gas detection. *Opt. Lett.* **2023**, *48*, 4508–4511. [CrossRef]
16. Musa-Veloso, K.; Likhodii, S.S.; Cunnane, S.C. Breath acetone is a reliable indicator of ketosis in adults consuming ketogenic meals. *Am. J. Clin. Nutr.* **2002**, *76*, 65–70. [CrossRef] [PubMed]
17. Chen, B.; Li, H.; Zhao, X.; Gao, M.; Cheng, K.; Shao, X.; Wu, H.; Dong, L.; Yin, X. Trace photoacoustic SO_2 gas sensor in SF6 utilizing a 266 nm UV laser and an acousto-optic power stabilizer. *Opt. Express* **2023**, *31*, 6974–6981. [CrossRef] [PubMed]
18. Peng, J.; Cao, Y.; Wang, R.; Wang, G.; Mei, J.; Liu, K.; Chen, W.; Gao, X. Simultaneous Detection of Major Greenhouse Gases with Multiresonator Photoacoustic Spectroscopy. *Anal. Chem.* **2024**, *96*, 14877–14883. [CrossRef]
19. Fu, L.; Zhang, J.; Pan, Y.; Lu, P. Differential photoacoustic spectroscopy for flow gas detection based on single microphone. *Photoacoustics* **2024**, *38*, 2213–5979. [CrossRef]
20. Kentaro, N.; Keisuke, H.; Naoko, N.; Ryuichiro, E.; Mariko, T.; Mai, N.; Hitomi, S.; Misaki, M.; Satoko, T.; Hisako, F.; et al. Ketogenic effects of medium chain triglycerides containing formula and its correlation to breath acetone in healthy volunteers: A randomized, double-blinded, placebo-controlled, single dose-response study. *Front. Nutr.* **2023**, *10*, 1224740.
21. Jiang, C.; Sun, M.; Wang, Z.; Chen, Z.; Zhao, X.; Yuan, Y.; Li, Y.; Wang, C. A portable Real-Time Ringdown Breath Acetone Analyzer: Toward Potential Diabetic Screening and Management. *Sensors* **2016**, *16*, 1199. [CrossRef]
22. Wang, A.; Li, J.; Xu, W.; Yuan, X.; Li, B.; Sun, K.; Liu, J. Air-mediated phosphoric acid activation strategy for preparation of oxygenated functional groups rich activated carbon for capacitive deionization. *Ind. Crops Prod.* **2024**, *215*, 118657. [CrossRef]
23. O'Keefe, A.; Deacon, D.A.G. Cavity ring-down optical spectrometer for absorption measurements using pulsed laser sources. *Rev. Sci. Instrum.* **1988**, *59*, 2544–2551. [CrossRef]
24. Busch, K.W.; Busch, M.A. *Cavity Ringdown Spectroscopy: An Ultra-Trace Absorption Measurement Technique*; ACS Symposium Series; Oxford University Press: New York, NY, USA, 1999; pp. 1–720.
25. Romanini, D.; Lehmann, K.K. Ring-down cavity absorption spectroscopy of the very weak HCN overtone bands with six, seven, and eight stretching quanta. *J. Chem. Phys.* **1993**, *99*, 6287–6301. [CrossRef]
26. Sun, J.; Yang, S.; Ai, J.; Yang, C.; Jia, Q.; Cao, B. Hierarchical Porous Activated Carbon Obtained by a Novel Heating-Rate-Induced Method for Lithium-Ion Capacitor. *Chem. Sel.* **2019**, *4*, 5300. [CrossRef]
27. Sun, J.; Yang, S.H.; Li, S.S.; Cao, B.Q. Double-activated porous carbons for high-performance supercapacitor electrodes. *Rare Met.* **2017**, *36*, 449–456. [CrossRef]
28. Sun, M.; Chen, Z.; Gong, Z.; Zhao, X.; Jiang, C.; Yuan, Y.; Wang, Z.; Li, Y.; Wang, C. Determination of breath acetone in 149 type 2 diabetic patients using a ringdown breath-acetone analyzer. *Anal. Bioanal. Chem.* **2015**, *407*, 1641–1650. [CrossRef] [PubMed]
29. Sun, M.; Zhao, X.; Yin, H.; Wang, Z.; Jiang, C.; Liu, W.; Chen, Z.; Yuan, Y.; Li, Y.; Wang, C. Study of breath acetone and its correlations with blood glucose and blood beta-hydroxybutyrate using an animal model with lab-developed type 1 diabetic rat. *RSC Adv.* **2015**, *5*, 71002–71010. [CrossRef]
30. Cheng, L.S.; Yang, R.T. Improved Horvath-Kawazoe equations including spherical pore models for calculating micropore size distribution. *Chem. Eng. Sci.* **1994**, *49*, 2599–2609. [CrossRef]
31. Finqueneisel, G.; Zimny, T.; Weber, J.V. On the prediction of adsorption isotherms of methanol/water vapour mixtures on microporous activated carbon. *Carbon* **2005**, *43*, 1093–1095. [CrossRef]

32. Samuel, M.S.; Mevada, C.; Mukhopadhyay, M. Hydrophilic Carbon Cloth (Chemically Activated) as an Electrode Material for Energy Storage Device. *Arab. J. Sci. Eng.* **2022**, *47*, 5949–5958. [CrossRef]
33. Tangsathitkulchai, C.; Ngernyen, Y.; Tangsathitkulchai, M. Surface modification and adsorption of eucalyptus wood-based activated carbons: Effects of oxidation treatment, carbon porous structure and activation method. *Korean J. Chem. Eng.* **2009**, *26*, 1341–1352. [CrossRef]
34. Hur, J.; Hwang, B.; Hong, L.; Yoo, S.J.; Chun, S.-E. Fabrication of hydrophilic porous carbon from polyvinylidene chloride-resin via synergetic activation of ZnO and tetrahydrofuran for aqueous supercapacitors. *Surf. Interfaces* **2023**, *40*, 103127. [CrossRef]
35. Canete, S.J.P.; Zhang, Z.Z.; Kong, L.M.; Schlegel, V.L.; Plantz, B.A.; Dowben, P.A.; Lai, R.Y. Application of synchrotron FT-IR microspectroscopy for determination of spatial distribution of methylene blue conjugated onto a SAM via "click" chemistry. *Chem. Commun.* **2011**, *47*, 11918–11920. [CrossRef]
36. Horikawa, T.; Kitakaze, Y.; Sekida, T.; Hayashi, J.; Katoh, M. Characteristics and humidity control capacity of activated carbon from bamboo. *Bioresour. Technol.* **2010**, *101*, 3964–3969. [CrossRef]
37. Long, C.; Li, Y.; Yu, W.H.; Li, A.M. Adsorption characteristics of water vapor on the hypercrosslinked polymeric adsorbent. *Chem. Eng. J.* **2012**, *180*, 106–112. [CrossRef]
38. Brennan, J.K.; Bandosz, T.J.; Thomson, K.T.; Gubbins, K.E. Water in porous carbons. *Colloid. Surf. A* **2001**, *187*, 539–568. [CrossRef]

Disclaimer/Publisher's Note: The statements, opinions and data contained in all publications are solely those of the individual author(s) and contributor(s) and not of MDPI and/or the editor(s). MDPI and/or the editor(s) disclaim responsibility for any injury to people or property resulting from any ideas, methods, instructions or products referred to in the content.

chemosensors

MDPI

Article

Effects of Base Materials (α-Alumina and/or γ-Alumina) on Volatile Organic Compounds (VOCs)-Sensing Properties of Adsorption/Combustion-Type Microsensors

Takeo Hyodo [1,*], Yuma Matsuura [2], Genki Inao [2], Takahiko Sasahara [3], Yasuhiro Shimizu [2] and Taro Ueda [1]

1 Graduate School of Integrated Science and Technology, Nagasaki University, 1-14 Bunkyo-machi, Nagasaki 852-8521, Nagasaki, Japan; taroueda@nagasaki-u.ac.jp
2 Graduate School of Engineering, Nagasaki University, 1-14 Bunkyo-machi, Nagasaki 852-8521, Nagasaki, Japan; shimizu@nagasaki-u.ac.jp (Y.S.)
3 Gas Equipment R&D Center, Yazaki Energy System Corporation, 23 Minami Kajama, Futamata-cho, Tenryu-ku, Hamamatsu-shi 431-3393, Shizuoka, Japan; takahiko.sasahara@jp.yazaki.com
* Correspondence: hyodo@nagasaki-u.ac.jp

Abstract: The sensing properties of adsorption/combustion-type microsensors using 5 wt% Pt-loaded aluminas, which consist of two kinds of alumina (α-Al_2O_3 and γ-Al_2O_3), as sensing (catalytic) materials for ethanol and toluene, were investigated in air, and the mixing effects of α-Al_2O_3 with γ-Al_2O_3 on the dynamic and static responses of the sensors were discussed in this study. The mixing of 50 wt% α-Al_2O_3 with γ-Al_2O_3 was the most effective in enhancing the dynamic responses to ethanol, which originated from the flash combustion behavior of ethanol and/or their partially decomposed products adsorbed on the sensing films from 150 °C to 450 °C, while further mixing of α-Al_2O_3 with γ-Al_2O_3 tended to increase the dynamic responses to toluene. On the other hand, the static responses to both ethanol and toluene, which arise from their catalytic combustion at elevated temperatures (450 °C), mainly increased with an increase in the addition of α-Al_2O_3 in the 5 wt% Pt-loaded aluminas. These results indicate that the synergistic effects of the catalytic activity and the thermal conductivity of the 5 wt% Pt-loaded aluminas are the most important for the sensing properties of these sensors to ethanol and toluene.

Keywords: adsorption/combustion-type gas sensor; MEMS; dynamic response; static response; thermal conductivity; catalytic activity; alumina; platinum; ethanol; toluene

1. Introduction

The large number of volatile organic compounds (VOCs) generated from various building materials, such as paints, adhesives, and resins, has been a serious problem for a long time, and it is widely recognized that these substances have a variety of adverse effects on human health [1–3]. In order to reduce these symptoms, so-called "sick house syndrome" and "chemical sensitivity", it is essential to develop specific VOC-sensing devices that can detect VOCs in indoor environments with high sensitivity. On the other hand, various kinds of VOCs are also contained in human biogases, such as exhaled breath and skin gas, and monitoring changes in their concentrations can help determine the health status and presence or absence of disease [4–6]. For example, acetone has been reported to be a biomarker for diabetes and metabolic activity, isoprene and pentane for heart disease, and toluene and 1-nonanal for lung cancer [7–11], and the development of a wearable device that can sensitively and selectively detect these VOCs is significantly attractive in the healthcare and medical fields. Recently, various gas sensors have been developed for the sensitive and selective detection of different types of VOCs [12–24]. Our group has developed adsorption/combustion-type gas sensors that combine the functions of "VOC adsorption" and "flash VOC combustion" for many years, by dynamically operating kinds of catalytic combustion-type gas sensors, in order to detect such ultra-trace amounts of VOCs [25,26]. After VOCs are adsorbed on the surface of the catalytic materials of the sensors at low temperatures, they are burned by flash heating and the large amount of combustion heat generated becomes output as a sensor signal. We previously reported that the VOC response of adsorption/combustion-type gas sensors was drastically improved by impregnating Pd and Au nanoparticles in an optimal ratio on γ-alumina (γ-Al_2O_3) or by highly dispersing core/shell (Au/Pd) nanoparticles by ultrasonic reduction on γ-Al_2O_3 [27,28]. We also confirmed that the VOC response was improved by co-loading Pt and a metal oxide on γ-Al_2O_3 and that the VOC-sensing characteristics were highly dependent on the type of metal oxide [29,30]. In the process of clarifying the VOC-response mechanism of adsorption/combustion-type gas sensors while closely investigating the VOC oxidation activity and adsorption/desorption characteristics, it was suggested that the thermal conductivity of the catalytic material also had a significant effect on the VOC-response characteristics [31]. In fact, our previous research revealed that the VOC-response characteristics of the sensor were improved when α-alumina (α-Al_2O_3), which has a small specific surface area but high crystallinity, was mixed with γ-Al_2O_3 and that they were loaded with Pd and Au [26]. In this case, it was inferred that the higher thermal conductivity of α-Al_2O_3 compared with γ-Al_2O_3 contributed to the improved VOC-response characteristics. In this study, therefore, we focused on Pt nanoparticles as a noble-metal catalyst for efficient VOC combustion and aimed to clarify the effect of mixing α-Al_2O_3 with γ-Al_2O_3 as the base material of the sensing film on the VOC-sensing properties of adsorption/combustion-type microsensors for two kinds of VOCs, ethanol and toluene.

2. Experimental Section

2.1. Preparation of Sensing Materials and Characterization

γ-Al_2O_3 powder (JGC Catalysts and Chemicals Ltd., Kanagawa, Japan) and α-Al_2O_3 powder (TM-DA, Taimei Chemical Chemicals Co., Ltd., Tokyo, Japan) were mechanically mixed in an agate mortar, and then Pt particles were loaded onto their surfaces by general impregnation with hydrogen hexachloroplatinate(IV) hexahydrate (FUJIFILM Wako Pure Chem. Corp., Osaka, Japan), followed by heat treatment at 500 °C for 1 h. The obtained powder is denoted as nPt/γ(r)α(t)-Al_2O_3 [n: the amount of Pt loaded (wt%), r: the weight ratio of γ-Al_2O_3 to all aluminas (wt%), t: the weight ratio of α-Al_2O_3 to all aluminas (wt%)], and nPt/γ(*100*)α(*0*)-Al_2O_3 and nPt/γ(*0*)α(*100*)-Al_2O_3 are abbreviated and simply expressed as nPt/γ-Al_2O_3 and nPt/α-Al_2O_3, respectively. The pore size distribution and the specific surface area (SSA) of the obtained powders were measured by Barrett-Joyner-Halenda (BJH) and Brunauer-Emmett-Teller (BET) methods using nitrogen adsorption-desorption isotherms (TriStar 3000, Micromeritics Inst. Corp., Norcross, GA, USA), respectively [32–34]. The crystal phases of the powders were characterized by X-ray diffraction analysis (XRD; Minflex 600-DX, Rigaku Corp., Tokyo, Japan) using Cu Kα radiation (40 kV, 40 mA), and their crystallite sizes were calculated by using the Scherrer equation. The chemical states of Al, O, and Pt on the surface of the representative powders were characterized by X-ray photoelectron spectroscopy using Al Kα radiation (XPS; ACIS-TLATRA DLD, Kratos Analytical, Manchester, UK). The binding energy was calibrated using the C 1s level (285.0 eV) of the usual contamination.

The dispersibility, surface area, and average particle size of the Pt nanoparticles on the surface of typical 5Pt/γ(*r*)α(*t*)-Al_2O_3 powders were calculated using the CO-pulse method (BELCAT II, MicrotracBEL Corp., Osaka, Japan). The nanostructure of the typical 5Pt/γ(*r*)α(*t*)-Al_2O_3 powder was observed by transmission electron microscopy (TEM; JEM-ARM200F, JEOL Ltd., Tokyo, Japan), and scanning transmission electron microscopy with energy dispersive X-ray spectroscopy for elemental mapping (STEM-EDS; JEM-ARM200F and JED-2300T, JEOL Ltd., Tokyo, Japan). The thermal conductivity of representative *n*Pt/γ(*r*)α(*t*)-Al_2O_3 powders was measured by using a modified transient plane source sensor (TRIDENT, C-Therm Technologies, Fredericton, NB, Canada). The *n*Pt/γ(*r*)α(*t*)-Al_2O_3 powders were pelletized by compressing at 1000 kg for 8 min, followed by annealing at 500 °C for 2 h in ambient air, and the obtained disks with a diameter of ca. 2 cm and a thickness of ca. 1 mm were used for the thermal conductivity measurement. The heat treatment condition was the same as that of the sensor fabrication described in the following section. The catalytic combustion behavior of ethanol and toluene over typical 5Pt/γ(*r*)α(*t*)-Al_2O_3 powders was measured using a handmade catalytic activity evaluation system. The 5Pt/γ(*r*)α(*t*)-Al_2O_3 powders were pelletized by compressing at 1000 kg for 8 min, followed by annealing at 500 °C for 2 h in ambient air, and the obtained disks with a diameter of ca. 1 cm were crushed to obtain irregularly shaped agglomerates with a size of 250–850 μm. They were packed into the region of ca. 12 mm in width in the center of a fixed bed reactor with an inner diameter of 2.0 cm, and the oxidation activity of ethanol and toluene over the typical *n*Pt/γ(*r*)α(*t*)-Al_2O_3 powders was measured in synthetic air at a flow rate of 20 cm^3 min^{-1} (space velocity: 8159 h^{-1}) and in compressed air at a flow rate of 20 cm^3 min^{-1} (space velocity: 40,795 h^{-1}).

2.2. Fabrication of Sensors and Measurement of Sensing Properties to Ethanol and Toluene

Figure 1a shows a stereomicroscope photograph of a microsensor platform with a couple of Pt microheaters, which was fabricated by MEMS technology, and Figure 1b shows a SEM photograph of a Pt microheater, which was covered with a thin alumina film, on a thin insulating substrate (ca. 70 × 70 μm^2) with a diaphragm structure. The Pt microheater also served as a detector. The microsensor platform, which was recently downsized from the previous one [16–23], was designed by Yazaki Energy System Corp. The *n*Pt/γ(*r*)α(*t*)-Al_2O_3 and unloaded γ(*r*)α(*t*)-Al_2O_3 powders were used as the sensing and reference materials, respectively. They were mixed with an appropriate amount of an organic vehicle consisting of polyvinyl butyral resin (mean polymerization degree: 700)], di-*n*-butyl phthalate, and terpineol by ball milling for 30 min. The obtained *n*Pt/γ(*r*)α(*t*)-Al_2O_3 and unloaded γ(*r*)α(*t*)-Al_2O_3 pastes were applied over both the Pt microheaters, as a sensing film or a reference film, respectively, by drop coating employing an air-pulse fluid dispenser (MS-10DX, Musashi Eng., Inc., Tokyo, Japan) with a suitable syringe. The microsensor chips attached to these films were then heat-treated at 500 °C for 2 h in ambient air. The obtained sensors were referred to as sensing materials (*n*Pt/γ(*r*)α(*t*)-Al_2O_3). Figure 1c shows the SEM photographs of a typical sensing 5Pt/γ(*r*)α(*t*)-Al_2O_3 film on a substrate with a Pt microheater. The thickness at the center of the film was approximately 30–40 μm, and the thickness gradually decreased toward the edge of the film. In addition, the surface of the film was relatively smooth, and the morphology of the film was formed with relatively good reproducibility.

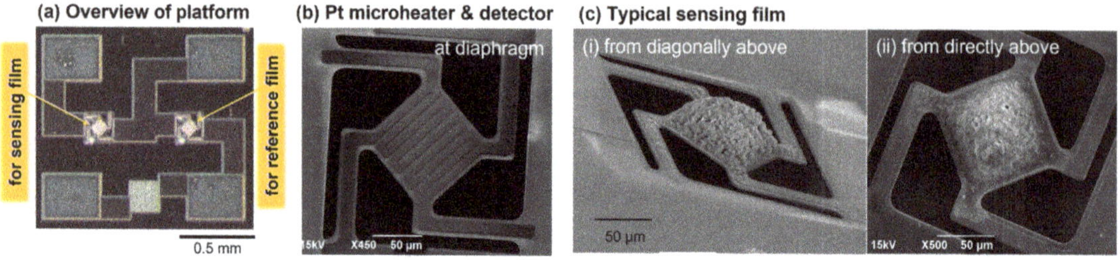

Figure 1. (**a**) Stereomicroscope photograph of microsensor platform with a couple of Pt microheaters, and SEM photographs of (**b**) Pt microheater and detector at diaphragm and (**c**) typical sensing film (5Pt/γ-Al_2O_3).

2.3. Measurement of Sensing Properties to Ethanol and Toluene

The sensing properties of ethanol and toluene were measured in an acrylic chamber (internal volume: approximately 50 dm^3) in which 10, 100, and 1000 ppm of ethanol or toluene were vaporized just on a compact ceramic heater. The nPt/γ(r)α(t)-Al_2O_3 sensors consist of sensing (nPt/γ(r)α(t)-Al_2O_3) and reference (γ(r)α(t)-Al_2O_3) films with Pt microheaters, and these films were assembled into a bridge circuit, as shown in Figure S1a. All sensors were consecutively pulse-heated during a cycle of 10 s from low temperature (LT, 150 °C, 9.6 s) to high temperature (HT, 450 °C, 0.4 s) in this study. A typical signal profile of the sensor is shown in Figure S1b. Ethanol, toluene, and their related components (intermediates), which are produced by partial oxidation and/or condensation, are adsorbed on the oxide surfaces of both films at 150 °C in the target gas. These adsorbates burn mainly on the oxide surface of the sensing film the moment the sensor is heated to 450 °C. At that time, the sensor-signal profile typically has one dynamic signal originating from the flash catalytic combustion of all the adsorbates mainly on the oxide surface of the sensing film. A static signal originating from the general catalytic combustion of ethanol or toluene mainly on the oxide surface of the sensing film is constantly confirmed after the instantaneous dynamic signal disappears during pulse heating at 450 °C. The magnitude of the general response (ΔV_{MAX}) was defined as the largest difference between the output voltage in air containing ethanol or toluene and that in air (ΔV). Two responses, which were calculated by integrating the dynamic and static signals of the sensor-signal profiles, were defined as the integrated dynamic response (IDR) and integrated static response (ISR), respectively (see Figure S1b).

3. Results and Discussion

3.1. Characterization of Sensor Materials

Figure S2 shows the nitrogen adsorption/desorption isotherms and pore size distributions of all the 5Pt/γ(r)α(t)-Al_2O_3 powders. Figure 2 shows the dependence of the SSA of the 5Pt/γ(r)α(t)-Al_2O_3 powders on the amount of α-Al_2O_3 (t). The amount of nitrogen adsorbed on the 5Pt/γ-Al_2O_3 powder was the largest among all the 5Pt/γ(r)α(t)-Al_2O_3 powders, and thus both the pore volume and the SSA (ca. 219 $m^2\ g^{-1}$) of the 5Pt/γ-Al_2O_3 powder were also the largest among them. In addition, the large hysteresis of the nitrogen adsorption-desorption isotherms was clearly confirmed under a relative pressure of 0.6–1.0, indicating that they were classified mainly as type-IV isotherms in the models of IUPAC [32]. Thus, the mesopore distribution arising from nitrogen adsorption was significantly different from that of nitrogen desorption. The SSA monotonically decreased with an increase in the amount of α-Al_2O_3, as shown in Figure 2, and the ratio of the small mesopores (diameter: ≤ca. 5 nm) to large mesopores (diameter: ≥ca. 5 nm) gradually increased with the decrease in the SSA. The SSA of the 5Pt/α-Al_2O_3 powder was the smallest among them (ca. 12 m^2 g^{-1}), while the adsorption and desorption isotherms had little hysteresis and the diameters of the mesopores were mainly less than 5 nm. Figure S3 shows the XRD spectra of all the 5Pt/γ(*r*)α(*t*)-Al_2O_3 powders, and Figure 3 shows the dependence of the ratio of the normalized integral intensity of the (400) peak of γ-Al_2O_3 to that of the (116) peak of α-Al_2O_3 ($I_{400,\gamma}/I_{116,\alpha}$) on the amount of α-$Al_2O_3$ in the 5Pt/γ(*r*)α(*t*)-Al_2O_3 powders. All the peaks of the 5Pt/γ-Al_2O_3 powder were assigned to γ-Al_2O_3 (JCPDS No. 10-0425), while all the peaks of the 5Pt/α-Al_2O_3 powder were assigned to either Pt (JCPDS No. 04-0802) or α-Al_2O_3 (JCPDS No. 71-1123). In addition, the peaks of both γ-Al_2O_3 and α-Al_2O_3 were confirmed in the XRD spectra of the other 5Pt/γ(*r*)α(*t*)-Al_2O_3 powders. On the other hand, the small peaks of Pt were detected only in the 5Pt/γ(75)α(25)-Al_2O_3 powder among them, which indicates that the crystallinity and/or crystallite size of Pt nanoparticles loaded to the γ(*r*)α(*t*)-Al_2O_3 powders increased with an increase in the amount of α-Al_2O_3 in the 5Pt/γ(*r*)α(*t*)-Al_2O_3 powders. Furthermore, the $I_{400,\gamma}/I_{116,\alpha}$ value monotonically decreased with an increase in the amount of α-Al_2O_3 in 5Pt/γ(*r*)α(*t*)-Al_2O_3. The dependences of both SSA (Figure 2) and $I_{400,\gamma}/I_{116,\alpha}$ (Figure 3) values on the amount of α-Al_2O_3 in the 5Pt/γ(*r*)α(*t*)-Al_2O_3 obviously show that α-Al_2O_3 was uniformly mixed with γ-Al_2O_3 in the 5Pt/γ(*r*)α(*t*)-Al_2O_3 powders.

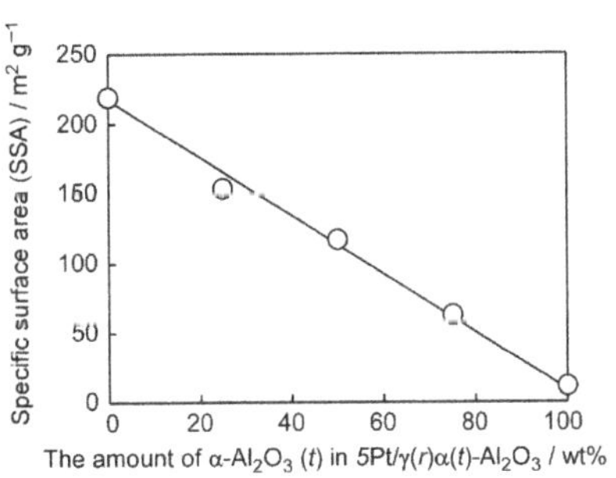

Figure 2. Dependence of specific surface area (SSA) of 5Pt/γ(*r*)α(*t*)-Al_2O_3 powders on the amount of α-Al_2O_3 (*t*).

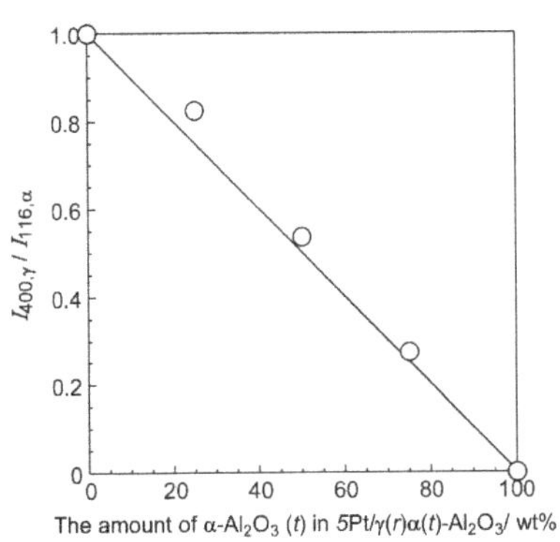

Figure 3. Dependence of the ratio of normalized integral intensity of (400) peak of γ-Al_2O_3 to that of (116) peak of α-Al_2O_3 ($I_{400,\gamma}/I_{116,\alpha}$) on the amount of α-$Al_2O_3$ (*t*) in 5Pt/γ(*r*)α(*t*)-Al_2O_3 powders.

Figures S4 and S5 show the XPS spectra of Pt 4f, Al 2s, and O 1s of all 5Pt/γ(*r*)α(*t*)-Al_2O_3 powders, and variations in the ratio of Pt or Al to all Pt and Al species and the ratio of each Pt component to all Pt species with the amount of α-Al_2O_3 in the 5Pt/γ(*r*)α(*t*)-Al_2O_3 powders are shown in Figure 4. All the ratios were calculated by the deconvolution of their XPS spectra. The half-widths of both the Al 2s and O 1s spectra gradually narrowed with an increase in the amount of α-Al_2O_3 ("from ca. 2.7 eV to ca. 2.0 eV" and "from ca. 2.8 eV to ca. 1.8 eV", respectively). In addition, the binding energy of the Al 2s spectra was hardly dependent on the amount of α-Al_2O_3, whereas the binding energy of the O 1s spectra was negatively shifted with an increase in the amount of α-Al_2O_3. On the other hand, the amount of Pt^0 (Pt metal) increased and the amount of Pt^{2+} (such as PtO) decreased with the amount of α-Al_2O_3, while the amount of Pt^{4+} (such as PtO_2) remained almost unchanged. Nevertheless, the binding energy of Pt positively shifted with the amount of α-Al_2O_3, while the amount of Pt^{2+} was much larger than those of Pt^0 and Pt^{4+}. These results indicate that the electron density in all the Pt components decreased with an increase in the amount of α-Al_2O_3 and that the electrons were transferred to the oxygen species of the Al_2O_3-components due to the interaction between Pt and Al_2O_3. Moreover, the ratio of Pt to all Pt and Al species on the 5Pt/γ(*r*)α(*t*)-Al_2O_3 surface increased with an increase in the amount of α-Al_2O_3, even though the amount of Pt loaded onto the mixed aluminas was only 5 wt% for all 5Pt/γ(*r*)α(*t*)-Al_2O_3 powders. In other words, the amount of Pt loaded per unit SSA increased with an increase in the amount of α-Al_2O_3 because the SSA of α-Al_2O_3 was much smaller than that of γ-Al_2O_3, as shown in Figure 2.

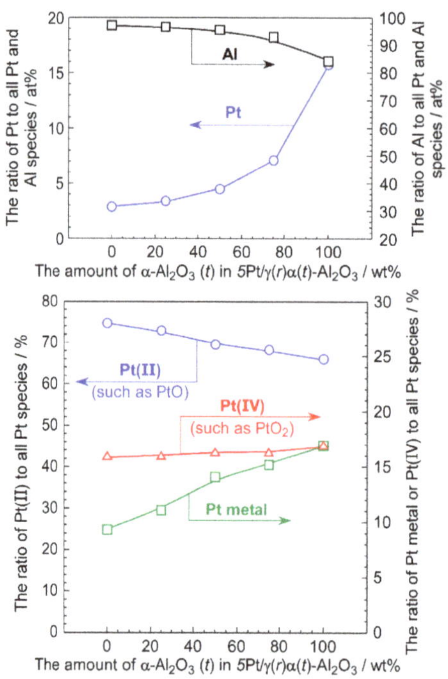

Figure 4. Variations in the ratio of Pt or Al to all Pt and Al species and the ratio of each Pt component to all Pt species with the amount of α-Al_2O_3 in 5Pt/γ(*r*)α(*t*)-Al_2O_3 powders.

The dispersibility, surface area, and average particle size of Pt loaded onto the aluminas of representative 5Pt/γ(*r*)α(*t*)-Al_2O_3 (5Pt/γ-Al_2O_3, 5Pt/γ(*50*)α(*50*)-Al_2O_3, and 5Pt/α-

Al_2O_3) powders were estimated from the CO adsorption/desorption properties, as shown in Table 1. The dispersibility and surface area of the Pt nanoparticles decreased and then the average particle size of Pt increased with an increase in the amount of α-Al_2O_3. These results indicate that the Pt nanoparticles, which were smaller than those on the surface of α-Al_2O_3, were loaded in a highly dispersed state on the surface of γ-Al_2O_3. The dispersibility and surface area of 5Pt/γ(50)α(50)-Al_2O_3 were relatively close to the average values of α-Al_2O_3 and γ-Al_2O_3 (dispersibility: ca. 31%, surface area: ca. 76 m^2 g^{-1}-Pt). This suggests that the Pt nanoparticles loaded on the surface of α-Al_2O_3 and γ-Al_2O_3 of 5Pt/γ(50)α(50)-Al_2O_3 existed in a relatively similar state to the Pt nanoparticles loaded on the surface of α-Al_2O_3 and γ-Al_2O_3 alone. Therefore, the average particle size of the Pt nanoparticles of 5Pt/γ(50)α(50)-Al_2O_3 calculated from the CO adsorption/desorption properties is not very meaningful when discussing the loading state. However, just the value is also very reasonable because it lies between the Pt nanoparticles on the 5Pt/γ-Al_2O_3 surface, which has a small average particle size, and the Pt nanoparticles on the 5Pt/α-Al_2O_3 surface, which has a large average particle size.

Table 1. Dispersibility, surface area, and average particle size of Pt loaded onto the aluminas of typical 5Pt/γ(*r*)α(*t*)-Al_2O_3 powders.

Sample	5Pt/γ-Al_2O_3	5Pt/γ(50)α(50)-Al_2O_3	5Pt/α-Al_2O_3
Dispersibility/%	47.8	32.0	14.0
Surface area/m^2 g^{-1}-Pt	118	80.0	34.5
Average particle size/nm	2.37	3.55	8.12

Figure 5 shows TEM photographs of representative 5Pt/γ(*r*)α(*t*)-Al_2O_3 (5Pt/γ-Al_2O_3, 5Pt/γ(50)α(50)-Al_2O_3, and 5Pt/α-Al_2O_3) powders. The γ-Al_2O_3 powder consisted of very fine particles, which were quite aggregated. In contrast, the α-Al_2O_3 powder was formed from particles with a diameter of approximately 200 nm. Furthermore, it was confirmed that the particles of both γ-Al_2O_3 and α-Al_2O_3 powders were uniformly mixed in the TEM photograph of 5Pt/γ(50)α(50)-Al_2O_3. Many Pt nanoparticles were confirmed with both γ-Al_2O_3 and α-Al_2O_3 particles in all highly magnified TEM photographs. The sizes of the Pt nanoparticles in the 5Pt/γ-Al_2O_3 and 5Pt/γ(50)α(50)-Al_2O_3 powders were smaller than those of the Pt nanoparticles in the 5Pt/α-Al_2O_3 powder, but their sizes seem to be comparatively smaller than the average sizes of the Pt nanoparticles, which were calculated from the amount of CO adsorption (Table 1).

Figure S6 shows the annular dark-field (ADF)-STEM images and EDS maps (Al K and Pt M) of representative 5Pt/γ(*r*)α(*t*)-Al_2O_3 (5Pt/γ-Al_2O_3, 5Pt/γ(50)α(50)-Al_2O_3, and 5Pt/α-Al_2O_3) powders, together with their TEM photographs. In all the samples, the EDS signal of Al K almost overlapped with the morphology of the alumina particles, as confirmed in the TEM photographs. The EDS signal of Pt M was also observed at almost the same location as the EDS signal of Al M in the TEM and ADF-STEM images of 5Pt/γ-Al_2O_3 and 5Pt/γ(50)α(50)-Al_2O_3, which indicates that Pt nanoparticles were uniformly loaded on both γ-Al_2O_3 and α-Al_2O_3 particles in these samples. On the other hand, on the basis of the ADF-STEM images and the EDS maps of the Pt M signal, it was confirmed that not only Pt nanoparticles randomly loaded on the surface of the α-Al_2O_3 particles, but also that a large amount of Pt nanoparticles was agglomerated at the interfaces (grain boundaries) of the α-Al_2O_3 particles. The large average size of the Pt nanoparticles in the 5Pt/α-Al_2O_3 powder (Table 2), which was calculated from the amount of CO adsorbed, probably originates from their agglomeration. Such agglomeration of Pt nanoparticles was not observed in the 5Pt/γ(50)α(50)-Al_2O_3 powder, because the γ-Al_2O_3 particles were well mixed with the α-Al_2O_3 particles and the γ-Al_2O_3 particles were well inserted among the α-Al_2O_3 particles, as shown in the TEM and ADF-STEM images. The amounts of Pt nanoparticles loaded in *5*Pt/γ-Al_2O_3, *5*Pt/γ(*50*)α(*50*)-Al_2O_3, and *5*Pt/α-Al_2O_3, which were evaluated by EDS analysis, were ca. 5.1 wt%, ca. 4.7 wt%, and ca. 4.9 wt%, respectively, and these values were almost the same as the amount of Pt added during the preparation process.

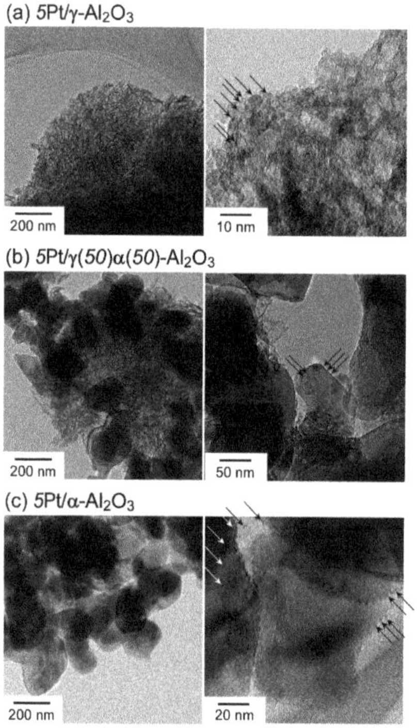

Figure 5. TEM photographs of representative *5*Pt/γ(*r*)α(*t*)-Al_2O_3 powders. Black and white arrows show Pt nanoparticles.

Table 2. Thermal conductivities of typical samples.

Sample	Thermal Conductivity/10^{-2} W m^{-1} K^{-1}
γ-Al_2O_3	9.32
*5*Pt/γ-Al_2O_3	9.63
*10*Pt/γ-Al_2O_3	9.97
γ(*50*)α(*50*)-Al_2O_3	14.9
α-Al_2O_3	37.7

The thermal conductivity of the sensing materials is one of the most important factors for adsorption/combustion-type gas sensors, because the thermal energy generated by the oxidation of VOCs on the catalyst surface needs to be effectively transferred to the Pt microheater (detector) to enhance the magnitude of the VOC response. Table 2 shows the thermal conductivities of representative sample disks fabricated under the same conditions as the sensor manufacturing conditions (heat treatment at 500 °C for 2 h in ambient air). The thermal conductivity of the γ-Al_2O_3 disk was small, and the thermal conductivity largely increased with an increase in the amount of α-Al_2O_3 added to γ-Al_2O_3 and the thermal conductivity of the α-Al_2O_3 disk was very large. On the other hand, the thermal conductivity of the nPt/γ-Al_2O_3 disks also tended to increase with an increase in the amount of Pt loaded, but the increase ratio was small in the range of 0–10 wt%. These results indicate that the addition of α-Al_2O_3 is effective in increasing the thermal conductivity of the sensing film.

3.2. Catalytic Activities

Figure 6a shows the catalytic combustion behavior of 100 ppm ethanol and 100 ppm toluene over representative 5Pt/γ(*r*)α(*t*)-Al_2O_3 (5Pt/γ-Al_2O_3, 5Pt/γ(*50*)α(*50*)-Al_2O_3, and 5Pt/α-Al_2O_3) powders in dry air at a flow rate of 20 cm^3 min^{-1}. The oxidation activities of 5Pt/γ-Al_2O_3 and 5Pt/γ(*50*)α(*50*)-Al_2O_3 for ethanol were extremely high, achieving nearly 100% ethanol conversion even at room temperature. However, there was a large difference in the amount of acetaldehyde produced as an intermediate product, and the amount of acetaldehyde produced over 5Pt/γ-Al_2O_3 was clearly smaller than that over 5Pt/γ(*50*)α(*50*)-Al_2O_3. The amount of CO_2 produced over both samples did not increase until relatively high temperatures were reached, and the CO_2 concentration reached 200 ppm, which is the concentration produced when ethanol is completely oxidized, at 250 °C. This was because other partially decomposed products except for acetaldehyde were produced over both samples. In fact, previous studies confirmed the production of ethylene and diethyl ether as partially decomposed products at incomplete medium temperatures for ethanol oxidation [22]. No partially decomposed products other than acetaldehyde were confirmed because of the low ethanol concentration used in this study. The catalytic combustion behavior of 100 ppm ethanol over representative 5Pt/γ(*r*)α(*t*)-Al_2O_3 (5Pt/γ-Al_2O_3, 5Pt/γ(*50*)α(*50*)-Al_2O_3, and 5Pt/α-Al_2O_3) powders in dry air was also investigated at a higher flow rate of 100 cm^3 min^{-1}, as shown in Figure S7a. It was not possible to quantify the amount of CO_2 produced in this experiment because compressed air was used as the base gas. Increasing the gas flow rate obviously clarified that the oxidation activity of 5Pt/α-Al_2O_3 for ethanol was higher than that of 5Pt/γ(*50*)α(*50*)-Al_2O_3. These results confirmed that the ethanol oxidation activity decreased with increasing amounts of α-Al_2O_3 added to 5Pt/γ(*r*)α(*t*)-Al_2O_3, owing to the large specific surface area of 5Pt/γ-Al_2O_3, as well as the high dispersibility of Pt nanoparticles and the large Pt surface area.

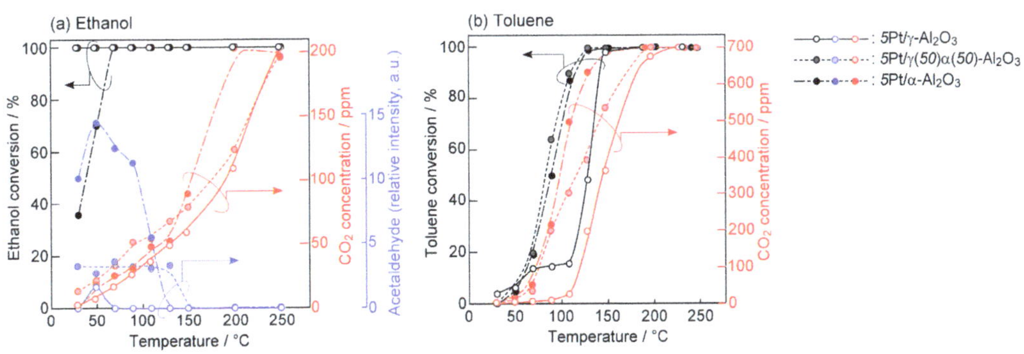

Figure 6. Catalytic combustion behavior of (**a**) 100 ppm ethanol and (**b**) 100 ppm toluene over representative 5Pt/γ(*r*)α(*t*)-Al_2O_3 powders in dry air (flow rate: 20 cm^3 min^{-1}).

Figure 6b shows the catalytic combustion behavior of 100 ppm toluene over representative 5Pt/γ(*r*)α(*t*)-Al_2O_3 (5Pt/γ-Al_2O_3, 5Pt/γ(*50*)α(*50*)-Al_2O_3, and 5Pt/α-Al_2O_3) powders in dry air at a flow rate of 20 cm^3 min^{-1}. The toluene conversion property of 5Pt/γ(*50*)α(*50*)-Al_2O_3 was the highest among them, but 5Pt/α-Al_2O_3 also showed a relatively high toluene conversion property. The amount of CO_2 generated, which is an indicator of complete oxidation, over 5Pt/α-Al_2O_3 was larger than that over 5Pt/γ(50)α(50)-Al_2O_3, even at low temperatures, confirming that partially decomposed products of toluene are not easily generated on the surface of 5Pt/α-Al_2O_3. On the other hand, the oxidation activity of 5Pt/γ-Al_2O_3 for toluene was much lower than that of the others, whereas the oxidation activity of 5Pt/γ-Al_2O_3 was comparable to that of the others only at lower temperatures. The catalytic combustion behavior of 100 ppm toluene over representative 5Pt/γ(*r*)α(*t*)-Al_2O_3 (5Pt/γ-Al_2O_3, 5Pt/γ(50)α(50)-Al_2O_3, and 5Pt/α-Al_2O_3) powders in dry air was also investigated at a higher flow rate of 100 cm^3 min^{-1}, as shown in Figure S7b. The amount of CO_2 produced was not able to be monitored in this experiment, too. The difference in the catalytic activity between 5Pt/γ(50)α(50)-Al_2O_3 and 5Pt/α-Al_2O_3 was clarified by increasing the gas flow rate, and it was confirmed that the toluene oxidation activity of 5Pt/γ(50)α(50)-Al_2O_3 was higher than that of 5Pt/α-Al_2O_3. It is also noteworthy that the oxidation activity of 5Pt/γ-Al_2O_3 for toluene became comparable to that of 5Pt/α-Al_2O_3 when the gas flow rate increased from 20 cm^3 min^{-1} to 100 cm^3 min^{-1}, which means that the gas flow rate (the rate at which toluene is supplied to the catalyst surface) is one of the most important factors for the relative comparison of the oxidation activities of the samples.

3.3. VOC-Sensing Properties

Figure 7 shows the sensor-signal profiles of representative 5Pt/γ(*r*)α(*t*)-Al_2O_3 (5Pt/γ-Al_2O_3, 5Pt/γ(50)α(50)-Al_2O_3, and 5Pt/α-Al_2O_3) sensors to ethanol and toluene in ambient air. These sensor-signal profiles largely depended on the kind of VOCs and their concentration, as well as the composition of the aluminas, that is, the amount of α-Al_2O_3, in the 5Pt/γ(*r*)α(*t*)-Al_2O_3 powders. The 5Pt/γ-Al_2O_3 sensor shows "medium dynamic responses and relatively small static responses to ethanol" and "quite small dynamic responses and medium static responses to toluene". The 5Pt/γ(50)α(50)-Al_2O_3 sensor showed "quite large dynamic responses and relatively small static responses to ethanol" and "small dynamic response and medium static response to toluene", but the dynamic responses of the 5Pt/γ(50)α(50)-Al_2O_3 sensor to toluene were much larger than those of the 5Pt/γ-Al_2O_3 sensor. On the other hand, the 5Pt/α-Al_2O_3 sensor showed small dynamic responses and large static responses to both gases. As mentioned above, the mixing of α-Al_2O_3 with γ-Al_2O_3 can significantly change the sensor-signal profiles to ethanol and toluene. Figure 8 summarizes the composition dependencies of the three kinds of responses of all the 5Pt/γ(*r*)α(*t*)-Al_2O_3 sensors to ethanol and toluene. The general response, ΔV_{MAX}, and integrated dynamic response, IDR, to ethanol were the largest when the amount of α-Al_2O_3 was 50 wt%, irrespective of the concentration of ethanol, while the maximum of ΔV_{MAX} and IDR to toluene tended to shift toward a larger amount of α-Al_2O_3. Therefore, the magnitudes of ΔV_{MAX} and IDR to ethanol were basically larger than those of toluene, but the differences were relatively smaller when the amount of α-Al_2O_3 was larger. The integrated static response, ISR, to 1000 ppm ethanol increased with an increase in the amount of α-Al_2O_3, while those to 10 and 100 ppm ethanol seemed to be the largest at the amount of α-Al_2O_3 of 50 wt%, as with their ΔV_{MAX} and IDR. On the other hand, the ISR to toluene increased with an increase in the amount of α-Al_2O_3 in all toluene concentration ranges. In addition, the magnitude of ISR to ethanol was smaller than that of toluene, irrespective of the toluene concentration and the amount of α-Al_2O_3.

Figure 9 summarizes the concentration dependencies of the three kinds of responses of all 5Pt/γ(*r*)α(*t*)-Al_2O_3 sensors to ethanol and toluene. The ΔV_{MAX} and IDR, especially the IDR to both ethanol and toluene, tend to be saturated with an increase in their concentration, which indicates that ethanol, toluene, and their partially decomposed products produced during the low-temperature period (150 °C) were adsorbed on the sensing-material surface and they burned during the flash heating-up period from 150 °C to 450 °C. On the other hand, the ISR to both ethanol and toluene increased approximately linearly with an increase in their concentration. This is because ethanol and toluene directly burned over the sensing-material surface during the static operation at 450 °C, and thus the ISR behavior of these sensors is quite similar to the behavior of general catalytic combustion-type gas sensors.

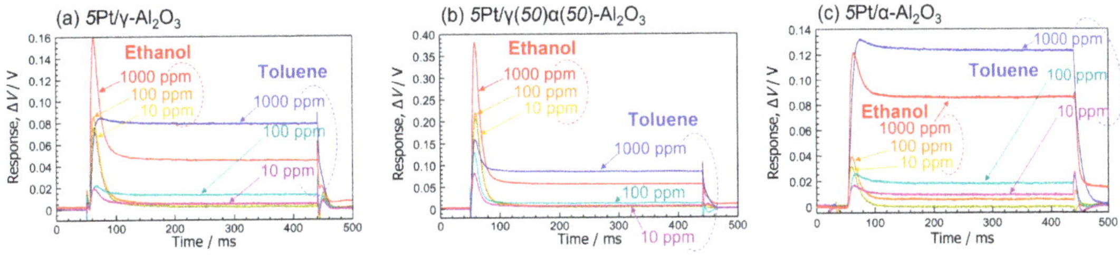

Figure 7. Sensor-signal profiles of (**a**) 5Pt/γ-Al_2O_3, (**b**) 5Pt/γ(*50*)α(*50*)-Al_2O_3, and (**c**) 5Pt/α-Al_2O_3 sensors to ethanol and toluene in ambient air.

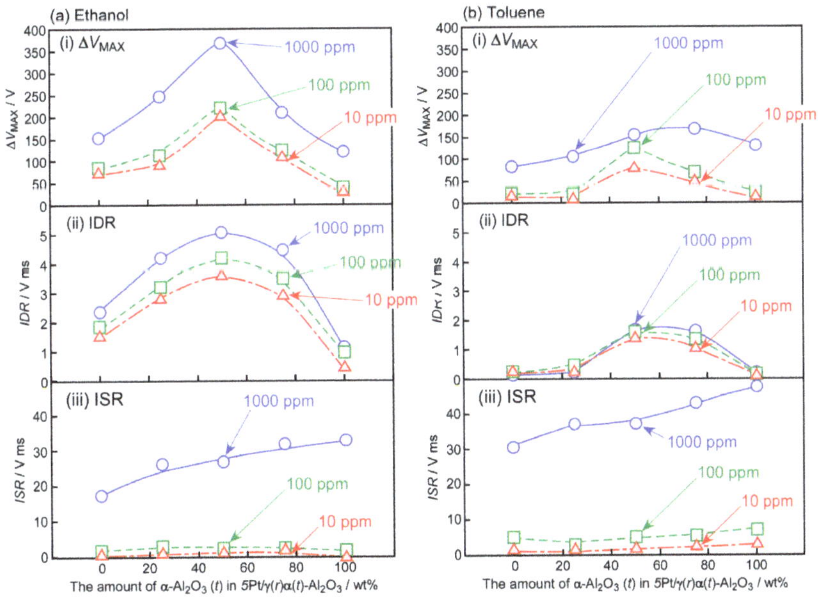

Figure 8. Composition dependencies of three kinds of responses of all 5Pt/$\gamma(r)\alpha(t)$-Al_2O_3 sensors to (**a**) ethanol and (**b**) toluene.

The variations in the three kinds of responses with the amount of α-Al_2O_3 seem to be influenced by the various characteristics of 5Pt/$\gamma(r)\alpha(t)$-Al_2O_3. The ethanol oxidation activity decreased with an increase in the amount of α-Al_2O_3. In addition, the amount of ethanol and/or the partially decomposed products generally decreases with an increase in the amount of α-Al_2O_3, because the specific surface area of 5Pt/$\gamma(r)\alpha(t)$-Al_2O_3 powder drastically decreased with an increase in the amount of α-Al_2O_3. However, the thermal conductivity increased significantly with an increase in the amount of α-Al_2O_3. The synergetic effects of these factors must determine the largest ΔV_{MAX} and IDR to ethanol when the amount of α-Al_2O_3 was 50 wt%. The shift of the largest ΔV_{MAX} and IDR to toluene toward a larger amount of α-Al_2O_3 in comparison with those to ethanol probably arises from the largest toluene oxidation activity of 5Pt/γ(*50*)α(*50*)-Al_2O_3. On the other hand, ISRs to both ethanol and toluene increased with an increase in the amount of α-Al_2O_3, which was almost the same as the dependence of thermal conductivity on the amount of α-Al_2O_3. The highly thermally conductive α-Al_2O_3 effectively transfers the thermal energy generated by the oxidation of ethanol and toluene toward the Pt microheater of the 5Pt/γ(*r*)α(*t*)-Al_2O_3 sensors during the high-temperature period.

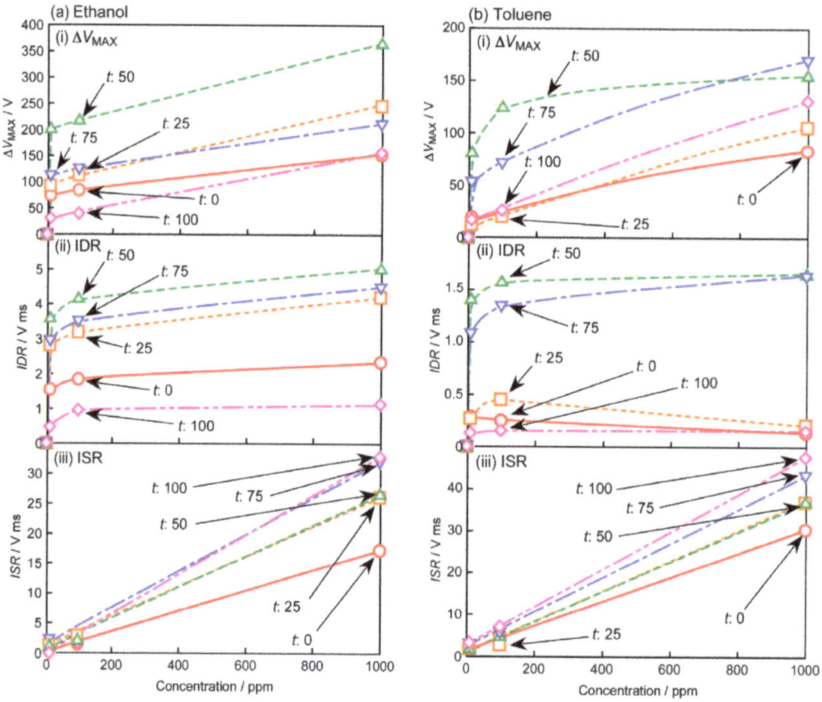

Figure 9. Concentration dependencies of three kinds of responses of all 5Pt/γ(*r*)α(*t*)-Al_2O_3 sensors to (**a**) ethanol and (**b**) toluene.

As mentioned above, the 5Pt/γ(*r*)α(*t*)-Al_2O_3 sensors have two important sensing-output information in one signal: dynamic response and static response. Therefore, we expect that these attractive characteristics of adsorption/combustion-type VOC sensors will bring great benefits to highly sensitive and selective VOC detection through accurate analysis using adequate computing technologies, such as machine learning and AI technology.

4. Conclusions

The sensing properties of adsorption/combustion-type 5Pt/γ(*r*)α(*t*)-Al_2O_3 sensors to ethanol and toluene were investigated in ambient air, and the effects of mixing α-Al_2O_3 and γ-Al_2O_3 as the base material of the sensing film on the sensing properties were discussed in this study. The α-Al_2O_3 powder was well mixed with the γ-Al_2O_3 powder, and the Pt nanoparticles were basically dispersed on the surface of these aluminas (some of the Pt nanoparticles were agglomerated at the grain boundaries only in the γ-Al_2O_3 powder), which was confirmed by the evaluation of the adsorption/desorption characteristics of N_2

and CO, as well as by the analyses using XRD, XPS, TEM, and STEM-EDS. The mixing of 50 wt% α-Al_2O_3 with γ-Al_2O_3 was the most effective in enhancing both ΔV_{MAX} and IDR to ethanol, while the maximum ΔV_{MAX} and IDR to toluene tended to shift toward a larger amount of α-Al_2O_3. The ethanol oxidation activity of 5Pt/γ(*r*)α(*t*)-Al_2O_3 decreased with an increase in the amount of α-Al_2O_3, whereas the thermal conductivity increased with an increase in the amount of α-Al_2O_3. These results indicate that the synergistic effects of catalytic activity and thermal conductivity of 5Pt/γ(*r*)α(*t*)-Al_2O_3 are quite important in enhancing the ethanol-sensing properties of the sensors. On the other hand, the ISR to both ethanol and toluene mainly increased with an increase in the addition of α-Al_2O_3 in 5Pt/γ(*r*)α(*t*)-Al_2O_3, even though the oxidation activities of α-Al_2O_3 for ethanol and toluene were lower than those of the other samples. Therefore, the relatively high thermal conductivity of α-Al_2O_3 must be of great importance in the effective transfer of thermal energy generated by the oxidation of ethanol and toluene toward the Pt microheater of the 5Pt/γ(*r*)α(*t*)-Al_2O_3 sensors during the high-temperature period.

Supplementary Materials: The following supporting information can be downloaded at https://www.mdpi.com/article/10.3390/chemosensors13010009/s1, Figure S1: Measurement circuit and typical sensor-signal profile with definition of three kinds of responses; Figure S2: Nitrogen adsorption-desorption isotherms and pore size distributions of all 5Pt/γ(*r*)α(*t*)-Al_2O_3 powders; Figure S3: XRD spectra of all 5Pt/γ(*r*)α(*t*)-Al_2O_3 powders; Figure S4: XPS spectra of Pt 4f of all 5Pt/γ(*r*)α(*t*)-Al_2O_3 powders; Figure S5: XPS spectra of Al 2s and O 1s of all 5Pt/γ(*r*)α(*t*)-Al_2O_3 powders; Figure S6: (i) TEM photographs, (ii) ADF-STEM images, and EDS maps of (iii) Al K and Pt M of representative 5Pt/γ(*r*)α(*t*)-Al_2O_3 (5Pt/γ-Al_2O_3, 5Pt/γ(*50*)α(*50*)-Al_2O_3, and 5Pt/α-Al_2O_3) powders; Figure S7: Catalytic combustion behavior of (a) 100 ppm ethanol and (b) 100 ppm toluene over representative 5Pt/γ(*r*)α(*t*)-Al_2O_3 powders in dry air (flow rate: 100 cm^3 min^{-1}).

Author Contributions: Conceptualization, T.H., T.S. and Y.S.; methodology, T.H.; software, T.S.; validation, T.H. and T.U.; investigation, Y.M., G.I. and T.U.; data curation, T.H. and T.U.; writing—original draft preparation, T.H.; writing—review, and editing, T.H. and T.U.; supervision, T.H., Y.S. and T.U.; project administration, T.H. and T.S.; funding acquisition, T.H. and T.S. All authors have read and agreed to the published version of the manuscript.

Funding: This research was partly funded by JSPS KAKENHI, grant number JP21H01626 and The Murata Science Foundation, grant number H31-076.

Institutional Review Board Statement: Not applicable.

Informed Consent Statement: Not applicable.

Data Availability Statement: The data that support the findings of this study are available from the corresponding author upon reasonable request.

Conflicts of Interest: The author, Takahiko Sasahara, was employed by the Gas Equipment R&D Center, Yazaki Energy System Corporation. The remaining authors declare that the research was conducted in the absence of any commercial or financial relationships that could be construed as potential conflicts of interest.

References

1. Rumchev, K.; Brouwn, H.; Spickett, J. Volatile Organic Compounds: Do they present a risk to our health? *Rev. Environ. Health* **2007**, *22*, 67–82. [CrossRef] [PubMed]
2. Bhoonah, R.; Maury-Micolier, A.; Jolliet, O. Integrated empirical and modelled determination of the human health impacts of building material VOCs. *Build. Environ.* **2023**, *242*, 110523. [CrossRef]
3. Haug, H.; Klein, L.; Sauerwald, T.; Poelke, B.; Beauchamp, J.; Alexander, R. Sampling Volatile organic compound emissions from consumer products: A review. *Crit. Rev. Anal. Chem.* **2024**, *54*, 1895–1916. [CrossRef] [PubMed]

4. Sekine, Y.; Kimura, K.; Umezawa, K. What does human skin gas analysis work for? *J. Jpn. Assoc. Odor Environ.* **2017**, *48*, 410–417. [CrossRef]
5. Moura, P.C.; Raposo, M.; Vassilenko, V. Breath volatile organic compounds (VOCs) as biomarkers for the diagnosis of pathological conditions: A review. *Biomed. J.* **2023**, *46*, 10623. [CrossRef]
6. Kim, C.; Kang, M.S.; Raja, I.S.; Oh, J.-W.; Joung, Y.K.; Han, D.-W. Current issues and perspectives in nanosensors-based artificial olfactory systems for breath diagnostics and environmental exposure monitoring. *Trends Anal. Chem.* **2024**, *174*, 117656. [CrossRef]
7. Tassopoulos, C.N.; Barnett, D.; Russell Fraseer, T. Breath-acetone and blood-sugar measurements in diabetes. *Lancet* **1969**, *28*, 1282–1286. [CrossRef]
8. Hunt, J.; Gaston, B. Airway nitrogen oxide measurements in asthma and other pediatric respiratory diseases. *J. Pediatr.* **2000**, *137*, 14–20. [CrossRef]
9. Cikach, F.S., Jr.; Dweik, R.A. Cardiovascular biomarkers in exhaled breath. *Prog. Cardiovasc. Dis.* **2012**, *55*, 34–43. [CrossRef]
10. Choueiry, F.; Barham, A.; Zhu, J. Analyses of lung cancer-derived volatiles in exhaled breath and in vitro models. *Exp. Biol. Med.* **2022**, *247*, 1179–1190. [CrossRef]
11. Schmidt, F.; Kohlbrenner, D.; Malesevic, S.; Huang, A.; Klein, S.D.; Puhan, M.A.; Kohler, M. Mapping the landscape of lung cancer breath analysis: A scoping review. *Lung Cancer* **2023**, *175*, 131–140. [CrossRef] [PubMed]
12. Güntner, A.T.; Weber, I.C.; Schon, S.; Pratsinis, S.E.; Gerber, P.A. Monitoring rapid metabolic changes in health and type-1 diabetes with breath acetone sensors. *Sens. Actuators B* **2022**, *367*, 132182. [CrossRef]
13. Mori, M.; Itagaki, Y.; Sadaoka, Y. VOC detection by potentiometric oxygen sensor based on YSZ and modified Pt electrodes. *Sens. Actuators B* **2012**, *161*, 471–479. [CrossRef]
14. Itoh, T.; Nakashima, T.; Akamatsu, T.; Izu, N.; Shin, W. Nonanal gas sensing properties of platinum, palladium, and gold-loaded tin oxide VOCs sensors. *Sens. Actuators B* **2013**, *187*, 135–141. [CrossRef]
15. Kim, N.-H.; Choi, S.-J.; Yang, D.-J.; Bae, J.; Park, J.; Kim, I.-D. Highly sensitive and selective hydrogen sulfide and toluene sensors using Pd functionalized WO_3 nanofibers for potential diagnosis of halitosis and lung cancer. *Sens. Actuators B* **2014**, *193*, 574–581. [CrossRef]
16. Ye, M.; Chien, P.-J.; Toma, K.; Arakawa, T.; Mitsubayashi, K. An acetone bio-sniffer (gas phase biosensor) enabling assessment of lipid metabolism from exhaled breath. *Biosens. Bioelectron.* **2015**, *73*, 208–213. [CrossRef]
17. Hyodo, T.; Kaino, T.; Ueda, T.; Izawa, K.; Shimizu, Y. Acetone-sensing properties of WO_3-based gas sensors operated in dynamic temperature modulation mode —Effects of loading of noble metal and/or NiO onto WO_3. *Sens. Mater.* **2016**, *28*, 1179–1189.
18. Suematsu, K.; Harano, W.; Oyama, T.; Shin, Y.; Watanabe, K.; Shimanoe, K. Pulse-driven semiconductor gas sensors toward ppt level toluene detection. *Anal. Chem.* **2018**, *90*, 11219–11223. [CrossRef]
19. Choi, H.-J.; Chung, J.-H.; Yoon, J.-W.; Lee, J.-H. Highly selective, sensitive, and rapidly responding acetone sensor using ferroelectric ε-WO_3 spheres doped with Nb for monitoring ketogenic diet efficiency. *Sens. Actuators B* **2021**, *338*, 129823. [CrossRef]
20. Sun, Y.; Zhao, Z.; Suematsu, K.; Zhang, W.; Zhang, W.; Zhuiykov, S.; Shimanoe, K.; Hu, J. MOF-derived Au-NiO/In_2O_3 for selective and fast detection of toluene at ppb-level in high humid environments. *Sens. Actuators B* **2022**, *360*, 131631. [CrossRef]
21. Shinkai, T.; Masumoto, K.; Iwai, M.; Inomata, Y.; Kida, T. Study on sensing mechanism of volatile organic compounds using Pt-loaded ZnO nanocrystals. *Sensors* **2022**, *22*, 6277. [CrossRef] [PubMed]
22. Ueda, T.; Oide, N.; Kamada, K.; Hyodo, T.; Shimizu, Y. Improved toluene response of mixed-potential type YSZ-based gas sensors using CeO_2-added Au electrodes. *ECS Sens. Plus* **2022**, *1*, 013604. [CrossRef]
23. Minami, K.; Zhou, Y.; Imamura, G.; Shiba, K.; Yoshikawa, G. Sorption kinetic parameters from nanomechanical sensing for discrimination of 2-nonenal from saturated aldehydes. *ACS Sens.* **2024**, *9*, 689–698. [CrossRef] [PubMed]
24. Schmitt, E.A.; Krott, M.; Epifani, M.; Suematsu, K.; Weimar, U.; Barsan, N. Volatile organic compound sensing with WO_3-based gas sensors: Surface chemistry basics. *J. Phys. Chem. C* **2024**, *128*, 1633–1643. [CrossRef]
25. Sasahara, T.; Kido, A.; Sunayama, T.; Uematsu, S.; Egashira, M. Identification and quantification of alcohol by a micro gas sensor based on adsorption and combustion. *Sens. Actuators B* **2004**, *99*, 532–538. [CrossRef]
26. Hyodo, T.; Shimizu, Y. Adsorption/combustion-type micro gas sensors: Typical VOC-sensing properties and material-design approach for highly sensitive and selective VOC detection. *Anal. Sci.* **2020**, *36*, 401–411. [CrossRef]
27. Yuzuriha, Y.; Hyodo, T.; Sasahara, T.; Shimizu, Y.; Egashira, M. Mesoporous Al_2O_3 co-loaded with Pd and Au as a combustion catalyst for adsorption/combustion-type gas sensors. *Sens. Lett.* **2011**, *9*, 409–413. [CrossRef]
28. Hyodo, T.; Yuzuriha, Y.; Nakgoe, O.; Sasahara, T.; Tanabe, S.; Shimizu, Y. Adsorption/combustion-type gas sensors employing mesoporous γ-alumina loaded with core(Au)/shell(Pd) nanoparticles synthesized reduction by sonochemical reduction. *Sens. Actuators B* **2014**, *202*, 748–757. [CrossRef]
29. Hyodo, T.; Hashimoto, T.; Ueda, T.; Nakgoe, O.; Sasahara, T.; Tanabe, S.; Shimizu, Y. Adsorption/combustion-type VOC sensors employing mesoporous γ-alumina co-loaded with noble-metal and oxide. *Sens. Actuators B* **2015**, *220*, 1091–1104. [CrossRef]

30. Hyodo, T.; Nagae, K.; Ueda, T.; Sasahara, T.; Shimizu, Y. Sensing behavior of adsorption/combustion-type gas microsensors to various alcoholic vapors. *Sens. Mater.* **2023**, *35*, 3851–3861. [CrossRef]
31. Hyodo, T.; Hiura, T.; Nagae, K.; Ueda, T.; Shimizu, Y. Effects of catalytic combustion behavior and adsorption/desorption properties on ethanol-sensing characteristics of adsorption/combustion-type gas sensors. *J. Asian Ceram. Soc.* **2021**, *9*, 1015–1030. [CrossRef]
32. Sing, K.S.W. Reporting physisorption data for gas/solid systems. *Pure Appl. Chem.* **1982**, *54*, 2201–2218. [CrossRef]
33. Brunauer, S.; Emmett, P.H.; Teller, E. Adsorption of gases in multimolecular layers. *J. Am. Chem. Soc.* **1938**, *60*, 309–319. [CrossRef]
34. Barrett, E.P.; Joyner, L.G.; Halenda, P.H. The determination of pore volume and area distributions in porous substances. I. Computations from nitrogen isotherms. *J. Am. Chem. Soc.* **1951**, *73*, 373–380. [CrossRef]

Disclaimer/Publisher's Note: The statements, opinions and data contained in all publications are solely those of the individual author(s) and contributor(s) and not of MDPI and/or the editor(s). MDPI and/or the editor(s) disclaim responsibility for any injury to people or property resulting from any ideas, methods, instructions or products referred to in the content.

chemosensors

MDPI

Article

SnO_2-Based CMOS-Integrated Gas Sensor Optimized by Mono-, Bi-, and Trimetallic Nanoparticles

Larissa Egger [1,*], Florentyna Sosada-Ludwikowska [1], Stephan Steinhauer [1,2], Vidyadhar Singh [3], Panagiotis Grammatikopoulos [4,5] and Anton Köck [1]

[1] Materials Center Leoben Forschung GmbH, 8700 Leoben, Austria
[2] Nanoparticles by Design Unit, Okinawa Institute of Science and Technology (OIST), Onna-son, Okinawa 904-0495, Japan
[3] Department of Physics, Jai Prakash University, Chapra 841301, Bihar, India
[4] Materials Science and Engineering, Guangdong Technion–Israel Institute of Technology, Shantou 515063, China
[5] Guangdong Provincial Key Laboratory of Materials and Technologies for Energy Conversion, Guangdong Technion–Israel Institute of Technology, Shantou 515063, China
* Correspondence: larissa.egger@mcl.at

Abstract: Chemical sensors, relying on electrical conductance changes in a gas-sensitive material due to the surrounding gas, have the (dis-)advantage of reacting with multiple target gases and humidity. In this work, we report CMOS-integrated SnO_2 thin film-based gas sensors, which are functionalized with mono-, bi-, and trimetallic nanoparticles (NPs) to optimize the sensor performance. The spray pyrolysis technology was used to deposit the metal oxide sensing layer on top of a CMOS-fabricated micro-hotplate (μhp), and magnetron sputtering inert-gas condensation was employed to functionalize the sensing layer with metallic NPs, Ag-, Pd-, and Ru-NPs, and all combinations thereof were used as catalysts to improve the sensor response to carbon monoxide and to suppress the cross-sensitivity toward humidity. The focus of this work is the detection of toxic carbon monoxide and a specific hydrocarbon mixture (HC_{mix}) in a concentration range of 5–50 ppm at different temperatures and humidity levels. The use of CMOS chips ensures low-power, integrated sensors, ready to apply in cell phones, watches, etc., for air quality-monitoring purposes.

Keywords: metal oxide; metallic nanoparticles; CMOS; tin dioxide

1. Introduction

Gas sensing has become a vital necessity for ensuring high air quality (AQ) indoors and outdoors. Since people in Europe spend 90% of their time indoors, environmental monitoring is of tremendous importance for private homes, offices, and vehicles. Poor indoor air quality has a great impact on human health (allergies, asthma, cancer, etc.) and also on employee productivity (15% reduction by low indoor air quality) [1]. Air pollution is considered a "silent killer", causing an estimated 7 million premature deaths every year [2]. Hence, it is crucial to monitor precisely and improve the quality of the air inside buildings. The State of the Art in the identification of gas species is lab-class analytical techniques, such as infrared spectroscopy, fluorescence spectroscopy, chemiluminescence, gas chromatography, or mass spectroscopy. While these lab-based technologies are very effective, all of these techniques are based on bulky and expensive tools, which are limited to stationary use and not suited for applications such as miniaturized handheld technologies, consumer electronic devices, or Internet-of-Things-based sensor networks [3].

In practice, air quality monitoring can be performed by gas sensor devices, which can be categorized into five main types according to their operational principle [4]: electrochemical (redox reactions on the electrodes), catalytic (burning of gas molecules), non-dispersive infrared (absorption of IR radiation), photo-ionization (gas ionization by UV-light), and solid-state (absorption/desorption of oxygen molecules from metal oxide (MO_x) surface) [5]. The last type is a conductometric MO_x-based gas sensor which has many advantages: it is highly sensitive to many dangerous gases, stable in an oxygen atmosphere, and has a long lifetime [6]. Over the past decades, various types of MO_x-based sensors have been developed [7–10]. Among those, sensors based on MO_x, such as SnO_2, ZnO, CuO, TiO_2, and WO_x [11,12], are the most promising candidates for inexpensive, compact implementations with industrial relevance [13–18]. Nevertheless, MO_x-based gas sensors have significant drawbacks, because they suffer from cross-sensitivity issues and require a comparatively high operation temperature, which results in high power consumption being necessary to operate the sensor [19]. These drawbacks have prevented their mass market penetration in consumer electronics, in particular, for ultralow-power applications (e.g., energy autonomous IoT devices) [20,21].

There are several approaches to overcome these disadvantages and improve the performance of gas sensor devices: Sensitivity can be improved by the implementation of nanomaterials, such as nanocrystalline thin films, nanowires, or NPs, to maximize the surface to volume ratio [22–24]. The use of specific MO_x nanostructures has demonstrated a general improvement in the gas-sensing performance. Selectivity can be improved by dedicated functionalization of MO_x nanostructures with metallic NPs [25–29], which provide a catalytic effect. Moreover, this kind of functionalization can lead to improved sensitivity, which enables a lower sensor operation temperature as compared to the basic device. High power consumption can be minimized by using micro-hotplate devices; integration of sensing materials on CMOS technology-fabricated micro-hotplates allows us to fabricate advanced gas-sensing devices with low power consumption and give us the potential to integrate them into everyday electronic devices [20,21,30].

In the literature, the enhanced influence of metallic nanoparticles on the sensing performance of MO_x-based sensors has already been widely investigated. However, a systematic study of different types of metallic nanoparticles and all their bi- and tri-metallic combinations has never been reported, especially not on CMOS-integrated chips. In this work, we demonstrate that the performance of CMOS-integrated SnO_2 thin film-based gas sensors can be significantly improved via functionalization with mono-, bi-, and trimetallic nanoparticles (NPs). Ag-, Pd-, and Ru-NPs and all combinations thereof have been deposited by magnetron sputtering inert-gas condensation on the SnO_2 films. This article focuses on the efficient detection of the toxic gas carbon monoxide (CO) in a humid atmosphere (50% relative humidity level) in a concentration regime between 5 ppm and 50 ppm, which is in the range of the maximum allowed concentration in the workplace in Germany (30 ppm [31]), and hence, it highly relevant for practical safety applications. In order to mimic cross-sensitivity toward Volatile Organic Compounds (VOCs), a specific mixture of hydrocarbon gases (HC_{mix}, mixture of acetylene, ethane, ethene, and propene, each of them 500 ppm) was investigated as well. We demonstrate that surface functionalization with mono-, bi-, and trimetallic NPs strongly improves the sensor in particular in the low operation temperature range between 100 °C and 200 °C, where the bare SnO_2 film exhibits no significant response. The idea is to find materials for optimizing the selectivity of a gas sensor system by increasing and decreasing the sensitivity of SnO_2 toward CO and HC_{mix}, respectively.

2. Materials and Methods

2.1. Fabrication of Sensor Devices

For this work, a CMOS-based sensor chip (ams OSRAM AG, Premstätten, Austria) with 8 micro-hotplates (μhps) [30] was used. Each μhp has two sensors, leading to a total of 16 sensors on one sensor chip. The μhps are based on a poly-Si heater and are etched through the CMOS stack down to the bulk Si. The sensors are contacted with Ti (5 nm)/Pt (150 nm) electrodes for a 4-point measurement.

The base sensing material is a SnO_2 thin film (thickness, 50 nm), which was deposited with spray pyrolysis technology at atmospheric pressure. The precursor solution was a tin chloride pentahydrate ($SnCl_4{\cdot}5H_2O$) in ethyl acetate. The solution was sprayed with a N_2 carrier gas on the chips, which were kept at 400 °C during the spraying process. The chips were annealed at 400 °C for 60 min for a more homogenous surface structure. The sensing films were structured by a photolithography process and dry Ar ion etching. Next, the films were coated with a resist protection layer before the μhps were under-etched by an isotropic XeF_2 dry-etching process. The finished chips were cleaned from the protective layer and further processed by stealth dicing into single dies (4.7 × 5.1 mm^2), which were glued and wire-bonded to printed circuit boards for the gas measurements. Figure 1a shows the final CMOS-integrated chip with 8x μhp array, and Figure 1b shows a single μhp with two SnO_2 sensor structures.

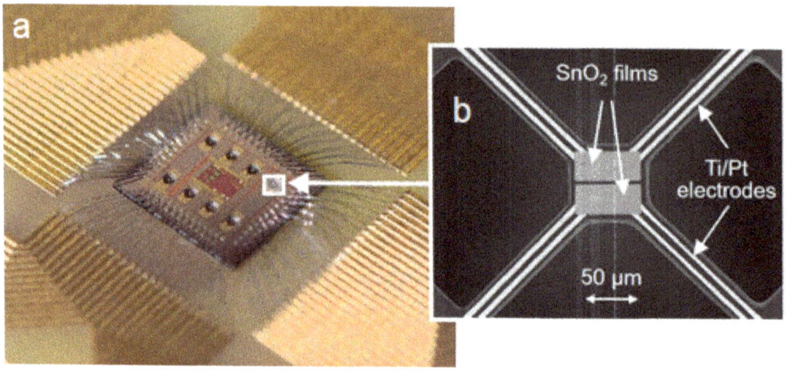

Figure 1. (**a**) Micro-hotplate chip and (**b**) single μhp.

The next step was the functionalization of the SnO_2-based chips with NPs; it was achieved by magnetron sputtering inert-gas condensation in a high-vacuum system using a cluster beam source, followed by an in-line quadrupole mass filter (QMF) for size selection [32]. For the generation of NPs, the NanoGen Trio deposition system (Mantis Deposition Ltd., Thame, UK), was employed, which features the possibilities of either sequential sputtering or co-sputtering in parallel up to three independent targets. Since the current study already contained a great number of variable parameters (metallic element, number of elements, detected gases, gas concentration range, humidity level, and temperature), we used sequential sputtering of the individual elements to keep it simple. The sputtering sequence should have no impact on the sensing performance, because the amount of each metal was controlled by the time of the individual sputtering steps. Hence, the analysis is focused on size distributions and coverages (NP number/area) for the multielement cases. High-resolution TEM images of the individual-element NPs are presented, which should be representative NP structures for all cases (no alloying, no segregation, and only minor surface coalescence at this coverage). The base pressure was ~10^{-8} mbar before deposition, whereas the pressures were ~10^{-4} mbar and ~10^{-1} mbar in the main chamber and the aggregation zone, respectively, during NP deposition. The argon-gas flow rate was set at 70 sccm, and the aggregation zone length was fixed at 120 mm. A 5 W power was applied to the Ag, Pd, and Ru targets during sputtering in order to produce mono-metallic (Ag, Ru, and Pd), mixed bi-metallic (AgRu, AgPd, and RuPd), and mixed tri-metallic (AgRuPd) NPs. The substrate was rotated at 2 rpm to obtain a more uniform coverage. NPs were characterized in terms of crystalline structure and size using a FEI Titan G2 Environmental TEM (operation voltage 300 kV):

- The monometallic Ru-NPs exhibit a monocrystalline hcp structure, and some degree of surface faceting is observed—see Figure 2a. The size distribution, as determined from 540 NPs, is in the range from 1 to 5 nm, with an average size of 2.5 nm—as shown in Figure 3d;
- The monometallic Ag-NPs exhibit an icosahedral-like multiple twin fcc structure, and some surface faceting is observed—see Figure 2b. The size distribution, as determined from 384 NPs, is in the range from 1 to 6 nm, with an average size of 3 nm, as shown in Figure 3e;

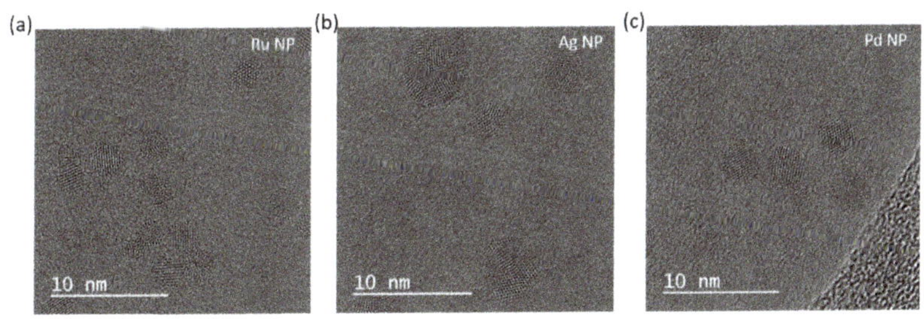

Figure 2. High-resolution TEM investigation of (**a**) Ru NP, (**b**) Ag NP, and (**c**) Pd NP.

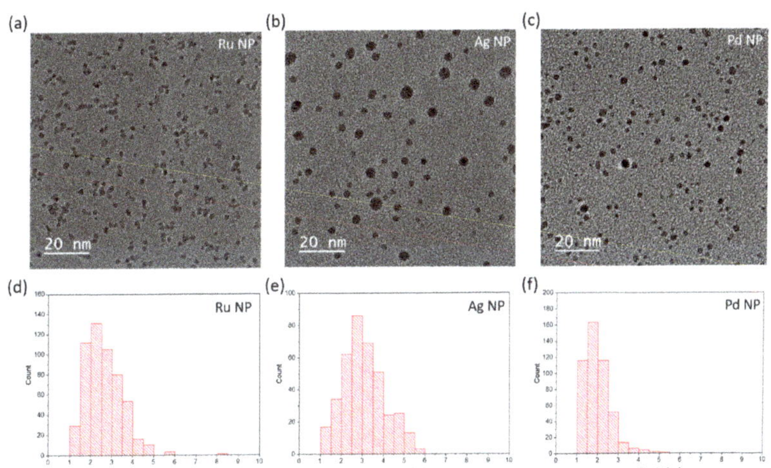

Figure 3. TEM investigation of (**a**) Ru NP, (**b**) Ag NP, and (**c**) Pd NP and the size distribution of (**d**) Ru NP, (**e**) Ag NP, and (**f**) Pd NP.

- The monometallic Pd-NPs exhibit a monocrystalline fcc structure, and some surface faceting is observed—see Figure 2c. The size distribution, as determined from 473 NPs, is in the range from 1 to 4 nm, with an average size of 2 nm, as shown in Figure 3f;
- The size distribution for bimetallic PdRu-NPs, as determined from 613 NPs, is in the range from 1 to 5 nm, with an average size of 2 nm, as shown in Figure 4d;
- The size distribution of bimetallic AgPd-NPs, as determined from 390 NPs, is in the range from 1 to 5 nm, with an average size of 3.5 nm, as shown in Figure 4e;
- The size distribution of bimetallic AgRu-NPs, as determined from 423 NPs, is in the range from 1 to 6 nm, with an average size of 2.5 nm, as shown in Figure 4f;

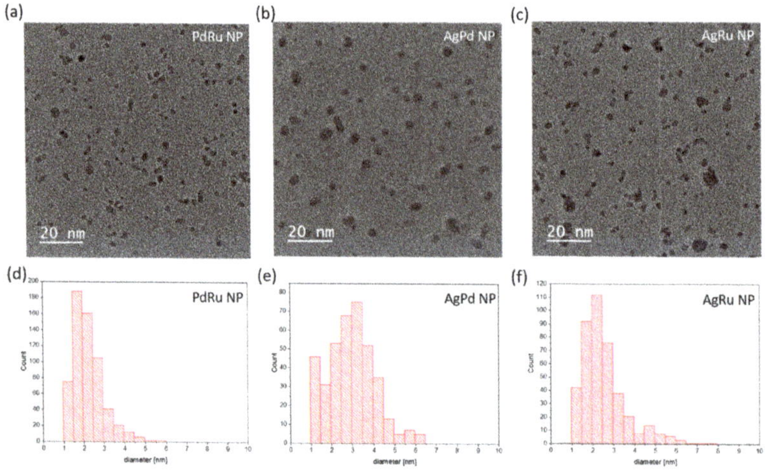

Figure 4. TEM investigation of (**a**) PdRu NP, (**b**) AgPd NP, and (**c**) AgRu NP and the size distribution of (**d**) PdRu NP, (**e**) AgPd NP, and (**f**) AgRu NP.

- The size distribution of trimetallic AgRuPd-NPs, as determined from 423 NPs, is in the range from 1 to 6 nm, with an average size of 2.5 nm, as shown in Figure 5b.

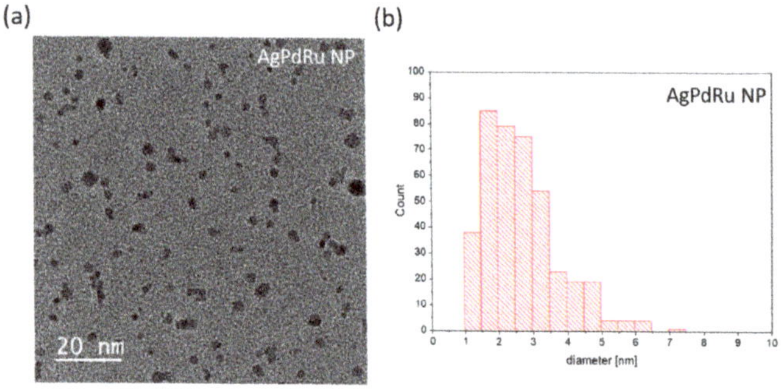

Figure 5. TEM investigation of (**a**) AgPdRu NP and the size distribution of (**b**) AgPdRu NP.

2.2. *Characterization of the CMOS-Based Sensors*

The performance of the CMOS-based sensors employing bare SnO_2 films and SnO_2 films functionalized with mono-, bi-, and trimetallic NPs was analyzed in an automated gas measurement setup. The sensing layers were not annealed at elevated temperatures (e.g., 400 °C) before the gas measurements to prevent potential migration and agglomeration or even melting of the NPs on the surface of the SnO_2 films [33]. Usually, we test our sensors also at higher temperatures up to 400 °C, because their response is usually maximum around 350 °C. However, many previous experiments have shown that the NPs can significantly change at temperatures above 150 °C in terms of morphology, agglomeration, and diffusion into the substrate, which has a negative impact on sensor response. Therefore, to ensure reproducibility and comparability between the different types of NPs, we strictly limited the operation temperature to a maximum of 200 °C. The gas chamber was a stainless-steel box with a volume of 80 cm^3 for gas flow and additional integrated electronics for operating the sensors. The operation temperature of the gas sensors was controlled by thermocouples, which are integrated into the CMOS stack of each µhp. The total gas flow was controlled by mass flow controllers and set to 1 l/min, and the humidity level was achieved by a water bubbler and feedback loop controlled by a commercial humidity sensor (AFK-E, KOBOLD Messing GmbH, Hofheim am Taunus, Germany). A python program was implemented to enable the automatic control of the gas measurement setup.

In this study, we focused on the toxic gas CO and the HC_{mix}. The HC_{mix}, consisting of four different gas components (acetylene, ethane, ethene, and propene), is a representative mixture of VOCs. Both test gases are indicators of air pollution. The main focus here is the influence of metallic nanoparticles on the sensitivity and selectivity toward CO and HC_{mix}. Synthetic air (80% nitrogen and 20% oxygen mixture) was used as background gas, whereas the CO and HC_{mix} test gases are mixtures in N_2. All gases are ready-to-use mixtures from Linde Gas Austria. Test gas concentrations were 5 ppm, 25 ppm, and 50 ppm. The sensor response was calculated as follows:

$$S = \frac{R_{air} - R_{gas}}{R_{air}} \quad (1)$$

where R_{gas} represents the resistance measured at the saturation point of the gas pulse, and R_{air} denotes the resistance in pure synthetic air. Consequently, the resistance trend in the background gas could be determined, and R_{air} was established based on this trend.

3. Results and Discussion

The different functionalized sensors are tested toward CO and HC_{mix} (mixture of acetylene, ethane, ethene, and propene, each of them 500 ppm). Our focus was on assessing the sensors' sensitivity toward the two test gases and exploring the potential for suppressing the cross-sensitivity to humidity through NP functionalization. The gas-sensing measurements were performed at a 200 °C operation temperature. Figure 6 shows a typical resistance measurement of a SnO_2 sensor at a 200 °C operation temperature and different humidity levels (25%, 50%, and 75%) for three different concentrations of CO (5 ppm, 25 ppm, and 50 ppm) gas pulses. The gas pulse duration was 5 min, and the recovery time after the gas pulse was set to 15 min. As is obvious from Figure 6, the resistance of the bare SnO_2 sensors is strongly influenced by the humidity of the background gas. The highest resistance is measured at the lowest (25%) humidity level. The resistance at 75% is at the same magnitude as for 50% humidity. The response of the bare SnO_2 sensors to CO significantly depends on the humidity, too (Figure 7). The highest response is observed at a 25% humidity level. The response decreases for 50% humidity, and it increases slightly for 75% humidity.

3.1. CO Measurements

In the initial phase of our investigation, we focused on the functionalization of the SnO_2 sensor using monometallic nanoparticles (Ag, Pd, and Ru). The responses of these sensors to varying concentrations of CO were examined under different humidity conditions, with the bare SnO_2 sensor serving as the reference. As illustrated in Figure 8, the response to CO increases with increasing CO concentration; however, the response does not show linear behavior. The sensitivity of sensors functionalized with monometallic NPs demonstrates a behavior closer to linearity compared to the bare sensor. Notably, both Pd and Ru monometallic NPs enhance the individual response to each CO concentration. The difference in the response between the functionalized and bare sensors becomes more pronounced with higher CO concentrations.

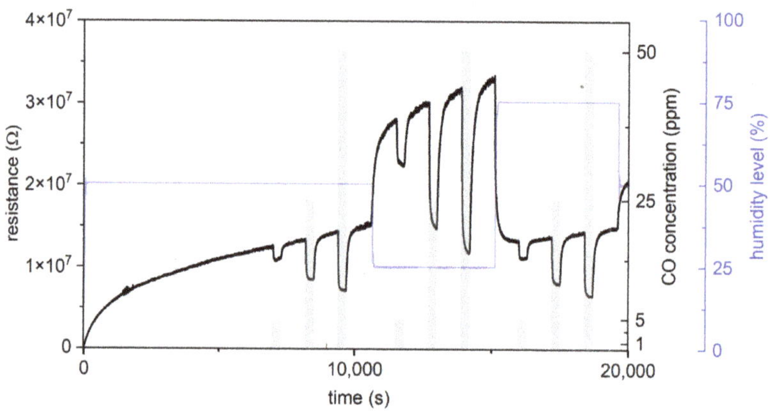

Figure 6. Resistance measurement of a bare SnO_2 sensor operated at 200 °C. The background gas has three different humidity levels: 50%, 25%, and 75%. At each humidity level, three CO gas pulses (duration 5 min) with 5 ppm, 25 ppm, and 50 ppm concentrations are introduced.

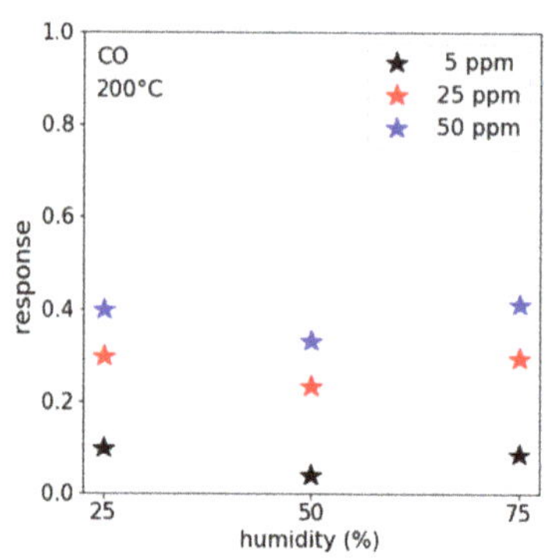

Figure 7. Response of a bare SnO_2 sensor at 200 °C operation temperature to three CO test gas pulses with 5 ppm (black), 25 ppm (red), and 50 ppm concentration (blue) at humidity levels of 25%, 50%, and 75%.

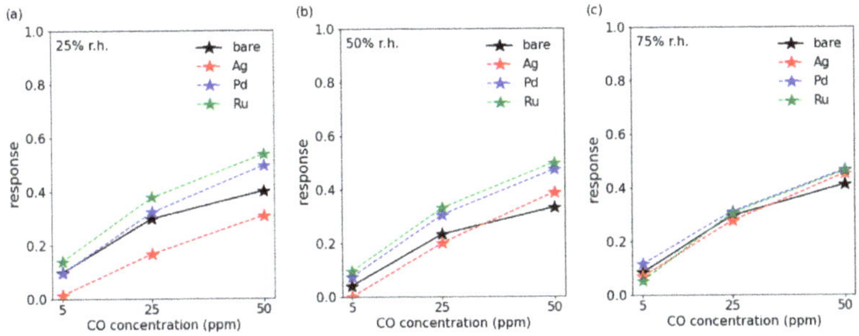

Figure 8. Response of a bare SnO_2 sensor (black) and SnO_2 functionalized with monometallic Ag (red), Pd (blue), and Ru (green) NPs to different CO concentrations (5 ppm, 25 ppm, and 50 ppm) at different humidity levels ((**a**) 25%, (**b**) 50%, and (**c**) 75%).

The response of the sensor functionalized with Ag NPs at 25% rh is lower than the response of the bare SnO_2 sensor but shows a linear increase with the CO concentration. Additionally, we observed that with increasing humidity, the response of the Ag-functionalized sensor converges with the response of bare SnO_2 sensors.

Figure 9 illustrates the responses to CO of sensors functionalized with bimetallic AgPd-, RuPd-, and AgRu-NPs. Notably, the sensors functionalized with AgPd- and AgRu-NPs show the same nonlinear behavior as the bare SnO_2 sensor. For 25% relative humidity, the RuPd-functionalized sensor shows a nearly linear behavior with increasing concentration of CO. For RuPd-NPs, the sensor response decreases with increasing humidity level. The AgPd-NP- and AgRu-NP-functionalized sensors exhibit a significantly reduced dependence on the humidity level, and the sensor response is almost independent of the humidity.

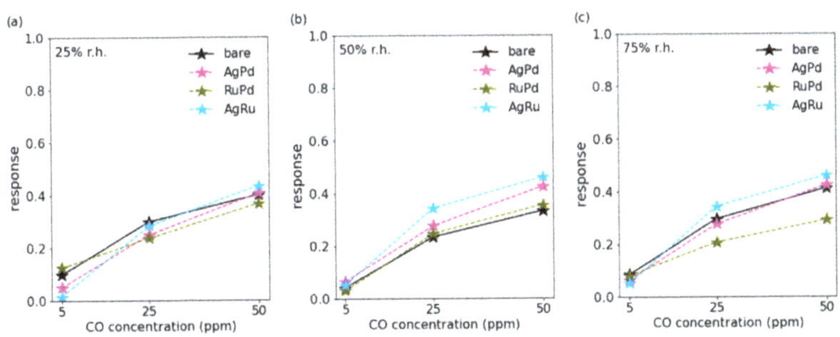

Figure 9. Response of a bare SnO_2 sensor (black) and SnO_2-functionalized bimetallic AgPd, RuPd, and AgRu NPs to different CO concentrations (5 ppm, 25 ppm, and 50 ppm) at different humidity levels ((**a**) 25%, (**b**) 50%, and (**c**) 75%).

Figure 10 compares the best sensors from the monometallic (Ru), bimetallic (AgRu), and trimetallic (PdAgRu) NP-functionalized sensors. The sensitivity of the trimetallic NP shows a pronounced increase in the response already at low concentrations. The sensor functionalized with trimetallic NPs shows a nonlinear increase in the response as the concentration increases, representing the most notable enhancement observed in these sensors.

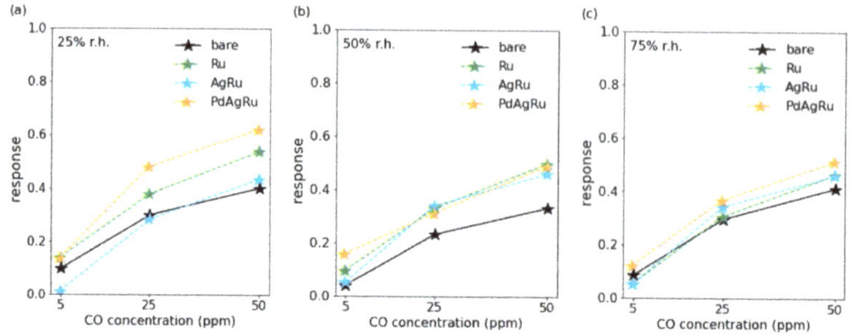

Figure 10. Response of a bare SnO_2 sensor (black) and SnO_2 functionalized with Ru, AgRu, and PdAgRu NPs to different CO concentrations (5 ppm, 25 ppm, and 50 ppm) at different humidity levels ((**a**) 25%, (**b**) 50%, and (**c**) 75%).

In Figure 11, we focus on the cross-sensitivity to humidity of the bare SnO_2 and the functionalized sensors for 50 ppm CO exposure. The bare SnO_2 sensor shows consistent response at 25% and 75%, but a significantly lower response at 50% rh. In contrast, the sensors functionalized with monometallic NPs show a distinct linear dependence on the humidity. For Pd- and Ru-NPs, the responses decrease, and for Ag-NPs, the response increases with increasing humidity. The response of AgRu- and AgPd-NP-functionalized sensors is almost independent of the humidity; for RuPd-NPs, the response decreases with increasing humidity, and this observation is more pronounced at high humidity levels. The PdAgRu-NP-functionalized sensor exhibits the highest response of all sensors, as it decreases with increasing humidity, a response more pronounced for low humidity levels, and reaches saturation around 50%. The Ag-NP-functionalized sensor is unique, as its response increases with rising humidity. Most promising are the AgPd- and AgRu-NP-functionalized sensors, since the responses of these sensors appear nearly independent of the humidity level.

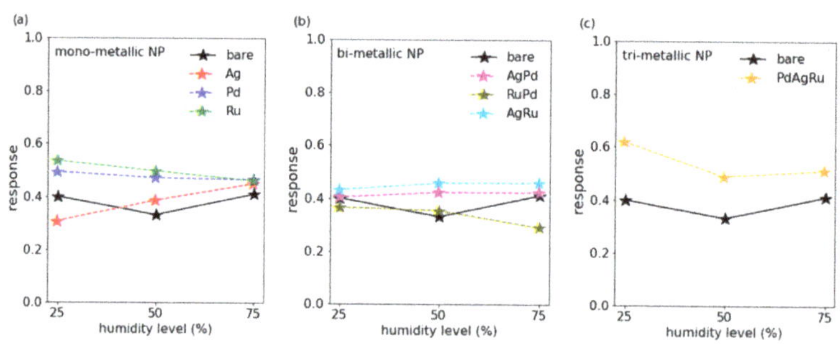

Figure 11. Response of a bare SnO_2 sensor (black) and SnO_2 functionalized with (**a**) monometallic, (**b**) bimetallic, and (**c**) trimetallic NP-functionalized sensors to 50 ppm CO concentrations at different humidity levels (25%, 50%, and 75%).

3.2. HC_{mix} Measurements

In the latter part of the work, our investigation focuses on the response of the same sensors to HC_{mix}.

As depicted in Figure 12, both the bare SnO_2 and all monometallic-functionalized sensors exhibited a clear increase in response correlating with the concentration of the HC_{mix}. Notably, sensors with Ag- and Pd-NPs demonstrated a stronger increase in the response within the low concentration range up to 25 ppm, plateauing toward 50 ppm. This observation suggests saturation, which is seen across all three humidity levels. Conversely, the behavior of the Ru-functionalized sensors is close to one of the bare sensors.

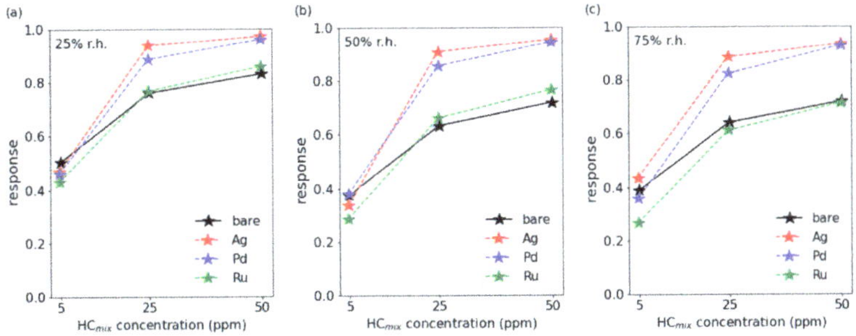

Figure 12. Response of a bare SnO_2 sensor (black) and SnO_2 functionalized with monometallic NP to different HC_{mix} concentrations (5 ppm, 25 ppm, and 50 ppm) at different humidity levels ((**a**) 25%, (**b**) 50%, and (**c**) 75%).

In Figure 13, the performance of bimetallic NP-functionalized sensors against HC_{mix} concentration is shown. The responses of the bimetallic NP-functionalized sensors show the same behavior as the monometallic ones. However, the response increase is more pronounced with increasing concentration compared with that of the bare SnO_2 sensor. In contrast to the monometallic ones, the individual responses are lower than the response of the bare sensor at low humidity levels. With increasing humidity levels, AgPd NP-functionalized sensors show a higher response than the bare sensor. The responses of the other bimetallic NPs increase, but they are still lower than for the bare SnO_2 sensor.

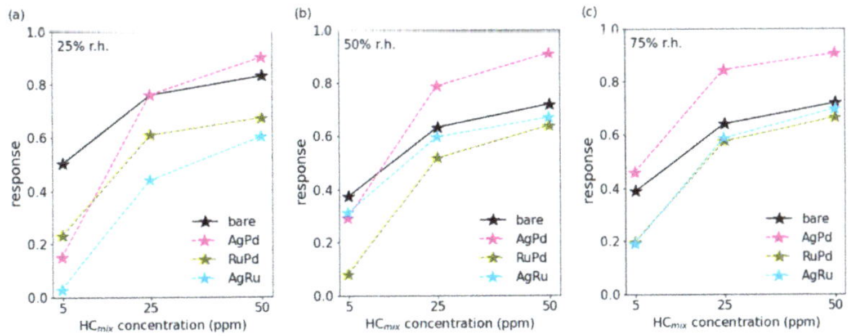

Figure 13. Response of a bare SnO_2 sensor (black) and SnO_2 functionalized with bimetallic NPs to different HC_{mix} concentrations (5 ppm, 25 ppm, and 50 ppm) at different humidity levels ((**a**) 25%, (**b**) 50%, and (**c**) 75%).

In Figure 14, we compare the best performing sensors from monometallic (Ag) and bimetallic (AgPd) with trimetallic (PdAgRu) NP-functionalized sensors across all three tested humidity levels. Both mono- and bimetallic sensors show a strong increase in response with concentration compared to the bare sensor and higher responses than the bare sensor. Conversely, the trimetallic NPs display a consistent dependence on concentration, even though they show slightly higher individual responses compared to the bare sensor.

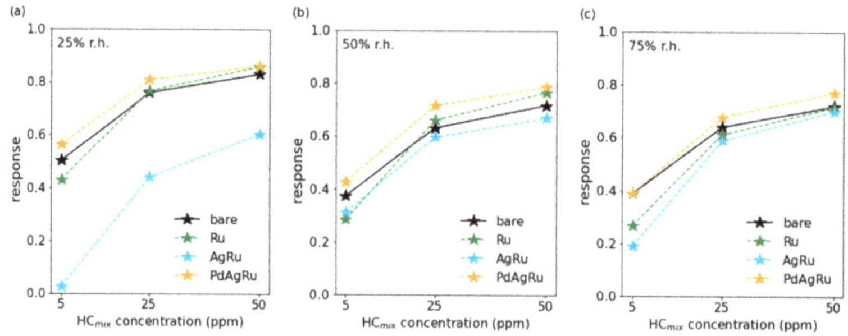

Figure 14. Response of a bare SnO_2 sensor (black) and SnO_2 functionalized with Ru, AgRu, and PdAgRu NPs to different HC_{mix} concentrations (5 ppm, 25 ppm, and 50 ppm) at different humidity levels ((**a**) 25%, (**b**) 50%, and (**c**) 75%).

Not only is the response to HC_{mix} crucial for an optimized sensor, but also for ensuring minimal cross-sensitivity to humidity. In Figure 15, we illustrate the responses of optimized sensors for 50 ppm HC_{mix} across three different humidity levels. The response of the bare sensor decreases from 25% to 50% humidity before it reaches saturation. The sensors with monometallic NPs display a decrease in response with increasing humidity, with slight decreases observed for Ag and Pd, and a more pronounced one for Ru. Sensors with RuPd-NPs show similar responses at 25% and 75% humidity levels, with a smaller response at 50%. Meanwhile, the response of the AgRu-NP-functionalized sensor increases with rising humidity, while the sensor with AgPd-NPs appears unaffected by humidity. Trimetallic-functionalized sensors have a response pattern similar to the bare sensor, indicating similar dependence on humidity.

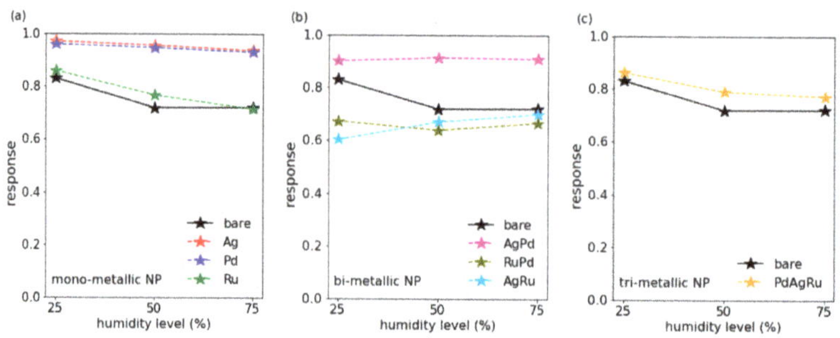

Figure 15. Response of a bare SnO_2 sensor (black) and SnO_2 functionalized with (**a**) monometallic, (**b**) bimetallic, and (**c**) trimetallic NPs to 50 ppm HC_{mix} concentrations at different humidity levels (25%, 50%, and 75%).

3.3. Discussion

The functionalization of sensors aimed at not only enhancing sensitivity toward specific test gases but also distinguishing between them. In Figure 16, we compare the performance improvement of the most promising NP-functionalized sensors with respect to the performance of the bare ones for both CO and HC_{mix} at different humidity levels. The Ag-functionalized sensor exhibits a significant enhancement for HC_{mix} across all three humidity levels (for 25% rh: increase about 20%; and for 50% rh and 75% rh: increase about 30%). On the other hand, for CO detection, Ag NPs reduce the response at a lower humidity level by about 20% but enhance it with higher humidity by about 20% (50%

rh) and 10% (75% rh). Pd NPs amplify the response to both test gases, with Ru more prominently increasing the response to CO.

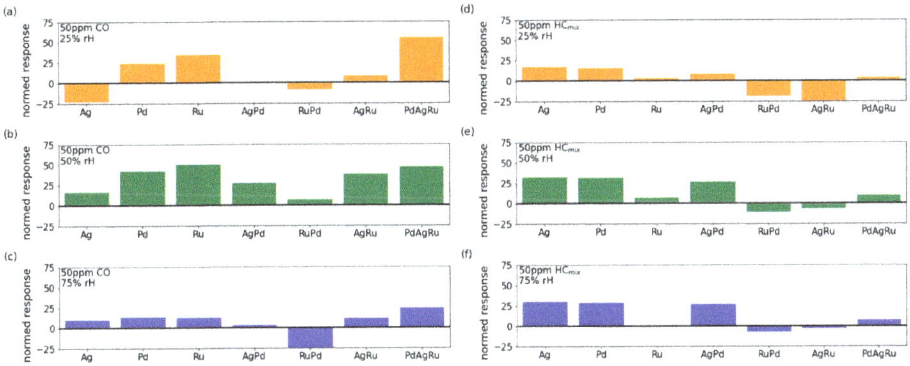

Figure 16. Relative change in the responses of the functionalized sensors compared to the bare SnO_2 for (**a–c**) 50 ppm CO and (**d–f**) 50 ppm HC_{mix} concentrations at different humidity levels ((**a**,**d**) 25%, (**b**,**e**) 50%, and (**c**,**f**) 75%).

Among the bimetallic NPs, those with Ag show the most substantial improvement for CO and HC_{mix}, respectively. The AgRu-functionalized sensors exhibit increased sensitivity to CO but markedly reduced sensitivity to HC_{mix} compared to the bare sensor. For the AgPd-functionalized sensor, the improvement for CO and HC_{mix} compared to the bare sensor is very similar. Trimetallic NPs enhance the response to CO, while the response to HC_{mix} remains close to that of the bare sensor.

In our case, the Ag-NPs showed the most promising results regarding distinguishing CO and HC_{mix}. The Ag NP-functionalized sensors significantly increase the response to the HC_{mix}, but they have only a small impact on the CO response. A similar behavior was observed for an acetylene (part of our HC_{mix}) sensor, which is based on Ag-loaded ZnO-reduced graphene; the sensor exhibited enhanced sensing toward C_2H_2 and decreased sensing toward CO at a 200 °C operating temperature. This behavior has been attributed to potential barriers formed by the chemisorbed oxygen ions on the ZnO surface, which prevents the C_2H_2 molecules from reacting. Ag catalytically activates the dissociation of molecular oxygen. Due to the spillover effect, more acetylene-sensing sites on the surface are created, consequently enhancing the sensor's response. A similar behavior was reported for an acetylene sensor fabricated on a µhp chip using In_2O_3 as the sensing material with Pd–Ag core–shell NPs on top as catalysts, where the Ag triggers an increased response to acetylene. We assume that the Ag plays the same role in the case of our SnO_2-based sensors and increases the sensor response to at least one of the four HC_{mix} components. It is plausible that a similar mechanism also increases the response toward the other three HC_{mix} components.

4. Conclusions

By combining different NP-functionalized sensors with a humidity sensor into a multisensor device, one can identify and detect concentrations ranging from 5 up to 50 ppm for both test gases, CO and HC_{mix}. Regarding sensitivity to humidity, some promising candidates, like Pd and AgPd, demonstrate low cross-sensitivity to humidity while detecting both test gases. In contrast, the Ag- and AgRu-functionalized samples exhibit different responses to humidity and the two test gases. Therefore, this behavior can be used to identify the observed test gas by comparing the response of the different sensor materials. Hence, in the next step, we will functionalize the 8 μhps of the CMOS device with different NPs, which can be achieved by using a shadow mask. Comparing the differently functionalized sensor arrays will allow for a clear discrimination of the target gases with a single micro-chip. Furthermore, sensor tests toward different other polluting target gases (e.g., NO, NO_2, formaldehyde, acetone, toluene, CO_2, and NH_3) and a detailed long-term stability study of the sensor behavior are planned for the future.

Author Contributions: Conceptualization, S.S. and A.K.; methodology, S.S., P.G., and V.S.; validation, F.S.-L. and L.E.; formal analysis, L.E.; investigation, S.S., P.G., and V.S.; writing—original draft preparation, L.E.; writing—review and editing, S.S., P.G., V.S., F.S.-L., and A.K. All authors have read and agreed to the published version of the manuscript.

Funding: This research was partially funded by ECSEL Joint Undertaking (JU), grant number 876362; and by FFG Austrian Research Promotion Agency, grant number FO999899024.

Informed Consent Statement: Not applicable.

Data Availability Statement: The data presented in this study are available upon request from the corresponding author.

Conflicts of Interest: Authors Larissa Egger, Florentyna Sosada-Ludwikowska, Stephan Steinhauer and Anton Köck were employed by the company Materials Center Leoben Forschung GmbH, The remaining authors declare that the research was conducted in the absence of any commercial or financial relationships that could be construed as a potential conflict of interest.

References

1. González-Martín, J.; Kraakman, N.J.R.; Pérez, C.; Lebrero, R.; Muñoz, R. A State–of–the-Art Review on Indoor Air Pollution and Strategies for Indoor Air Pollution Control. *Chemosphere* **2021**, *262*, 128376. [CrossRef]
2. World Health Organization. *World Health Organization (WHO) Global Air Quality Guidelines: Particulate Matter (PM2.5 and PM10), Ozone, Nitrogen Dioxide, Sulfur Dioxide and Carbon Monoxide*; World Health Organization: Geneva, Switzerland, 2021.
3. Lara-Ibeas, I.; Rodríguez Cuevas, A.; Le Calvé, S. Recent Developments and Trends in Miniaturized Gas Preconcentrators for Portable Gas Chromatography Systems: A Review. *Sens. Actuators B Chem.* **2021**, *346*, 130449. [CrossRef]
4. Nikolic, M.V.; Milovanovic, V.; Vasiljevic, Z.Z.; Stamenkovic, Z. Semiconductor Gas Sensors: Materials, Technology, Design, and Application. *Sensors* **2020**, *20*, 6694. [CrossRef] [PubMed]
5. Yi, W.Y.; Lo, K.M.; Mak, T.; Leung, K.S.; Leung, Y.; Meng, M.L. A Survey of Wireless Sensor Network Based Air Pollution Monitoring Systems. *Sensors* **2015**, *15*, 31392–31427. [CrossRef]
6. Pathania, A.; Dhanda, N.; Verma, R.; Sun, A.-C.A.; Thakur, P.; Thakur, A. Metal Oxide Chemoresistive Gas Sensing Mechanism, Parameters, and Applications. *ECS Sens. Plus* **2024**, *3*, 13401. [CrossRef]
7. Degler, D.; Weimar, U.; Barsan, N. Current Understanding of the Fundamental Mechanisms of Doped and Loaded Semiconducting Metal-Oxide-Based Gas Sensing Materials. *ACS Sens.* **2019**, *4*, 2228–2249. [CrossRef]
8. Schröder, S.; Ababii, N.; Brînză, M.; Magariu, N.; Zimoch, L.; Bodduluri, M.T.; Strunskus, T.; Adelung, R.; Faupel, F.; Lupan, O. Tuning the Selectivity of Metal Oxide Gas Sensors with Vapor Phase Deposited Ultrathin Polymer Thin Films. *Polymers* **2023**, *15*, 524. [CrossRef] [PubMed]
9. Li, T.; Yin, W.; Gao, S.; Sun, Y.; Xu, P.; Wu, S.; Kong, H.; Yang, G.; Wei, G. The Combination of Two-Dimensional Nanomaterials with Metal Oxide Nanoparticles for Gas Sensors: A Review. *Nanomaterials* **2022**, *12*, 982. [CrossRef] [PubMed]
10. Dadkhah, M.; Tulliani, J.-M. Green Synthesis of Metal Oxides Semiconductors for Gas Sensing Applications. *Sensors* **2022**, *22*, 4669. [CrossRef] [PubMed]
11. Korotcenkov, G. Metal Oxides for Solid-State Gas Sensors: What Determines Our Choice? *Mater. Sci. Eng. B* **2007**, *139*, 1–23. [CrossRef]
12. Xavier, R.; Sivaperuman, K. Review on the of Physical Vapor Deposition on Imminent Chemiresistive Metal Oxide Gas Sensors and Their Future Scope. *Mater. Today Commun.* **2024**, *38*, 107831. [CrossRef]
13. Zhang, J.; Liu, X.; Neri, G.; Pinna, N. Nanostructured Materials for Room-Temperature Gas Sensors. *Adv. Mater.* **2016**, *28*, 795–831. [CrossRef] [PubMed]

14. Zhang, D.; Yang, Z.; Yu, S.; Mi, Q.; Pan, Q. Diversiform Metal Oxide-Based Hybrid Nanostructures for Gas Sensing with Versatile Prospects. *Coord. Chem. Rev.* **2020**, *413*, 213272. [CrossRef]
15. Korotcenkov, G.; Cho, B.K. Engineering Approaches for the Improvement of Conductometric Gas Sensor Parameters: Part 1. Improvement of Sensor Sensitivity and Selectivity (Short Survey). *Sens. Actuators B Chem.* **2013**, *188*, 709–728. [CrossRef]
16. Kolmakov, A. Some Recent Trends in the Fabrication, Functionalisation and Characterisation of Metal Oxide Nanowire Gas Sensors. *Int. J. Nanotechnol.* **2008**, *5*, 450–474. [CrossRef]
17. Kim, K.W.; Cho, P.S.; Kim, S.J.; Lee, J.H.; Kang, C.Y.; Kim, J.S.; Yoon, S.J. The Selective Detection of C_2H_5OH Using SnO_2-ZnO Thin Film Gas Sensors Prepared by Combinatorial Solution Deposition. *Sens. Actuators B Chem.* **2007**, *123*, 318–324. [CrossRef]
18. Sharma, S.; Madou, M. Review Article: A New Approach to Gas Sensing with Nanotechnology. *Philos. Trans. R. Soc. A Math. Phys. Eng. Sci.* **2012**, *370*, 2448–2473. [CrossRef]
19. Liu, H.; Zhang, L.; Li, K.H.H.; Tan, O.K. Microhotplates for Metal Oxide Semiconductor Gas Sensor Applications—Towards the CMOS-MEMS Monolithic Approach. *Micromachines* **2018**, *9*, 557. [CrossRef] [PubMed]
20. Köck, A.; Wimmer-Teubenbacher, R.; Sosada-Ludwikovska, F.; Rohracher, K.; Wachmann, E.; Herold, M.; Welden, T.V.; Kim, J.M.; Ali, Z.; Poenninger, A.; et al. 3D-Integrated Multi-Sensor Demonstrator System for Environmental Monitoring. In Proceedings of the 2019 20th International Conference on Solid-State Sensors, Actuators and Microsystems & Eurosensors XXXIII (Transducers & Eurosensors XXXIII), Berlin, Germany, 23–27 June 2019; pp. 1136–1139.
21. Mallires, K.R.; Wang, D.; Tipparaju, V.V.; Tao, N. Developing a Low-Cost Wearable Personal Exposure Monitor for Studying Respiratory Diseases Using Metal--Oxide Sensors. *IEEE Sens. J.* **2019**, *19*, 8252–8261. [CrossRef] [PubMed]
22. Bhati, V.S.; Hojamberdiev, M.; Kumar, M. Enhanced Sensing Performance of ZnO Nanostructures-Based Gas Sensors: A Review. *Energy Rep.* **2020**, *6*, 46–62. [CrossRef]
23. Ahemad, M.J.; Le, T.D.; Kim, D.-S.; Yu, Y.-T. Bimetallic AgAualloy@ ZnO Core-Shell Nanoparticles for Ultra-High Detection of Ethanol: Potential Impact of Alloy Composition on Sensing Performance. *Sens. Actuators B Chem.* **2022**, *359*, 131595. [CrossRef]
24. Isaac, N.A.; Pikaar, I.; Biskos, G. Metal Oxide Semiconducting Nanomaterials for Air Quality Gas Sensors: Operating Principles, Performance, and Synthesis Techniques. *Microchim. Acta* **2022**, *189*, 196. [CrossRef] [PubMed]
25. Zhu, L.Y.; Ou, L.X.; Mao, L.W.; Wu, X.Y.; Liu, Y.P.; Lu, H.L. *Advances in Noble Metal-Decorated Metal Oxide Nanomaterials for Chemiresistive Gas Sensors: Overview*; Springer Nature: Singapore, 2023; Volume 15, ISBN 0123456789.
26. Ren, Y.; Xie, W.; Li, Y.; Ma, J.; Li, J.; Liu, Y.; Zou, Y.; Deng, Y. Noble Metal Nanoparticles Decorated Metal Oxide Semiconducting Nanowire Arrays Interwoven into 3D Mesoporous Superstructures for Low-Temperature Gas Sensing. *ACS Cent. Sci.* **2021**, *7*, 1885–1897. [CrossRef] [PubMed]
27. Zhang, S.; Zhang, M.; Guo, Y. Highly Sensitive Ethanol Gas Sensor Based on Ag Nanoparticles Decorated In_2O_3. *Chem. Res. Chin. Univ.* **2024**, *40*, 1033–1040. [CrossRef]
28. Sosada-Ludwikowska, F.; Reiner, L.; Egger, L.; Lackner, E.; Krainer, J.; Wimmer-Teubenbacher, R.; Singh, V.; Steinhauer, S.; Grammatikopoulos, P.; Köck, A. Adjusting Surface Coverage of Pt Nanocatalyst Decoration for Selectivity Control in CMOS-Integrated SnO_2 Thin Film Gas Sensors. *Nanoscale Adv.* **2024**, *6*, 1127–1134. [CrossRef] [PubMed]
29. Wawrzyniak, J. Advancements in Improving Selectivity of Metal Oxide Semiconductor Gas Sensors Opening New Perspectives for Their Application in Food Industry. *Sensors* **2023**, *23*, 9548. [CrossRef] [PubMed]
30. Lackner, E. *Optimization of CMOS Integrated Tin Oxide Gas Sensors Using Metallic and Bimetallic Nanoparticles*; TU Graz: Styria, Austria, 2017.
31. Deutsche Forschungsgemeinschaft. *List of MAK and BAT Values 2022 (Maximum Concentrations and Biological Tolerance Values at the Workplace)*; Deutsche Forschungsgemeinschaft: Bonn, Germany, 2022; Report 58.
32. Grammatikopoulos, P.; Steinhauer, S.; Vernieres, J.; Singh, V.; Sowwan, M. Nanoparticle Design by Gas-Phase Synthesis. *Adv. Phys. X* **2016**, *1*, 81–100. [CrossRef]
33. Aguado, A.; Jarrold, M.F. Melting and Freezing of Metal Clusters. *Annu. Rev. Phys. Chem.* **2011**, *62*, 151–172. [CrossRef]

Disclaimer/Publisher's Note: The statements, opinions and data contained in all publications are solely those of the individual author(s) and contributor(s) and not of MDPI and/or the editor(s). MDPI and/or the editor(s) disclaim responsibility for any injury to people or property resulting from any ideas, methods, instructions or products referred to in the content.

chemosensors

MDPI

Review

The Screening and Diagnosis Technologies Towards Pneumoconiosis: From Imaging Analysis to E-Noses

Yuqian Zhang [1], Wufan Xuan [2,*], Shuai Chen [1,3,*], Mingna Yang [1] and Huakun Xing [4]

1 Jiangxi Provincial Engineering Research Center for Waterborne Coatings, School of Chemistry and Chemical Engineering, Jiangxi Science & Technology Normal University, Nanchang 330013, China; zhangyuqian202412@163.com (Y.Z.); yangmingna1023@163.com (M.Y.)
2 School of Safety Science, Tsinghua University, Beijing 100084, China
3 Jiangxi Provincial Key Laboratory of Flexible Electronics, Nanchang 330013, China
4 Institute of Energy Materials and Nanotechnology, School of Civil Engineering and Architecture, Nanchang Jiaotong Institute, Nanchang 330100, China; xhk1235@126.com
* Correspondence: xuanwf@mail.tsinghua.edu.cn (W.X.); shuaichen@jxstnu.edu.cn (S.C.); Tel./Fax: +86-791-83832345 (S.C.)

Abstract: Pneumoconiosis, as the most widely distributed occupational disease globally, poses serious health and social hazards. Its diagnostic techniques have evolved from conventional imaging and computer-assisted analysis to emerging sensor strategies covering biomarker analysis, routine breath sensing, integrated electronic nose (E-nose), etc. All of them both have special advantages and face shortcomings or challenges in practical application. In recent years, the emergence of advanced data analysis technologies, including artificial intelligence (AI), has provided opportunities for large-scale screening of pneumoconiosis. On the basis of a deep analysis of the characteristics of the technologies for screening and diagnosis of pneumoconiosis, this paper comprehensively and systematically reviews the current development of these technologies, especially focusing on the research progress of emerging sensor technologies, and provides a forecast for their future development.

Keywords: occupational disease; pneumoconiosis; diagnostic technique; early screening; imaging; breath analysis; E-nose

1. Introduction

As one of the most common occupational diseases worldwide, pneumoconiosis has an extremely wide range of occupational exposures. In various types of dust-filled workplaces, workers may suffer from more than 10 types of pneumoconiosis, including coal workers' pneumoconiosis and silicosis. It is a chronic preventable but currently incurable interstitial occupational lung disease that can lead to serious and long-term health and social problems [1–3]. The situation is particularly concerning because the disease can continue to develop even after exposure to dust has stopped. Moreover, it can trigger or worsen other comorbidities or complications, particularly chronic obstructive pulmonary disease (COPD) and tuberculosis (TB) [4,5]. Regarding COPD, individuals with pneumoconiosis are at higher risk of developing this condition compared to the general population due to occupational exposure and subsequent recurrent respiratory infections. The lung damage caused by pneumoconiosis further impairs the normal structure and function of airways and alveoli, exacerbating the characteristic airflow limitation in COPD. The coexistence of pneumoconiosis and COPD often leads to more severe respiratory symptoms, accelerated decline in lung function, and increased risk of respiratory failure. TB represents another significant comorbidity because the immunological changes in the lungs induced by pneumoconiosis increase susceptibility to Mycobacterium TB infection. The compromised lung tissue and reduced immunity create a favorable environment for the growth and spread of TB bacilli. This coexistence creates a vicious cycle: pneumoconiosis increases the risk of TB infection, while TB exacerbates the existing lung damage caused by pneumoconiosis.

In addition, as the middle-aged and elderly male labor force is the main affected group, and there is a trend of younger onset, the burden and harm to families and society are even more severe. The currently widely accepted pathogenesis is mainly that after the inhalable dust enters the lungs and is phagocytosed by macrophages, it cannot be expelled and digested, leading to retention and causing lesions mainly characterized by inflammation and diffuse fibrosis of lung tissue [6]. Due to the high morbidity and disability rate, and the lack of effective treatment and cure methods, as well as the diverse and complex manifestations of the lesions, early screening and diagnosis have long been the focus and difficulty of occupational disease research and are also the key to early detection and complete control of the patient's condition. The World Health Organization (WHO) and the International Labour Organization (ILO) jointly established the Joint Commission on Occupational Health. In 1995, this commission launched the "International Plan for the Global Elimination of Silicosis". The aim of this plan was to significantly reduce the incidence rate of pneumoconiosis by 2010 and completely eliminate pneumoconiosis by 2030 [7]. However, the achievement of this ultimate goal may now need to be postponed considerably, especially given the different stages of development in different countries.

As an occupational disease, pneumoconiosis is diagnosed based on both medical and policy considerations, making it an attribution diagnosis. Over the past several decades, considerable advancements have been achieved in this area, but many challenges persist. Previously, there have been some reviews focusing on the progress of imaging technologies used in this field [8–10]. Unlike them, this paper offers a systematic evaluation of the development situation of existing technologies from traditional imaging to pulmonary function tests (PFTs), tissue invasion, etc., as well as of advanced sensing technologies for pneumoconiosis markers (from biomarking to routine breath sensing and integrated E-noses) (Figure 1). Lastly, this paper provides a forward-looking perspective, including the optimization direction of existing technologies, the great value of artificial intelligence (AI)-assisted analysis and diagnosis, etc.

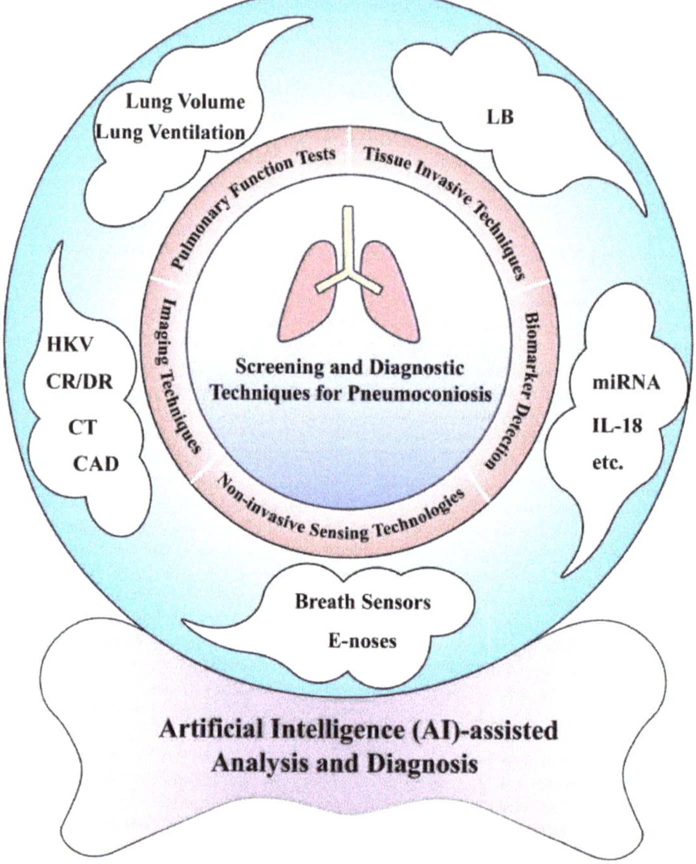

Figure 1. Screening and diagnostic techniques for pneumoconiosis include imaging techniques (high-kV (HKV) X-ray imaging, computed radiography (CR), direct digital radiography (DR), computed tomography (CT), computer-aided diagnosis (CAD)), PFT, tissue-invasive techniques (lung biopsy (LB)), biomarker detection (miRNA, IL-18, etc.), and non-invasive sensing technologies (breath sensors and E-noses).

2. Established Techniques for Screening and Diagnosis of Pneumoconiosis

2.1. Imaging Techniques

2.1.1. HKV X-Ray Imaging

Chest X-ray films utilize the different absorption amounts of X-rays during the process of penetrating the human body, forming grayscale images of lung tissues due to differences in density [11,12]. Then, indicators such as the profusion of small opacities and the distribution in lung regions are used to indirectly display pathological differences. Conventional low-kilovolt chest X-ray films adopt a tube voltage of 60 to 70 kilovolts (kV). Although they can be used for diagnosis, they have relatively low sensitivity to the early manifestations of pneumoconiosis. There are deficiencies, such as the overlap of anterior and posterior chest images, low image density resolution, and unclear display of the internal structures of small and large opacities, which easily lead to misdiagnosis and missed diagnosis [13]. However, HKV chest X-ray films adopt a tube voltage of $\geq$120 kV with an exposure of 5 mAs to 7 mAs. The load on the X-ray tube is low. As the absorption coefficient of substances for X-rays decreases with the increase in tube voltage, the exposure time is significantly shortened [14]. This results in advantages, such as a reduced radiation dose for the examinees, rich image layers, high image quality, and higher resolution for small opacities and fine tissues. Currently, the chest X-ray of HKV is an authoritative diagnostic method in the national diagnostic standards for pneumoconiosis in China (such as GBZ 70-2015). While HKV chest X-rays optimize lung visualization by reducing bone artifacts and enhancing penetration through thick tissues, the image contrast can be influenced by several factors. These include technical parameters (such as tube current and exposure time), patient-specific conditions (e.g., body habitus), equipment performance, and post-processing techniques [15,16]. These variables may alter the final image quality and contrast, potentially affecting diagnostic accuracy. However, in the differentiation of inflammatory masses, pulmonary tuberculosis, lung abscesses, lung cancer, and so on, it still has an advantage over chest CT [8,17,18].

2.1.2. CR and DR

Compared with HKV, both CR and DR are digital imaging techniques, which combine computer digital image processing technology with traditional X-ray imaging technology and have many similarities and differences in image quality and post-processing [19–22]. CR is an indirect digital imaging method. It requires an imaging plate as an information intermediary and goes through steps such as recording, reading, processing, and displaying image information before an image can be formed. On the other hand, DR is a direct digital imaging method [23–25]. It relies on amorphous selenium flat-panel detectors and analog-to-digital converters to directly convert X-ray information into digital images. Compared with CR, DR has a faster imaging speed, higher image contrast and clarity, richer image layers, and higher density resolution, thus ensuring image quality. Moreover, it exposes the examinees to less radiation, has a faster photography speed, is convenient for quality control, can display images instantly, and is more efficient. Therefore, the DR technique is used for the screening and staging diagnosis of pneumoconiosis [9,26]. However, it is required that image post-processing techniques, such as noise reduction and edge enhancement, are not used because they will lead to problems such as insufficiently clear display of small opacities on chest X-ray films and the loss of some detailed image information. In addition, DR equipment has a high cost and requires a high level of technical expertise. It needs to be operated and maintained by professional technicians to ensure image quality and the normal operation of the equipment [27].

Both CR and DR digital imaging possess powerful post-processing functions. They can change the grayscale by adjusting the window width and window level, which increases the tolerance of the instrument's technical parameter settings, reduces the noise caused by inaccurate selection of exposure conditions, enables the acquisition of more information about different tissues, and allows for direct annotation on the computer. Moreover, they can be remotely transmitted to realize remote expert consultations. However, digital radiography requires different program controls in all aspects, such as photographic imaging, image display, processing, and printing. And the technical differences between the equipment of different manufacturers must also be taken into account [28,29]. Therefore, at present, their application in the diagnosis of pneumoconiosis is still greatly restricted. There is still no unified diagnostic standard formulated, and there is a lack of standard films for imaging classification. Nevertheless, with the continuous improvement of relevant technologies and the continuous reduction in the cost of equipment and its use, it is believed that they will completely replace analog imaging techniques in the future and become the main tools for clinical diagnosis and remote consultations.

2.1.3. CT

CT is a widely used imaging technique that converts X-rays passing through the body into digital signals, which are then processed to generate detailed chest images for lung examination. Compared to traditional X-ray imaging, CT offers higher accuracy and resolution, enabling the detection of subtle pathological changes in the lungs, such as characteristic small opacities, which are critical for early pneumoconiosis diagnosis [30,31]. However, CT has limitations, including exposure to ionizing radiation, the potential toxicity of iodinated contrast agents, increased diagnostic costs, and the lack of standardized evaluation protocols.

High-resolution CT (HRCT) further enhances diagnostic capabilities by providing sub-millimeter slice reconstruction and finer morphological details of lung parenchyma. Despite its improved sensitivity and specificity, HRCT remains an auxiliary tool for early screening and staging due to the non-specificity of imaging changes and clinical symptoms in pneumoconiosis [32,33]. When combined with advanced image post-processing techniques, such as 3D reconstruction and multiplanar reconstruction (MPR), HRCT can improve diagnostic accuracy and aid in the identification of complications. HRCT scanning risks inducing other diseases from radiation exposure. Thus, low-dose CT (LDCT) scanning for pneumoconiosis CT examinations has drawn special attention and is widely used, aiming to balance radiation dose and diagnostic image quality [33]. However, LDCT image quality is lower than routine CT and HRCT, with higher noise and unclear fine structures, potentially affecting lesion detail observation and analysis [34,35]. Dual-energy spectrum CT, another advanced technique, quantifies dust content in lung tissues, improving staging accuracy, though its high technical requirements limit widespread adoption [36]. Thin-section CT with multiplanar reconstruction can clearly show bronchial conditions, local masses, and tissue invasion, helping distinguish stage III pneumoconiosis large opacities from lung cancer masses [37]. But its thin slice thickness and limited range mean more slices, longer scan time, and higher radiation dose when scanning a larger area. Multislice spiral CT (MSCT) scans and acquires data fast, has high temporal and spatial resolution, better image quality, and obvious 3D effects. It is advantageous in observing lung small opacities, overcoming respiratory interference, and detecting other diseases [38]. However, its radiation dose is relatively high, especially in large-range or multiple scans. As CT technology becomes more accessible and affordable, it is expected to play an increasingly routine role in pneumoconiosis diagnosis. Table 1 shows the advantages and limitations of diagnostic techniques such as HKV X-ray imaging, CR, DR, HRCT, LDCT, dual-energy spectrum CT, thin-section CT, and MSCT.

Table 1. Advantages and limitations of different diagnostic techniques.

Diagnostic Techniques	Advantages	Limitations
HKV X-ray Imaging	Short exposure time; less radiation dose	Multifactorial
CR	Indirect digital imaging; reduces the exposure dose	Low image resolution and clarity, slow imaging process vs. DR
DR	Direct digital imaging; high image resolution and clarity, fast imaging process vs. CR	High cost; complex lesions inferior to CT; specialized technical staff required
HRCT	High spatial resolution	Risk of radiation exposure
LDCT	Low radiation dose; universal adoption	Increased image noise vs. HRCT
Dual Energy Spectrum CT	High accuracy	Complicated operation
Thin-Section CT	High clarity	Limited scanning range
MSCT	Fast scanning and data acquisition speed; high temporal and spatial resolution; higher image quality; obvious 3D effects	High radiation dose

2.1.4. CAD

CAD, a technology that assists physicians in interpreting medical data, plays a crucial role in processing and analyzing diverse medical data, particularly in the diagnosis of pneumoconiosis [39]. In recent years, AI, especially deep learning (DL)—a subset of AI that enables autonomous feature learning through computer algorithms—has seen significant advancements in medical imaging research. Since 2012, DL has been increasingly applied to analyze complex medical data, offering improved accuracy over traditional machine learning models that rely on manual feature extraction [40]. Chest X-ray examination is a common preliminary screening method for pneumoconiosis. DL models, trained on large datasets of chest X-ray images (including normal lungs and pneumoconiosis cases at various stages), can automatically identify disease-related features, such as changes in lung texture and the appearance and distribution of nodules. By integrating their expertise with CAD-generated reports, physicians can make more accurate diagnostic decisions. This collaborative approach enhances diagnostic accuracy and efficiency and alleviates the workload on physicians while lowering healthcare costs [41]. In addition to DL, models constructed using algorithms such as decision trees, support vector machines (SVM), and artificial neural networks (ANN) have achieved relatively satisfactory results in the diagnosis of lung diseases [42,43]. For instance, SVM models incorporating multiple biomarkers (e.g., TGF-β1, CTGF, and PDGF) have shown clinical value in pneumoconiosis screening [44,45]. However, these models face challenges, including limited specificity and small sample sizes.

Despite these advancements, the application of AI in pneumoconiosis diagnosis faces several challenges. The need for large-scale, high-quality datasets and lengthy training cycles limits the widespread adoption of AI technologies [46]. Additionally, the lack of standardized evaluation criteria for CT diagnosis of pneumoconiosis hinders the application of AI in CT image analysis. While AI cannot yet replace manual diagnosis entirely, the "AI + physician" dual-reading mechanism reduces diagnostic errors, alleviates physician workload, and bridges gaps in expertise, particularly benefiting primary hospitals and less experienced physicians [47,48]. However, this approach requires careful validation to avoid over-reliance on AI-generated results.

Recently, the integration of large language models (LLMs) with visual encoders, as demonstrated by PneumoLLM, has enabled feature extraction and accurate classification of pneumoconiosis from chest X-ray images [49]. Unlike traditional DL methods, LLMs combined with visual encoders can leverage natural language processing capabilities to enhance feature extraction and classification. This approach not only improves diagnostic accuracy but also reduces reliance on extensive training data, offering new perspectives and methods for medical imaging analysis (Figure 2).

2.2. PFT

Since imaging and biomarker techniques cannot assess the functional status of patients, PFT methods, such as lung volume and lung ventilation, have been used as important supplementary approaches to evaluate the disease severity of patients with pneumoconiosis [50,51]. PFT is a universal method for diagnosing airflow limitation, as it can assess a patient's respiratory function and help physicians understand the lung's ventilation capacity and the elasticity of lung tissue. Patients with pneumoconiosis may experience symptoms such as dyspnea, shortness of breath, and chest tightness. Therefore, PFTs are used as important supplementary methods for evaluating the severity of pneumoconiosis. They are also important bases for identifying the labor capacity, judging the prognosis, and thus confirming the disability grade [52]. The study by Huang et al. indicates that pulmonary function abnormalities in patients with pneumoconiosis are closely related to occupational dust exposure, and obstructive and mixed ventilatory dysfunction have been observed in patients [53]. PFT has been used in occupational health examinations for decades and is an important auxiliary diagnostic technique for pneumoconiosis. For dust-exposed workers, a significant decline in pulmonary function test results within the normal range should raise concerns. A limitation of this technique is the lack of a direct and specific correlation between the decline in pulmonary function and the pathological changes of pneumoconiosis, as changes in pulmonary function can be influenced by various factors, such as age, lifestyle habits (e.g., smoking), other pulmonary diseases, and the individual's overall health status.

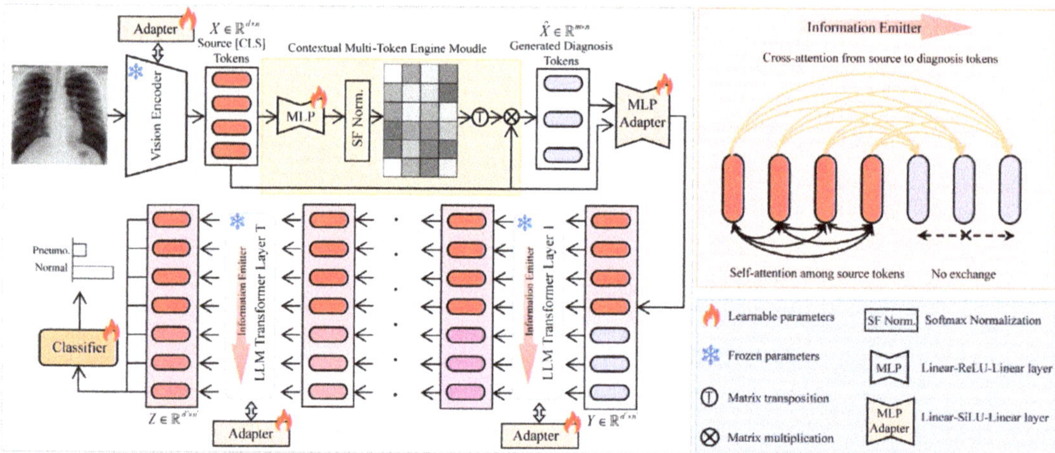

Figure 2. Schematic diagram of the proposed PneumoLLM for pneumoconiosis diagnosis [49].

2.3. *Tissue-Invasive Techniques*

LB is a technique to obtain lung tissue samples for pathological examination. When the imaging manifestations are atypical or in the early stage of the development of pneumoconiosis, LB techniques can detect and provide evidence of early pathological changes caused by dust in lung tissue specimens, improving diagnostic accuracy and reducing the rates of misdiagnosis and missed diagnosis [54,55]. Currently, the main LB techniques include surgical lung biopsy (SLB), transbronchial lung biopsy (TBLB), percutaneous lung biopsy (PCLB), transbronchial cryobiopsy (TBCB), etc. Among them, TBCB is the safest and most effective, and it can obtain high-quality large-area tissue specimens, which is suitable for diagnosing diffuse lung diseases [56,57]. LB can provide direct evidence of dust exposure and pathological changes in the lung tissue, thereby assisting in the diagnostic process of pneumoconiosis [58]. However, as lung tissue biopsy is an invasive procedure, it can cause certain degrees of physical trauma [59,60]. Additionally, due to the stringent technical requirements and medical conditions necessary for its performance, it is currently less frequently used in the diagnosis of pneumoconiosis and is not suitable as a routine or population-based screening method for detecting pathological changes.

3. Advanced Sensing Technologies for Screening and Diagnosis of Pneumoconiosis

3.1. *Biomarker Detection*

Biomarkers, encompassing proteins, genes, metabolites, and other molecular indicators, are critical for identifying structural or functional changes in systems, organs, tissues, cells, and subcellular components [61–63]. In pneumoconiosis, various biomarkers are associated with its pathogenesis, including inflammatory responses, pulmonary fibrosis, oxidative stress, and immune dysfunction. For instance, neopterin (NPT), an oxidative stress marker released by activated mononuclear macrophages, has been extensively studied. Research shows that serum NPT levels are significantly elevated in silicosis patients compared to healthy controls, highlighting its diagnostic potential [64]. However, biomarkers in serum proteins, cytokines, and apoptosis-related factors may suffer from limited sensitivity and specificity.

Epigenetic mechanisms, such as DNA methylation, histone modification, and miRNA regulation, have also gained attention for their role in maintaining genetic stability while responding to environmental stimuli [65]. Serum-specific miRNAs, which are stable, detectable, and resistant to freeze-thaw cycles, are particularly promising for clinical applications. Studies have linked the expression levels of certain miRNAs (e.g., miR-155 and miR-4516) to the severity of pneumoconiosis, suggesting their utility in early diagnosis [66–68]. However, these studies are often limited by small sample sizes and a lack of exploration into metabolic pathways, necessitating further validation [69]. Additionally, the complexity, cost, and time-intensive nature of current detection technologies hinder their widespread clinical adoption. miRNAs, which serve as early biomarkers for pneumoconiosis, are traditionally detected using methods that suffer from high temperature requirements, contamination risks, fluorescence bleaching, and primer design challenges. To overcome these limitations, surface-enhanced Raman spectroscopy (SERS) has emerged as a groundbreaking alternative. SERS offers advantages such as room-temperature detection, non-contact operation, label-free analysis, and no need for RNA primers, making it an ideal tool for rapid and ultrasensitive miRNA detection. Cui et al. [70] utilized the label-free SERS technique to conduct detection and analysis on miRNA biomarkers related to pneumoconiosis based on 3D gold-coated zinc oxide nanorod arrays (Au-ZnO NRA) (Figure 3a). The miRNA biomarkers (miR-19a, miR-149, miR-146a, and miR-155) used in the study are associated with the occurrence of pneumoconiosis. The presence of these miRNAs in blood and bronchoalveolar lavage fluid has been screened as biomarkers for early pneumoconiosis. Detecting these miRNAs through the label-free SERS method can provide a rapid, reliable, and ultrasensitive alternative approach for the early diagnosis of pneumoconiosis.

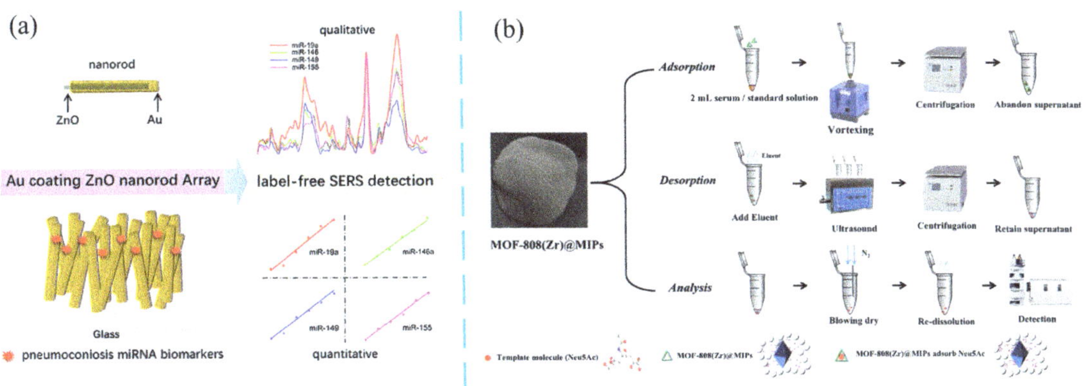

Figure 3. (**a**) Label-free SERS detection of miRNA [70]. (**b**) Technical route for Neu5Ac determination [71].

In addition to miRNAs, free N-acetylneuraminic acid (Neu5Ac) in serum can also serve as a biomarker for the early diagnosis of pneumoconiosis (Figure 3b). Lin et al. [71] took metal-organic frameworks (MOFs) as carriers and synthesized a type of molecularly imprinted polymer (MIPs), which was used for the specific adsorption of free Neu5Ac in human serum. And they adopted liquid chromatography-tandem mass spectrometry (LC-MS/MS) to determine the free Neu5Ac in serum samples. The linear range of this detection system is 50–10,000 ng mL^{-1}. The limit of detection (LOD) and the limit of quantification (LOQ) are 3 and 10 ng mL^{-1}, respectively. Under these conditions, the recovery rates of serum samples are between 83.54% and 92.21%. This method has been successfully applied to the analysis of free Neu5Ac in the serum of pneumoconiosis patients without protein precipitation. Furthermore, Wang et al. [72] applied non-targeted metabolomics and lipidomics techniques to conduct characterization and analysis on the serum of patients with pneumoconiosis and silicosis. Four metabolites were discovered, namely 1,2-dioctanoyl-sn-glycero-3-phosphocholine, phosphatidylcholine (O-18:1/20:1), indole-3-acetamide, and L-homoarginine. In addition, kynurenine, N-tetracosanoylsphingosine 1-phosphate, 5-methoxyethanol, and phosphatidylethanolamine (22:6/18:1) can be used to predict the staging of pneumoconiosis.

Beyond epigenetic markers, other biomarkers, such as pulmonary surfactants, cytokines (e.g., IL-18, TGF-β1), and long non-coding RNAs (e.g., lncRNA-ATB) have been explored. The upregulation of plasma LncRNA-ATB is closely related to the genetic targets in pneumoconiosis. LncRNA-ATB can be activated by TGF-β and is significantly upregulated in patients with pneumoconiosis, closely correlating with the expression levels of TGF-β1 [73]. TGF-β1 plays a key role in the fibrosis process of pneumoconiosis, and its genetic polymorphisms (such as +869T/C) are associated with susceptibility to pneumoconiosis and the degree of fibrosis [74]. As pneumoconiosis progresses, the degree of fibrosis in lung tissues increases, and the expression levels of TGF-β1 usually rise accordingly. Therefore, the upregulation of LncRNA-ATB may be a result of the activation of the TGF-β1 signaling pathway, further promoting the fibrosis process in pneumoconiosis and increasing the disease risk for patients [75]. Similarly, dysregulation of the respiratory microbiome, characterized by altered abundances of specific bacterial genera (e.g., Prevotella, Actinobacillus, Leptotrichia), has been linked to inflammatory and fibrotic indicators in pneumoconiosis. These microbial shifts, alongside elevated levels of inflammatory markers such as TNF-α and hydroxyproline (HYP), provide insights into the dynamic progression of pulmonary lesions and offer potential avenues for microbial-based diagnostic approaches [76,77].

B cell-derived immunoglobulins are routinely used in clinical practice for the diagnosis of coal workers' pneumoconiosis, as they provide crucial information on the humoral immune status. Previous studies have shown that the concentrations of IgA and IgG increase in patients with coal workers' pneumoconiosis, but little is known about the role of serum IgG subclasses in the diagnosis of coal workers' pneumoconiosis [6]. Li et al. [6] found that compared with dust-exposed workers without pneumoconiosis and healthy controls (HCs), the levels of serum IgG1, IgG2, IgM, and IgA were elevated in patients with coal workers' pneumoconiosis. In particular, the IgG2/IgG3 ratio provides a feasible alternative approach for the diagnosis of coal workers' pneumoconiosis. The study involved the biomarkers IgG1, IgG2, IgG3, IgG4, IgA, and IgM (Figure 4a). It analyzed the ROC curves of immunoglobulins in coal workers' pneumoconiosis and dust-exposed workers (Figure 4b), as well as the ratios of IgG1/IgG3 and IgG2/IgG3 (Figure 4c). Additionally, ROC analysis was performed on pulmonary function parameters and the IgG2/IgG3 ratio to distinguish coal workers' pneumoconiosis from workers exposed to dust (Figure 4d). In coal workers' pneumoconiosis patients, the serum concentrations of IgG1, IgG2, IgA, and IgM are increased, while the concentration of IgG3 is decreased. The IgG2/IgG3 ratio shows a certain value in distinguishing between coal workers' pneumoconiosis and dust-exposed workers, with a relatively large area under the curve. Its diagnostic performance is better than that of IgG1 or IgG2 alone, and there is no significant difference compared with the lung function index FEV1/FVC, suggesting that it can be regarded as a potential biomarker for the diagnosis of coal workers' pneumoconiosis.

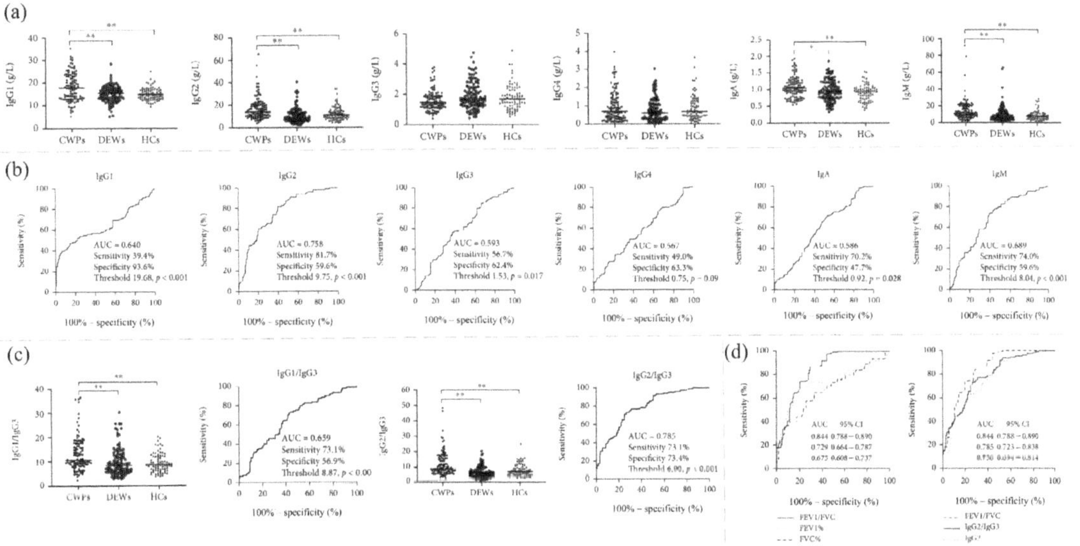

Figure 4. (**a**) Comparison of serum concentrations of IgG1, IgG2, IgG3, IgG4, IgA, and IgM among the study groups, * $p < 0.05$, ** $p < 0.01$. (**b**) The ROC curve analysis of immunoglobulin test results between coal workers' pneumoconiosis and dust-exposed workers. (**c**) Comparison of the IgG1/IgG3 ratio and IgG2/IgG3 ratio among the study groups. (**d**) ROC analysis of the spirometry parameters and the IgG2/IgG3 ratio for distinguishing coal workers' pneumoconiosis and dust-exposed workers [6].

In summary, the exploration of biomarkers for pneumoconiosis diagnosis has yielded promising yet incomplete results. While oxidative stress markers (e.g., NPT), epigenetic regulators (e.g., miRNAs, DNA methylation), B cell-derived immunoglobulins, and inflammatory cytokines (e.g., TNF-α, IL-1β) provide valuable insights, challenges related to sensitivity, specificity, and technological limitations persist. Future research should focus on multi-omics approaches, combining multiple biomarkers and advanced detection technologies to enhance diagnostic accuracy and facilitate early intervention in pneumoconiosis.

3.2. Breath Analysis

3.2.1. Monitoring of Exhaled Components

For early pneumoconiosis screening, both sensitivity and accuracy are indispensable. However, the convenience and non-invasiveness of the tests are also of great importance. Existing technologies based on large-scale instruments and equipment, such as X-ray, CT, MRI, and so on, are all costly, invasive, or may have side effects related to radiation exposure that can affect health, be time-consuming, or require well-trained personnel. All these factors are not conducive to the popularization of the equipment and the implementation of extensive population screening.

Thousands of volatile organic compounds (VOCs) have been found in exhaled breath, and they are related to the internal biochemical processes of the human body, either directly or indirectly [78–80]. For example, lipid peroxidation plays an important role in the pathogenesis of pneumoconiosis. Pentane and methylated alkanes such as C5–C7, as metabolites of lipid peroxides, constitute the main VOCs in the breath of patients with pneumoconiosis [81]. Compared with the technologies based on high-end instruments and professional operators mentioned above, breath sensors and their arrays (electronic noses) are cutting-edge and hot sensing technologies that have attracted extensive attention in recent years. They also represent a kind of diagnostic technology that is rapid, simple, non-contact, safe, non-invasive, capable of continuous sampling, and easy to repeat [81–87]. Some patients may not have obvious organic lesions in the early stage, but their exhaled breath already contains specific VOCs with disease characteristics [88]. Breath analysis technology, leveraging VOCs as disease biomarkers, demonstrates significant potential for non-invasive diagnostics. While current applications in pulmonary disorders (e.g., lung cancer, tuberculosis), infectious diseases, and urological conditions have advanced, their clinical adoption remains limited due to unresolved pathophysiological mechanisms governing VOC metabolic alterations. Research limitations include small sample sizes, confounding variables from environmental exposures (e.g., smoking) and host metabolic processes, and nascent exploration in pneumoconiosis detection. As an adjunctive tool, breath analysis could prioritize high-risk individuals for confirmatory imaging studies, thereby optimizing resource allocation and transforming disease management paradigms. Its non-invasive nature and capacity for continuous monitoring position it as a revolutionary approach in early disease detection and personalized healthcare, particularly for occupational lung diseases like pneumoconiosis.

Respiratory sampling may not require advanced and specialized techniques and knowledge, but it still needs to utilize high-end technologies such as spectroscopy to detect and analyze the VOCs exhaled in diseases [89]. Commonly used techniques include gas chromatography, gas chromatography-mass spectrometry, proton transfer reaction mass spectrometry, selected ion flow tube-mass spectrometry, ion mobility spectrometry, and laser spectroscopy [90,91]. These devices are powerful, yet they have relatively high requirements in terms of instrument cost and size, operational professionalism and proficiency, data analysis level, sample pretreatment, etc., and they are also relatively time-consuming. Therefore, it is quite necessary to develop miniaturized and direct respiratory sensor analysis platforms. However, the real market-oriented application of this technology still depends on the improvement of material technology, sensor technology, machine learning methods, and disease-specific reference libraries and databases, especially the search and identification of potential biomarkers for specific respiratory diseases.

3.2.2. Respiratory Physiological Parameters

Pneumoconiosis poses life-threatening risks through respiratory failure and other severe complications, necessitating comprehensive respiratory physiological parameter monitoring throughout diagnosis and treatment. Real-time continuous breath sensor-based surveillance plays a critical role in the early detection of pathophysiological deteriorations, including respiratory arrhythmias and hypoxemic exacerbations, particularly given the compromised pulmonary parenchyma and fragile respiratory function in affected patients. Such monitoring enables dynamic risk stratification for complication prevention. Furthermore, the heterogeneous nature of pneumoconiosis (variable disease severity, comorbidities, and treatment responses) mandates precision medicine approaches, where sensor-derived physiologic data support evidence-based adjustments to individualized therapeutic regimens. The detailed information provided by breath sensors gives doctors a basis for accurately evaluating patients' responses to existing treatment measures [82]. For example, by analyzing the data recorded by breath sensors, doctors can determine whether drug treatment has achieved the expected results and whether oxygen therapy meets the actual needs of patients. Based on these accurate evaluation results, doctors can make targeted fine adjustments to drug dosages and reasonably adjust key treatment parameters such as the duration and intensity of oxygen therapy. This not only improves the precision and effectiveness of treatment but also significantly enhances patients' treatment experience, making them more comfortable during the treatment process. Meanwhile, it helps to improve the prognosis and brings greater hope for patients' recovery.

In the field of early diagnosis of the disease, breath sensors also have a unique value. They can continuously monitor key parameters such as patients' breathing patterns, respiratory frequencies, and depths of breathing. In the early stage of pneumoconiosis, patients' bodies often undergo some subtle changes, and these slight alterations in these parameters may be important signals in the early stage of the disease [92]. Thanks to their high sensitivity, breath sensors can detect these early changes in a timely manner, providing doctors with valuable diagnostic clues and thus enabling the early diagnosis of pneumoconiosis. Early diagnosis is crucial for the treatment of pneumoconiosis, as it can buy more treatment time for patients and increase the success rate of treatment. Although imaging examinations, such as X-rays and CT scans, are important means for the diagnosis of pneumoconiosis, breath sensors can provide additional physiological information and serve as a powerful supplement to imaging examinations. For example, by monitoring the gas exchange efficiency during breathing, doctors can indirectly infer the degree of pulmonary fibrosis. Pulmonary fibrosis is an important pathological feature of pneumoconiosis, and understanding its degree is of great significance for accurately assessing patients' lung conditions and formulating reasonable treatment plans [93]. The combination of breath sensors and imaging examinations is like providing doctors with a pair of "X-ray eyes" that can comprehensively and accurately understand patients' conditions, enabling them to provide higher-quality medical services for patients.

3.2.3. Key Factors and Strategies for Enhancing Breath Analysis Performance

High-quality breath analysis samples also rely on the control of individual breathing patterns, breathing depths, sampling time methods, working temperatures, etc., as well as the progress of gas sample storage technologies. In addition, the lack of selectivity and the reliability of working in high-humidity environments (the exhaled breath of a normal human body has a high relative humidity of 90%) remain the key factors limiting the use of most sensors. Moreover, when the concentration of a single VOC component is relatively high, problems such as sensor drift and the inability to accurately calibrate the sensors may also occur [90]. It is also necessary to establish standardized methods to utilize different datasets and consider the reproducibility of instruments and classification models among different sensors, that is, repeatability. The responses of sensors to VOCs can be analyzed by pattern recognition algorithms to classify different situations separately. Among them, principal component reduction and subsequent discriminant analysis pattern recognition are the most commonly used types of raw data analysis in their responses. Other AI technologies can also be used for data analysis, such as machine learning algorithms and ANN [94], but the diversity of analysis techniques may hinder the standardization of sensor array technologies.

Therefore, by selecting materials with high specific surface area, unique crystal structures, and favorable electronic properties—such as metal oxide semiconductors, graphene and its derivatives, and metal-organic frameworks (MOFs)—high-performance breath sensors can be developed [95–98]. By employing chemical methods to introduce specific functional groups or active sites on the surface of the sensing material, it can interact specifically with target volatile organic compound (VOC) molecules, thereby significantly enhancing the sensor's selectivity [99]. Furthermore, constructing an array of sensors with different sensitivities and combining it with pattern recognition algorithms (such as principal component analysis and artificial neural networks) to process and analyze the response signals enables accurate identification and concentration measurement of various VOCs, thus greatly improving the sensor's selectivity and anti-interference capability [100]. Choosing appropriate packaging materials is also crucial. Materials such as polydimethylsiloxane (PDMS) and cellulose nanocrystals (CNC), which have good gas permeability and chemical stability, can effectively isolate the sensor's sensitive elements from the external environment, preventing interference from moisture, dust, and other impurities while ensuring that target VOCs gases can reach the sensing surface [101,102]. Integrating environmental compensation components within the sensor package allows for real-time monitoring of changes in environmental parameters, such as temperature and humidity, and compensating the sensor's response through circuitry or algorithms, thereby enhancing the measurement accuracy of the sensor under different environmental conditions [103].

3.3. E-Noses Technology

Compared with breath sensors that have a single structure and lack practicality, E-noses utilize gas sensor arrays to achieve sensing and data analysis directly after sampling and realize pattern recognition of diseases, food, pharmaceuticals, explosives, drugs, etc. on site. The E-nose is an advanced device that simulates the human olfactory system. Its working principle is based on the precise detection of gas chemical components by an array of sensors. These sensors can sensitively identify and respond to various VOCs, efficiently converting the captured chemical signals into electrical signals that are easy to analyze. The VOCs in the exhaled breath of patients with pneumoconiosis are significantly different from those of healthy individuals. The electronic nose leverages this characteristic by analyzing the unique patterns of VOCs in exhaled breath to identify biomarkers associated with pneumoconiosis, thereby providing a strong basis for its diagnosis [104,105]. The detection using the electronic nose has many advantages. First, the detection process only requires the collection of exhaled breath, eliminating the need for invasive procedures. This greatly enhances the patient experience, making it more comfortable and convenient. This feature also makes the electronic nose suitable for large-scale screening. Second, the detection is rapid and capable of being completed in a short time, meeting the need for efficient testing. Moreover, the electronic nose has excellent detection capabilities for low-concentration VOCs, which is beneficial for the early diagnosis of pneumoconiosis, providing patients with valuable treatment time [106].

However, the application of the electronic nose in the detection of pneumoconiosis still faces some challenges. On the one hand, research on VOC biomarkers related to pneumoconiosis is still ongoing and not yet fully clarified, which, to some extent, limits the accuracy and reliability of electronic nose detection. On the other hand, VOCs in exhaled breath are easily influenced by factors such as diet and environment. Effective measures need to be taken during the detection process to exclude these interferences and ensure the validity of the detection results. The application of the electronic nose in pneumoconiosis detection is still in the developmental stage. More research is needed to fully validate its clinical value and promote its widespread use in the diagnosis of pneumoconiosis [107]. The research on E-noses involves many interdisciplinary fields and generally includes three structural units, namely the sensor unit array, the signal processing unit, and the pattern recognition unit. The sensor and its array technology, as well as the pattern recognition system (mathematical and statistical algorithms such as discriminant factor analysis and partial least squares method) and AI technologies, such as ANNs, are the two construction bases of electronic noses. Electronic and computer technologies are the preparation bases, while neurophysiology and mathematics are the theoretical bases and point out the direction for their development [108]. It not only means real-time monitoring to help doctors achieve the detection and treatment of patients but also can be combined with in-vehicle analysis, wearable analysis, and so on.

The sensing array units of the E-noses also rely on the development of sensing materials and technologies. Currently, the gas-sensitive sensing units are still mainly based on several relatively mature types, such as metal oxide semiconductors [109], field-effect transistors [110], quartz oscillators [111], optical methods [112], and surface acoustic waves [113]. They involve materials like metal oxides, conductive polymers, carbon-based materials such as carbon nanotubes (CNTs), and two-dimensional (2D) materials [114–116]. Each of these materials has its own advantages and disadvantages, and the selectivity of a single material is limited. For example, the E-nose based on the CNT-TiO_2 composite structure developed by Shooshtari et al. can distinguish acetone, ethanol, butanol, and propanol vapors with an accuracy rate of 97.5%, which can be used for rapid and efficient monitoring of VOCs [117]. With the development of sensing technologies, microfabrication technologies, nanotechnologies, advanced signal processing algorithms, etc., and the development of sensor arrays with strong selectivity and high sensitivity, together with the significant optimization of data processing algorithms, it is believed that the E-nose technology will eventually play a significant role in the screening, diagnosis, and treatment of pneumoconiosis [85,115,118].

As an endeavor, Xuan et al. [87] developed an E-nose based on an array of 16 organic nanofiber sensors (Figure 5a). It combines machine learning, pattern recognition algorithms, big data analysis techniques, etc. They constructed an exhaled breath screening and diagnosis model for pneumoconiosis and an early warning model for pulmonary fibrosis lesions, with an accuracy rate of over 85%. The organic nanofiber sensors in the E-nose system can respond to VOCs present in exhaled breath. When people suffer from diseases such as silicosis, the concentrations of these VOCs may change. Metabolites generated by lesions in the alveolar-capillary membrane are directly released into the alveolar space and can be detected in the breath. The sensor array is directly exposed to the breath mixture, and the composite profile (breath fingerprint) generated by the responses of all 16 nanofibers can distinguish between diseased and healthy control groups, reflecting metabolic changes in the breath components. This method is similar to the olfactory system of mammals, where a large number of olfactory receptors (sensors) work as a cooperative array to generate specific patterns for different odors or mixtures without the need to know the detailed information of individual components. In this way, the E-nose can quickly and non-invasively detect diseases, providing an ideal technology for large-scale disease screening (Figure 5b).

In addition to detecting pneumoconiosis, E-noses can also detect lung cancer. For example, Tirzīte et al. [119] applied the E-nose combined with logistic regression analysis (LRA) in the detection of lung cancer. The study involved 252 lung cancer patients and 223 non-lung cancer patients. Exhaled breath samples were collected and analyzed through the Cyranose 320 sensor device, and LRA was used for data analysis to distinguish between lung cancer patients, patients with other lung diseases, and healthy individuals. The sensitivities for cancer detection among smokers and non-smokers were 95.8% and 96.2%, respectively, and the specificities were 92.3% and 90.6%, respectively. This indicates that the combination of E-nose and LRA can effectively identify lung cancer patients, thus assisting doctors in diagnosing lung cancer at an early stage and providing more timely treatment for patients. Of course, in the medical field, a single detection method often has limitations, while multimodal detection can improve the diagnostic accuracy of diseases by integrating the advantages of multiple detection means. For example, combining breath analysis, imaging examinations, and blood tests can evaluate patients' health conditions more comprehensively, especially for complex diseases such as lung cancer, chronic obstructive pulmonary disease (COPD), and pneumoconiosis [120].

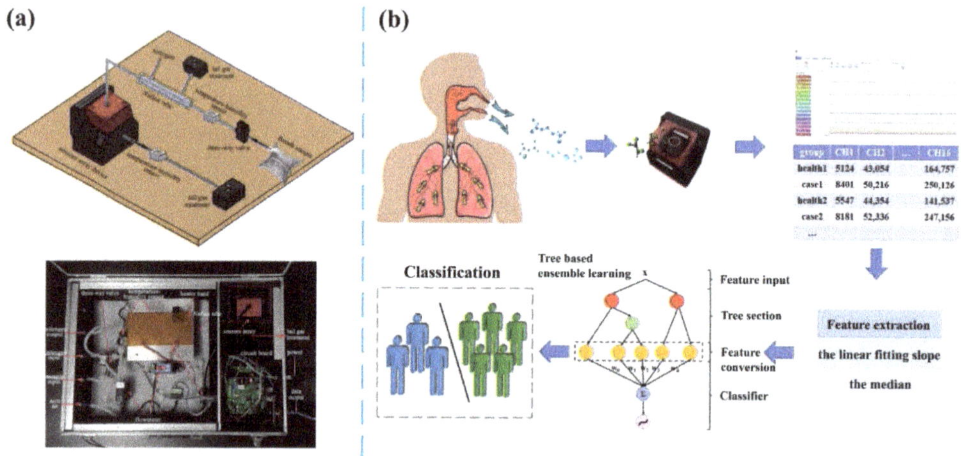

Figure 5. (**a**) E-nose system. (**b**) Schematic diagram of the process of screening for pneumoconiosis with an E-nose system [87].

4. Conclusions and Outlook

In summary, this paper provides a comprehensive overview of various screening and diagnostic technologies for pneumoconiosis, ranging from the dominant and most widely used clinical imaging analysis, PFT, tissue-invasive techniques, and biomarker detection to E-nose technology. When these technologies are used in combination, their strengths are fully realized. Imaging analysis offers macroscopic structural information about the lungs. PFT provides quantitative assessment information on lung function, while tissue-invasive techniques provide precise pathological diagnoses and biomarker technologies offer diagnostic evidence at the biomolecular level. The E-nose, on the other hand, provides an ideal non-invasive chemical sensing technology by analyzing the chemical composition of exhaled breath. Together, these technologies complement and validate each other, allowing for a comprehensive assessment of the patient's condition from multiple dimensions. This significantly improves the accuracy and reliability of pneumoconiosis diagnosis.

However, currently, the situation of preventing and controlling pneumoconiosis remains extremely severe. Early prevention and control are still crucial tasks in this work, which still demand the coordinated cooperation and long-term efforts of the government, enterprises, workers, and researchers. They should not only consider the sensitivity, specificity, and the employed technology itself but also take into account factors like convenience, cost-effectiveness, and safety. There is still a long way to go to optimize mainstream imaging technologies, especially considering their reliance on the doctors' experience and the non-negligible misdiagnosis rate. Thus, the development of other technologies, such as biomarkers and respiratory sensing, are given great expectations, combined with the rapid progress in technologies like mobile devices, remote consultations, and digital imaging. The continuous reduction in the costs of equipment and usage will further offer more opportunities for their widespread implementation, especially in developing countries and specific industries with high pneumoconiosis incidence. However, the development of biomarker technology is still restricted by issues such as high technical barriers, a relatively small sample size in clinical studies, and the complexity of excluding interfering factors, so its future promotion will be greatly limited. Currently, although some potential biomarkers (such as inflammatory factors) may be of significant importance for the early diagnosis and assessment of disease progression in pneumoconiosis, their clinical application value has not yet been verified through large-scale clinical trials. In the future, with a deeper understanding of the pathogenesis of pneumoconiosis and continuous advancements in detection technologies, new biomarkers are expected to play an important role in clinical diagnosis and treatment. However, before large-scale clinical application, their sensitivity, specificity, and clinical utility must be rigorously validated through clinical trials.

In contrast, as a type of chemosensor for gaseous analytes detection, non-invasive diagnosis of pneumoconiosis through expiratory VOC sensing is quite ideal and suitable for large-scale, low-cost, and easily accepted screening. It can safely and frequently collect a wide range of samples and is expected to become the best option for screening the whole population and for daily monitoring. However, developing high-performance breath sensors or even E-nose devices remains a complex challenge that awaits persistent endeavors and in-depth research. For example, the production of VOCs is associated with a variety of pathophysiological processes, including inflammation, oxidative stress, and microbial infections. Due to the complex pathological mechanisms of pneumoconiosis, changes in VOCs may overlap with other pulmonary diseases, thus greatly limiting their specificity and sensitivity. The strong interference of the moisture co-existing with VOCs in the expiratory air with their detection also needs to be taken seriously, compared with other detection scenarios. To address these challenges, existing experience with general VOC sensors should be actively used: (a) enhancing selectivity through material innovation and sensor array design, etc.; (b) realizing humidity interference through constructing hydrophobic surfaces, using humidity-resistant materials (e.g., porous silicon), employing AI-based humidity compensation algorithms, etc.; (c) enhancing sensitivity through material innovation and functional design, etc.; (d) AI-driven data processing and pattern recognition, etc. These will help address the challenges and drive breakthroughs in the field of pneumoconiosis detection using breath analysis and E-noses. In addition, significant progress may be made to produce wearable devices, which could lay a solid foundation for the application of pneumoconiosis-related pathogenic environmental monitoring simultaneously. Combining sensors with the Internet of Things (IoT) and communication technologies can also minimize the possibility of workplace accidents to the greatest extent. This will help to reduce the incidence or severity of pneumoconiosis at the source.

Moreover, animal models and related methods, as powerful scientific research tools, deserve more attention. Their establishment and application can not only provide a basis for clinical treatment but also lay a solid foundation for research on the occurrence, development, and mechanism of pneumoconiosis. But there are ethical and other limitations to real animals; in contrast, in vitro biomimetic architectures like that using microfluidic chips have shown great application prospects as a model for the study of pneumoconiosis pathophysiology.

No matter which strategy is used, the explosive development of big data, AI, and other technical principles has greatly promoted the progress of this field [41]. In particular, the explosive development of AI-generated content (AIGC) technology is making the "AI doctor" a reality. It not only fills the technical gap caused by the difference in medical conditions and doctor experience but also makes remote diagnosis and larger population screening possible and will greatly improve the efficiency and accuracy of diagnosis. It also may partially eliminate the diagnostic errors caused by a single technology and enable collaboration across territories, disciplines, and industries so as to better benefit pneumoconiosis patients or potential populations in the future. Most importantly, it could make up for the imbalance between different countries and regions, ultimately, for the benefit of all mankind.

Author Contributions: Y.Z.: investigation, methodology, software, and writing—original draft.; W.X.: writing—original draft, writing—review and editing, and supervision; S.C.: conceptualization, project administration, resources, writing—original draft, writing—review and editing, and supervision.; M.Y. and H.X.: formal analysis, software, writing—review and editing. All authors have read and agreed to the published version of the manuscript.

Funding: The financial support from the Academic Development Project of TongXin Funds (No. 2024161817) is gratefully acknowledged.

Conflicts of Interest: The authors declare no conflicts of interest.

References

1. Han, B.; Liu, H.; Zhai, G.; Wang, Q.; Liang, J.; Zhang, M.; Cui, K.; Shen, F.; Yi, H.; Li, Y.; et al. Estimates and predictions of coal workers' pneumoconiosis cases among redeployed coal workers of the Fuxin mining industry group in China: A historical cohort study. *PLoS ONE* **2016**, *11*, e0148179. [CrossRef] [PubMed]
2. Iijima, Y.; Bando, M.; Yamasawa, H.; Moriyama, H.; Takemura, T.; Niki, T.; Sugiyama, Y. A case of mixed dust pneumoconiosis with desquamative interstitial pneumonia-like reaction in an aluminum welder. *Respir. Med. Case Rep.* **2017**, *20*, 150–153. [CrossRef] [PubMed]
3. Peng, Y.; Li, X.; Cai, S.; Chen, Y.; Dai, W.; Liu, W.; Zhou, Z.; Duan, J.; Chen, P. Prevalence and characteristics of COPD among pneumoconiosis patients at an occupational disease prevention institute: A cross-sectional study. *BMC Pulm. Med.* **2018**, *18*, 22. [CrossRef] [PubMed]
4. Fan, Y.; Xu, W.; Wang, Y.; Wang, Y.; Yu, S.; Ye, Q. Association of occupational dust exposure with combined chronic obstructive pulmonary disease and pneumoconiosis: A cross-sectional study in China. *BMJ Open* **2020**, *10*, e038874. [CrossRef]
5. Hu, M.; Wang, Z.; Hu, X.; Wang, Y.; Wang, G.; Ding, H.; Bian, M. High-resolution computed tomography diagnosis of pneumoconiosis complicated with pulmonary tuberculosis based on cascading deep supervision U-Net. *Comput. Methods Programs Biomed.* **2022**, *226*, 107151. [CrossRef]
6. Li, Z.-G.; Li, B.-C.; Li, Z.-W.; Hu, H.-Y.; Ma, X.; Cao, H.; Yu, Z.-H.; Dai, H.-P.; Wang, J.; Wang, C. The potential diagnostic biomarkers for the IgG subclass in coal workers' pneumoconiosis. *J. Immunol. Res.* **2023**, *2023*, 9233386. [CrossRef]
7. Martínez, C.; Prieto, A.; García, L.; Quero, A.; González, S.; Casan, P. Silicosis, una enfermedad con presente activo. *Arch. Bronconeumol.* **2010**, *46*, 97–100. [CrossRef]
8. Alam, M.S.; Wang, D.; Sowmya, A. AMFP-net: Adaptive multi-scale feature pyramid network for diagnosis of pneumoconiosis from chest X-ray images. *Artif. Intell. Med.* **2024**, *154*, 102917. [CrossRef]
9. Akima, R.; JP, N.A.; Ito, K.; Nogami, S.; Nishimori, M.; Oogi, K.; Hayashi, N.; Suganuma, N.; Yamagami, T. Perceptual and objective physical quality of chest images: A comparison between digital radiographic chest images processed for cancer screening and pneumoconiosis screening in Japan. *Ind. Health* **2023**, *61*, 260–268. [CrossRef]
10. Hayashi, H.; Ashizawa, K.; Takahashi, M.; Kato, K.; Arakawa, H.; Kishimoto, T.; Otsuka, Y.; Noma, S.; Honda, S. The diagnosis of early pneumoconiosis in dust-exposed workers: Comparison of chest radiography and computed tomography. *Acta Radiol.* **2022**, *63*, 909–913. [CrossRef]
11. Foley, R.W.; Nassour, V.; Oliver, H.C.; Hall, T.; Masani, V.; Robinson, G.; Rodrigues, J.C.; Hudson, B.J. Chest X-ray in suspected lung cancer is harmful. *Eur. Radiol.* **2021**, *31*, 6269–6274. [CrossRef] [PubMed]
12. Bosdelekidis, V.; Ioakeimidis, N.S. Lung field segmentation in chest X-rays: A deformation-tolerant procedure based on the approximation of rib cage seed points. *Appl. Sci.* **2020**, *10*, 6264. [CrossRef]
13. Sauter, A.P.; Andrejewski, J.; Frank, M.; Willer, K.; Herzen, J.; Meurer, F.; Fingerle, A.A.; Makowski, M.R.; Pfeiffer, F.; Pfeiffer, D. Correlation of image quality parameters with tube voltage in X-ray dark-field chest radiography: A phantom study. *Sci. Rep.* **2021**, *11*, 14130. [CrossRef] [PubMed]
14. Wang, Z.; Hu, M.; Zeng, M.; Wang, G. Intelligent image diagnosis of pneumoconiosis based on wavelet transform-derived texture features. *Comput. Math. Method Med.* **2022**, *2022*, 2037019. [CrossRef]
15. Zhao, C.-Y.; Jia, R.-S.; Liu, Q.-M.; Liu, X.-Y.; Sun, H.-M.; Zhang, X.-L. Chest X-ray images super-resolution reconstruction via recursive neural network. *Multimed. Tools Appl.* **2021**, *80*, 263–277. [CrossRef]
16. Sarapata, A.; Stayman, J.; Finkenthal, M.; Siewerdsen, J.; Pfeiffer, F.; Stutman, D. High energy X-ray phase contrast CT using glancing-angle grating interferometers. *Med. Phys.* **2014**, *41*, 021904. [CrossRef]

17. Devnath, L.; Luo, S.; Summons, P.; Wang, D. Automated detection of pneumoconiosis with multilevel deep features learned from chest X-ray radiographs. *Comput. Biol. Med.* **2021**, *129*, 104125. [CrossRef]
18. Alam, M.S.; Wang, D.; Sowmya, A. DLA-Net: Dual lesion attention network for classification of pneumoconiosis using chest X-ray images. *Sci. Rep.* **2024**, *14*, 11616. [CrossRef]
19. Tucker, D.; Souto, M.; Barnes, G. Scatter in computed radiography. *Radiology* **1993**, *188*, 271–274. [CrossRef]
20. Soneji, S.; Yang, J.; Tanner, N.T.; Dang, R.; Silvestri, G.A.; Black, W. Underuse of chest radiography versus computed tomography for lung cancer screening. *Am. J. Public Health* **2017**, *107*, 1248–1250. [CrossRef]
21. Yu, H.; Guo, H.; Wang, Y.; Wang, Y.; Zhang, L. Bismuth nanomaterials as contrast agents for radiography and computed tomography imaging and their quality/safety considerations. *WIREs Nanomed. Nanobiotechnol.* **2022**, *14*, e1801. [CrossRef] [PubMed]
22. Penenberg, B.L.; Samagh, S.P.; Rajaee, S.S.; Woehnl, A.; Brien, W.W. Digital radiography in total hip arthroplasty: Technique and radiographic results. *J. Bone Jt. Surg.* **2018**, *100*, 226–235. [CrossRef] [PubMed]
23. Chung, N.; Cheng, Y.-H.; Po, H.-L.; Ng, W.-K.; Cheung, K.-C.; Yung, H.-Y.; Lai, Y.-M. Spinal phantom comparability study of cobb angle measurement of scoliosis using digital radiographic imaging. *J. Orthop. Translat.* **2018**, *15*, 81–90. [CrossRef] [PubMed]
24. Doyle, P.; Martin, C. Calibrating automatic exposure control devices for digital radiography. *Phys. Med. Biol.* **2006**, *51*, 5475. [CrossRef]
25. Scott, A.W.; Zhou, Y.; Allahverdian, J.; Nute, J.L.; Lee, C. Evaluation of digital radiography practice using exposure index tracking. *J. Appl. Clin. Med. Phys.* **2016**, *17*, 343–355. [CrossRef]
26. Li, X.; Xu, M.; Yan, Z.; Xia, F.; Li, S.; Zhang, Y.; Xing, Z.; Guan, L. Deep convolutional network-based chest radiographs screening model for pneumoconiosis. *Front. Med.* **2024**, *11*, 1290729. [CrossRef]
27. Marincowitz, C.; Gravesteijn, B.; Sheldon, T.; Steyerberg, E.; Lecky, F. Performance of the hull Salford Cambridge decision rule (HSC DR) for early discharge of patients with findings on CT scan of the brain: A CENTER-TBI validation study. *Emerg. Med. J.* **2022**, *39*, 213–219. [CrossRef]
28. Moore, C.S.; Wood, T.J.; Jones, S.; Saunderson, J.R.; Beavis, A.W. A practical method to calibrate and optimise automatic exposure control devices for computed radiography (CR) and digital radiography (DR) imaging systems using the signal-to-noise ratio (SNR) metric. *Biomed. Phys. Eng. Express* **2019**, *5*, 035027. [CrossRef]
29. Thomas, M.A.; Rowberg, A.H.; Langer, S.G.; Kim, Y. Interactive image enhancement of CR and DR images. *J. Digit. Imaging* **2004**, *17*, 189–195. [CrossRef]
30. Petsonk, E.L.; Rose, C.; Cohen, R. Coal mine dust lung disease. New lessons from an old exposure. *Am. J. Respir. Crit. Care Med.* **2013**, *187*, 1178–1185. [CrossRef]
31. Gao, X.; Qian, Y.; Gao, A. COVID-VIT: Classification of COVID-19 from CT chest images based on vision transformer models. *arXiv* **2021**, arXiv:2107.01682.
32. Catapano, F.; Marchitelli, L.; Cundari, G.; Cilia, F.; Mancuso, G.; Pambianchi, G.; Galea, N.; Ricci, P.; Catalano, C.; Francone, M. Role of advanced imaging in COVID-19 cardiovascular complications. *Insights Imaging* **2021**, *12*, 28. [CrossRef] [PubMed]
33. Walkoff, L.; Hobbs, S. Chest imaging in the diagnosis of occupational lung diseases. *Clin. Chest Med.* **2020**, *41*, 581–603. [CrossRef] [PubMed]
34. Khanna, S.A.; Nance, J.W.; Suliman, S.A. Detection and monitoring of interstitial lung disease in patients with systemic sclerosis. *Curr. Rheumatol. Rep.* **2022**, *24*, 166–173. [CrossRef]
35. da Silva Teles, G.B.; Macedo, A.C.S.; Chate, R.C.; Valente, V.A.T.; de Gusmao Funari, M.B.; Szarf, G. LDCT lung cancer screening in populations at different risk for lung cancer. *BMJ Open Respir. Res.* **2020**, *7*, e000455. [CrossRef]
36. Chang, S.; Li, M.; Yu, H.; Chen, X.; Deng, S.; Zhang, P.; Mou, X. Spectrum estimation-guided iterative reconstruction algorithm for dual energy CT. *IEEE Trans. Med. Imaging* **2019**, *39*, 246–258. [CrossRef]
37. Sui, X.; Huang, Y.; Song, W.; Zheng, F.; Wang, X.; Xu, X.; Wang, Z.; Jiang, J.; Jin, Z. Clinical features of pulmonary cryptococcosis in thin-section CT in immunocompetent and non-AIDS immunocompromised patients. *Radiol. Med.* **2020**, *125*, 31–38. [CrossRef]
38. Leineman, I.; Faizenberg, R.; Golimbievskaya, T.; Shostak, M.; Mazurov, V. OP0120 post-processing techniques for enhanced visualization of pulmonary vasculitis and pulmonary thrombosis by multislice spiral computed tomography in patients with systemic lupus erythematosus and antiphospholipid syndrome. *Ann. Rheum. Dis.* **2016**, *75*, 100–101. [CrossRef]
39. Hamid, N.; Portnoy, J.M.; Pandya, A. Computer-assisted clinical diagnosis and treatment. *Curr. Allergy Asthma Rep.* **2023**, *23*, 509–517. [CrossRef]
40. Young, C.; Barker, S.; Ehrlich, R.; Kistnasamy, B.; Yassi, A. Computer-aided detection for tuberculosis and silicosis in chest radiographs of gold miners of South Africa. *Int. J. Tuberc. Lung Dis.* **2020**, *24*, 444–451. [CrossRef]
41. Mushtaq, F.; Bhattacharjee, S.; Mandia, S.; Singh, K.; Chouhan, S.S.; Kumar, R.; Harjule, P. Artificial intelligence for computer aided detection of pneumoconiosis: A succinct review since 1974. *Eng. Appl. Artif. Intell.* **2024**, *133*, 108516. [CrossRef]
42. Wang, X.; Huang, F.; Cheng, Y. Computational performance optimization of support vector machine based on support vectors. *Neurocomputing* **2016**, *211*, 66–71. [CrossRef]

43. Okumura, E.; Kawashita, I.; Ishida, T. Computerized classification of pneumoconiosis on digital chest radiography artificial neural network with three stages. *J. Digit. Imaging* **2017**, *30*, 413–426. [CrossRef] [PubMed]
44. Okamoto, M.; Tominaga, M.; Shimizu, S.; Yano, C.; Masuda, K.; Nakamura, M.; Zaizen, Y.; Nouno, T.; Sakamoto, S.; Yokoyama, M. Dental technicians' pneumoconiosis. *Intern. Med.* **2017**, *56*, 3323–3326. [CrossRef] [PubMed]
45. Rosenbloom, J.; Ren, S.; Macarak, E. New frontiers in fibrotic disease therapies: The focus of the Joan and Joel Rosenbloom Center for fibrotic diseases at Thomas Jefferson University. *Matrix Biol.* **2016**, *51*, 14–25. [CrossRef]
46. Zhang, L.; Rong, R.; Li, Q.; Yang, D.M.; Yao, B.; Luo, D.; Zhang, X.; Zhu, X.; Luo, J.; Liu, Y. A deep learning-based model for screening and staging pneumoconiosis. *Sci. Rep.* **2021**, *11*, 2201. [CrossRef]
47. Chen, M.; Wang, Y.; Wang, Q.; Shi, J.; Wang, H.; Ye, Z.; Xue, P.; Qiao, Y. Impact of human and artificial intelligence collaboration on workload reduction in medical image interpretation. *NPJ Digit. Med.* **2024**, *7*, 349. [CrossRef]
48. Wu, Y.; Xia, S.; Liang, Z.; Chen, R.; Qi, S. Artificial intelligence in COPD CT images: Identification, staging, and quantitation. *Resp. Res.* **2024**, *25*, 319. [CrossRef]
49. Song, M.; Wang, J.; Yu, Z.; Wang, J.; Yang, L.; Lu, Y.; Li, B.; Wang, X.; Wang, X.; Huang, Q. PneumoLLM: Harnessing the power of large language model for pneumoconiosis diagnosis. *Med. Image Anal.* **2024**, *97*, 103248. [CrossRef]
50. Dankert, A.; Dohrmann, T.; Löser, B.; Zapf, A.; Zöllner, C.; Petzoldt, M. Pulmonary function tests for the prediction of postoperative pulmonary complications: A systematic review. *Deutsch. Ärztebl Int.* **2022**, *119*, 99. [CrossRef]
51. Presti, T.P.; Johnson, D.C. Improving pulmonary function test interpretation. *Eur. Respir. J.* **2023**, *61*, 2201858. [CrossRef] [PubMed]
52. Hou, X.; Wei, Z.; Jiang, X.; Wei, C.; Dong, L.; Li, Y.; Liang, R.; Nie, J.; Shi, Y.; Qin, X. A comprehensive retrospect on the current perspectives and future prospects of pneumoconiosis. *Front. Public Health* **2025**, *12*, 1435840. [CrossRef] [PubMed]
53. Huang, C.; Liu, R.; Cai, C.; Huang, L.; Xia, T.; Luo, S.; Wang, S.; Gan, Y.; Cai, J.; Peng, X. Investigation of factors influencing abnormal pulmonary ventilation function in occupational exposed populations and the establishment of a risk prediction model. *Sci. Rep.* **2024**, *14*, 25215. [CrossRef] [PubMed]
54. Carney, J.; McAdams, P.; McCluskey, J.; Roggli, V.L. Aluminum-induced pneumoconiosis confirmed by analytical scanning electron microscopy: A case report and review of the literature. *Ultrastruct. Pathol.* **2016**, *40*, 155–158. [CrossRef]
55. Raj, R.; Raparia, K.; Lynch, D.A.; Brown, K.K. Surgical lung biopsy for interstitial lung diseases. *Chest* **2017**, *151*, 1131–1140. [CrossRef]
56. Govender, P.; Keyes, C.M.; Hankinson, E.A.; O'Hara, C.J.; Sanchorawala, V.; Berk, J.L. Transbronchial biopsies safely diagnose amyloid lung disease. *Amyloid* **2017**, *24*, 37–41. [CrossRef]
57. Mukhopadhyay, S.; Mehta, A.C. Utility of core needle biopsies and transbronchial biopsies for diagnosing nonneoplastic lung diseases. *Arch. Pathol. Lab. Med.* **2018**, *142*, 1054–1068. [CrossRef]
58. Yoon, H.-Y.; Kim, Y.; Park, H.S.; Kang, C.-W.; Ryu, Y.J. Combined silicosis and mixed dust pneumoconiosis with rapid progression: A case report and literature review. *World J. Clin. Cases* **2018**, *6*, 1164. [CrossRef]
59. Chao, B.; Sage, A.; Cypel, M.; Liu, M.; Yeung, J.; Bai, X.; Van Raemdonck, D.; Ceulemans, L.; Neyrinck, A.; Verleden, S. The reliability and validity of donor tissue biopsies in lung transplantation. *J. Heart Lung Transpl.* **2021**, *40*, S347. [CrossRef]
60. Guo, Y.; Li, H.; Chen, H.; Li, Z.; Ding, W.; Wang, J.; Yin, Y.; Jin, L.; Sun, S.; Jing, C. Metagenomic next-generation sequencing to identify pathogens and cancer in lung biopsy tissue. *EBioMedicine* **2021**, *73*, 103639. [CrossRef]
61. Liu, G.; Xing, R.; Zhu, Z.; Wan, C. Changes in serum protein expression in patients with pneumoconiosis fibrosis treated with Vernonia anthelmintica Willd. injection. *Asian J. Surg.* **2022**, *45*, 1285–1286. [CrossRef] [PubMed]
62. Chair, S.Y.; Chan, J.Y.W.; Law, B.M.H.; Waye, M.M.Y.; Chien, W.T. Genetic susceptibility in pneumoconiosis in China: A systematic review. *Int. Arch. Occup. Environ. Health* **2023**, *96*, 45–56. [CrossRef] [PubMed]
63. Qian, Q.; Cao, X.; Wang, B.; Qu, Y.; Qian, Q.; Sun, Z.; Feng, F. Retracted: TNF-α-TNFR signal pathway inhibits autophagy and promotes apoptosis of alveolar macrophages in coal worker's pneumoconiosis. *J. Cell. Physiol.* **2019**, *234*, 5953–5963. [CrossRef] [PubMed]
64. Pingle, S.K.; Tumane, R.G.; Jawade, A.A. Neopterin: Biomarker of cell-mediated immunity and potent usage as biomarker in silicosis and other occupational diseases. *Indian J. Occup. Environ.* **2008**, *12*, 107–111. [CrossRef]
65. Hakim, M.; Broza, Y.Y.; Barash, O.; Peled, N.; Phillips, M.; Amann, A.; Haick, H. Volatile organic compounds of lung cancer and possible biochemical pathways. *Chem. Rev.* **2012**, *112*, 5949–5966. [CrossRef]
66. Pan, X.; Shi, X.; Zhang, H.; Chen, Y.; Zhou, J.; Shen, F.; Wang, J.; Jiang, R. Exosomal miR-4516 derived from ovarian cancer stem cells enhanced cisplatin tolerance in ovarian cancer by inhibiting GAS7. *Gene* **2024**, *927*, 148738. [CrossRef]
67. Kiszałkiewicz, J.; Majewski, S.; Piotrowski, W.; Górski, P.; Pastuszak-Lewandoska, D.; Migdalska-Sęk, M.; Brzeziańska-Lasota, E. Evaluation of selected IL6/STAT3 pathway molecules and miRNA expression in chronic obstructive pulmonary disease. *Sci. Rep.* **2021**, *11*, 22756. [CrossRef]
68. Yarali, E.; Eksin, E.; Torul, H.; Ganguly, A.; Tamer, U.; Papakonstantinou, P.; Erdem, A. Impedimetric detection of miRNA biomarkers using paper-based electrodes modified with bulk crystals or nanosheets of molybdenum disulfide. *Talanta* **2022**, *241*, 123233. [CrossRef]

69. Huang, R.; Yu, T.; Li, Y.; Hu, J. Upregulated has-miR-4516 as a potential biomarker for early diagnosis of dust-induced pulmonary fibrosis in patients with pneumoconiosis. *Toxicol. Res.* **2018**, *7*, 415–422. [CrossRef]
70. Cui, J.; Guan, Q.; Lv, H.; Fu, K.; Fu, R.; Feng, Z.; Chen, F.; Zhang, G. Three-dimensional nanorod array for label-free surface-enhanced Raman spectroscopy analysis of microRNA pneumoconiosis biomarkers. *Spectrochim. Acta Part A Mol. Biomol. Spectrosc.* **2021**, *261*, 120015. [CrossRef]
71. Lin, R.; Chen, B.; Luo, D.; Wu, Y.; Huang, L.; Huang, L.; Liao, Q.; Zheng, L. Molecularly imprinted polymers coated on the surface of metal-organic frameworks combined with liquid chromatography-tandem mass spectrometry for the detection of free N-acetylneuraminic acid in serum of pneumoconiosis patients. *Microchem. J.* **2022**, *181*, 107821. [CrossRef]
72. Wang, H.; Zhou, S.; Liu, Y.; Yu, Y.; Xu, S.; Peng, L.; Ni, C. Exploration study on serum metabolic profiles of Chinese male patients with artificial stone silicosis, silicosis, and coal worker's pneumoconiosis. *Toxicol. Lett.* **2022**, *356*, 132–142. [CrossRef] [PubMed]
73. Hamdy, N.M.; Zaki, M.B.; Abdelmaksoud, N.M.; Elshaer, S.S.; Abd-Elmawla, M.A.; Rizk, N.I.; Fathi, D.; Doghish, A.S.; Abulsoud, A.I. Comprehensive insights and in silico analysis into the emerging role of LincRNAs in lung diseases pathogenesis; a step toward ncRNA precision. *Funct. Integr. Genom.* **2025**, *25*, 34. [CrossRef] [PubMed]
74. Bahia, W.; Soltani, I.; Abidi, A.; Mahdhi, A.; Mastouri, M.; Ferchichi, S.; Almawi, W.Y. Structural impact, ligand-protein interactions, and molecular phenotypic effects of TGF-β1 gene variants: In silico analysis with implications for idiopathic pulmonary fibrosis. *Gene* **2024**, *922*, 148565. [CrossRef]
75. Deng, C.-W.; Zhang, X.-X.; Lin, J.-H.; Huang, L.-F.; Qu, Y.-L.; Bai, C. Association between genetic variants of transforming growth factor-β1 and susceptibility of pneumoconiosis: A meta-analysis. *Chin. Med. J.* **2017**, *130*, 357–364. [CrossRef]
76. Zhang, Y.; Sun, D.; Song, Y.; Ye, Q. Candidate gene polymorphisms associated with silicosis and coal workers' pneumoconiosis: A systematic review and meta-analysis. *BMC Pulm. Med.* **2024**, *24*, 580. [CrossRef]
77. Sullivan, D.E.; Ferris, M.; Pociask, D.; Brody, A.R. The latent form of TGFβ1 is induced by TNFα through an ERK specific pathway and is activated by asbestos-derived reactive oxygen species in vitro and in vivo. *J. Immunotoxicol.* **2008**, *5*, 145–149. [CrossRef]
78. Wojnowski, W.; Dymerski, T.; Gębicki, J.; Namieśnik, J. Electronic noses in medical diagnostics. *Curr. Med. Chem.* **2019**, *26*, 197–215. [CrossRef]
79. Behera, B.; Joshi, R.; Vishnu, G.A.; Bhalerao, S.; Pandya, H.J. Electronic nose: A non-invasive technology for breath analysis of diabetes and lung cancer patients. *J. Breath Res.* **2019**, *13*, 024001. [CrossRef]
80. Scarlata, S.; Finamore, P.; Meszaros, M.; Dragonieri, S.; Bikov, A. The role of electronic noses in phenotyping patients with chronic obstructive pulmonary disease. *Biosensors* **2020**, *10*, 171. [CrossRef]
81. Hsiao-Yu, Y.; Shie, R.-H.; Che-Jui, C.; Pau-Chung, C. Development of breath test for pneumoconiosis: A case-control study. *Resp. Res.* **2017**, *18*, 178.
82. Hashoul, D.; Haick, H. Sensors for detecting pulmonary diseases from exhaled breath. *Eur. Respir. Rev.* **2019**, *28*, 190011. [CrossRef] [PubMed]
83. Marzorati, D.; Mainardi, L.; Sedda, G.; Gasparri, R.; Spaggiari, L.; Cerveri, P. A review of exhaled breath: A key role in lung cancer diagnosis. *J. Breath Res.* **2019**, *13*, 034001. [CrossRef] [PubMed]
84. Park, S.Y.; Kim, Y.; Kim, T.; Eom, T.H.; Kim, S.Y.; Jang, H.W. Chemoresistive materials for electronic nose: Progress, perspectives, and challenges. *InfoMat* **2019**, *1*, 289–316. [CrossRef]
85. Guntner, A.T.; Abegg, S.; Konigstein, K.; Gerber, P.A.; Schmidt-Trucksass, A.; Pratsinis, S.E. Breath sensors for health monitoring. *ACS Sens.* **2019**, *4*, 268–280. [CrossRef]
86. Adiguzel, Y.; Kulah, H. Breath sensors for lung cancer diagnosis. *Biosens. Bioelectron.* **2015**, *65*, 121–138. [CrossRef]
87. Xuan, W.; Zheng, L.; Bunes, B.R.; Crane, N.; Zhou, F.; Zang, L. Engineering solutions to breath tests based on an e-nose system for silicosis screening and early detection in miners. *J. Breath Res.* **2022**, *16*, 036001. [CrossRef]
88. Broza, Y.Y.; Mochalski, P.; Ruzsanyi, V.; Amann, A.; Haick, H. Hybrid volatolomics and disease detection. *Angew. Chem. Int. Ed.* **2015**, *54*, 11036–11048. [CrossRef]
89. Hu, W.; Wu, W.; Jian, Y.; Haick, H.; Zhang, G.; Qian, Y.; Yuan, M.; Yao, M. Volatolomics in healthcare and its advanced detection technology. *Nano Res.* **2022**, *15*, 8185–8213. [CrossRef]
90. Song, X.; Qian, L.; Fu, L.; Cao, R.; Wang, X.; Chen, M. Real-time mildew detection and gradation in simulated containerized soybeans: Insights from GC-IMS analysis of mVOCs and VOCs. *Food Sci. Nutr.* **2024**, *12*, 6772–6788. [CrossRef]
91. Yang, C.; Jiao, L.; Dong, C.; Wen, X.; Lin, P.; Duan, D.; Li, G.; Zhao, C.; Fu, X.; Dong, D. Long-range infrared absorption spectroscopy and fast mass spectrometry for rapid online measurements of volatile organic compounds from black tea fermentation. *Food Chem.* **2024**, *449*, 139211. [CrossRef] [PubMed]
92. Xiong, H.; Zhang, X.; Sun, J.; Xue, Y.; Yu, W.; Mou, S.; Hsia, K.J.; Wan, H.; Wang, P. Recent advances in biosensors detecting biomarkers from exhaled breath and saliva for respiratory disease diagnosis. *Biosens. Bioelectron.* **2025**, *267*, 116820. [CrossRef] [PubMed]
93. Kołodziej, M.; de Veer, M.J.; Cholewa, M.; Egan, G.F.; Thompson, B.R. Lung function imaging methods in cystic fibrosis pulmonary disease. *Respir. Res.* **2017**, *18*, 96. [CrossRef] [PubMed]

94. Zhou, Z.; Xu, T.; Zhang, X. Empowerment of AI algorithms in biochemical sensors. *TrAC Trends Anal. Chem.* **2024**, *173*, 117613. [CrossRef]
95. Ni, Y.; Zang, X.; Yang, Y.; Gong, Z.; Li, H.; Chen, J.; Wu, C.; Huang, J.; Lai, Y. Environmental stability stretchable organic hydrogel humidity sensor for respiratory monitoring with ultrahigh sensitivity. *Adv. Func. Mater.* **2024**, *34*, 2402853. [CrossRef]
96. Lagopati, N.; Valamvanos, T.-F.; Proutsou, V.; Karachalios, K.; Pippa, N.; Gatou, M.-A.; Vagena, I.-A.; Cela, S.; Pavlatou, E.A.; Gazouli, M. The role of nano-sensors in breath analysis for early and non-invasive disease diagnosis. *Chemosensors* **2023**, *11*, 317. [CrossRef]
97. Lourenço, C.; Turner, C. Breath analysis in disease diagnosis: Methodological considerations and applications. *Metabolites* **2014**, *4*, 465–498. [CrossRef]
98. Najafabadi, S.A.N.; Huang, C.; Betlem, K.; van Voorthuizen, T.A.; de Smet, L.C.; Ghatkesar, M.K.; van Dongen, M.; van der Veen, M.A. Advancements in inkjet printing of metal and covalent organic frameworks: Process design and ink optimization. *ACS Appl. Mater. Interfaces* **2025**, *17*, 11469–11494. [CrossRef]
99. Emmanuel, A.O.; Zine, N.; Raffin, G.; Jaffrezic-Renault, N.; Elaissari, A.; Errachid, A. Integrated breath analysis technologies: Current advances and future prospects. *TrAC Trend. Anal. Chem.* **2024**, *181*, 118048.
100. Zuliani, C.; Luque, J.; Falco, C.; Gardner, E.; De Luca, A.; Vincent, T.; Tripathy, S.; Ali, Z.; Udrea, F. Flow compensated gas sensing array for improved performances in breath-analysis applications. *IEEE Sens. Lett.* **2020**, *4*, 5500504. [CrossRef]
101. Janeiro, R.; Flores, R.; Viegas, J. Silicon photonics waveguide array sensor for selective detection of VOCs at room temperature. *Sci. Rep.* **2019**, *9*, 17099. [CrossRef] [PubMed]
102. Li, Q.; He, C.; Wang, C.; Huang, Y.; Yu, J.; Wang, C.; Li, W.; Zhang, X.; Zhang, F.; Qing, G. Sustainable, insoluble, and photonic cellulose nanocrystal patches for calcium ion sensing in sweat. *Small* **2023**, *19*, 2207932. [CrossRef]
103. Singh, S.; Varma, P.; Sreelekha, G.; Adak, C.; Shukla, R.P.; Kamble, V.B. Metal oxide-based gas sensor array for VOCs determination in complex mixtures using machine learning. *Microchim. Acta* **2024**, *191*, 196. [CrossRef] [PubMed]
104. Li, Y.; Wang, Z.; Zhao, T.; Li, H.; Jiang, J.; Ye, J. Electronic nose for the detection and discrimination of volatile organic compounds: Application, challenges, and perspectives. *TrAC Trend. Anal. Chem.* **2024**, *180*, 117958. [CrossRef]
105. Yu, K.-L.; Yang, H.-C.; Lee, C.-F.; Wu, S.-Y.; Ye, Z.-K.; Rai, S.K.; Lee, M.-R.; Tang, K.-T.; Wang, J.-Y. Exhaled breath analysis using a novel electronic nose for different respiratory disease entities. *Lung* **2025**, *203*, 14. [CrossRef]
106. Cai, M.; Xu, S.; Zhou, X.; Lu, H. Electronic nose humidity compensation system based on rapid detection. *Sensors* **2024**, *24*, 5881. [CrossRef]
107. Togbe, D.; Benmerzoug, S.; Jacobs, M.; Ryffel, B.; Quesniaux, V. Silica-related diseases in the modern world: A role for self-DNA sensing in lung inflammatory diseases. *Allergy* **2020**, *75*, 3009–3010. [CrossRef]
108. Hendrick, H.; Hidayat, R.; Horng, G.-J.; Wang, Z.-H. Non-invasive method for tuberculosis exhaled breath classification using electronic nose. *IEEE Sens. J.* **2021**, *21*, 11184–11191. [CrossRef]
109. Kononov, A.; Korotetsky, B.; Jahatspanian, I.; Gubal, A.; Vasiliev, A.; Arsenjev, A.; Nefedov, A.; Barchuk, A.; Gorbunov, I.; Kozyrev, K. Online breath analysis using metal oxide semiconductor sensors (electronic nose) for diagnosis of lung cancer. *J. Breath Res.* **2019**, *14*, 016004. [CrossRef]
110. Wu, Y.; Yuan, J.; Li, X.; Zhang, X. GeP_3 monolayer as a promising 2D sensing materials in detecting SO_2, H_2S, SOF_2 and SO_2F_2. *Phys. Scr.* **2024**, *99*, 085956. [CrossRef]
111. Ohnishi, M.; Ishimoto, C.I.C.; Aoki, M.A.M. Fundamental properties and characteristics of a gas-sensitive device, made of a surface acoustic wave device and Langmuir-Blodgett film, used in a recognition system for odoriferous gases. *Jpn. J. Appl. Phys.* **1994**, *33*, 5987. [CrossRef]
112. Burgués, J.; Hernández, V.; Lilienthal, A.J.; Marco, S. Gas distribution mapping and source localization using a 3D grid of metal oxide semiconductor sensors. *Sens. Actuators B Chem.* **2020**, *304*, 127309. [CrossRef]
113. Yang, J.; Wang, T.; Zhu, C.; Yin, X.; Dong, P.; Wu, X. A composite gas-phase explosives sensor with frequency-resistance dual-signal display based on surface acoustic wave. *ACS Appl. Electron. Mater.* **2023**, *5*, 1526–1535. [CrossRef]
114. Ji, H.; Zeng, W.; Li, Y. Gas sensing mechanisms of metal oxide semiconductors: A focus review. *Nanoscale* **2019**, *11*, 22664–22684. [CrossRef]
115. Krishna, K.G.; Parne, S.; Pothukanuri, N.; Kathirvelu, V.; Gandi, S.; Joshi, D. Nanostructured metal oxide semiconductor-based gas sensors: A comprehensive review. *Sens. Actuators A Phys.* **2022**, *341*, 113578. [CrossRef]
116. Zhou, X.; Cheng, X.; Zhu, Y.; Elzatahry, A.A.; Alghamdi, A.; Deng, Y.; Zhao, D. Ordered porous metal oxide semiconductors for gas sensing. *Chin. Chem. Lett.* **2018**, *29*, 405–416. [CrossRef]
117. Shooshtari, M.; Salehi, A. An electronic nose based on carbon nanotube-titanium dioxide hybrid nanostructures for detection and discrimination of volatile organic compounds. *Sens. Actuators B Chem.* **2022**, *357*, 131418. [CrossRef]
118. Amoah, N.A.; Xu, G.; Kumar, A.R.; Wang, Y. Calibration of low-cost particulate matter sensors for coal dust monitoring. *Sci. Total Environ.* **2023**, *859*, 160336. [CrossRef]

119. Tirzïte, M.; Bukovskis, M.; Strazda, G.; Jurka, N.; Taivans, I. Detection of lung cancer with electronic nose and logistic regression analysis. *J. Breath Res.* **2018**, *13*, 016006. [CrossRef]
120. Kovacs, D.; Bikov, A.; Losonczy, G.; Murakozy, G.; Horvath, I. Follow up of lung transplant recipients using an electronic nose. *J. Breath Res.* **2013**, *7*, 017117. [CrossRef]

Disclaimer/Publisher's Note: The statements, opinions and data contained in all publications are solely those of the individual author(s) and contributor(s) and not of MDPI and/or the editor(s). MDPI and/or the editor(s) disclaim responsibility for any injury to people or property resulting from any ideas, methods, instructions or products referred to in the content.

chemosensors MDPI

Review

Progress in Layered Double Hydroxide-Based Materials for Gas and Electrochemical Sensing Applications

Waseem Raza [1], Khursheed Ahmad [2,*] and Tae Hwan Oh [2,*]

[1] Department of Chemistry and Biochemistry, Facultad de Farmacia, Universidad San Pablo-CEU, CEU Universities, Urbanización Montepríncipe, Boadilla del Monte, 28668 Madrid, Spain
[2] School of Chemical Engineering, Yeungnam University, Gyeongsan 38541, Republic of Korea
* Correspondence: khursheed@yu.ac.kr (K.A.); taehwanoh@ynu.ac.kr (T.H.O.)

Abstract: In the current scenario, it is considered that environmental pollution is one of the significant challenges for the global world. Various toxic and hazardous substances such as hydrazine, phenolic compounds, and pharmaceutical waste significantly contribute to environmental pollution. Exposure to such substances and compounds increases the chances of negative effects on human health as well as the environment. Therefore, it is considered that monitoring toxic gases and hazardous substances/compounds is of great significance. In the past few years, layered double hydroxide (LDH)-based materials have received significant interest for gas sensing and electrochemical sensing studies. The presence of layered structured, larger surface area, decent conductivity, and electrochemical properties makes them a suitable material for sensing applications. This motivates us to summarize the recent progress in the development of LDH material-based gas and electrochemical sensors for the detection of toxic and hazardous gases/compounds. It was observed in previous reports that LDH-based materials are promising candidates for gas sensing as well as electrochemical sensing applications. It was found that LDH and its composites may exhibit larger surface areas and high electrical conductivity when combined with other materials such as metal oxides, MXenes, polymers, and metal sulfides. Thus, researchers prepared hybrid composites of LDH-based materials for gas and electrochemical sensing applications. It is worth mentioning that many solvents which have negative impacts on the environment could not be detected by electrochemical methods, while some toxic compounds/substances could not be determine by gas sensing methods. This may create a gap between the determinations of different kinds of pollutants that exist in the environment. Thus, it is required to find a bi-functional material which can be used for kind of sensing technology. In addition, it may also overcome the limitations or gap between the two sensing techniques. LDH-based materials have demonstrated excellent performance in gas and electrochemical sensing technologies. Thus, it would be of great significance to employ the single LDH-based materials for gas as well as electrochemical sensing applications. In this review article, we have tried our best to compile the progress in the various LDH-based materials for gas sensing and electrochemical sensing applications towards the detection of hazardous compounds.

Keywords: gas sensors; layered double hydroxides (LDHs); hazardous compounds; sensors

1. Introduction

In the past few years, it was observed that environmental pollution has been increasing significantly due to urbanization and industrialization [1]. The increase in industrial activities may led to a substantial increase in the burning of fossil fuels, which releases harmful pollutants such as carbon dioxide (CO_2), sulfur dioxide (SO_2), and particulate matters into the atmosphere, which may contribute to air pollution and climate change [2,3]. Industrial processes may also produce a large amount of waste water-containing chemicals such as hydrazine, hydrogen peroxide (H_2O_2), phenolic compounds, heavy metals, etc., which are often discharged into water bodies, and cause negative impacts on the environment, aquatic life, and human health [4–7]. In addition, exposure to toxic gases and hazardous compounds may cause several health issues such as chronic diseases and respiratory diseases [8]. Thus, it is clear that monitoring toxic gases and hazardous compounds would be of great significance in order to reduce their negative impacts on human health and the environment.

Sensor technology is one of the promising approaches to monitor toxic gases and compounds [9]. Gas sensing is an efficient technique for the monitoring of various toxic gases and has been widely used for the detection of various gases [10,11]. According to the reported studies, gas sensors have promising features for the detection of various organic volatile gases (VOCs) [12], whereas the electrochemical method is the most promising technique for the sensing of various toxic/hazardous compounds [13]. In previous reports, various materials such as reduced graphene oxide [14], carbon nitride [15], metal–organic-frameworks [16], MXenes [17], polymer [18], metal oxides [19], and layered double hydroxides (LDHs) [20] were used in sensing applications.

Recently, LDH-based materials have received enormous attention because of their excellent optoelectronic properties [20]. LDH materials, which are also known as metal alloy hydroxides or hydrotalcite, are inorganic materials with a two-dimensional structure (2D) [21]. Generally, LDH materials consists of a mixture of metal hydroxides with the presence of different anions (anionic layer) and oxidation states (cationic layer), which are present next to each other [22], and LDH materials possess excellent structural and chemical properties. The high specific surface area, acceptable stability, and layered structure of LDH materials makes them suitable for various optical and electrical applications. In previous reports, copper (Cu)-, nickel (Ni)-, aluminum (Al)-, cobalt (Co)-, magnesium (Mg)-, iron (Fe)-, and manganese (Mn)-based LDH materials were prepared by various synthetic procedures [23–30]. The LDH materials were also combined with various metal oxides and polymers to further improve their properties for various optoelectronic applications [29]. The LDH-based hybrid composites attracted the scientific community due to their unique functional, structural properties, and synergistic interactions. LDH-based composites also exhibited high electrical conductivity, larger surface areas, and various active sites for adsorption/catalytic reactions [31]. For gas sensing studies, LDH and its composites facilitated electron transfer and enabled the sensors to detect the toxic gases rapidly. Similarly for electrochemical sensing studies, LDH-based hybrid materials may enhance sensitivity, selectivity, signal response, and stability for the determination of heavy metals and environmental pollutants.

Herein, we report on the use of LDH-based materials for gas and electrochemical sensing of environmental pollutants. We believe that recent articles may be valuable to material scientists who are sincerely engaged in the development of cost-effective LDH-based materials for these sensing applications.

2. LDH-Based Materials in Gas Sensors

With the rapid industrial development and growing living standards, air population has become a serious global problem and poses serious risks to human and animal health, ecosystems, and the environment [32]. According to a World Health Organization (WHO) report, more than seven million people die worldwide every year due to air pollution [31].

Major contributors to air pollution are mobile vehicles, oil refineries, power plants, and chemical industries. Therefore, monitoring and controlling combustion-related emissions are of prime importance to protect public health, assure food safety, support medical growth, enhance air quality, and improve environmental sustainability. A gas sensor is a chemical sensing device which detects gas molecules in the environment by converting chemical information into an electronic signal such as current, frequency, or voltage change and reduces risks to human health at an early stage [33–36]. Hence, there is an urgent need to develop advanced, high quality, efficient energy gas sensors for environmental monitoring and society protection for the real time detection and monitoring of toxic and flammable gases, as well as volatile organic compounds (VOCs) [37]. With the development of the internet and wireless technology, gas-sensing intelligent systems have been securely integrated with smart living, becoming a critical part of our daily life and modern industries [35]. Gas sensors are widely applied in biochemical, chemical, and food industries for applications in diagnostics, air quality monitoring, food testing, and the detection of toxic, flammable, and explosive gases [38]. For example, ethanol sensors are widely explored for various applications including alcohol detection on human breath for drunk driving, to assess the quality of wine, to monitor industrial leakages, and to ensure food and biomedical safety [39]. The advancement in science and technology has motivated the way for the development of gas sensors with high sensitivity, good selectivity, and long-term stability [40]. As a result, great efforts have been carried out to design novel sensing materials since the first chemresisitive metal oxide-based gas sensor was introduced. So far, in addition to conventional semiconductor oxide, a variety of gas-sensing materials such as conductive polymer, organic compounds, and carbon nanotubes have been utilized for gas detection using techniques like catalytic, optical, electrochemical, and acoustic gas sensors [41,42]. However, the performance of the sensors is mainly dependent on some basic properties such as sensitivity, selectivity, detection limit, response, and recovery time. In addition, factors such as low power consumption, shape, size, and wireless functionality play a critical role in boosting gas-sensing performance and making sensors more appealing [43]. Figure 1 exhibits the classification of gas-sensing materials into two categories based on electrochemical and other principles, associated with fundamentals of the gas-sensing and gas-sensitive materials [44]. Nanomaterials have attracted significant attention due to their outstanding properties, such as large surface area, and superior mechanical, chemical, physical, as well as electrical, characteristics. The large surface area is critical for improving sensing performance by providing a greater number of active sites. Moreover, nanomaterials can be further functionalized. To enhance the sensitivity, selectivity, and stability of sensors, scientists have devoted substantial effort into modifying electrodes with various nanomaterials, such as multiwalled carbon nanotubes (MWCNTs), gold nanostars, metal–organic frameworks (MOFs), covalent organic frameworks (COFs), and layered double hydroxides (LDHs) [42].

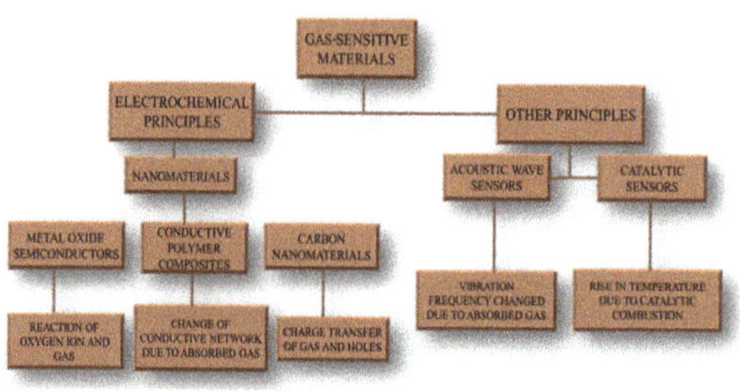

Figure 1. Classification of gas-sensing materials [9,10,34–36,44].

Among the various electrode modification nanomaterials, LDHs have attracted significant attention and have become a rapidly increasing area of research in recent years. LDHs are exciting materials characterized as two-dimensional (2D) anionic lamellar clay materials [40]. LDHs demonstrate various chemical and physical properties due to their unique molecular structures [45]. These include a high surface area, catalytic activity, chemical and thermal stability, anion exchange capability, tunable and flexible interlayer spacing, and cost-effectiveness [46]. LDHs, also referred to as hydrotalcites or metal hydroxide alloys, are inorganic materials made up of layers of brucite-like sheets stacked together [46]. These 2D layered structures are aligned parallel and are held together by electrostatic interactions such as hydrogen bonds or Vander Walls forces, along the vertical axis as shown in Figure 2 [47]. LDHs are made up of a combination of metal hydroxides with different oxidation states (forming cationic layer) and anions (forming anionic layers) ordered next to each other (Figure 2) [47]. The natural inorganic hydrotalcite was first discovered in Sweden around 1842, and approximately a century later, Feitknecht successfully synthesized it in the laboratory in 1942 [48,49]. It was then registered as the first patent under the name of Brocker in 1970 [47]. However, research on using LDHs as gas sensors began back in 2006. The general formula of an LDH is $[M_{(1-x)}{}^{2+} M_x{}^{3+} (OH)_2]^{x+}(A^{n-})_{x/n}{\cdot}mH_2O$, where M^{2+} represents a divalent cation (Mg^{2+}, Zn^{2+}, Ni^{2+}, Co^{2+}, Ca^{2+}...), M^{3+} is a trivalent cation (Al^{3+}, Ga^{3+}, Cr^{3+}, Fe^{3+}...), A^{n-} denotes the interlayer anion of valance n ($CO_3{}^{2-}$, $SO_4{}^{2-}$, polyoxometalates, porphyrins, etc.), and X is the molar ration of $M^{3+}/(M^{2+} + M^{3+})$, typically ranging from 0.2 and 0.4 due to structural stability factors (Figure 2) [47,50–52].

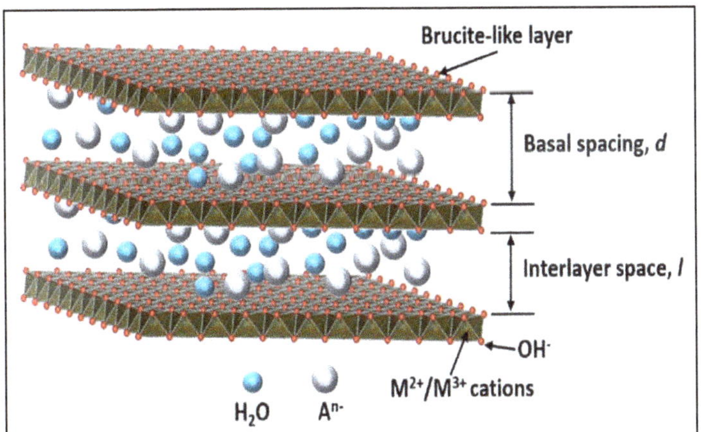

Figure 2. Schematic illustration of typical LDH structure and its chemical components [52].

Gas sensors composed of organic–inorganic hybrids offer several advantages, such as molecular-scale thickness control, low cost, and high reproducibility. However, their design is challenging due to difference in vapor pressure and decomposition temperatures between the organic and inorganic components using standard physical techniques. Taking LDH nanosheets into account, they can serve as 2D building blocks for producing ultrathin films (UTFs) with superior functionalities. Therefore, $(\text{calcein-ZnS})_{30}$ UTF was fabricated using two-step layer by layer (LBL) assembly process using calcein and exfoliation Zn_2Al layered LDH nanosheets, followed by an in situ gas/solid reaction with H_2S [50]. The synthesized $(\text{calcein-ZnS})_{30}$ UTF composite, made up of organic molecules and inorganic semiconductors, facilitates the development of an innovative ethanol sensor with high sensitivity, fast response, as well as quick recovery time, operating at a relatively low working temperature (90 °C). Xiao et al. [53] reported the synthesis of a $\{\text{pyrenetetrasulfonate(PyTS)/ZnS}\}_n$ UTF sensor using an LBL electronic deposition technique. This process consists of an alternating assembly of 1,3,6,8-PyTS and exfoliated Zn_2Al LDH nanosheets, followed by an effective in situ gas/solid sulfurization reaction of H_2S. The UV/Vis spectroscopy was used to observe the assembly process, which indicates the regular stepwise growth of $(PyTS/LDH)_n$ UTFs with each deposition cycle. The performance of both synthesized $(PyTS/LDH)_n$ UTFs and sulfurized $(PyTS/LDH)_n$ UTFs was investigated in the presence of various gases such as ammonia, hydrogen, carbon monoxide, methane, acetylene, and ethanol. While both sensors exhibited sensitivity to all tested gases, the sulfurized $(PyTS/LDH)_n$ UTFs showed a much better response to ethanol, even at a relatively low operating temperature. They reported that the enhanced performance of the sensor towards ethanol could be attributed to the synergistic interactions between the inorganic ZnS and organic pyrene components. Morandi et al. reported the synthesis of Pt/Zn/Al layered double hydroxides as gas-sensing materials [54]. They employed two simple synthesis routes, classical co-precipitation and the sol–gel method, for the synthesis of a Pt/Zn/Al LDH gas sensor. The Pt/Zn/Al LDH sensor synthesized with the help of the sol–gel method was found to be pure, and the results are consistent with previously reported X-ray diffraction (XRD) results (Figure 3a). However, the same Pt/Zn/Al LDH sensor synthesized via the co-prepetition method exhibited trace amounts of the ZnO phase in its XRD pattern (Figure 3b). The sensor's performance was then investigated using different VOCs such as CO (100 ppm), CH_4 (500 ppm), and benzene (C_6H_6, 10 ppm) with and without the addition of Pt. These measurements were performed at 450 °C using a conductance measurement method. As shown in Figure 3c, the addition of Pt improves the electrical response, with the highest sensitivity found for CH_4 in dry air. However, it is difficult to distinguish the response of different gases in dry air. Therefore, the sensing performance of all three VOCs was also investigated in the presence of wet air. As illustrated in Figure 3d, carbon monoxide exhibits the highest response under these conditions.

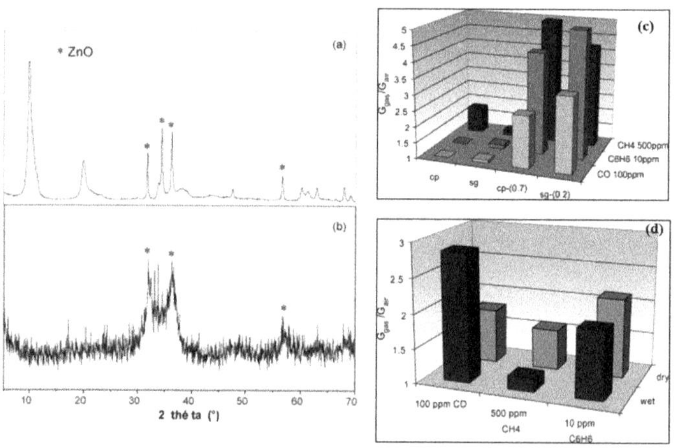

Figure 3. (**a**) XRD pattern of co-precipitated LDH (**a**) with 0.7% of Pt and (**b**) cp-2 mixed oxide. (**c**) Responses to CO, C_6H_6, and CH_4 of two Pt-containing samples (cp-(0.7), sg-(0.2)) and two correspondent samples without Pt (cp, sg) at 450 °C in dry air. (**d**) Responses to CO, CH_4, and C_6H_6 in dry and wet air at 400 °C of cp-(2) thick films [54].

In another study, Guan et al. reported the synthesis of a $ZnO/ZnAl_2O_4$ sensor via the calcination method [55]. In a typical synthesis approach, all starting reagents were dissolved in deionized water using the co-precipitation method. The resulting slurry was then transferred to a Teflon-lined autoclave and aged at 100 °C for 12 h. Finally, the products were centrifuged and washed several times with water and ethanol followed by calcination at different temperatures in air. The sensing performances of the sensors calcined at various temperatures were investigated against ethanol. They reported that the sensor calcined at 1000 °C exhibited a better response to ethanol compared to other samples calcined at 600 and 800 °C. Additionally, the sample calcined at 1000 °C presents fine repeatability and good selectivity for ethanol. They reported that the improved performance of the sample calcined at 1000 °C could be attributed to better crystallinity of $ZnAl_2O_4$ than that calcined at 800 or 600 °C, respectively. Liu et al. [55] reported the preparation of a $ZnO/ZnFe_2O_4$ composite sensor with a hexagonal nanostructure via a calcination method. For this firstly, Zn_2Fe-LDH with sodium dodecyl sulfate (SDS) was synthesized using a combination of co-precipitation and hydrothermal methods at 100 °C for 10 h. The resulting powder was filtered and washed several times with water and dried at 60 °C for 12 h in electric oven. Finally, the $ZnO/ZnFe_2O_4$ composite sensor was calcined at various temperatures to obtain the desired hexagonal structure as shown in Figure 4a. They focused on the typical problem tackled by metal oxide sensors, which usually need high operating temperatures around 200 °C. They found that by applying light irradiation, the operating temperature could be significantly lowered (Figure 4a). They observed that under light illumination, the gas-sensing performance of the $ZnO/ZnFe_2O_4$ composite, calcined at 600 °C, was significantly enhanced for detecting triethylamine (TEA) at a reduced operating temperature of 80 °C. The gas-sensing performance of the $ZnO/ZnFe_2O_4$ sensor towards TEA was investigated through independent resistance change measurements conducted at an operating temperature of 80 °C. In a typical gas-sensing experiment, liquid TEA was first injected followed by gas evaporation. The chamber was then irradiated with a light wavelength of $\lambda > 320$ nm and a power of 150 mW/cm^2. It can be seen from Figure 4b that the $ZnO/ZnFe_2O_4$ sensor exhibited a gradual increase in R_o/R_t upon light irradiation for varying TEA concentrations ranging from 5 to 1000 ppm. Initially, the resistance (R_t) for TEA decreased means R_o/R_t continues to increase upon introducing TEA gas into the chamber. This might be the chemical reaction of TEA with surface absorbed oxide ions (O_2^-) resulting in electron transfer back to the sensor, leading to a decrease in resistance (R_t). However, the sensor's response (R_o/R_t) decreases to the initial value after switching off the light and being exposed to air (Figure 4b). As shown in Figure 4c, a comparative evaluation of the $ZnO/ZnFe_2O_4$ sensor calcined at 600 °C was investigated to assess its performance in the presence and absence of light illumination, while maintaining all other experimental conditions the same. The results indicate that a noticeable difference in response across all TEA concentrations clearly show that light illumination significantly enhances the gas-sensing performance of the sensor. Hong et al. reported the synthesis of hierarchical flower-like Ni-Al-LDH and Ni-Fe-Al-LDH intercalation sensors using a facile one-step hydrothermal approach for NO_x sensing at room temperature (RT) [56]. In another study, Sun et al. reported the preparation of three-dimensional hierarchical flower-like a Mg-Al-LDH sensor using a simple hydrothermal method for sensing NO_x at RT [57]. They reported that the hierarchical porous nanostructure provides natural channels for efficient and fast carrier transportation, resulting in fast response and recovery times to 100 ppm NO_x at RT. Qu et al. [58] designed a porous double-shelled nanocage $ZnO/Ni_{0.9}Zn_{0.1}O$ sensor using a unique metal–organic framework route. They reported that the designed $ZnO/Ni_{0.9}Zn_{0.1}O$ sensor demonstrates high sensitivity and selectivity towards xylene compared to pristine ZnO nanocages. The outstanding gas-sensing performance of the $ZnO/Ni_{0.9}Zn_{0.1}O$ sensor is ascribed to the large surface area, high porosity, and synergistic effect of ZnO and $Ni_{0.9}Zn_{0.1}O$.

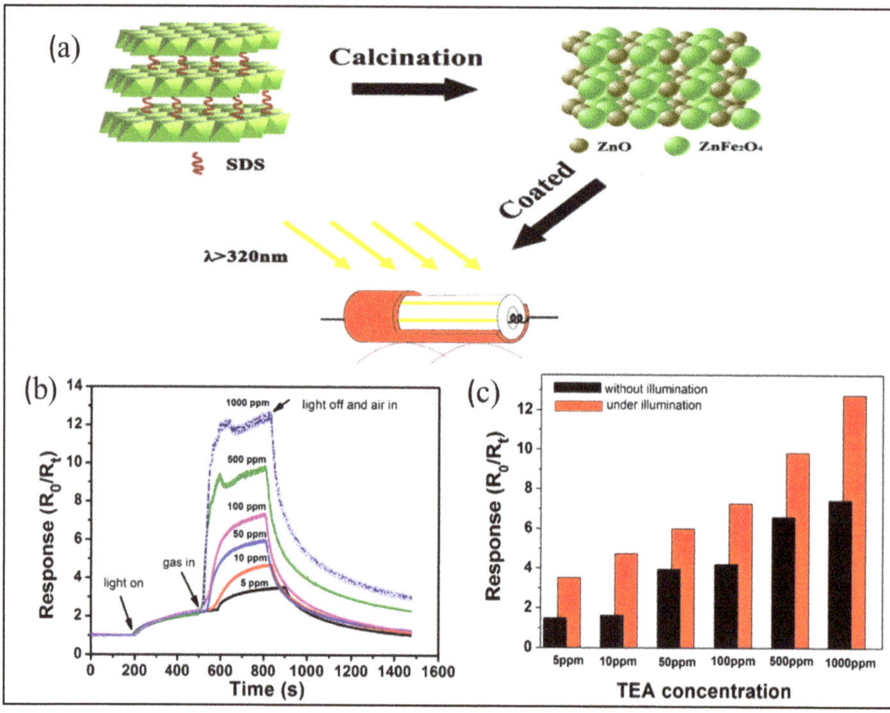

Figure 4. (**a**) Pictorial presentation for synthesis of gas sensor based on $ZnO/ZnFe_2O_4$ composite. (**b**) Sensing response towards various TEA concentrations as function of time under light illumination at working temperature of 80 °C. (**c**) Comparative sensing response for TEA in presence and absence of light illumination at 80 °C [55].

Zhang et al. developed a highly sensitive, durable, and active sensor for sensing o VOCs using a combination of p-n heterojunction [59]. The sensor was prepared using a facile thermal conversion of hierarchical CoTi LDHs precursors at 300–400 °C [59]. The prepared optimized mesoporous hierarchical Co_3O_4–TiO_2 nanocomposite demonstrated excellent sensing performance for toluene and xylene at an operating temperature of 115 °C. In another study, Kang et al. designed a novel Ni-Al–LDH sensor using a hydrothermal method for detecting different gases including ozone, H_2, NO_2, and ethanol [60]. Their investigations indicate that the synthesized Ni-Al–LDH sensor exhibited exceptional and selective detection of ozone, even in the presence of other interfering gases such as H_2, NO_2, and C_2H_5OH. Moreover, the developed sensor exhibited an outstanding response and recovery times, along with excellent selectivity and stability towards ozone. A phenomenological sensing performance of the chlorine intercalated LDH sensor was investigated by Polese et al. [48] for five different VOCs (CO, CO_2, NO, NO_2, and CH_4). Their study highlighted that LDHs are a class of nanomaterials that exhibit outstanding properties such as large surface area, high porosity, and excellent chemical interaction capabilities with a wide range of analytes. Additionally, these materials are easy to synthesize and tailor for sensing performance. In another study, Qu et al. [61] used a MOF-based synthesis approach to develop a $Co_3O_4/NiCo_2O_4$ double-shelled nanocage for acetone sensing. To design a MOF-derived $Co_3O_4/NiCo_2O_4$, zeolite imidazolate framework-67 (ZIF-67) was first synthesized. Afterwards, the synthesized ZIF-67 was then dispersed into an ethanol solution containing $Ni(NO_3)_2{\cdot}6H_2O$ during continuous stirring. Finally, the mixture was annealed at 350 °C for 2 h, as shown in Figure 5a. To confirm the morphology, crystallinity, and elemental composition of the synthesized $Co_3O_4/NiCo_2O_4$ double-shelled nanocages sensor, SEM, TEM, XRD, and elemental analyses were carried out. It could be seen from the XRD analysis (Figure 5b,e) that the synthesized sensor is pure and crystalline in nature, which is consistent with the reported literature. The scanning electron microscopic (SEM) image of ZIF-67 and Co_3O_4 are displayed in Figure 5c and 5d, respectively. Surface analysis exhibits the formation of rhombo-dodecahedron-shaped structures with an average size of 500 nm. The SEM results for ZIF-67/Ni-Co LDH and $Co_3O_4/NiCo_2O_4$ are depicted in Figure 5f and 5h, respectively. On the other hand, transmission electron microscopic (TEM) images of the ZIF-67/Ni-Co LDH and $Co_3O_4/NiCo_2O_4$ are presented in Figure 5g and 5i, respectively. The SEM results exhibited the formation of a rougher surface texture of $Co_3O_4/NiCo_2O_4$ compared to the ZIF-67/Ni-Co LDH.

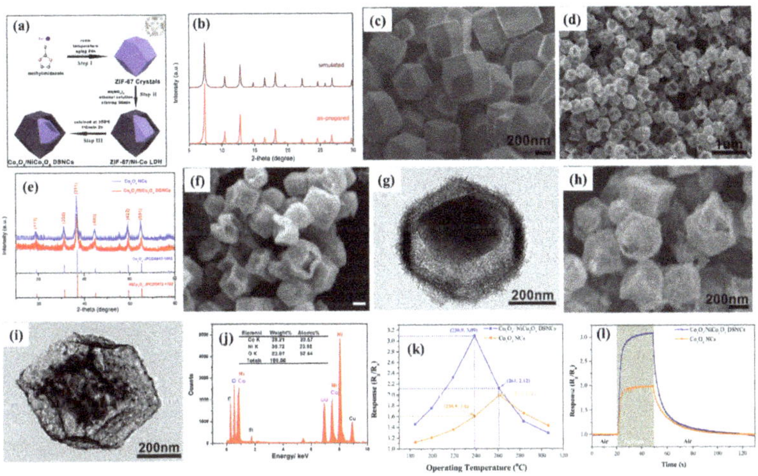

Figure 5. (**a**) Schematic illustration for synthesis of $Co_3O_4/NiCo_2O_4$ double-shelled nanocages. (**b**) XRD pattern of synthesized ZIF-67 and simulated sample. SEM image of (**c**) ZIF-67 and (**d**) Co_3O_4. XRD patterns of (**e**) Co_3O_4 and $Co_3O_4/NiCo_2O_4$. SEM (**f**,**g**) TEM images of ZIF-67/Ni-Co LDH. SEM (**h**,**i**) TEM images of $Co_3O_4/NiCo_2O_4$. EDX spectrum (**j**) of $Co_3O_4/NiCo_2O_4$. (**k**) Sensing performance of Co_3O_4 and $Co_3O_4/NiCo_2O_4$ sensor for 100 ppm acetone as function of operating temperature, and (**l**) response of Co_3O_4 and $Co_3O_4/NiCo_2O_4$ sensor for 100 ppm acetone at 238.9 °C. Reproduced with permission [61].

TEM analysis further confirmed the formation of a double-shelled nanocage structure. In order to confirm the elemental composition of the $Co_3O_4/NiCo_2O_4$ double-shelled nanocage sensor, Energy Dispersive X-ray Spectroscopic (EDX) analysis was carried out, which indicated the presence of Co, Ni, and O elements, as shown in Figure 5j. The sensing performance of the synthesized sensors was examined for acetone at different temperatures to study the impact of operating temperature, a key parameter influencing the performance of semiconducting sensors. Therefore, the sensing performance of the $Co_3O_4/NiCo_2O_4$ sensor was tested for 10 ppm acetone as a function of temperature to obtain the optimum temperature, as shown in Figure 5k. The optimum temperature for $Co_3O_4/NiCo_2O_4$ was found to be 238.9 °C. Therefore, the rest of the sensing experiments were performed at the optimum temperature, which indicated that the $Co_3O_4/NiCo_2O_4$ sensor revealed fast response and recovery times for 100 ppm acetone (Figure 5l).

Li et al. [62] demonstrated the development of core–shell of polystyrene spheres combined with cobalt-based LDHs for the detection of ethanol. In another study, Qu et al. [63] fabricated the porous hollow $Co_3O_4/ZnCo_2O_4$ nanostructure composite using a self-sacrificing template method for acetone sensing. Zhang et al. [64] established an ultra-sensitive and selective NO_x gas sensor with an ultra-low detection limit by prepared expanded graphite/NiAl LDH nanowires using a hydrothermal method. It was reported in the literature that nanocomposites made up of combinations of organic and inorganic materials have attracted significant interest in the field of sensors due to the synergistic effects between respective components. In this contest, LDHs have emerged as promising host-matrices, providing a confined and stable environment for uniform composite formation. In this contest, Qin et al. proposed that LDH-based sensing devices could be a possible solution due to their exclusive structural properties [65]. To explore this possibility, a 3D nanocomposite of polyaniline (PANI) and ZnTi-LDHs was developed using a hydrothermal method, where PANI was uniformly joined into the ZnTi-LDH matrices. The optimized ZnTi-LDH sensor exhibited outstanding sensing performance, confirmed by its high response and notable long-term stability at RT. In order to detect the nitrogen dioxide (NO_2) gas, which plays a key role for the formation of photochemical smog, PM2.5, acid rain, and ozone formation in the atmosphere, an LDH-based sensor was developed by Liu et al. [66]. They used a simple one-step hydrothermal method for the preparation of a 3D flower-like CoAl-LDH nanocomposite sensor. To improve gas sensitivity of a 3D flower-like CoAl-LDH sensor, fluoride ion was used as a functional template agent. The functionalized optimized sensor demonstrated excellent sensing performance towards NO_2 detection, as confirmed by ultrafast response and recovery times. Zhang et al. [67] reported the synthesis of 3D flower-like NiZnAL ternary sensor with multi-metal oxide and ultra-thin porous nanosheets for the detection of NO_x. In another study, a 3D flower-like Zn and Al sodium dodecyl sulfate-LDH sensor intercalated by anions was synthesized using a hydrothermal method. In this system, urea was used a precipitant and sodium dodecyl sulfate as a functional templating agent for the detection of NO_2 [68]. Lang et al. [69] highlighted that an ideal gas sensor must exhibit fast response and recovery times, along with excellent repeatability and long-term stability, to contribute effectively to environmental protection. To fulfill this purpose, they developed the 3D flower-like layered Ni-Co-LDH composite sensor using a simple hydrothermal method. In a typical procedure, $Ni(NO_3)_2.6H_2O$ and $Co(NO_3)_2.6H_2O$ in ratios of 3:1, 1:1, and 1:3 were dissolved in distilled water along with hexamethylenetetramine (HMT) under continuous stirring for 4 h. The resulting suspension then transferred to Teflon-lined autoclave and was heated at 95 °C for 12 h. the obtained powder was washed several times with water and ethanol and dried at 60 °C in oven. The samples were denoted as Ni3Co1, Ni1Co1, and Ni1Co3 (Figure 6a).

In order to confirm the flower-like structure of the sensors, SEM analysis was conducted. The authors found that synthesized materials are composed of uniform 2D sheets making a 3D flower-like structure. Moreover, the nanosheets were interconnected and intertwined similar to petals and almost perpendicular to the outer surface, providing a diameter of 3–4 μm. This may be attributed to the fact that during hydrothermal synthesis, the nanosheets gradually deposited on each other, producing a shape similar to a flower-like structure. The performance of the synthesized 3D flower-like NiCo-LDH sensor was investigated to detect NO_2 gas at RT. Figure 6b exhibits the sensing response of the optimized Ni1Co1 sensor against different concentrations (0.01–100 ppm) of NO_2 at RT. As shown in Figure 6b, as the concentration of NO_2 increases, the reaction of the Ni1Co1 sensor also increases. The lowest concentration detected by Ni1Co1 and Ni3Co1 was found to be 0.1 ppm, whereas the lowest concentration detected by Ni1Co3 was found to be 0.05 ppm. Figure 6c exhibits the response and recovery times of the Ni3Co1, Ni1Co1, and Ni1Co3 sensors towards NO_2 sensing. Among all the sensors, Ni1Co1 indicates the highest sensing response and recovery times for NO_2 sensing, as shown in Figure 6c. The calibration curve for Ni1Co1 was obtained from the logarithm of sensor response (log S) and the logarithm of gas concentration (log NO_2). It can be seen from Figure 6d that the calibration curve shows a linear relationship. They also measure the stability and selectivity, which are very important parameters for sensors to be used at an industrial level. The stability of the Ni1Co1 sensor was performed by repeatability using the sensor for 14 consecutive cycles, as shown in Figure 6e. The selectivity of the sensor was investigated in the presence of other interfering gases such as CH_4, CO, NH_3, H_2S, H_2, and NO_2 at the same experimental conditions at RT. As shown in Figure 6f, the optimized Ni1Co1 sensor is more selective towards NO_2 gas. Moreover, the long-term stability of the Ni1Co1 sensor was subjected to 100 ppm NO_2 every 5 days at RT. As shown in Figure 6f, after 60 days of stability measurement, the sensor response is excellent. The performance of the gas sensors based on LDH materials are shown in Table 1.

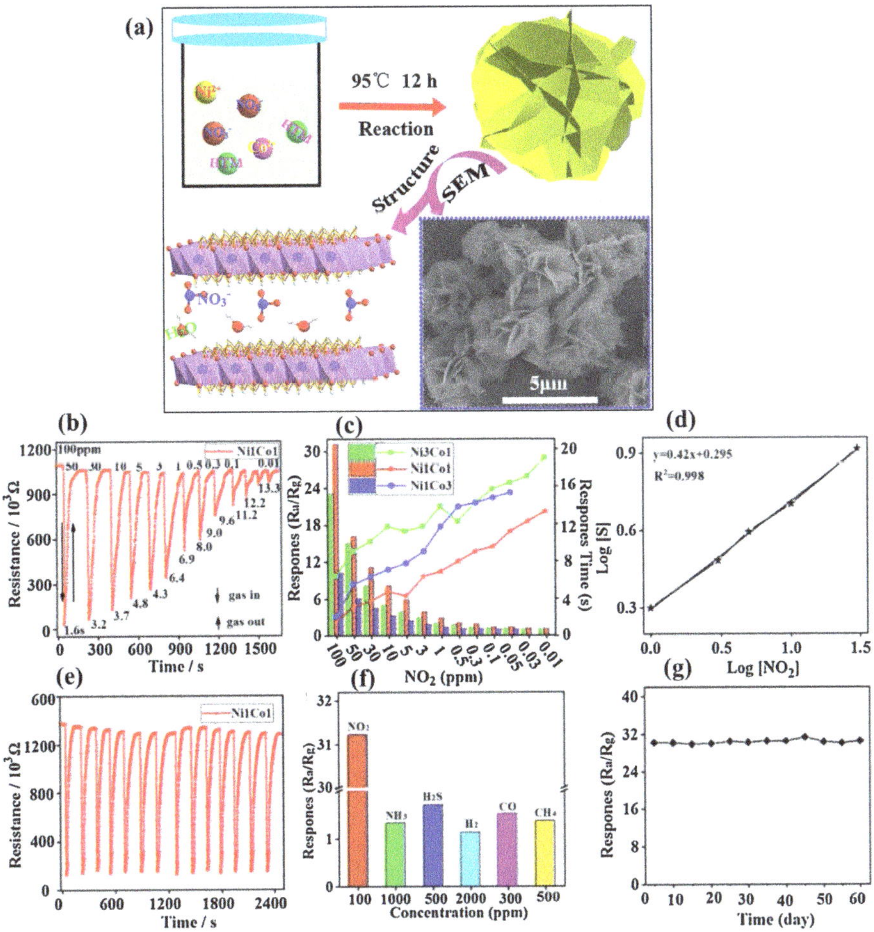

Figure 6. (**a**) Schematic illustration of preparation of 3D flower-like NiCo-LDH gas sensor. (**b**) Response and recovery times of optimized Ni1Co1 sensor towards varying concentrations of NO_2 gas, (**c**) gas response and recovery times of NiCo-LDH sensor at different molar concertation for NO_2 at RT, (**d**) NO_2 calibration curve of Ni1Co1 composite, (**e**) dynamic response of Ni1Co1 sensor 100 ppm NO_2 for 14 cycles, (**f**) selectivity of Ni1Co1 sensor in presence of different interfering gases, and (**g**) stability of Ni1Co1 sensor for 100 ppm NO_2 for 60 days. Reproduced with permission [69].

Table 1. Sensing activity and comparison of gas sensors based on LDH [70–85].

Sensors	Analytes	Detection Range (ppm)	Response Time (min)	Recovery Time (min)	Response (S)	References
Ni-Cr-AL-LDH	Acetone, ethanol	500–5000	6, 1	3, 1	$R_g - R_a/R_a \times 100$	[70]
$NiO/NiGa_2O_4$-LDH	Toluene	0.1–5	4.7	6.3	R_g/R_a	[71]
ZnAl−Cl, ZnFe−Cl, ZnAl−NO_3, and MgAl−NO_3-LDH	Acetone, ethanol, ammonia, chlorine	Fixed concentrations of 10% of saturated vapors	3.1, 3.1, 3.9, 2.3	2.3, 4.05, 3.7, 3.2	$R_g - R_a/R_a \times 100$	[72]
CuCr-, ZnCr-, and ZnTi-LDH	Methanol, ethanol, acetone	1–40	---	0.61, 0.71, 1.48	$R_g - R_a/R_a$	[73]
ZnTi-LDHs/rGO	NO_2	0.05–20	0.033	0.86	$R_g - R_a/R_a \times 100$	[74]
CoNi- LDH on GO (GO@LDH)	2,4-Dimethyl benzaldehyde	10–300 μg/L	----	-----	-----	[75]
zinc–chromium-LDH	SO_2	0.1–100	0.33	5.12	$R_g - R_a/R_a \times 100$	[76]
zinc–chromium-LDH	SO_2	0.1–100	0.33	2.8	$R_g - R_a/R_a \times 100$	[77]
NiCo-LDH	Methylene Blue, Methylene orange	0.005–10	4.5, 2.2	---	$R_g - R_a/R_a$	[78]
Zn–Al-LDH	SO_2	-	-	-	R_{gas}/R_{air}	[79]
Cu/Fe -LDH	Imidacloprid	-	-	-	$R_g - R_a/R_a$	[80]
$(CeO_2)x$/Ni-Al LDH	Methanol, ethanol, and acetone	25–300	1.01, 0.666, 0.42	3.45, 0.91, 2.13	R_{air}/R_{gas}	[81]
CeO_2/Ni–Al-LDH	Ethanol	25–300	0.666	1.27	R_{air}/R_{gas}	[82]
Ni-Al LDH	Methanol, ethanol, and acetone	20- 300	0.2, 0.83, 0.33	2.25, 1.95, 1.9	R_{air}/R_{gas}	[83]
$(CeO_2)x$/Ni-Al–LDH	Methanol, ethanol, and acetone	25–300	1.01, 0.666, 0.41	3.45, 0.92, 2.13	R_{air}/R_{gas}	[84]
ZnCo-LDHs	Ethanol	50–250	---	----	R_g/R_a	[85]

3. LDH-Based Electrochemical Sensors

Electrochemical sensors are promising voltammetric sensing technology for the detection of biomolecules, hazardous compounds, drugs, etc. In this report, we have summarized the progress in LDH-based materials for the sensing of toxic and hazardous substances. It is known that endocrine-disrupting substances or compounds such as bisphenol A (BPA; 2,2-bis(4-hydroxyphenyl)propane) are one of the environmental pollutants that may have negative impacts on aquatic life and environment. BPA is widely used in various products and can be easily leached to water, food, or soil, and it can further migrate to aquatic organisms and human bodies [86]. BPA may be responsible for various diseases such as sexual dysfunction, neurotoxicity problems, cardiovascular, lower sperm quality, and various cancers such as breast cancer. Thus, determination and monitoring of BPA is necessary; Zhan et al. [86] reported the preparation of 1-aminopropyl-3-methylimidzaolium tetrafluoroborate-modified zinc-aluminum (Zn-Al) LDH (ILs-LDH) by co-precipitation method. The authors found that ILs-LDH has a disk-like surface morphology by SEM-based analysis. The glassy carbon electrode, i.e., GCE, was coated with ILs-LDH towards the sensing of BPA. The authors used cyclic voltammetry (CV), differential pulse voltammetry (DPV), and electrochemical impedance spectroscopy (EIS) techniques for the sensing of BPA. The ILs-LDH-coated GCE exhibits improved electrochemical performance for the sensing of BPA and an acceptable linear range of 0.02 to 3 μM with a limit of detection (LOD) of 4.6 nM and excellent selectivity, stability, and reproducibility. The ILs-LDH-coated GCE shows that the proposed BPA sensor can be employed for real sample studies with a recovery range of 94.9% to 102%. Hydrogen peroxide (H_2O_2) has been known as a ubiquitous compound in the processes and as an intermediate product for biological reactions and aerobic organisms. H_2O_2 is also used in various industrial applications such as textile, medical, packaging, and waste water treatment. Despite various applications, H_2O_2 has negative influences on human and aquatic life. Thus, the sensing of H_2O_2 is of great significance. Heli et al. [87] synthesized cobalt-aluminum (CoAl) LDH nanoshells integrated multiwalled carbon nanotube (MWCNT) composites via a reflux assisted method. The synthesized CoAl-LDH/MWCNT composite was characterized by XRD, which revealed the presence of a rhombohedral structure of LDH. The formation of the composite was also confirmed by XRD and suggested that LDH material was successfully deposited on MWCNTs. The SEM and TEM analyses indicated the formation of nanoshel-integrated MWCNTs. The carbon paste electrode (CPE) was modified with a CoAl-LDH/MWCNTs composite (denoted as MCPE) and adopted as a non-enzymatic H_2O_2 sensor. Chronoamperometry (CA) and CV analyses show excellent electrochemical behavior of the MCPE and stability with good selectivity in the presence of various interfering substances. Heavy metals are major threats for human beings and the environment due to their highly toxic nature. Mercury (Hg) is one of the heavy metals and possesses hazardous properties and it may be responsible for various health problems such as kidney failure, lung diseases, and nervous system disorders. Thus, a novel Hg sensor was developed using mercaptocarboxylic acid (thioglycolic acid; TGA) intercalated magnesium (Mg)-Al LDH as the electrode modifier. An anion exchange method was adopted for the preparation of Mg–Al–TGA LDH, which was further coated on GCE and explored as a Hg sensor. The authors achieved an LOD of 0.8 nM by employing square wave anodic stripping voltammetry (SWASV) as a sensing technology and Mg–Al–TGA LDH/GCE as the working electrode [88]. In another report [89], co-precipitation and hydrothermal routs methods were used for the fabrication of copper oxide (CuO)@manganese (Mn)Al LDH. The fabrication mechanism for the formation of CuO@MnAl isdepicted in Figure 7a. The GCE was modified with the prepared CuO@MnAl and this electrode (CuO@MnAl/GCE) was used for H_2O_2 sensing applicatiosn by employing amperometry analysis. Figure 7b shows the ameprometric responses of CuO@MnAl/GCE for various concentrations of H_2O_2. The potential for the sensing of H_2O_2 was fixed as −0.85 V. It was observed that current responses increase with increasing concentrations of H_2O_2 and a calibration plot between the current values and concentrations of H_2O_2 is depicted in Figure 7c. The current response was linearly increased, and authors achieved a decent LOD of 0.126 μM and a linear range of 6 μM to 22 mM was obtained. Selectivity is one of the challenging tasks; therefore, the authors studied the selectivity of CuO@MnAl/GCE for H_2O_2 via amperometry analysis. It is clear from Figure 7d that CuO@MnAl/GCE has excellent selectivity for H_2O_2 in the presence of various interfering substances. The long-term stability and reproducibility of CuO@MnAl/GCE were also checked by measuring the current response with 5 mM H_2O_2, as shown in Figure 7e. The authors observed that CuO@MnAl/GCE has acceptable stability for 30 days and decent reproducibility (inset of Figure 7e).

Li et al. [90] proposed the fabrication of ternary LDH electrode material for heavy metal ion sensing applications. An iron(Fe)/magnesium(Mg)/nickel(Ni) ternary LDH was synthesized via a co-precipitation method. The fabricated ternary Fe/Mg/Ni LDH was deposited on GCE and the SWASV technique was utilized for the determination of lead (Pb(II)). A remarkably good LOD of 0.032 μM was obtained for Pb(II) sensing. Excellent selectivity and stability was also obtained for Fe/Mg/Ni LDH-modified GCE towards Pb(II) sensing. Polycyclic aromatic hydrocarbons, i.e., PAHs, are globally known as environmental contaminants and receive significant concern because of their higher toxicity and bioaccumulative properties [91]. Cadmium (Cd)/Al LDH was obtained using a green electrochemical synthetic method. The XRD results revealed that prepared Cd/Al LDH has a poor crystalline nature while SEM analysis suggested the presence of a regular and small sheet structure. Cd/Al LDH further coated GCE and its sensing activity for anthracene was studied by employing EIS, CV, and DPV methods. The constructed Cd/Al/GCE electrode shows an LOD of 0.5 fM with decent selectivity. The NiFe LDH was fabricated on a nickel foam electrode using a hydrothermal method [92]. Their phase purity, formation, and crystalline nature were evaluated by XRD while SEM analysis was adopted for the study of morphological characteristics. It can be seen that NiFe LDH has decent crystalline properties with good phase purity and XRD results confirmed the growth of NiFe LDH on the Ni foam electrode (Figure 8a). The SEM-based results clearly show that NiFe LDH has hierarchical spheres (Figure 8b). The fabricated NiFe LDH-based nickel foam electrode was employed for H_2O_2 sensing. The NiFe LDH-based nickel foam electrode shows that an LOD of 0.5 μM was achieved using an amperometric method. This proposed electrode also shows good selectivity in the presence of potassium chloride, sodium chloride, dopamine, glucose, and uric acid (Figure 8c).

In a previous study, MgFe LDH was also reported by employing a one-step hydrothermal method [93]. MgFe LDH was immobilized on graphene sheets and the resulting composite (MgFe LDH/graphene/GCE) demonstrated higher electrochemical activity, which may be ascribed to the presence of synergism in the fabricated MgFe LDH/graphene composite. The authors investigated simultaneously the determination of Cd (II) and Pb (II). The LOD of 5.9 nM and 2.7 nM were reported for the detection of Cd (II) and Pb (II), respectively. The MgFe LDH/graphene composite-modified GCE also displayed good recovery for real sample studies in lake water. Zhan et al. [94] fabricated a novel BPA sensor by utilizing exfoliated Ni_2P/Al LDH as an electrode modifier. Ni_2P/Al was prepared by using l-asparagine as a pre-intercalator and synthesized Ni2P/Al LDH was exfoliated, as demonstrated in Figure 9. The exfoliated Ni_2P/Al LDH-modified GCE was explored as a BPA sensor and the DPV technique was utilized as voltametric approach towards the determination of BPA. The mechanism of BPA sensing is illustrated in the DPV graph displayed in Figure 9. This proposed sensor shows a linear range of 0.02 to 1.51 μM, an LOD of 6.8 nM, and good selectivity, reproducibility, and stability.

Nitrite is widely used in various industries including the food industry, agriculture (as a fertilizer), and corrosion science (as an inhibitor). Unfortunately, nitrite has toxic properties and can interact with amines to transform to carcinogenic N-nitrosamines. Additionally, it can have negative effects if it is present in drinking water with higher concentrations. Mg/Al LDH was prepared on carbon paper (CP) using a hydrothermal method. Surface morphological investigations revealed that Mg/Al LDH has a flower-like structure and grew on CP. This electrode demonstrated decent electrochemical performance for the sensing of nitrite with decent selectivity and stability [95]. Metronidazole (MNZ; 2-(2-methyl-5-nitroimidazole-1-yl) ethanol) is the derivative of nitroimidazole which is used for the treatment of protozoal diseases such as giardiasis and trichomoniasis [96]. MNZ may cause some health-related issues such as genotypic, mutagenic, genotoxic, and carcinogenic side effects. Thus, Vilian et al. [96] fabricated a novel MNZ sensor by using NiCo LDH as the electrode material and the authors prepared NiCo LDH on carbon nanofibers (CNFs) via a hydrothermal method (Figure 10a). SEM analysis revealed that CNF–NiCo-LDH with a dense hierarchical nanowire structure was successfully anchored on the surface of CNFs (Figure 10b). The synthesized CNF–NiCo-LDH-based electrode was used as a MNZ sensor by using a DPV--based sensing method. The authors observed that the current value increases with increasing concentrations of MNZ (Figure 10c) and a calibration plot between the current value and concentration of MNZ (Figure 10d) indicated that current values linearly increase and a decent linear range of 3 to 57 nM was obtained with an LOD of 0.13 nM. The authors also stated in this article that the proposed CNF–NiCo-LDH-GCE for the sensing of MNZ has excellent anti-interfering properties and can be used for the selective detection studies of MNZ in pharmaceutical industries.

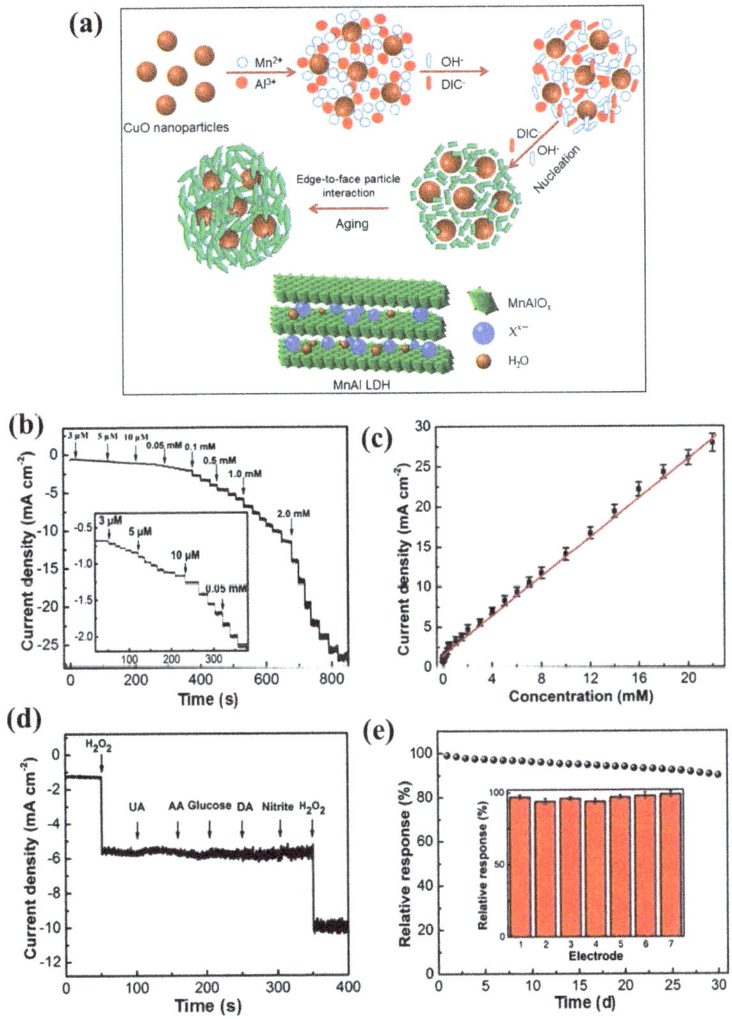

Figure 7. (**a**) Schematic picture for synthesis mechanism of CuO@MnAl NSs. (**b**) Amperometric response of CuO@MnAl/GCE with different concentrations of H_2O_2 at −0.85 V and (**c**) calibration curve between current versus concentrations. (**d**) Selectivity test and (**e**) variation in current responses for 5 mM H_2O_2 with time (s). Inset shows current values for seven electrodes for 5 mM H_2O_2. Reproduced with permission [89].

Wang et al. [97] fabricated zeolitic imidazole framework (ZIF-67)/LDH nanosheets (NSs) for the sensing of α-naphthol and β-naphthol. Synthetic protocols for the preparation of ZIF-67/LDH composites can be seen in Figure 11a. XRD results shos the presence of good phase purity and a decent crystalline nature of the obtained samples (Figure 11b). ZIF-67/LDHNS (Co/Al LDH nanosheets) was coated on GCE and electrochemical studies revealed the excellent electrochemical behavior of the ZIF-67/LDHNS-modified GCE. DPV results demonstrated decent LOD of 62 and 94 nM for the sensing of α-naphthol and β-naphthol, respectively. The decent linear range of 0.3 to 150 μM with high stability, reproducibility, and selectivity were also highlighted by the authors. It was observed that LDHNS@ZIF-67/GCE may be used as an efficient sensor for the simultaneous determination of α-naphthol and β-naphthol (Figure 11c) with good selectivity.

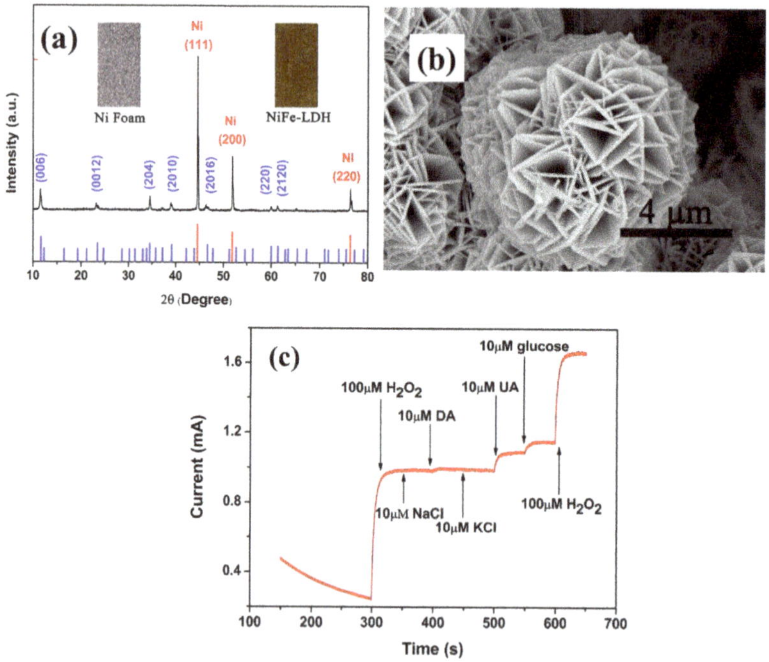

Figure 8. (**a**) XRD and (**b**) SEM results for Cd/Al LDH. (**c**) Selectivity test of Cd/Al LDH/GCE for H_2O_2. Reproduced with permission [92].

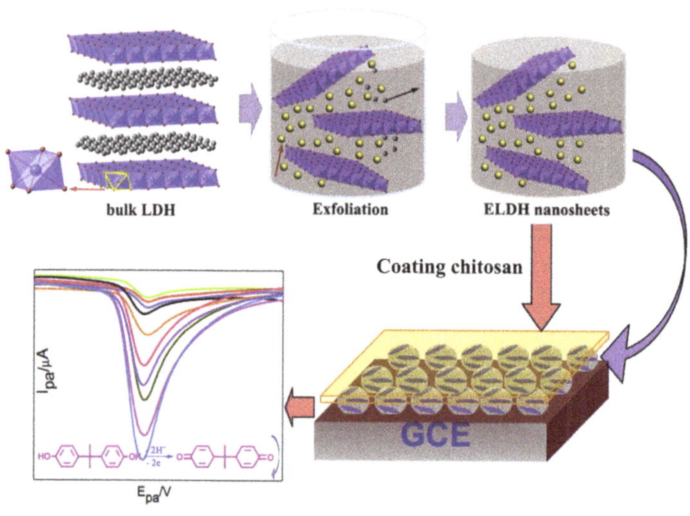

Figure 9. Schematic illustration of fabrication of BPA sensor. Reproduced with permission [94].

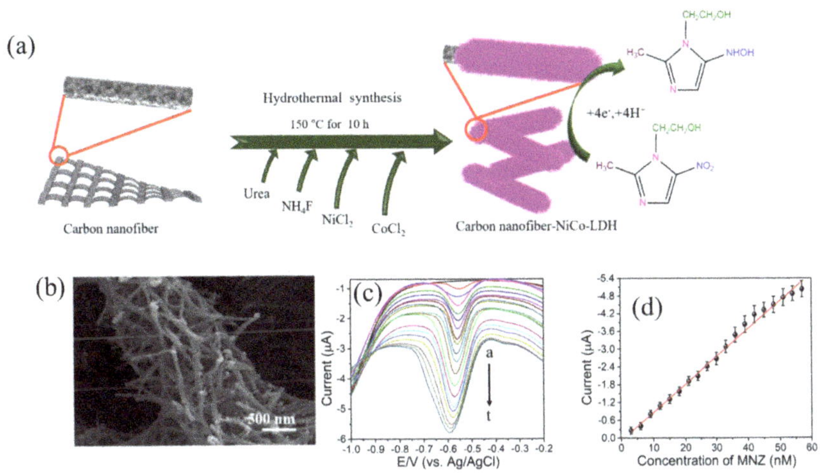

Figure 10. (**a**) Schematic graph shows formation of CNF-NiCo LDH and SEM image (**b**) of CNF-NiCo LDH. (**c**) DPVs of CNF-NiCo LDH/GCE for MNZ sensing under various concentrations and (**d**) calibration curve of current versus concentrations of MNZ. Reproduced with permission [96].

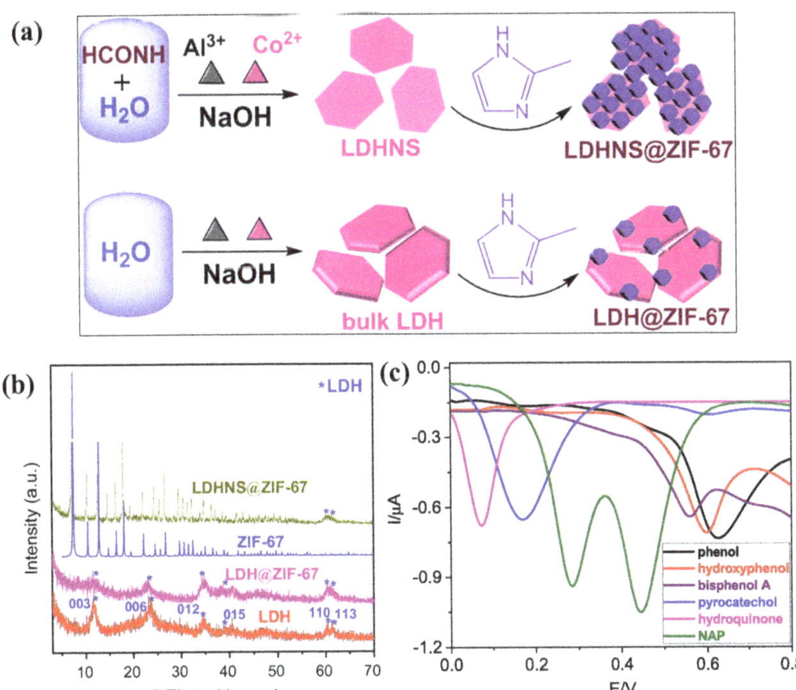

Figure 11. (**a**) Schematic illustration of formation of LDHNS@ZIF-67. (**b**) XRD patterns of LDH, ZIF-67, LDH@ZIF-67, and LDHNS@ZIF-67. (**c**) Selectivity test using DPV method. Reproduced with permission [97].

In a different report [98], a unique reaction–diffusion framework (RDF) was adopted for the formation of CoAl-LDHNS (CoAl-LDHNS@ZIF-67). The proposed CoAl-LDHNS@ZIF-67/GCE shows LOD of 54 and 82 nM for the sensing of α-naphthol and β-naphthol, respectively, with linear range of 0.3 to 150 μM. It was believed that the presence of hierarchically structured functional LDH materials may exhibit better electrochemical activity and can be explored for sensing applications. Joseph et al. [99] reported the formation of a novel FeMn LDH entrapped tungsten carbide (WC) composite for the determination of diphenylamine (DPA) via a hydrothermal method. An FeMn LDH/WC-based electrode showed a wide linear range of 0.01 to 183.34 μM with an LOD of 1.1 nM and good selectivity via the DPV method. In other work [100], a NiCo LDH/WC composite (WC@NiCo-LDH) was also formed by employing the simple and efficient hydrothermal method. The WC@NiCo-LDH-modified electrode was able to maintain excellent electrochemical activity for the sensing of norfloxacin (NRF) using CV, DPV, and amperometric methods. The authors reported LOD of 0.005 μM and 0.002 μM using DPV and amperometry methods, respectively. The linear range of 0.02–83.4 μM and 0.002–346 μM were reported for the DPV- and amperometry-based methods, respectively. This improved performance was attributed to the high surface area, electrical conductivity, and synergism between the WC and LDH materials. Karuppiah et al. [101] reported the preparation of two-dimensional (2D) hydrogen ammonium ZnMo LDH (AZnMo LDH) via a co-precipitation method and integrated it with 1D vapor grown carbon fiber (VGCF) using an ultrasonication method. AZnMo-LDHs@VGCF was explored as the sensing material for the determination of dimetridazole (DMZ) and an LOD of 0.021 μM, sensitivity of 1 μA μM^{-1} cm^{-2}, and linear range of 0.25 to 520.75 μM were obtained using the DPV method.

Oxygen vacancy-based Co-Al LDH (OV-LDH) was integrated with hydroxylated MWCNTs (H-MWCNTs) by a simple self-assembly approach, as demonstrated in Figure 12 [102]. It is believed that O-vacancies may enhance the electrochemical performance of Co-A LDH and the presence of H-functional groups in H-MWCNTs can enable the composite to form H-bonds with O-vacancies and improve the stability. The resulting OV-LDHs/H-MWCNT composite-based electrode showed excellent LOD of 0.074 and 0.076 μM for hydroquinone, i.e., HQ, and catechol, i.e., CC, respectively. The proposed sensor also has excellent recovery in industrial waste water and suggested its potential for practical applications.

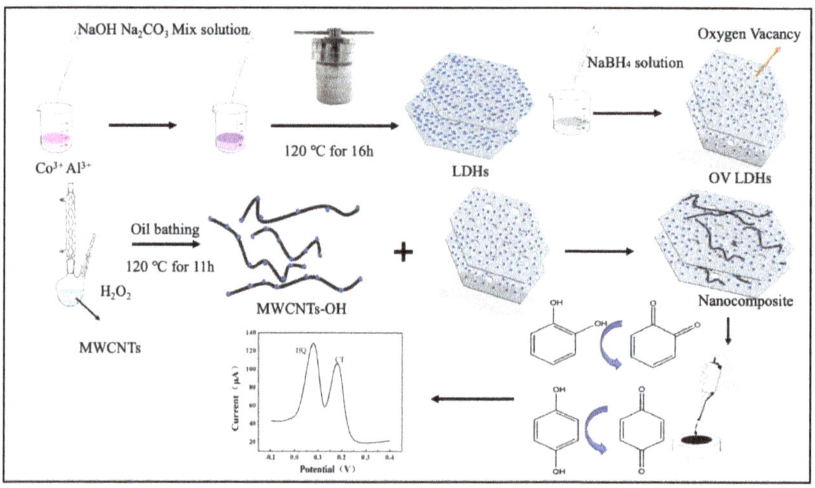

Figure 12. Schematic picture for formation of OV LDH/H-MWCNTs and their application for CC and HQ sensing. Reproduced with permission [102].

The NiCo LDH/functionalized halloysite nanotubes (NiCo-LDH/F-HNTs) were proposed as efficient electrode modifiers by Kokulnathan et al. [103] and showed an LOD of 0.002 μM, linear range of 0.01 to 33.4 μM, and sensitivity of 13.0 μA μM^{-1} cm^{-2} towards the determination of parathion (PT). It was stated that the presence of synergism, higher conductivity, various active sites, rapid electron transport, and high surface area are the reasons for the enhanced sensing activity of the NiCo-LDH/F-HNT-based electrode for the sensing of PT. Lv et al. [104] also reported the fabrication of CoAl LDH decorated hematite (CoAl-LDH/α-Fe_2O_3) for the sensing of H_2O_2, which demonstrated an LOD of 0.04 μM and linear range of 1 to 2000 μM including a sensitivity of 132.0 μA/mM·cm^2. A Co-based LDH integrated with gold nanoparticles (Au NPs) was fabricated as an electrode modifier for the sensing of H_2O_2 [105]. The AuNPs/Co-LDH-based electrode demonstrated an LOD of 0.19 μM and a sensitivity of 406.61 μA mM^{-1} cm^{-2} for the determination of H_2O_2 using an amperometry method. Diethofencarb (DFC) is used to fight fungal attacks in the agriculture industries to improve crop production [106]. However, the negative influences of DFC motivated the electrochemists to fabricate a sensor for monitoring DFC. A Zn-chromium (Cr) LDH/vanadium carbide (VC) was prepared by using a hydrothermal method, as shown in Figure 13. It was observed that ZnCr LDH has a nanoflower-shaped structure, which is attached on the VC surface. ZnCr-LDH/VC was coated on a screen-printed carbon electrode, i.e., SPCE, for electrochemical sensing applications. ZnCr-LDH/VC/SPCE exhibited an LOD of 2 nM for the monitoring of DFC via the DPV method. This sensor was also efficient in detecting the DFC in a real sample analysis in water and tomato samples.

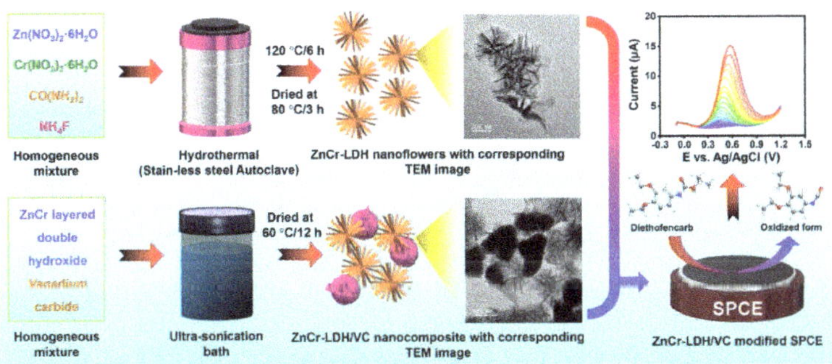

Figure 13. Schematic picture illustrating ZnCr LDH/VC composite for sensing of DFC. Reproduced with permission [106].

Carmoisine is an azo dye which has negative environmental impacts and its monitoring is of great significance [107]. Ni-Co LDH was obtained under facile conditions and was coated on a screen-printed graphite electrode (SPGE) for the sensing of carmoisine in the presence of tartrazine via the DPV method. The LOD of 0.09 μM was achieved for the detection of carmoisine with a linear range of 0.3 to 125 μM. Kokulnathan et al. [108] reported the construction of carbendazim (CDM) in the residues of the environment. For the sensing of CDM, the authors prepared a NiCo LDH and coated it on GCE and EIS, CV, and DPV techniques were used to determine CDM. The authors achieved an LOD of 0.001 μM and a linear range of 0.006 to 14.1 μM, with a sensitivity of 3.38 μA μM^{-1} cm^{-2}. Singh et al. [109] reported the fabrication oxidized graphitic carbon nitride (O-g-CN)/copper (Cu)-Al LDH composite for the construction of a diclofenac sodium (DS) sensor. It was found that Cu-Al LDH was homogenously deposited on the O-g-CN surface. The obtained O-g-CN/Cu-Al LDH was deposited on the surface of GCE, which demonstrated an acceptable LOD of 0.38 μM with a linear range of 0.5 to 60 μM via the DPV method. It is understood that phenolic compounds, for example, benzenediol (BD), have poor biodegradability and toxicity and pose threats to the environment and human life [110]. In this regard, AZnMo LDH was prepared with carbon black (CB) and proposed as a working electrode for the sensing of BD. AZnMo LDH/CB showed low electrical resistance and higher electrochemical reactivity with improved electron transport. The presence of synergistic interactions between AZnMo LDH and CB enhanced the sensing mechanism of HQ, CC, and resorcinol (RC) via DPV. The authors reported LOD of 0.0054 μM, 0.0018 μM, and 0.075 μM towards HQ, CC, and RC, respectively. The proposed electrode was also validated towards real sample investigations in water, soil, and environmental samples. In 2023, a simple and cost-effective strategy was applied for the formation of ternary MnFeZn LDH and deposited on the GCE surface towards the sensing of flutamide (FLA) [111]. The formation of MnFeZn LDH is illustrated in Figure 14. A linear range of 0.019 to 2735.79 μM and LOD of 12.9 nM were obtained by using the DPV method. The proposed sensor also showed stability of 50 cycles with real sample sensing in river water. Thus, this type of sensor can be applied for practical applications.

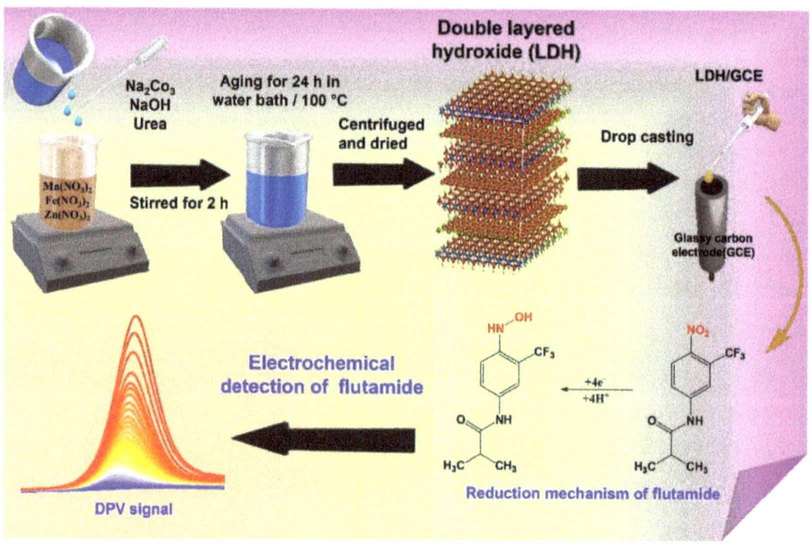

Figure 14. Schematic graph shows formation of MnFeZn LDH for FLA sensing. Reproduced with permission [111].

He et al. [112] reported the sensing of glucose and H_2O_2 using the CoZn-LDH@CuO-modified electrode via an amperometry method. An interesting LOD of 0.17 μM and high sensitivity of 4585 μA mM^{-1} cm were obtained via CoZn-LDH@CuO NSA/CF. Tajik et al. [113] fabricated a Ni-Co LDH/MWCNT composite and modified it with CPE to develop the 4-aminophenol (4-AP) sensor. The authors adopted CV, DPV, and chronoamperometry techniques for the electrochemical characterization of the Ni-Co-LDH/MWCNTs/CPE towards the sensing of 4-AP. The electrochemical studies exhibited a wide linear range of 0.02 to 700 μM, a sensitivity of 0.076 μA/μM, and an LOD of 0.01 μM. Ni-Co-LDH/MWCNTs/CPE also worked well for selectivity studies and real sample investigations in water samples. In 2024, a hydrazine (N_2H_4) sensor was developed by Zhou et al. [114] by employing an LDH-modified electrode. The authors obtained NiCo-LDH elf-assembled hollow nanocages (HNCs) by employing an in situ etching approach via a Cu_2O nanocube template (Figure 15a). Furthermore, the authors found that this unique structure is beneficial for electrochemical sensing applications and constructed a NiCo-LDH HNCs/CP electrode for the determination of N_2H_4. The pH was optimized by recording CVs of NiCo-LDH HNCs/CP in different pH conditions for the sensing of N_2H_2 (Figure 15b). The higher activity of NiCo-LDH HNCs/CP was observed for N_2H_4 under a pH of 14. However, oxygen evolution at pH 14 may interfere with the sensing of N_2H_4;

therefore, the authors used pH 13 for further electrochemical studies. Furthermore, the authors observed that chronoamperometry is highly sensitive compared to the CV and further studies were carried out using chronoamperometry. The potential for the determination of N_2H_4 was also optimized and 0.6 V was found to be suitable and efficient (Figure 15c). The amperometric response of the NiCo-LDH HNCs/CP electrode shows that current response increases with the addition of N_2H_4 via chronoamperometry at 0.6 V (Figure 15d). This rapid increase in the current response was found to be linear by calibrating current responses versus the concentration of N_2H_4, as shown in Figure 15e. An interesting LOD of 0.135 μM and a linear range of 3×10^{-3} mM to 2 mM and 3 μM to 6 mM were observed with good selectivity and stability.

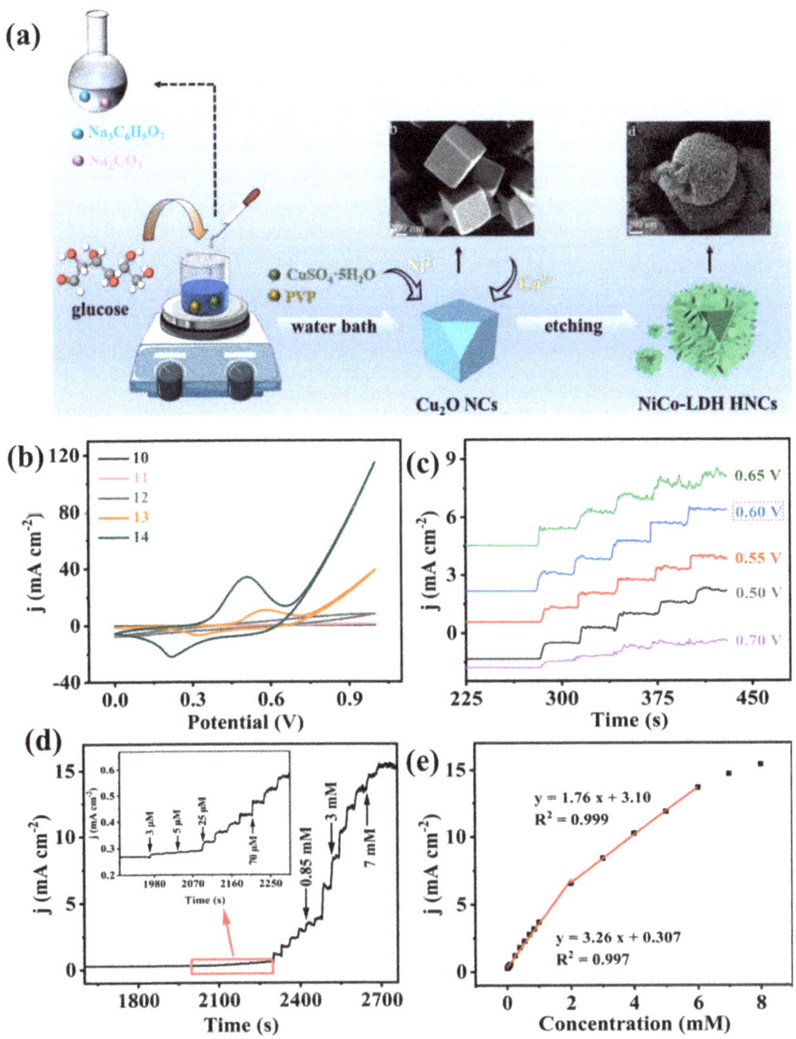

Figure 15. (**a**) Schematic illustration of formation of NiCo-LDH HNCs. (**b**) CV graphs of NiCo-LDH HNCs/CP for 1 mM N_2H_4 (under different pH conditions). (**c**) Amperometric responses of NiCo-LDH HNCs/CP with successive spike of 1 mM N_2H_4 (at different potentials). (**d**) Amperometric graph of NiCo-LDH HNCs/CP for N_2H_4 at 0.6 V. Inset enlarged graph. (**e**) Calibration curve between current versus concentration of N_2H_4. Reproduced with permission [114].

Khan et al. [115] obtained a decent LOD of 0.004 μM and a linear range of 0.05 to 50 μM towards the detection of pentachlorophenol (PCP) by fabricating a Ni-Al LDH-modified GCE. EIS investigations revealed that Ni-Al LDH-modified GCE has good electrical conductivity and electrochemical ability for the detection of PCP. Ashry et al. [116] obtained an LOD of 0.3 μM for the sensing of paroxetine by using a β-Cyclodextrin/Zn–Fe LDH/g-CN composite-based electrode whereas Ragumoorthy et al. [117] developed a Mesalazine (MLZ) sensor by utilizing Co-Al LDH/GCE, which exhibits an LOD of 0.029 μM and a linear range of 0.049 to 665.1 μM. In a recent report by Stanley et al. [118], a propyl gallate (PG) sensor was developed by fabricating a NiFeCo-LDH and graphene aerogel (NiFeCo-LDH/GA)-based electrode and an LOD of 0.87 nM with excellent sensitivity of 41.22 μA μM^{-1} cm^{-2} was achieved. The authors stated that the presence of high surface area and active sites enhanced the sensing of the proposed electrode for the determination of PG. In another report [119], NiFeCu-LDH/GA/SPCE was also explored as a PG sensor, which demonstrated an LOD of 0.004 μM and a linear range of 0.02 to 279.1 μM with excellent recovery in real samples. In another study [120], a Ni/Co LDH hollow cake (HC) was prepared by using a Co-based ZIF precursor via a benign ion etching process. The Ni/Co LDH HC-based electrode showed an LOD of 0.22 μM and a sensitivity of 7050 μA mM^{-1} cm^{-2} for the monitoring of H_2O_2. Guo et al. [121] reported the synthesis of $NiMoO_4$ nanorods@NiCo-LDH nanosheets for the sensing of H_2O_2. The $NiMoO_4$ NRs@NiCo-LDH NSs/CC was explored as an electrode and an LOD of 112 nM with a wide linear range of 1 μmol to 9.0 mmol was observed concerning the monitoring of H_2O_2 using amperometry. The performance of the various reported sensors are displayed in Table 2.

Table 2. Sensing performance of reported sensors using LDH-based electrodes.

Electrode Material	Analyte	Detection Limit (μM)	Linear Range (μM)	Sensitivity (μA/μM·cm^2)	Sensing Method	References
ILs-LDH modified GCE	BPA	0.0046	0.01–3.0	-	DPV	[86]
CoAl-LDH/MWCNTs	H_2O_2	5 μmol dm^{-3}	100–4000 μmol dm^{-3}	118 mA dm^3 mol^{-1} cm^{-2}	CA	[87]
Mg–Al–TGA LDH/GCE	Hg	0.8 nM	2–800 nM	-	SWASV	[88]
CuO@MnAl/GCE	H_2O_2	0.126	6 μM to 22 mM	-	Amperometry	[89]
Fe/Mg/Ni LDH/GCE	Pb	0.032	0.03–1	68.1 μA μM^{-1}	SWASV	[90]
Cd/Al/GCE	Anthracene	0.5 fM	0.1–100.0 pM	-	DPV	[91]
NiFe LDH/nickel foam	H_2O_2	0.5	5×10^{-4}–0.84 mM	-	Amperometry	[92]
MgFe LDH/GCE	Cd (II)	5.9 nM	0.1–1.0	-	SWASV	[93]
MgFe LDH/GCE	Pb (II)	2.7 nM	0.1–1.0	-	SWASV	[93]
Exfoliated Ni_2P/Al LDH/GCE	BPA	6.8 nM	0.02 to 1.51 μM	-	DPV	[94]
MgAl LDH/CP	Nitrite	0.03	14.8–222 μM	-	Amperometry	[95]
CNF–NiCo-LDH-GCE	MNZ	0.13 nM	3 to 57 nM	1.294 μA $nM^{-1}cm^{-2}$.	DPV	[96]
WC@NiCo-LDH	NRF	0.005	0.02–83.4	6.53 μA $\mu M^{-1}cm^2$	DPV	[97]
WC@NiCo-LDH	NRF	0.002	0.002–346	40.81 μA $\mu M^{-1}cm^2$	Amperometry	[98]
WC@FeMn-LDH	DPA	0.0011	0.01–183.34	-	DPV	[99]
WC@NiCo–LDH	NRF	0.005	0.02–83.4	-	DPV	[100]
AZnMo-LDHs@VGCF	DMZ	0.021	0.25–570	1	DPV	[101]
OV-LDHs/H-MWCNTs/GCE	CT	0.074	0.5–150	-	DPV	[102]

Table 2. *Cont.*

Electrode Material	Analyte	Detection Limit (μM)	Linear Range (μM)	Sensitivity (μA/μM·cm^2)	Sensing Method	References
OV-LDHs/H-MWCNTs/GCE	HQ	0.076	0.5–150	-	DPV	[102]
NiCo-LDH/F-HNTs	PT	0.003	0.012–24.5	-	DPV	[103]
CoAl-LDH/α-Fe_2O_3	H_2O_2	0.04	0.001–2	-	PEC	[104]
AuNPs/Co-LDH	H_2O_2	0.19	4 μM–16 mM	406.61 μA $mM^{-1}cm^{-2}$	Amperometry	[105]
ZnCr-LDH/VC	DFC	0.002	0.01–228	-	DPV	[106]
Ni–Co LDH NSs/SPGE	Carmoisine	0.09	0.3–125	-	DPV	[107]
NiCo-LDH	CDM	0.001	0.006–14.1	3.38	DPV	[108]
o-g-C_3N_4/CuAl-LDH/GCE	DS	0.64	0.5–60	-	DPV	[109]
AZnMo-LDHs/CB	HQ	0.0054	0.05–971	-	DPV	[110]
AZnMo-LDHs/CB	CC	0.0018	0.1–1036	-	DPV	[110]
AZnMo-LDHs/CB	RC	0.075	0.5–1408.5	-	DPV	[110]
MnFeZn-LDH	FLA	0.012	0.0199–2735.7	5.04	DPV	[111]
CoZn-LDH@CuO NSA/CF	H_2O_2	0.17	0.8 μM–3.5 mM	4585 μA $mM^{-1}cm$	Amperometry	[112]
Ni-Co-LDH/MWCNTs/CPE	4-AP	0.01	0.02–700	0.076 μA/μM	DPV	[113]
NiCo-LDH	Hydrazine	0.135	3–2 × 10^3	3260 μA $mM^{-1}cm^{-2}$	Amperometry	[114]
Ni-Al-LDH	PCP	0.004	0.05–50	-	DPV	[115]
CoAl-LDH/GCE	MLZ	0.029	0.049–665.1	-	DPV	[117]
NiFeCo-LDH/GA	PG	0.00087	0.001–297.43	-	DPV	[118]
NiFeCu-LDH/GA	FG	0.004	0.02–279.1	-	DPV	[119]
NiCo-LDH HC	H_2O_2	0.22	-	7050 μA $mM^{-1}cm^{-2}$	Amperometry	[120]
$NiMoO_4$ NRs@NiCo-LDH NSs/CC	H_2O_2	0.112	1–9000	-	Amperometry	[121]

4. Conclusions and Future Perspectives

This article reviewed the progress in the development of 2D LDH and its composite-based materials for gas sensing and electrochemical sensing applications. The progress in LDH materials based on Mn, Cu, Fe, Mg, Co, Ni Gd, and Al metals were compiled for sensing studies. The reported literature exhibits that LDH-based materials have promising features such as decent surface area and conductivity and physiochemical properties for the detection of various gases and hazardous compounds. The presence of synergy effects in the fabricated LDH-based composites provide active sites for catalytic reactions/adsorption of toxic gases and facilitate charge transport, which resulted in improved sensing performance. We believed that tuning the composition of the LDH-based materials may improve the selectivity and sensitivity of the gas sensors. In addition, LDH-based materials may offer various advantages such as the development of gas sensors at low temperatures, which makes them promising candidates for practical applications at a large scale. On the other side, LDH-based materials coated on different electrodes such as CPE, GCE, or SPCE showed excellent electrochemical activity for various electrochemical reactions. The LDH-based materials are found to be promising electrode modifiers for the construction of electrochemical sensors for the detection of environment pollutants. Unfortunately, despite the excellent sensing performance of the 2D LDH-based materials, some challenges still exist, which include structural instability and low electronic conductivity. The limited conductivity of LDHs may also restrict their applications as electrode modifiers in electrochemical sensors and sensing material for gas sensors. It can be noted that MXenes are highly conductive material and are widely used as conductive support for the preparation of hybrid composites for various optoelectronic applications. The presence of synergistic interactions in the MXene-based composites may further improve their physicochemical properties. To further enhance the performance of LDH-based materials, it is crucial to combine LDHs with appropriate conductive materials to boost their overall conductivity. Thus, we also believe that future investigations may focus on the design and fabrication of simple and benign synthetic methods, with a combination of LDH with MXenes, conductive materials, or other layered metal sulfides to further improve the electrical conductivity and charge transport properties of the LDH materials. The combination of MXene and LDH materials may be a great idea to develop a highly efficient hybrid material for gas sensing and electrochemical applications. LDH/MXene-based hybrid composites may be adopted as sensing materials which can be further used for commercial purposes such as monitoring hazardous compounds or toxic gases for environment-related fields.

Author Contributions: Conceptualization, K.A.; and W.R.; writing—original draft preparation, K.A. and W.R.; writing—review and editing, T.H.O.; supervision, T.H.O. All authors have read and agreed to the published version of the manuscript.

Funding: This research received no external funding.

Institutional Review Board Statement: Not applicable.

Informed Consent Statement: Not applicable.

Data Availability Statement: No new data were generated in this study.

Conflicts of Interest: Authors declare no conflicts of interest.

References

1. Chen, H.; Deng, G.; Liu, Y. Monitoring the Influence of Industrialization and Urbanization on Spatiotemporal Variations of AQI and $PM_{2.5}$ in Three Provinces, China. *Atmosphere* **2022**, *13*, 1377. [CrossRef]
2. Nakhjiri, A.; Kakroodi, A.A. Air pollution in industrial clusters: A comprehensive analysis and prediction using multi-source data. *Ecol. Inform.* **2024**, *80*, 102504. [CrossRef]
3. Afifa; Arshad, K.; Hussain, N.; Ashraf, M.H.; Saleem, M.Z. Air pollution and climate change as grand challenges to sustainability. *Sci. Total Environ.* **2024**, *928*, 172370. [CrossRef]
4. Cuerda-Correa, E.M.; Alexandre-Franco, M.F.; Fernández-González, C. Advanced Oxidation Processes for the Removal of Antibiotics from Water. *An Overview. Water* **2020**, *12*, 102. [CrossRef]
5. Hashim, H.S.; Fen, Y.W.; Omar, N.A.S.; Fauzi, N.I.M. Sensing Methods for Hazardous Phenolic Compounds Based on Graphene and Conducting Polymers-Based Materials. *Chemosensors* **2021**, *9*, 291. [CrossRef]
6. Silva, J.A. Wastewater Treatment and Reuse for Sustainable Water Resources Management: A Systematic Literature Review. *Sustainability* **2023**, *15*, 10940. [CrossRef]
7. Negahdary, M.; Ameku, W.A.; Santos, B.G.; Lima, I.S.; de Oliveira, T.G.; França, M.C.; Angnes, L. Recent electrochemical sensors and biosensors for toxic agents based on screen-printed electrodes equipped with nanomaterials. *Microchem. J.* **2023**, *185*, 108281. [CrossRef]
8. Nishida, C.; Yatera, K. The Impact of Ambient Environmental and Occupational Pollution on Respiratory Diseases. *Int. J. Environ. Res. Public Health* **2022**, *19*, 2788. [CrossRef]
9. Yuan, Z.; Li, R.; Meng, F.; Zhang, J.; Zuo, K.; Han, E. Approaches to Enhancing Gas Sensing Properties: A Review. *Sensors* **2019**, *19*, 1495. [CrossRef]
10. Liu, X.; Cheng, S.; Liu, H.; Hu, S.; Zhang, D.; Ning, H. A Survey on Gas Sensing Technology. *Sensors* **2012**, *12*, 9635–9665. [CrossRef]
11. Delfino, I.; Diano, N.; Lepore, M. Advanced Optical Sensing of Phenolic Compounds for Environmental Applications. *Sensors* **2021**, *21*, 7563. [CrossRef] [PubMed]
12. Kuş, E.; Altındemir, G.; Bostan, Y.K.; Taşaltın, C.; Erol, A.; Wang, Y.; Sarcan, F. A Dual-Channel MoS_2-Based Selective Gas Sensor for Volatile Organic Compounds. *Nanomaterials* **2024**, *14*, 633. [CrossRef] [PubMed]

13. Khan, M.A.H.; Rao, M.V.; Li, Q. Recent Advances in Electrochemical Sensors for Detecting Toxic Gases: NO_2, SO_2 and H_2S. *Sensors* **2019**, *19*, 905. [CrossRef] [PubMed]
14. Chakraborthy, A.; Nuthalapati, S.; Nag, A.; Afsarimanesh, N.; Alahi, M.E.E.; Altinsoy, M.E. A Critical Review of the Use of Graphene-Based Gas Sensors. *Chemosensors* **2022**, *10*, 355. [CrossRef]
15. Idris, A.O.; Oseghe, E.O.; Msagati, T.A.M.; Kuvarega, A.T.; Feleni, U.; Mamba, B. Graphitic Carbon Nitride: A Highly Electroactive Nanomaterial for Environmental and Clinical Sensing. *Sensors* **2020**, *20*, 5743. [CrossRef]
16. Dourandish, Z.; Tajik, S.; Beitollahi, H.; Jahani, P.M.; Nejad, F.G.; Sheikhshoaie, I.; Di Bartolomeo, A. A Comprehensive Review of Metal–Organic Framework: Synthesis, Characterization, and Investigation of Their Application in Electrochemical Biosensors for Biomedical Analysis. *Sensors* **2022**, *22*, 2238. [CrossRef]
17. Zhao, Z.; Cao, J.; Zhu, B.; Li, X.; Zhou, L.; Su, B. Recent Advances in MXene-Based Electrochemical Sensors. *Biosensors* **2025**, *15*, 107. [CrossRef]
18. Pan, Y.; Zhang, J.; Guo, X.; Li, Y.; Li, L.; Pan, L. Recent Advances in Conductive Polymers-Based Electrochemical Sensors for Biomedical and Environmental Applications. *Polymers* **2024**, *16*, 1597. [CrossRef]
19. Tereshkov, M.; Dontsova, T.; Saruhan, B.; Krüger, S. Metal Oxide-Based Sensors for Ecological Monitoring: Progress and Perspectives. *Chemosensors* **2024**, *12*, 42. [CrossRef]
20. Kim, A.; Varga, I.; Adhikari, A.; Patel, R. Recent Advances in Layered Double Hydroxide-Based Electrochemical and Optical Sensors. *Nanomaterials* **2021**, *11*, 2809. [CrossRef]
21. Nemček, L.; Hagarová, I.; Matúš, P. Layered Double Hydroxides as Next-Generation Adsorbents for the Removal of Selenium from Water. *Appl. Sci.* **2024**, *14*, 8513. [CrossRef]
22. Chubar, N.; Gilmour, R.; Gerda, V.; Mičušík, M.; Omastova, M.; Heister, K.; Man, P.; Fraissard, J.; Zaitsev, V. Layered double hydroxides as the next generation inorganic anion exchangers: Synthetic methods versus applicability. *Adv. Colloid Interface Sci.* **2017**, *245*, 62–80. [CrossRef] [PubMed]
23. Bukhtiyarova, M.V.; Bulavchenko, O.A.; Bukhtiyarov, A.V.; Nuzhdin, A.L.; Bukhtiyarova, G.A. Selective Hydrogenation of 5-Acetoxymethylfurfural over Cu-Based Catalysts in a Flow Reactor: Effect of Cu-Al Layered Double Hydroxides Synthesis Conditions on Catalytic Properties. *Catalysts* **2022**, *12*, 878. [CrossRef]
24. Li, C.; Bao, Y.; Liu, E.; Zhao, B.; Sun, T. Recent Advances of Modified Ni (Co, Fe)-Based LDH 2D Materials for Water Splitting. *Molecules* **2023**, *28*, 1475. [CrossRef]
25. Huang, S.-H.; Chen, Y.-J.; Huang, W.-F.; Uan, J.-Y. Electrodeposition of a Li-Al Layered Double Hydroxide (LDH) on a Ball-like Aluminum Lathe Waste Strips in Structured Catalytic Applications: Preparation and Characterization of Ni-Based LDH Catalysts for Hydrogen Evolution. *Catalysts* **2022**, *12*, 520. [CrossRef]
26. Li, Y.; Narducci, R.; Varone, A.; Kaciulis, S.; Bolli, E.; Pizzoferrato, R. Zn–Al Layered Double Hydroxides Synthesized on Aluminum Foams for Fluoride Removal from Water. *Processes* **2021**, *9*, 2109. [CrossRef]
27. Zhang, X.; Zhang, M.; Li, R.; Feng, X.; Pang, X.; Rao, J.; Cong, D.; Yin, C.; Zhang, Y. Active Corrosion Protection of Mg–Al Layered Double Hydroxide for Magnesium Alloys: A Short Review. *Coatings* **2021**, *11*, 1316. [CrossRef]
28. Wang, S.; Zhu, Y.q.; Zhang, Y.; Wang, B.; Yan, H.; Liu, W.; Lin, Y. Manganese-based layered double hydroxide nanoparticles as highly efficient ozone decomposition catalysts with tunable valence state. *Nanoscale* **2020**, *12*, 12817–12823. [CrossRef]
29. Mureseanu, M.; Cioatera, N.; Carja, G. Fe-Ce/Layered Double Hydroxide Heterostructures and Their Derived Oxides: Electrochemical Characterization and Light-Driven Catalysis for the Degradation of Phenol from Water. *Nanomaterials* **2023**, *13*, 981. [CrossRef]
30. Xia, G.; Zheng, Y.; Sun, Z.; Xia, S.; Ni, Z.; Yao, J. Fabrication of ZnAl-LDH mixed metal-oxide composites for photocatalytic degradation of 4-chlorophenol. Environ. *Sci. Pollut. Res.* **2022**, *29*, 39441–39450. [CrossRef]
31. Huang, Y.; Liu, C.; Rad, S.; He, H.; Qin, L. A Comprehensive Review of Layered Double Hydroxide-Based Carbon Composites as an Environmental Multifunctional Material for Wastewater Treatment. *Processes* **2022**, *10*, 617. [CrossRef]
32. Ahmad, K.; Raza, W.; Alsulmi, A.; Kim, H. Fabrication of Electrochemical Sensor for Metronidazole Using MoS_2/Graphite-like Carbon Nitride Composite Modified Glassy Carbon Electrode. *Diam. Relat. Mater.* **2023**, *138*, 110178. [CrossRef]
33. Raza, W.; Ahmad, K. Room Temperature Gas Sensor Based on Reduce Graphene Oxide for Environmental Monitoring. In *Handbook of Nanomaterials and Nanocomposites for Energy and Environmental Applications*; Springer International Publishing: Cham, Switzerland, 2020; pp. 1–19.
34. Fanget, S.; Hentz, S.; Puget, P.; Arcamone, J.; Matheron, M.; Colinet, E.D. Gas sensors based on gravimetric detection—A review. *Sens. Actuators B Chem.* **2011**, *160*, 804–821. [CrossRef]
35. Lin, T.; Lv, X.; Hu, Z.; Xu, A.; Feng, C. Semiconductor Metal Oxides as Chemoresistive Sensors for Detecting Volatile Organic Compounds. *Sensors* **2019**, *19*, 233. [CrossRef]
36. Ghidotti, M.; Fabbri, D.; Torri, C. Determination of linear and cyclic volatile methyl siloxanes in biogas and biomethane by solid-phase microextraction and gas chromatography-mass spectrometry. *Talanta* **2019**, *195*, 258–264. [CrossRef] [PubMed]

37. Fois, M.; Cox, T.; Ratcliffe, N.; de Lacy Costello, B. Rare Earth Doped Metal Oxide Sensor for the Multimodal Detection of Volatile Organic Compounds (VOCs). *Sens. Actuators B Chem.* **2021**, *330*, 129264. [CrossRef]
38. Saruhan, B.; Lontio Fomekong, R.; Nahirniak, S. Review: Influences of Semiconductor Metal Oxide Properties on Gas Sensing Characteristics. *Front. Sens.* **2021**, *2*, 657931. [CrossRef]
39. Baig, N.; Sajid, M. Applications of Layered Double Hydroxides Based Electrochemical Sensors for Determination of Environmental Pollutants: A Review. *Trends Environ. Anal. Chem.* **2017**, *16*, 1–15. [CrossRef]
40. Guo, Y.; Xiao, Y.; Zhang, L.; Song, Y.F. Fabrication of (Calcein-ZnS) n Ordered Ultrathin Films on the Basis of Layered Double Hydroxide and Its Ethanol Sensing Behavior. *Ind. Eng. Chem. Res.* **2012**, *51*, 8966–8973. [CrossRef]
41. Jiménez-Cadena, G.; Riu, J.; Rius, F.X. Gas Sensors Based on Nanostructured Materials. *Analyst* **2007**, *132*, 1083–1099. [CrossRef]
42. Wang, K.; Cao, Z.; Ding, Q.; Yoo, J.; Singh, N.; Kang, H.; Wang, L.; Xu, L.; Kim, J.S. Electronic Structure Tuning of Layered Double Hydroxides for Electrochemical Sensing and Its Biomedical Applications. *Sci. China Chem.* **2024**, *67*, 3614–3630. [CrossRef]
43. Yunusa, Z.; Hamidon, M.N.; Kaiser, A.; Awang, Z. Gas Sensors: A Review. *Sens. Transducers* **2014**, *168*, 61–75.
44. Feng, S.; Farha, F.; Li, Q.; Wan, Y.; Xu, Y.; Zhang, T.; Ning, H. Review on Smart Gas Sensing Technology. *Sensors* **2019**, *19*, 3760. [CrossRef] [PubMed]
45. Guan, M.Y.; Xu, D.M.; Song, Y.F.; Guo, Y. ZnO/$ZnAl_2O_4$ Prepared by Calcination of ZnAl Layered Double Hydroxides for Ethanol Sensing. *Sens. Actuators B Chem.* **2013**, *188*, 1148–1154. [CrossRef]
46. Tang, S.; Yao, Y.; Chen, T.; Kong, D.; Shen, W.; Lee, H.K. Recent Advances in the Application of Layered Double Hydroxides in Analytical Chemistry: A Review. *Anal. Chim. Acta* **2020**, *1103*, 32–48. [CrossRef]
47. Asif, M.; Aziz, A.; Azeem, M.; Wang, Z.; Ashraf, G.; Xiao, F.; Chen, X.; Liu, H. A Review on Electrochemical Biosensing Platform Based on Layered Double Hydroxides for Small Molecule Biomarkers Determination. *Adv. Colloid Interface Sci.* **2018**, *262*, 21–38. [CrossRef]
48. Polese, D.; Mattoccia, A.; Giorgi, F.; Pazzini, L.; Di Giamberardino, L.; Fortunato, G.; Medaglia, P.G. A Phenomenological Investigation on Chlorine Intercalated Layered Double Hydroxides Used as Room Temperature Gas Sensors. *J. Alloys Compd.* **2017**, *692*, 915–922. [CrossRef]
49. Zakaria, S.A.; Ahmadi, S.H.; Amini, M.H. Chemiresistive Gas Sensors Based on Layered Double Hydroxides (LDHs) Structures: A Review. *Sens. Actuators A Phys.* **2022**, *346*, 113827. [CrossRef]
50. Karmakar, A.K.; Hasan, M.S.; Sreemani, A.; Das Jayanta, A.; Hasan, M.M.; Tithe, N.A.; Biswas, P. A Review on the Current Progress of Layered Double Hydroxide Application in Biomedical Sectors. *Eur. Phys. J. Plus* **2022**, *137*, 801. [CrossRef]
51. Aziz, A.; Asif, M.; Ashraf, G.; Azeem, M.; Majeed, I.; Ajmal, M.; Wang, J.; Liu, H. Advancements in Electrochemical Sensing of Hydrogen Peroxide, Glucose and Dopamine by Using 2D Nanoarchitectures of Layered Double Hydroxides or Metal Dichalcogenides. A Review. *Microchim. Acta* **2019**, *186*, 671. [CrossRef]
52. Boumeriame, H.; Da Silva, E.S.; Cherevan, A.S.; Chafik, T.; Faria, J.L.; Eder, D. Layered Double Hydroxide (LDH)-Based Materials: A Mini-Review on Strategies to Improve the Performance for Photocatalytic Water Splitting. *J. Energy Chem.* **2021**, *64*, 406–431. [CrossRef]
53. Xiao, Y.P.; Zhang, L.M.; Guo, Y.; Song, Y.F. (Pyrenetetrasulfonate/ZnS)n Ordered Ultrathin Films with ZnAl Layered Double Hydroxide as Precursor and Ethanol-Sensing Properties. *Eur. J. Inorg. Chem.* **2013**, *2013*, 3348–3351. [CrossRef]
54. Morandi, S.; Prinetto, F.; Di Martino, M.; Ghiotti, G.; Lorret, O.; Tichit, D.; Malagù, C.; Vendemiati, B.; Carotta, M.C. Synthesis and Characterisation of Gas Sensor Materials Obtained from Pt/Zn/Al Layered Double Hydroxides. *Sens. Actuators B Chem.* **2006**, *118*, 215–220. [CrossRef]
55. Liu, S.R.; Guan, M.Y.; Li, X.Z.; Guo, Y. Light Irradiation Enhanced Triethylamine Gas Sensing Materials Based on ZnO/$ZnFe_2O_4$ Composites. *Sens. Actuators B Chem.* **2016**, *236*, 350–357. [CrossRef]
56. Hong, D.; Zhang, J.; Rehman, A.U.; Gong, L.; Zhou, J.; Kan, K.; Li, L.; Shi, K. One-Step Synthesis of Hierarchical Ni-Fe-Al Layered Double Hydroxide with Excellent Sensing Properties for NO:X at Room Temperature. *RSC Adv.* **2016**, *6*, 103192–103198. [CrossRef]
57. Sun, H.; Chu, Z.; Hong, D.; Zhang, G.; Xie, Y.; Li, L.; Shi, K. Three-Dimensional Hierarchical Flower-like Mg-Al-Layered Double Hydroxides: Fabrication, Characterization and Enhanced Sensing Properties to NOx at Room Temperature. *J. Alloys Compd.* **2016**, *658*, 561–568. [CrossRef]
58. Qu, F.; Zhang, B.; Zhou, X.; Jiang, H.; Wang, C.; Feng, X.; Jiang, C.; Yang, M. Metal-Organic Frameworks-Derived Porous ZnO/Ni0.9Zn0.1O Double-Shelled Nanocages as Gas Sensing Material for Selective Detection of Xylene. *Sens. Actuators B Chem.* **2017**, *252*, 649–656. [CrossRef]
59. Zhang, J.; Tang, P.; Liu, T.; Feng, Y.; Blackman, C.; Li, D. Facile Synthesis of Mesoporous Hierarchical Co_3O_4-TiO_2 p-n Heterojunctions with Greatly Enhanced Gas Sensing Performance. *J. Mater. Chem. A* **2017**, *5*, 10387–10397. [CrossRef]
60. Kang, G.; Zhu, Z.; Tang, B.H.; Wu, C.H.; Wu, R.J. Rapid Detection of Ozone in the Parts per Billion Range Using a Novel Ni–Al Layered Double Hydroxide. *Sens. Actuators B Chem.* **2017**, *241*, 1203–1209. [CrossRef]

61. Qu, F.; Jiang, H.; Yang, M. MOF-Derived $Co_3O_4/NiCo_2O_4$ Double-Shelled Nanocages with Excellent Gas Sensing Properties. *Mater. Lett.* **2017**, *190*, 75–78. [CrossRef]
62. Li, Y.; Zhou, F.; Gao, L.; Duan, G. Co_3O_4 Nanosheet-Built Hollow Spheres Containing Ultrafine Neck-Connected Grains Templated by PS@Co-LDH and Their Ppb-Level Gas-Sensing Performance. *Sens. Actuators B Chem.* **2018**, *261*, 553–565. [CrossRef]
63. Qu, F.; Thomas, T.; Zhang, B.; Zhou, X.; Zhang, S.; Ruan, S.; Yang, M. Self-Sacrificing Templated Formation of $Co_3O_4/ZnCo_2O_4$ Composite Hollow Nanostructures for Highly Sensitive Detecting Acetone Vapor. *Sens. Actuators B Chem.* **2018**, *273*, 1202–1210. [CrossRef]
64. Zhang, X.; Ikram, M.; Liu, Z.; Teng, L.; Xue, J.; Wang, D.; Li, L.; Shi, K. Expanded Graphite/NiAl Layered Double Hydroxide Nanowires for Ultra-Sensitive, Ultra-Low Detection Limits and Selective NO: X Gas Detection at Room Temperature. *RSC Adv.* **2019**, *9*, 8768–8777. [CrossRef] [PubMed]
65. Qin, Y.; Wang, L.; Wang, X. A High Performance Sensor Based on PANI/ZnTi-LDHs Nanocomposite for Trace NH3 Detection. *Org. Electron.* **2019**, *66*, 102–109. [CrossRef]
66. Liu, Z.; Teng, L.; Ma, L.; Liu, Y.; Zhang, X.; Xue, J.; Ikram, M.; Ullah, M.; Li, L.; Shi, K. Porous 3D Flower-like CoAl-LDH Nanocomposite with Excellent Performance for NO_2 Detection at Room Temperature. *RSC Adv.* **2019**, *9*, 21911–21921. [CrossRef]
67. Zhang, X.; Teng, L.; Liu, Y.; Liu, Z.; Xue, J.; Ikram, M.; Ullah, M.; Li, L.; Shi, K. 3D Flower-like NiZnAl Multimetal Oxide Constructed by Ultra-Thin Porous Nanosheets: A Long-Term and Stable Sensing Material for NOx at Room Temperature. *Sens. Actuators B Chem.* **2019**, *300*, 126899. [CrossRef]
68. Hong, Y.; Wang, D.; Lin, C.; Luo, S.; Pan, Q.; Li, L.; Shi, K. Room-Temperature Efficient NO_2 gas Sensors Fabricated by Porous 3D Flower-like ZnAl-Layered Double Hydroxides. *New J. Chem.* **2020**, *44*, 18469–18480. [CrossRef]
69. He, L.; Zhang, W.; Zhang, X.; Bai, X.; Chen, J.; Ikram, M.; Zhang, G.; Shi, K. 3D flower-like NiCo-LDH composites for a high-performance NO2 gas sensor at room temperature. *Colloids Surf. A Physicochem. Eng. Aspects* **2020**, *603*, 125142. [CrossRef]
70. Gheibi, S.O.; Fallah Shojaei, A.; Khorshidi, A.; Hosseini-Golgoo, S.M. Synthesis, characterization, and gas sensing properties of Ni-Cr-Al LDH. *Appl. Phys. A* **2021**, *127*, 614. [CrossRef]
71. Nie, L.; Fan, G.; Wang, A.; Zhang, L.; Guan, J.; Han, N.; Chen, Y. Finely Dispersed and Highly Toluene Sensitive $NiO/NiGa_2O_4$ Heterostructures Prepared from Layered Double Hydroxides Precursors. *Sens. Actuators B Chem.* **2021**, *345*, 130412. [CrossRef]
72. Vigna, L.; Nigro, A.; Verna, A.; Ferrari, I.V.; Marasso, S.L.; Bocchini, S.; Fontana, M.; Chiodoni, A.; Pirri, C.F.; Cocuzza, M. Layered Double Hydroxide-Based Gas Sensors for Voc Detection at Room Temperature. *ACS Omega* **2021**, *6*, 20205–20217. [CrossRef] [PubMed]
73. Qin, Y.; Xu, X.; Xia, Q.; Wang, X. Layered Double Oxides for Sensing Reducing Volatile Organic Compounds: The Effect of Local Charge Region Modulation. *Chempluschem* **2021**, *86*, 904–912. [CrossRef] [PubMed]
74. Qin, Y.; Ding, W.; Zhao, R. ZIF-8-Derived ZnTi-LDHs with Unique Self-Supported Architecture and Corresponding LDHs/RGO Hybrid for Gas Sensor Applications. *Chem. Phys. Lett.* **2021**, *781*, 138965. [CrossRef]
75. Zhong, M.; Lao, Z.; Tan, J.; Yu, G.; Liu, Y.; Liang, Y. Synthesis of CoNi-Layered Double Hydroxide on Graphene Oxide as Adsorbent and Construction of Detection Method for Taste and Odor Compounds in Smelling Water. *J. Hazard. Mater.* **2022**, *428*, 128227. [CrossRef]
76. Shinde, R.B.; Padalkar, N.S.; Sadavar, S.V.; Patil, A.S.; Kale, S.B.; Magdum, V.V.; Chitare, Y.M.; Kulkarni, S.P.; Patil, U.M.; Parale, V.G.; et al. Lattice Engineering Route for Self-Assembled Nanohybrids of 2D Layered Double Hydroxide with 0D Isopolyoxovanadate: Chemiresistive SO_2 Sensor. *Mater. Today Chem.* **2022**, *24*, 100801. [CrossRef]
77. Shinde, R.B.; Padalkar, N.S.; Sadavar, S.V.; Kale, S.B.; Magdum, V.V.; Chitare, Y.M.; Kulkarni, S.P.; Patil, U.M.; Parale, V.G.; Park, H.H.; et al. 2D–2D Lattice Engineering Route for Intimately Coupled Nanohybrids of Layered Double Hydroxide and Potassium Hexaniobate: Chemiresistive SO_2 Sensor. *J. Hazard. Mater.* **2022**, *432*, 128734. [CrossRef]
78. Sadrolhosseini, A.R.; Ghasemi, E.; Pirkarimi, A.; Hamidi, S.M.; Taheri Ghahrizjani, R. Highly Sensitive Surface Plasmon Resonance Sensor for Detection of Methylene Blue and Methylene Orange Dyes Using NiCo-Layered Double Hydroxide. *Opt. Commun.* **2023**, *529*, 129057. [CrossRef]
79. Gupta, N.K.; Leyva, C.; Viltres, H.; Dhavale, R.P.; Kim, K.S.; Romero-Galarza, A.; Park, H.H. Zinc-Aluminum Layered Double Hydroxide and Double Oxide for Room-Temperature Oxidation of Sulfur Dioxide Gas. *Chemosphere* **2023**, *338*, 139503. [CrossRef]
80. Selvaraj, A.K.; Suresh, I.; Nesakumar, N.; Rayappan, J.B.B.; Kulinich, S.A.; Kulandaiswamy, A.J. Highly Sensitive Cu/Fe Calcined Layered Double Hydroxide Sensor to Detect Imidacloprid, a Neurotransmission Blocker. *Colloids Surf. A Physicochem. Eng. Asp.* **2024**, *687*, 133540. [CrossRef]
81. Zakaria, S.A.; Ahmadi, S.H.; Amini, M.H. Investigation the AMOST Effect on the Sensitivity of CeO_2/Ni-Al Layered Double Hydroxide Composite Based Sensors. *Microchem. J.* **2024**, *207*, 112128. [CrossRef]
82. Zakaria, S.A.; Amini, M.H.; Ahmadi, S.H. CeO_2/Ni–Al Layered Double Hydroxide Composite Decorated with Ag Nanoparticles as a Gas Sensor. *Mater. Sci. Semicond. Process.* **2024**, *178*, 108391. [CrossRef]
83. Zakaria, S.A.; Ahmadi, S.H.; Amini, M.H. Investigation of Ag Nanoparticles Decorated Ni-Al Layered Double Hydroxide as a Gas Sensor. *Mater. Chem. Phys.* **2024**, *318*, 129216. [CrossRef]

84. Zakaria, S.A.; Ahmadi, S.H.; Amini, M.H. CeO_2/Ni-Al Layered Double Hydroxide Composite as a Gas Sensor for Volatile Organic Compounds Detection. *Mater. Today Commun.* **2024**, *40*, 109677. [CrossRef]
85. Fu, Z.; Lu, J.; Zhu, J.; Wang, R.; Feng, L. Effect of Laminate Structure on Morphology and Ethanol Sensing Properties of Hierarchical Zinc–Cobalt Layered Double Hydroxides. *New J. Chem.* **2024**, *48*, 8963–8973. [CrossRef]
86. Zhan, T.; Song, Y.; Li, X.; Hou, W. Electrochemical sensor for bisphenol A based on ionic liquid functionalized Zn-Al layered double hydroxide modified electrode. *Mater. Sci. Eng. C* **2016**, *64*, 354–361. [CrossRef]
87. Heli, H.; Pishahang, J.; Amiri, H.B. Synthesis of hexagonal CoAl-layered double hydroxide nanoshales/carbon nanotubes composite for the non-enzymatic detection of hydrogen peroxide. *J. Electroanal. Chem.* **2016**, *768*, 134–144. [CrossRef]
88. Asadpour-Zeynali, K.; Amini, R. A novel voltammetric sensor for mercury(II) based on mercaptocarboxylic acid intercalated layered double hydroxide nanoparticles modified electrode. *Sens. Actuators B Chem.* **2017**, *246*, 961–968. [CrossRef]
89. Asif, M.; Wang, H.; Dong, S.; Aziz, A.; Zhang, G.; Xiao, F.; Liu, H. Metal oxide intercalated layered double hydroxide nanosphere: With enhanced electrocatalyic activity towards H_2O_2 for biological applications. *Sens. Actuators B Chem.* **2017**, *239*, 243–252. [CrossRef]
90. Li, S.-S.; Jiang, M.; Jiang, T.-J.; Liu, J.-H.; Guo, Z.; Huang, X.-J. Competitive adsorption behavior toward metal ions on nano-Fe/Mg/Ni ternary layered double hydroxide proved by XPS: Evidence of selective and sensitive detection of Pb(II). *J. Hazard. Mater.* **2017**, *338*, 1–10. [CrossRef]
91. Qiao, X.; Wei, M.; Tian, D.; Xia, F.; Chen, P.; Zhou, C. One-step electrosynthesis of cadmium/aluminum layered double hydroxides composite as electrochemical probe for voltammetric detection of anthracene. *J. Electroanal. Chem.* **2018**, *808*, 35–40. [CrossRef]
92. Tao, Y.; Chang, Q.; Liu, Q.; Yang, G.; Guan, H.; Chen, G.; Dong, C. Highly sensitive nonenzymatic H_2O_2 sensor based on NiFe-layered double hydroxides nanosheets grown on Ni foam. *Surf. Interfaces* **2018**, *12*, 102–107. [CrossRef]
93. Ma, Y.; Wang, Y.; Xie, D.; Gu, Y.; Zhu, X.; Zhang, H.; Wang, G.; Zhang, Y.; Zhao, H. Hierarchical MgFe-layered double hydroxide microsphere/graphene composite for simultaneous electrochemical determination of trace Pb(II) and Cd(II). *Chem. Eng. J.* **2018**, *347*, 953–962. [CrossRef]
94. Zhan, T.; Song, Y.; Tan, Z.; Hou, W. Electrochemical bisphenol A sensor based on exfoliated Ni_2Al-layered double hydroxide nanosheets modified electrode. *Sens. Actuators B Chem.* **2017**, *238*, 962–971. [CrossRef]
95. Xiang, X.; Pan, F.; Du, Z.; Feng, X.; Gao, C.; Li, Y. MgAl-layered double hydroxide flower arrays grown on carbon paper for efficient electrochemical sensing of nitrite. *J. Electroanal. Chem.* **2019**, *855*, 113632. [CrossRef]
96. Vilian, A.T.E.; Ranjith, K.S.; Lee, S.J.; Umapathi, R.; Hwang, S.-K.; Oh, C.W.; Huh, Y.S.; Han, Y.-K. Hierarchical dense Ni−Co layered double hydroxide supported carbon nanofibers for the electrochemical determination of metronidazole in biological samples. *Electrochim. Acta* **2020**, *354*, 136723. [CrossRef]
97. Wang, R.; Zhang, P.; Zhan, T.; Yu, X.; Wen, Y.; Liu, X.; Gao, H.; Wang, P.; She, X. In situ growth of ZIF-67 on ultrathin CoAl layered double hydroxide nanosheets for electrochemical sensing toward naphthol isomers. *J. Colloid Interface Sci.* **2020**, *576*, 313–321. [CrossRef]
98. Liu, Y.; Wang, R.; Wang, H.; Yu, X.; Liu, X.; He, P.; She, X.; Zhan, T. Electrochemical sensing for naphthol isomers based on the in situ growth of zeolitic imidazole framework-67 on ultrathin CoAl layered double hydroxide nanosheets by a reaction–diffusion technique. *J. Colloid Interface Sci.* **2021**, *599*, 762–772. [CrossRef]
99. Joseph, X.B.; Sherlin, V.A.; Wang, S.-F.; George, M. Integration of iron–manganese layered double hydroxide/tungsten carbide composite: An electrochemical tool for diphenylamine $H^{\bullet+}$ analysis in environmental samples. *Environ. Res.* **2022**, *212*, 113291. [CrossRef]
100. Joseph, X.B.; Stanley, M.M.; Wang, S.-F.; George, M. Growth of 2D-layered double hydroxide nanorods heterojunction with 2D tungsten carbide nanocomposite: Improving the electrochemical sensing of norfloxacin. *J. Ind. Eng. Chem.* **2022**, *110*, 434–446. [CrossRef]
101. Karuppiah, C.; Babulal, S.M.; Chen, T.-W.; Chen, S.-M.; Hsu, L.-F.; Al Farraj, D.A.; Ramaraj, S.K.; Elshikh, M.S.; Yang, C.-C. A novel ammonium zinc molybdate layered double hydroxide nanoflakes/vapor grown carbon fibers nanomaterials based electrocatalyst for the monitoring of dimetridazole drug in real samples. *J. Environ. Chem. Eng.* **2022**, *10*, 108227. [CrossRef]
102. Liu, Z.; Liao, D.; Yu, J.; Jiang, X. An electrochemical sensor based on oxygen-vacancy cobalt–aluminum layered double hydroxides and hydroxylated multiwalled carbon nanotubes for catechol and hydroquinone detection. *Microchem. J.* **2022**, *175*, 107216. [CrossRef]
103. Kokulnathan, T.; Wang, T.-J.; Kumar, E.A.; Ahmed, F. Construction of nickel cobalt-layered double hydroxide/functionalized–halloysite nanotubes composite for electrochemical detection of organophosphate insecticide. *Chem. Eng. J.* **2022**, *433*, 133639. [CrossRef]
104. Lv, J.; Wu, M.; Fan, M.; Zhang, Q.; Chang, Z.; Wang, X.; Zhou, Q.; Wang, L.; Chong, R.; Zhang, L. Insights into the multirole CoAl layered double hydroxide on boosting photoelectrochemical activity of hematite: Application to hydrogen peroxide sensing. *Talanta* **2023**, *262*, 124681. [CrossRef]

105. Yuan, J.; Chen, Q.; Xiao, Y.; Li, D.; Jiang, X.; Wu, P. Cobalt-based layered double hydroxide nanosheet-supported AuNPs for high performance electrochemical H_2O_2 detection. *Appl. Surf. Sci.* **2023**, *630*, 157463. [CrossRef]
106. Joseph, X.B.; Baby, J.N.; Wang, S.-F.; George, M. Emerging carbonate anion intercalated- ZnCr-layered double hydroxide/vanadium carbide nanocomposite: A electrochemical sensor for diethofencarb fungicide monitoring. *Chemosphere* **2023**, *335*, 139099. [CrossRef]
107. Beitollahi, H.; Nejad, F.G.; Dourandish, Z.; Aflatoonian, M.R. Electrochemical detection of carmoisine in the presence of tartrazine on the surface of screen printed graphite electrode modified with nickel-cobalt layered double hydroxide ultrathin nanosheets. *Chemosphere* **2023**, *337*, 139369. [CrossRef]
108. Kokulnathan, T.; Wang, T.-J.; Ahmed, F.; Arshi, N. Fabrication of flower-like nickel cobalt-layered double hydroxide for electrochemical detection of carbendazim. *Surf. Interfaces* **2023**, *36*, 102570. [CrossRef]
109. Singh, A.K.; Gautam, R.K.; Agrahari, S.; Tiwari, I. Oxidized g-C_3N_4 decorated with Cu–Al layered double hydroxide as a sustainable electrochemical sensing material for quantification of diclofenac. *Mater. Chem. Phys.* **2023**, *294*, 127002. [CrossRef]
110. Karuppiah, C.; Babulal, S.M.; Chen, S.-M.; Palanisamy, S.; Hsu, L.-F.; Yang, C.-C.; Chiesa, M. Simultaneous electrochemical determination of benzenediol compounds in environmental samples using nano architectures of hydrogen ammonium zinc molybdate layered double hydroxides integrated with carbon black modified electrode. *J. Water Process Eng.* **2023**, *55*, 104202. [CrossRef]
111. Tharuman, S.; Nataraj, N.; Chen, S.-M.; Sasirekha, V.; Ragumoorthy, C. Ternary MnFeZn layered double hydroxides as an efficient material for the electrochemical sensing of flutamide. *Surf. Interfaces* **2023**, *41*, 103195. [CrossRef]
112. He, Z.; Jin, Y.; Yuan, X.; Xue, K.; Hu, J.; Xiong, X. ZIF in situ transformation hollow porous self-supporting CoZn-LDH@CuO electrode for electrochemical sensing of glucose and H_2O_2. *Microchem. J.* **2023**, *195*, 109457. [CrossRef]
113. Tajik, S.; Beitollahi, H.; Mohammadi, S.Z.; Nejad, F.G.; Dourandish, Z. Voltammetric sensor based on a Ni-Co-layered double hydroxide/multi-walled carbon nanotubes nanocomposite for 4-aminophenol determination. *Chem. Phys. Impact* **2024**, *8*, 100653. [CrossRef]
114. Zhou, Y.; Zhao, S.; Li, J.; Chen, H.; Gao, Y.; Xiong, X. Construction of hollow NiCo-LDH nanocages for electrochemical sensing of hydrazine based on template etching strategy. *Microchem. J.* **2024**, *207*, 111803. [CrossRef]
115. Khan, M.M.; Shaikh, H.; Al Souwaileh, A.; Khan, M.Y.; Batool, M.; Memon, S.Q.; Solangi, A.R. A highly selective nickel-aluminum layered double hydroxide nanostructures based electrochemical sensor for detection of pentachlorophenol. *Arab. J. Chem.* **2024**, *17*, 105604. [CrossRef]
116. Ashry, A.; Rabia, M.; Mostafa, S.M.; Korany, M.A.; Farghali, A.A.; Khalil, M.M. β-Cyclodextrin/Zn–Fe layered double hydroxides/graphitic carbon nitride nanomaterials based potentiometric sensor for paroxetine determination in environmental water samples. *RSC Adv.* **2024**, *14*, 34791–34803. [CrossRef]
117. Ragumoorthy, C.; Nataraj, N.; Chen, S.-M. Cobalt aluminum layered double hydroxide as a superior electrochemical probe for the sensitive detection of anti-inflammatory agent mesalazine. *J. Alloys Compd.* **2024**, *989*, 174352. [CrossRef]
118. Stanley, M.M.; Sriram, B.; Wang, S.-F.; Sherlin, A.; Kogularasu, S.; Lee, Y.-Y. Mary George, Rational confinement of ternary layered double hydroxide within three-dimensional graphene aerogel: An electrochemical tool for propyl gallate sensing. *Chem. Eng. J.* **2024**, *485*, 149682. [CrossRef]
119. Stanley, M.M.; Sriram, B.; Wang, S.-F.; Sherlin, V.A.; Kogularasu, S.; George, M. Preserving food quality: Electrochemical detection of synthetic food antioxidant, propyl gallate in processed foods using ternary component layered double hydroxide/graphene aerogel synergy. *Mater. Today Sustain.* **2025**, *29*, 101061. [CrossRef]
120. Wu, Y.; Xu, F.; Wu, C.; Hu, K.; Yang, Q.; Huang, K. Preparation of Ni/Co layered double hydroxide hollow cakes as an electrochemical sensing platform for monitoring the production of hydrogen peroxide in formulated beverages. *Microchem. J.* **2024**, *202*, 110793. [CrossRef]
121. Guo, J.; Wu, J.; Xu, L.; Yuan, X.; Tan, C.; Wang, Q.; Xiong, X. Microplasma-assisted construction of cross-linked network hierarchical structure of $NiMoO_4$ nanorods @NiCo-LDH nanosheets for electrochemical sensing of non-enzymatic H_2O_2 in food. *Food Chem.* **2024**, *461*, 140940. [CrossRef]

Disclaimer/Publisher's Note: The statements, opinions and data contained in all publications are solely those of the individual author(s) and contributor(s) and not of MDPI and/or the editor(s). MDPI and/or the editor(s) disclaim responsibility for any injury to people or property resulting from any ideas, methods, instructions or products referred to in the content.

chemosensors

MDPI

Article

Influence of P(V3D3-co-TFE) Copolymer Coverage on Hydrogen Detection Performance of a TiO_2 Sensor at Different Relative Humidity for Industrial and Biomedical Applications

Mihai Brinza [1,2], Lynn Schwäke [1], Lukas Zimoch [3], Thomas Strunskus [1,4,*], Thierry Pauporté [5], Bruno Viana [5], Tayebeh Ameri [4,6], Rainer Adelung [3,4], Franz Faupel [1,4,*], Stefan Schröder [1,4,6,*] and Oleg Lupan [1,2,3,5]

[1] Multicomponent Materials, Department of Materials Science, Kiel University, Kaiserstraße 2, D-24143 Kiel, Germany; mihai.brinza@mib.utm.md (M.B.); lysc@tf.uni-kiel.de (L.S.); oleg.lupan@mib.utm.md (O.L.)
[2] Center for Nanotechnology and Nanosensors, Department of Microelectronics and Biomedical Engineering, Technical University of Moldova, 168 Stefan cel Mare Av., MD-2004 Chisinau, Moldova
[3] Functional Nanomaterials, Department of Materials Science, Kiel University, Kaiserstr. 2, D-24143 Kiel, Germany; luzi@tf.uni-kiel.de (L.Z.); ra@tf.uni-kiel.de (R.A.)
[4] Kiel Nano, Surface and Interface Science (KiNSIS), Kiel University, Christian Albrechts-Platz 4, D-24118 Kiel, Germany; tam@tf.uni-kiel.de
[5] Institut de Recherche de Chimie Paris-IRCP, Chimie ParisTech, Université PSL, rue Pierre et Marie Curie 11, 75231 Paris Cedex 05, France; thierry.pauporte@chimieparistech.psl.eu (T.P.); bruno.viana@chimieparistech.psl.eu (B.V.)
[6] Composite Materials, Department of Materials Science, Kiel University, Kaiserstraße 2, D-24143 Kiel, Germany
* Correspondence: ts@tf.uni-kiel.de (T.S.); ff@tf.uni-kiel.de (F.F.); ssch@tf.uni-kiel.de (S.S.)

Abstract: The detection of hydrogen gas is crucial for both industrial fields, as a green energy carrier, and biomedical applications, where it is a biomarker for diagnosis. TiO_2 nanomaterials are stable and sensitive to hydrogen gas, but their gas response can be negatively affected by external factors such as humidity. Therefore, a strategy is required to mitigate these influences. The utilization of organic–inorganic hybrid gas sensors, specifically metal oxide gas sensors coated with ultra-thin copolymer films, is a relatively novel approach in this field. In this study, we examined the performance and long-term stability of novel TiO_2-based sensors that were coated with poly(trivinyltrimethylcyclotrisiloxane-co-tetrafluoroethylene) (P(V3D3-co-TFE)) co-polymers. The P(V3D3-co-TFE)/TiO_2 hybrid sensors exhibit high reliability even for more than 427 days. They exhibit excellent hydrogen selectivity, particularly in environments with high humidity. An optimum operating temperature of 300 °C to 350 °C was determined. The highest recorded response to H_2 was approximately 153% during the initial set of measurements at a relative humidity of 10%. The developed organic–inorganic hybrid structures open wide opportunities for gas sensor tuning and customization, paving the way for innovative applications in industry and biomedical fields, such as exhaled breath analysis, etc.

Keywords: sensors; organic–inorganic structures; hybrid gas sensor; tuning; hydrogen; copolymer; PTFE; PV3D3; P(V3D3-co-TFE)

1. Introduction

Gas sensors are of great interest in many applications. They enable monitoring and control in industrial applications, but also in areas such as medicine, food, automotives, agriculture, or domestic use [1–3]. In recent years, hydrogen monitoring has become very important. Hydrogen is a clean and versatile energy carrier [4], enabling affordable and sustainable modern energy [5,6]. As reported in the literature, there are a multitude of methods of hydrogen production [4] and storage [7], which typically involve high pressure and/or low temperature. These factors, in combination with inadequate long-term monitoring and non-rapid leak detection, pose a significant danger to humans and nature alike, limiting the safe usage of hydrogen for industrial applications. Beyond its industrial applications, hydrogen has been used in the pharmaceutical industry [5] for the manufacture of specific drugs [8] and for producing hydrogen peroxide, which is a key chemical and oxidant [9]. Furthermore, hydrogen use is surging as a therapeutic gas, e.g., as a treatment for reperfusion or re-oxygenation injury [10]. Hydrogen therapy is additionally recognized as a treatment for neurodegenerative diseases, rheumatoid arthritis, diabetes mellitus, brain stem infarction, cancer, as well as exercise- or sports-induced oxidative stress [11,12]. Several different modes of administration are known for these treatments [13,14], some of which lead to detectable changes in exhaled air [15]. In addition, hydrogen is a potential biomarker for irritable bowel syndrome starting from 12 ppm [16,17] and lactose intolerance if there is an increase of 20 ppm above the baseline [15,18,19], which extends the use of hydrogen gas sensors to the field of diagnostics. Thus, hydrogen is used in a wide variety of fields, especially in industry and in the medical field.

However, for monitoring and control in these application areas, long-term, stable, fast responding, and highly sensitive hydrogen gas sensors are required. In addition, the sensors should be essentially unaffected by external factors such as the presence of other gases or humidity. The latter property in particular poses a challenge for state-of-the-art sensor materials such as TiO_2, resulting in the introduction of polymer-coated hybrid sensors [15,20–25].

In this study, TiO_2 was selected as a sensor material due to its recognized stability over extended time periods [26–29], which has led to its utilization in hydrogen gas sensors that exhibit a time-independent gas response [26,30–32].

However, further investigation is essential to increase the stability and reduce the interference of relative humidity (RH) to improve reliable long-term operation, since RH significantly affects sensor performances and capabilities. For example, it is important to note that relative humidity is the main cause of the lack in performance of metal oxide sensors [33–35].

Although TiO_2 is sensitive to hydrogen [36], its gas sensing performance can be negatively affected by external factors too, like humidity, etc. [26]. The effects of humidity on the magnetic properties and electrical conductivity of semiconductors are well known and have been a major challenge for the last decades [37,38]. Hybrid structures open new opportunities for gas sensor tuning and customization, as well as immunization vs. humidity [39], paving the way for innovative applications in industry and biomedical fields, such as exhaled breath analysis.

Combining different nanomaterials with excellent mechanical and electronic properties creates composite materials of great technological interest, as adding a second phase can significantly enhance their capabilities, including electronic [40] and sensing performance, especially in humid ambients [33,40] and by mixing with ternary phases [41].

Researchers use different approaches to obtain practical promising compounds as electronic materials such as different mixes of metals and oxides [42]. While this research path takes into account the addition of a second phase [40] and considers the deviation of the original cations of the metals used [43] or even puts an emphasis on the effect of crystallite size and distribution [44], it is also possible to obtain promising composite materials by using different polymer and copolymer coatings [15,20,21,45,46].

The coating of semiconducting metal oxide gas sensing structures with polytetrafluoroethylene (PTFE) and poly(trivinyltrimethylcyclotrisiloxane) (PV_3D_3) thin films leading to organic–inorganic hybrid gas sensors appears to be of great interest for hydrogen detection. The former is a fluorine-based polymer that exhibits high hydrophobicity, which can have a positive effect on the long-term stability of sensor performance, especially in humid environments like in exhaled air, etc. In addition, PTFE coatings have shown an effect on sensor selectivity [46]. However, their operation temperature is limited due to the fact that PTFE shows phase transitions in its semicrystalline regions and its onset of degradation starts gradually at around 260 °C, followed by rapid degradation at 400 °C and higher [47–49]. Hybrid gas sensors with PV_3D_3 retain their hydrophobicity even after annealing at 400 °C [39]. While individual PTFE/TiO_2 and PV_3D_3/TiO_2 hybrid gas sensors have been investigated previously by us, a copolymer coating combining the promising properties of both polymers has not been analyzed for gas sensor applications. Furthermore, no information on their long-term stability with regard to gas sensing has been reported so far.

Consequently, this gap is addressed in the current work. Aiming for enhanced long-term stability, overall performance, and specifically tailored sensitivity [20,39,45] in a humid environment, we fabricated a sensor based on a TiO_2 nano-layer entirely coated with an ultra-thin P(V_3D_3-co-TFE) copolymer film via solvent-free initiated chemical vapor deposition (iCVD) [50], referred to as a P(V_3D_3-co-TFE)/TiO_2 hybrid gas sensor. Its gas sensing performance was studied and monitored over a total time span of 427 days to investigate its long-term performance. The hydrophobic nature of the hybrid sensors prevents the recorded data from being influenced by moisture and the environment, providing the next step towards the detection of hydrogen as a biomarker in exhaled air.

2. Materials and Methods

2.1. Sample Production

The hybrid gas sensor principle in this study is similar to the one described in our previous works [20,45]. A quartz or glass substrate (Thermo scientific, 2.6 × 7.6 cm, Menzel-Gläser, Braunschweig, Germany) was precleaned by dipping it into a HCl solution (10%), followed by rinsing it with deionized water in an ultrasonic bath. It was then placed on a heating plate at 445 °C for 15 min. TiO_2 was deposited on this substrate by spray pyrolysis as described by Pauporté et al. [51]. Subsequently, the samples were thermally treated at 450 °C for another 30 min and cooled down naturally. The resulting 20 nm thick TiO_2 sprayed layer was post treated thermally for 60 min at 610 °C in air to decompose all residue materials from the surface and stabilize the crystalline phase [52]. Lastly, Au-interdigital electrodes were sputtered on top through a meander shaped shadow mask with a gap of 1.0 mm [53]. To avoid confusion, all production details for the sensor are indicated directly in the figure captions as well as in the related discussion parts.

The gas sensors, consisting of a TiO_2 nanolayer with Au contacts, were subsequently coated with a copolymer thin film to protect them from the environment, moisture, and humidity, and to tailor their gas sensing selectivity and performances. The deposition of 25 nm thick P(V_3D_3-co-TFE) copolymer thin films was performed using a custom-built iCVD reactor. The technical details are described elsewhere [39]. V_3D_3 and HFPO monomers were combined with the initiator perfluorobutanesulfonyl fluoride (PFBSF) to produce the polymer thin films. This combination has already been demonstrated in the literature [54]. The flow rates for V_3D_3, HFPO, and PFSBF during the deposition process were 0.2 sccm, 0.2 sccm, and 0.1 sccm, respectively. The substrate temperature was 30 °C and the process pressure 40 Pa. The filament array inside the reactor was resistively heated by applying 3.1 A and 16.7 V to the filament. The deposition rate was 1.8 nm/min. We refer to as $P(V_3D_3$-co-TFE$)/TiO_2$/quartz hybrid gas sensors for the fabricated organic–inorganic structures in the following text.

After fabrication and between measurements, the samples were stored in opaque containers that had been lined with paraffine paper under ambient conditions. The heating of the sensors before usage after long-term storage can be beneficial for several reasons. During storage, the sensor surface (especially metal oxides or porous materials) can adsorb moisture, hydrocarbons, or other environmental contaminants. Long storage periods may cause the sensor's baseline signal to drift. Thermal treatment at 300 °C can help to stabilize the baseline before measurements begin, reducing initial response fluctuations.

2.2. Computational

A section of the molecular structure of the $P(V_3D_3$-co-TFE) copolymer structure was modelled in Avogadro version 1.99.0. Geometry optimization of the obtained structure was performed via universal force field (UFF) [55]. Molden version 6.4 was used to visualize the output.

2.3. Sample Characterization

The sensing performances to test gas were performed using a heterostructured detector connected to interdigital gold electrodes, as can be observed in Figure 1, where (a) represents a schematic of the studied sensor. Figure 1b shows a magnified image of the meander architecture and (c) the proposed schematic sensing structure. The measuring apparatus was linked to a set-up with gas flow as described in our previous works [56–59].

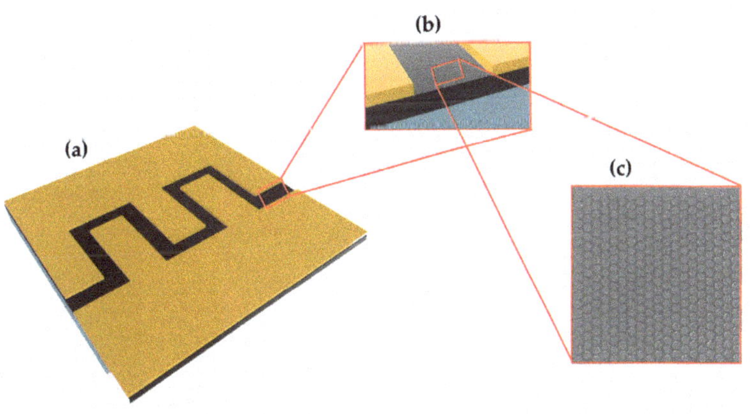

Figure 1. Schematic representation of the developed hybrid sensor with a gap of 1 mm between gold electrodes: (**a**) top view of the studied sample showing Au contacts/electrodes with meander shape on hybrid surface of sensor nanomaterial; (**b**) a magnified image of the meander architecture in cross-sectional view of the gas sensor and (**c**) top view of the proposed schematic sensing structure, namely of polymer/oxide coated gas sensing structure.

The gas response S for each dataset was calculated according to Equation (1), which was also mentioned in previous research [45], where a percentage-based ratio was obtained from electrical conductance, namely G_{gas} and G_{air}. This equation is based on the conductance G and calculated from Equation (2). R_{air} represents the resistance of the sample in air for G_{air} and in the same way R_{gas} for G_{gas} (influenced by applied gas).

$$S = \frac{G_{gas} - G_{air}}{G_{air}} 100\% \quad (1)$$

$$G_{air} = \frac{1}{R_{air}} \quad (2)$$

The gas response of the sample was measured using a custom setup and protocol based on a computer-controlled Keithley 2400 source-meter described in our previous works [20,26,33]. While varying the operating temperature (OPT), volatile organic compounds (VOCs) and hydrogen were used as test gases or vapors with a flow of 500 sccm (mL/min) of ambient gas (synthetic air) to obtain optimum sensor response. Furthermore, the described setup was utilized to assess dynamic responses, while maintaining a constant operation temperature. This enabled the determination of reaction times τ_r and recovery times τ_d, extracted from the first applied vapor pulse. Considering the lack of regulation, a feeding gas flow was supplied from a cylinder with a specific concentration and then adjusted using pre-calibrated mass flow controllers [58–60]. The required concentrations were obtained using Equations (3) and (4) to evaluate the sensor performances for the specimen [61–64]:

$$C\ (ppm) = \frac{C_1 \cdot F_{gas}}{F_{tot}} \quad (3)$$

where C is the required concentration of gas, C_1 is the initial concentration of the test gas, and F_{gas} is the gas flow. F_{tot} is the total flow of the gas–air mixture [33].

$$V_x = (Vol \cdot C \cdot M)/(22.4 \cdot d \cdot p) \cdot [(273 + T_r)/(273 + T_c)] \cdot 10^{-9} \quad (4)$$

where V_x is the volume of VOC injected into the test chamber volume *Vol*. C is the required VOC concentration (ppm), M is the molar mass, d is the density (g/cm^3), p is the purity, T_r is the room temperature, and T_c is the test chamber temperature (operating temperature).

In addition, the limit of detection (LoD) of the sensors for different temperatures and relative humidity was determined using Equation (5) [65]:

$$LoD = 3\frac{\sigma_B}{b} \quad (5)$$

The symbol σ_B is the population standard deviation of the blank signals, b is the slope of the signal/concentration functional relationship, and '3' is the chosen expansion factor. Origin software's linear fit function, in OriginLab 2024, was used to determine both the standard deviation and the slope.

Different levels of relative humidity (RH) were generated using a bubble humidification setup. This process entailed the passage of air through deionized water, followed by its continuous injection into the measurement chamber to generate the necessary RH value. The humidity was continuously monitored by a standard hygrometer. More details can be found in previous works [26,33,58,66]. In addition, relative humidity was measured continuously throughout the experiment using a specialized, calibrated sensor (SHT43; Digital Humidity Sensor with ISO17025 certification) placed next to the sample [33,58,66,67].

Fourier-transform infrared (FTIR) spectra of two P(V_3D_3-co-TFE) co-polymer thin films deposited onto Si substrates were measured using a FTIR spectrometer (Bruker Invenio R, Billerica, MA, USA). The first sample set was an as-deposited 260 nm thick film for reference, while the second sample set was heated to 350 °C for 15 min, simulating the sensor's measurement conditions. The spectra were recorded from 7500 to 368 cm^{-1} with 32 scans at 4 cm^{-1} resolution. A range from 4000 to 500 cm^{-1} was selected for further investigation. Baseline correction (spline), atmospheric compensation (CO_2), normalization, and smoothing (Savitzky–Golay) were performed on the recorded data using the Origin software (OriginLab 2024).

3. Results and Discussion

The fabricated $P(V_3D_3$-co-TFE) hybrid sensors were investigated with regard to their chemical and the gas sensing properties. All applied gases had a concentration of 100 ppm.

3.1. Chemical Characterization of the Fabricated Hybrid Sensors

During the iCVD process, V_3D_3 adsorbs at the substrate stage and HFPO and the initiator PFBSF are thermally decomposed to yield, e.g., difluorocarbene and fluorobutane radicals, respectively. Consequently, the difluorocarbene as well as the initiator radicals can initiate the polymerization via the vinyl groups of V_3D_3. Thus, using V_3D_3 and HFPO as co-monomers, we expected a molecular structure in which the CF_2 chains are connected by V_3D_3 crosslinks between the chains, resulting in sieve-like structures as schematically illustrated in Figure 2a.

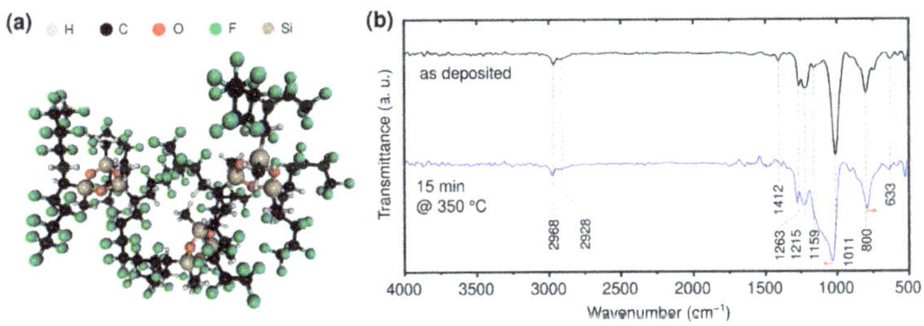

Figure 2. (**a**) Visualization of the molecular structure inside the deposited $P(V_3D_3$-co-TFE) thin films. (**b**) FTIR spectra of an as-deposited $P(V_3D_3$-co-TFE) thin film (black curve) and a $P(V_3D_3$-co-TFE) annealed at 350 °C in air for 15 min (blue curve). Red arrows indicate a change in position of the cyclotrisiloxane ring-associated band in the FTIR measurement.

iCVD as a solvent-free process allows the combination of co-monomers, which lack a common solvent needed in wet-chemical preparations. To confirm the functionality of both co-monomers inside our polymer films, we performed FTIR measurements shown in Figure 2b. The as-deposited 260 nm $P(V_3D_3$-co-TFE) (black curve) copolymer layer revealed C-H stretching bands between 3000 cm^{-1} and 2800 cm^{-1} for the formation of sp^3 hybridized carbon links and thus the successful incorporation of V_3D_3 units inside the polymer film. The cyclotrisiloxane rings of V_3D_3 were preserved during the polymerization, indicated by the band at 1011 cm^{-1} for the as-deposited film. Additional bands corresponding to $Si\text{-}CH_3$ symmetric as well as asymmetric bending and rocking vibrations can be observed at 1263, 1412 and 800 cm^{-1}, respectively [54]. The 1412 cm^{-1} band can also be associated with $Si\text{-}CH_2$ groups [68], which indicates the expected crosslinking via reacted vinyl groups.

Fluoropolymer functionalities are shown by the bands at 1215, 1159, and 633 cm^{-1} [54,69]. This reveals a successful formation of a $P(V_3D_3$-co-TFE) polymer film. The 1:1 ratio of V_3D_3 and HFPO flows should ideally result in an equal number of V_3D_3 and the TFE units in the copolymer. However, the strong bands related to V_3D_3 indicate a higher density of V_3D_3 units inside the film.

In order to simulate the conditions that are present during the gas sensing measurement, the sample was heated to 350 °C for a period of 15 min in ambient air. The purpose of this experiment was to investigate whether a chemical modification of the film occurs during the gas sensing measurement. After heat treatment at 350 °C, the deposited film is reduced in thickness, resulting in a reduced signal-to-noise ratio. Furthermore, a change in position of the cyclotrisiloxane ring-associated band can be noted in the FTIR measurement (blue curve in Figure 2b). A shoulder is formed, while the band shifts towards larger wavenumbers, indicated by the small red arrow in Figure 2b. Additionally, the Si-CH_3 rocking associated band shifts towards smaller wavenumbers. These phenomena can be attributed to structural changes within the thin film, such as an increase in Si-O-Si bond angle, indicating a change from a ring towards a cage structure of the crosslinking unit [70] or opening of the ring structure, which is beneficial for sensor applications. The PTFE-associated bands do not appear to undergo substantial alterations, suggesting that there is no significant decomposition of this component.

3.2. Gas Sensing Measurements and Evaluation

The first set of gas sensing measurement results of the P(V_3D_3-co-TFE)/TiO_2 hybrid gas sensor is shown in Figure 3, which shows an entire set (a) at 10% RH and the dynamic responses at the best response operating temperatures (b,c). The measurement conditions were set to 10% RH due to the potential relevance of gas detection in low RH environments to various application fields, particularly in biomedicine and food storage [71–74]. For instance, cleanrooms are frequently operated within the range of 10–20% RH to mitigate the risk of electrostatic discharge (ESD) [75,76] and/or to enhance virus attenuation [77,78]. Figure 3a shows the measured sensor response of our hybrid structure at different operating temperatures at 10% RH. This sensor structure shows the best results for hydrogen gas with a response value of ~153% at an operating temperature of 300 °C.

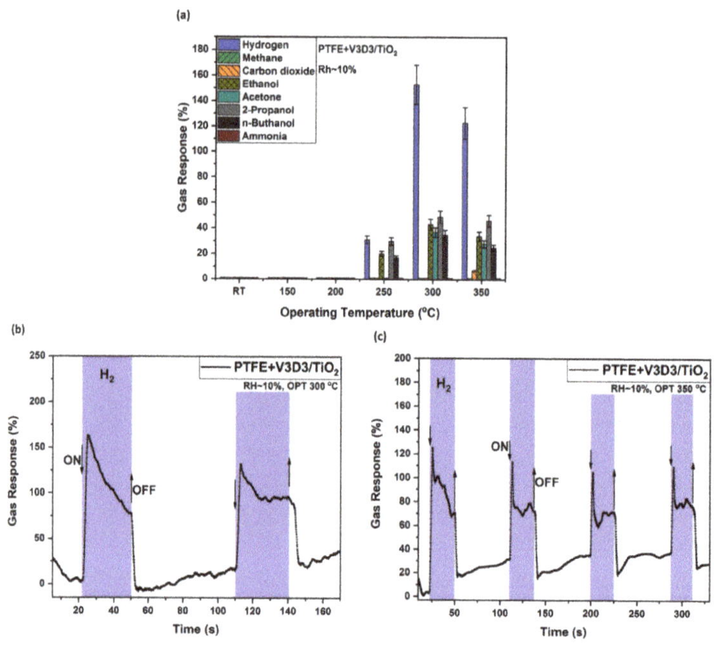

Figure 3. (**a**) Experimental measurements of the gas response of the P(V_3D_3-co-TFE)/TiO_2 hybrid structure to different gases and VOCs (hydrogen, methane, carbon dioxide, ethanol, acetone, acetone, 2-propanol, n-butanol, ammonia) at different operating temperatures. (**b**) Dynamic response of the same structure to hydrogen gas (100 ppm concentration) at 10% RH and operating temperatures of 300 °C and (**c**) 350 °C.

A similar selectivity is maintained at operating temperatures of 250 °C and 350 °C with reduced gas responses of ~30 and ~123%, respectively. At elevated temperatures, ethanol, n-butanol, 2-propanol, and acetone vapor were detected, with relatively low gas response values. Consequently, an operating temperature of 300 °C was selected for the analysis of the dynamic respons to hydrogen (Figure 3b). Our hybrid sensor exhibits a ~153% response in combination with a relatively promising reaction time of τ_r = 2.0 s and recovery time of τ_d = 2.1 s. For comparison, Figure 3c depicts the dynamic response at an elevated operating temperature of 350 °C. At this temperature, a slightly reduced gas response of ~123% is observed, yet faster reaction and recovery times of τ_r = 0.5 s and τ_d = 1.6 s are noted.

To investigate the long-term stability, we reevaluated the same sample 309 days after the first measurements for the same gases. We chose the same environment (10% RH) as well as an environment with higher relative humidity (45% RH) (Figure 4) in a similar way as in previous studies **[33,58]**. Before measurement, the hybrid sensors were heated/thermally treated to remove moisture, hydrocarbons, or other environmental contaminants from their surfaces, as described in Section 2.1. The hybrid sensors are fabricated to work in real-world environments. Thus, it is not necessary to store them in inert gas or a vacuum.

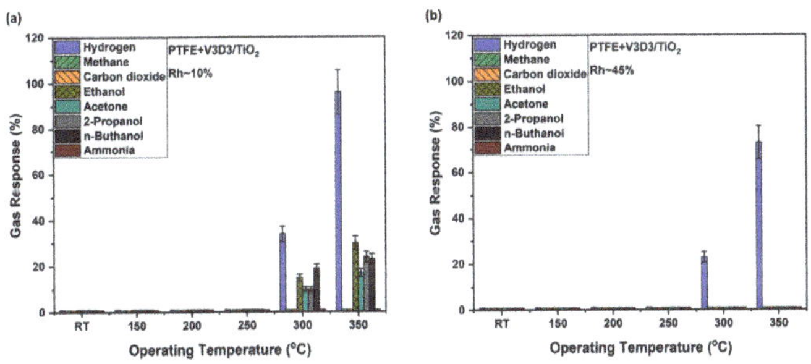

Figure 4. The gas response of the P(V_3D_3-co-TFE)/TiO_2 hybrid structure after 309 days from first measurements for different gases and VOCs (hydrogen, methane, carbon dioxide, ethanol, acetone, acetone, 2-propanol, n-butanol, ammonia) at different operating temperatures. (**a**) shows the results obtained in a relative low humidity of ~10% RH and (**b**) shows those obtained at ~45% ambient relative humidity.

Figure 4a shows that even after 309 days, a clear detection of ethanol, n-butanol, 2-propanol, acetone, and especially hydrogen gas is observed at elevated temperatures. Thereby, in comparison to Figure 3a, the optimal gas response for hydrogen gas shifted towards higher operating temperatures, with slightly reduced response values of 34% and 96% at 300 °C and 350 °C, respectively. A rise in relative humidity keeping all other measurement conditions constant results in a general decrease in gas response (Figure 4b), which could occur due to a specific gas sensing mechanism at higher operating temperature [34,79], which has to be investigated in more detail in the future. A hydrogen gas response of ~23% at 300 °C and ~73% at 350 °C is observed, indicating a decrease of about ~11% and ~23%, respectively, due to relative humidity increase. The gas response decrease for other test gases and VOCs is significantly higher, resulting in an improved hydrogen gas selectivity of the presented hybrid sensors in high humidity environments.

Again, an operating temperature of 300 °C as well as 350 °C was selected for the analysis of the dynamic response to hydrogen at 10% and 45% RH (Figure 5). A 34% hydrogen gas response at 10% RH and 300 °C was measured. In addition, reaction and recovery times of τ_r = 3.0 and τ_d = 4.9 s were determined (Figure 5a). An increase of relative humidity to ~45% RH, again, leads to a decreased hydrogen gas response value of ~23% and response and recovery times of τ_r = 4.2 s and τ_d = 4.4 s, respectively (Figure 5b). Furthermore, it can be noted that the signal became slightly unstable, which is assumed to be due to the abundance of water molecules on the top surface of the sensor structure. Figure 5c shows the dynamic response data for 10% RH and 350 °C operating temperature. The higher operation temperature had a positive influence on the hydrogen gas response value (approximately 96%). The response and recovery times are τ_r = 1.3 s and τ_d = 2.1 s, respectively. An increase in humidity (45% RH) leads again to a reduced hydrogen response value (approximately 73%) and response and recovery times of τ_r = 2.3 s and τ_d = 3.3 s (Figure 5d).

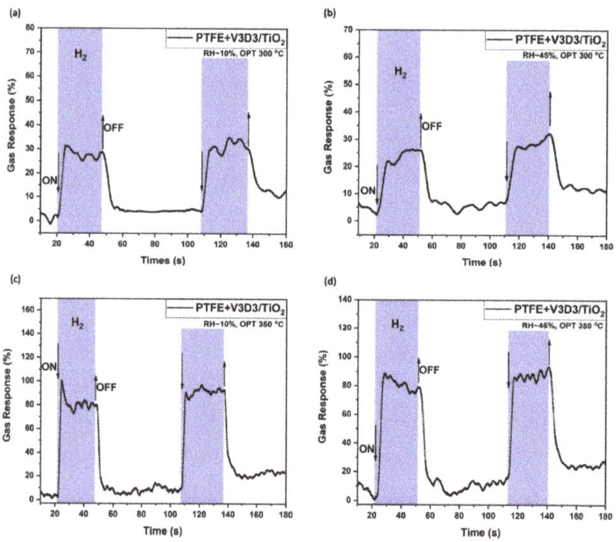

Figure 5. Dynamic response to hydrogen gas of a P(V_3D_3-co-TFE)/TiO_2 hybrid structure investigated 309 days after its first measurements. The individual subplots show the measurements at different operating temperatures and different relative humidity values: (**a**) 300 °C, ~10% RH; (**b**) 300 °C, ~45% RH; (**c**) 350 °C, ~10% RH, and (**d**) 350 °C, ~45% RH.

Unlike the overall gas responses, the reaction and recovery times seem to exhibit no significant changes with humidity or temperature. At higher relative humidity levels, the recorded signal shows slight perturbations manifested as different "noise hills" between the pulses. Nevertheless, the overall signal maintains a consistent shape. As noted in a recent review [80], low-frequency noise can stem from various sources and be impacted by numerous parameters, including, e.g., the sputtering power used during the fabrication process [81], etc. In our situation, the noise could originate from the fabrication process as well, specifically from the interface formed between the copolymer film and the gas sensing structure. This interface could cause vapor molecules to move chaotically, occasionally resulting in low-frequency noise. Due to the intricate nature of the noise phenomenon and the numerous potential causes, a definitive conclusion cannot be provided at this stage. Further research is necessary to understand the underlying mechanisms and identify the specific factors contributing to the noise phenomenon.

After an additional 118 days after the second measurement (after 427 days in total), we tested the hybrid structure again using the same gases, VOCs, and environments (10%

RH and 45% RH). The corresponding gas responses and dynamic hydrogen responses at 300 °C and 350 °C are shown in Figures S1 and S2. These data support the trend observed during the second measurement (Figure 5), highlighting the high hydrogen gas selectivity in high humidity environments. As time progresses, the overall hydrogen gas response of the analyzed hybrid sensor exhibits a decline, as shown in (Figure 6). After a period of 427 days, the hydrogen gas response is reduced to approximately 28% at 300 °C and 75% at 350 °C operating temperature.

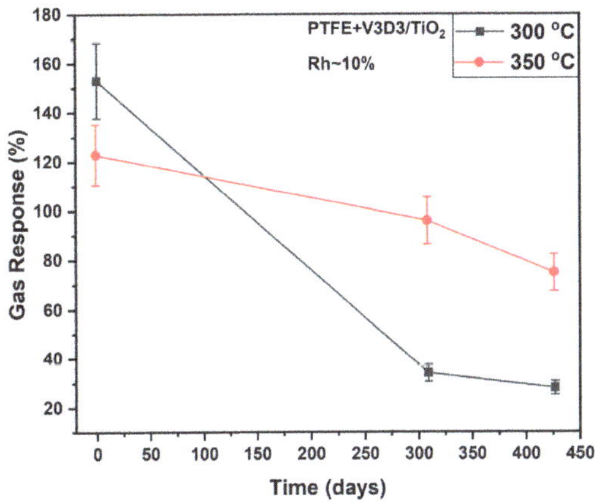

Figure 6. Gas response of the P(V_3D_3-co-TFE)/TiO_2 hybrid structure to hydrogen gas at 300 °C (gray symbols) and 350 °C (red symbols) and 10% RH measured over a time span of 427 days.

This behavior must be taken into account when integrating the sensor into devices for applications such as human breath testing. It creates opportunities for further research, particularly in the development of mathematical control logics. Apart from the hydrogen gas response values, the optimum operation temperature also experiences changes. The initial measurement (Figure 3) indicates an optimum operation temperature of 300 °C, while the second and third measurements show the highest response values and consequently better operation conditions at 350 °C. A similar trend can be observed for the determination of the LoD (Figure S3).

In this context, it should be noted that elevated temperatures for sensor operation could have an annealing effect on the presented hybrid sensors. As mentioned previously, PTFE usually cannot withstand operating temperatures higher than 320 °C, whereas PV_3D_3 usually retains its properties even at 400 °C. As indicated in Figure 2, the used copolymer does not show significant signs of decomposition, but rather slight changes of the internal ring structure. These structural changes occur during the first measurements at elevated temperatures, which could result in the observed shift of optimum operation temperature from 300 to 350 °C in all subsequent measurements. This phenomenon of enhanced performance at higher operation temperature can also be attributed to the underlying metal oxide sensing structure. However, conflicting observations have been made regarding the optimum temperature. While some studies suggest a peak performance around 300–350 °C [15,45], others indicate continuous performance enhancement with increasing temperature [20]. The parameters determined in this work, such as operating temperature and response value, enable a comparison of the TiO_2 sensor with additional P(V_3D_3-co-TFE) coating with different polymer-containing sensors used in the gas sensing industry. As seen in Table 1, scientists have experimented with different hybrid structures for the detection of a variety of different gases. The number of publications in this area is relatively limited and focusses particularly on the use of conductive copolymers, while non-conductive polymers have yet to be further explored.

Table 1. Copolymer coatings studied in the gas sensor industry.

No.	Sensing Material	Polymer	Target	OPT, °C	Concentration	Response	Ref.
1.	ZIF-8/PDMS	PDMS	CH_4	-	50%	16%	[82]
2.	PANI-co-PIN /Cu–Al_2O_3	PANI-co-PIN	Ammonia	RT	-	8 μΩ *	[83]
3.	(PANA-*Co*-PIN): Fe_3O_4-	(PANA-*Co*-PIN)	Ammonia	-	-	7 μΩ *	[84]
4.	TiO_2-g-P(SPEA-co-GMA)	P(SPEA-co-GMA)	Light Responsiveness	-	-	-	[85]
5.	PANI-*co*-PPy/Cu-Al_2O_3	(PANI-*co*-PPy)	Ammonia	RT	100 ppm	5 μΩ *	[86]
6.	PLANC-GOD	PLANC	Glucose	-	1; 3; 5; 7; 10^{-6} mol L	-	[87]
7.	(EMAA)/(MWCNT)	EMAA	Strain sensing	220	-	41.1 S m^{-1}	[88]
D1	P(V3D3-co-TFE)/TiO_2	P(V3D3-co-TFE)	H_2	300	100 ppm	153%	T.w.
D1	P(V3D3-co-TFE)/TiO_2	P(V3D3-co-TFE)	H_2	350	100 ppm	123%	T.w.

*—estimated from graphs; T.w. = this work; D1 = dataset from Figure 1; PDMS = poly (dimethylsiloxane); PANI = polyaniline; PIN = polyindole; PANA = poly(anthranilic acid); SPEA = poly(spiropyran ethylacrylate); PGMA = poly(glycidyl methacrylate); PPy = polypyrrole; PLANC = poly(luminol–aniline) nanowires composite; PEMAA = poly(ethylene-co-methacrylic acid); PV_3D_3 = poly(trivinyltrimethylcyclotrisiloxane); PTFE = polytetrafluoroethylene.

Table 2 compares the hybrid sensors reported in this study with selected hydrogen detectors coated with conventional polymers. Notably, the reaction and recovery times achieved are relatively fast in comparison with sensors from the literature. In addition, despite the very low concentrations of 100 ppm used in our experiment, high hydrogen response values were obtained, exceeding even those of sensors tested in higher initial concentrations. The relative humidity varies for the measurement in most test setups.

Table 2. Polymer-coated H_2 detectors studied in sensor industry and publications.

No.	Sensing Material	Polymer	H2 Concentration	RH, ~%	OPT, °C	Response	Time, s		Ref.
							Response	Recovery	
1.	TiO_2	PV4D4	100 ppm	-	300	100%	3	44	[15]
2.	TiO_2	PV4D4	100 ppm	-	350	709.07%	3.02	23.23	[20]
3.	PEDOT:PSS@Pd	PEDOT: PSS	4 %	-	RT	31.6%	19 (±4)	73 (±11)	[89]
4.	PANI/SnO_2	PANI	6000 ppm	-	RT	42%	11	7	[90]
5.	PANI/Al-SnO_2	PANI	1000 ppm	-	48	-	2	2	[91]
6.	PANI/Al-SnO_2	PANI	100 ppm	-	340	-	3	2	[91]
7.	PANI/SnO_2 + Pd	PANI	50 ppm	-	RT	19.2%	39	53	[92]
8.	PANI/SnO_2 + Pd	PANI	350 ppm	-	RT	353.7%	141	76	[92]
9.	PMMA/SnO_2:In_2O_3	PMMA	600 ppm	14	RT	1.05×10^3 Ω	196	282	[93]
10.	PMMA/SnO_2:In_2O_3	PMMA	600 ppm	65	RT	1.34×10^2 Ω	842	387	[93]
11.	Cytop/SnO_2:In_2O_3	Cytop	600 ppm	14	RT	1.49×10^1 Ω	1550	35	[93]
12.	Cytop/SnO_2:In_2O_3	Cytop	600 ppm	65	RT	7.52×10^1 Ω	322	44	[93]
13.	Fluoropel/SnO_2:In_2O_3	fluoropolymer	600 ppm	14	RT	1.55×10^2 Ω	134	30	[93]
14.	Fluoropel/SnO_2:In_2O_3	fluoropolymer	600 ppm	65	RT	9.79×10^1 Ω	356	56	[93]

Table 2. *Cont.*

No.	Sensing Material	Polymer	H2 Concentration	RH, ~%	OPT, °C	Response	Time, s		Ref.
							Response	Recovery	
D2	P(V3D3-co-TFE)/TiO_2	P(V3D3-co-TFE)	100 ppm	10	300	34%	3	4.9	T.w.
D2	P(V3D3-co-TFE)/TiO_2	P(V3D3-co-TFE)	100 ppm	10	350	96%	1.3	2.1	T.w.
D2	P(V3D3-co-TFE)/TiO_2	P(V3D3-co-TFE)	100 ppm	45	300	23%	4.2	4.4	T.w.
D2	P(V3D3-co-TFE)/TiO_2	P(V3D3-co-TFE)	100 ppm	45	350	73%	2.3	3.3	T.w.
D3	P(V3D3-co-TFE)/TiO_2	P(V3D3-co-TFE)	100 ppm	10	300	28%	1.5	3.1	T.w.
D3	P(V3D3-co-TFE)/TiO_2	P(V3D3-co-TFE)	100 ppm	10	350	75%	2.5	3.2	T.w.
D3	P(V3D3-co-TFE)/TiO_2	P(V3D3-co-TFE)	100 ppm	45	300	23%	1.7	2.1	T.w.
D3	P(V3D3-co-TFE)/TiO_2	P(V3D3-co-TFE)	100 ppm	45	350	48%	1.7	3.5	T.w.

T.w. = this work; D2 = dataset from Figure 2; D3 = dataset from Figure S1; PV4D4 = poly(1,3,5,7-tetramethyl-tetravinylcyclotetrasiloxane); PEDOT = poly(3,4-ethylenedioxythiophene); PSS = poly(styrene sulfonate); PANI = polyaniline; PMMA = polymethyl methacrylate; Cytop = poly-perfluorobutenyl vinyl ether; PV3D3 = poly (trivinyltrimethyltrimethylcyclotrisiloxane); PTFE = polytetrafluoroethylene.

On the other hand, in Table 3 one can see our previous research on the materials used in this work, which have been adapted through various improvement processes.

Table 3. Comparison of all our previous works results with different sensing structures and coatings for gas detection.

Sensing Structure	Polymer Coating	Enhancement	OPT, °C	Detected Gas	Gas Response, %	Ref.
TiO_2	Uncoated	TA 450 °C	250	Hydrogen	600	[22]
TiO_2	PTFE	-	350	2-propanol	64	[46]
TiO_2	PV4D4	-	400	2-propanol	225	[45]
TiO_2	PV4D4	TA 610 °C for TiO_2	300/RT	Hydrogen/Ammonia	100/52	[15]
TiO_2	PV4D4	TA 450 °C for PV4D4	350	Hydrogen	709	[20]
$CuO/Cu_2O/ZnO$:Fe	PV3D3	RTA 650 °C	350	Hydrogen	191	[39]
TiO_2	PV3D3	-	-	-	-	To be published
TiO_2	P(V3D3 + TFE)	Copolymer structure	300	Hydrogen	153	This work

TA—thermal annealing; RTA—rapid thermal annealing.

4. Conclusions

This study demonstrates that the proposed P(V_3D_3-co-TFE)/TiO_2 hybrid sensor is relatively reliable over an extended period of time. It exhibits hydrogen selectivity in different environments demanded for different application fields such as in cleanrooms, biomedical applications, or food storage. Tests at 10% RH are necessary to isolate the hydrogen response [94], because the sensor materials can adsorb water molecules which can influence the gas response. Even after a total of 427 days, similar trends were observed in all experimental measurements. An optimum operating temperature of 300 and 350 °C and a constant selectivity for hydrogen gas were determined. The elevated operating temperatures do not lead to a significant decomposition of the P(V_3D_3-co-TFE) film according to our experiments. A change in the internal structure was rather observed from FTIR studies. While the selectivity remains constant, the overall response values to hydrogen gas decrease slightly over time; namely, they were reduced to approximately 75% at 350 °C operating temperature. Nevertheless, it is possible to integrate this type of sensor structure with constant selectivity into smart devices where a response time curve is used to filter all the collected data. Hence, it can be applied not only as a simple detector, but rather as a potential exhaled breath detector for diagnostics or a feedback system for industrial applications. This provides new pathways to tune the gas sensor performance for specific applications. The applied copolymer thin film can introduce novel, e.g., sieve-like structures that enhance selectivity, sensitivity, and stability for preexisting sensor structures. Future research could investigate the potential tailoring of the additional polymer films to enhance the functionality in the context of molecular polymer–gas interactions. Different co-monomer combinations could give rise to sensors which are specifically tailored for certain gases and applications. Nevertheless, more research could be performed on this type of hybrid sensors, taking into consideration even other aspects, from an electronic device point of view, such as performing experiments at different magnetic fields [42], finding factors on how to improve self-healing and improve relationships between polymer breakdown strength and permittivity [40], adding ferromagnetic components [43], and further stoichiometry as in similar research [44]. In this regard, this research will and must continue to obtain reliable results and improve both the medical and industrial fields.

Supplementary Materials: The following supporting information can be downloaded at: https://www.mdpi.com/article/10.3390/chemosensors13040150/s1, Figure S1. The gas response of the TiO_2 structure, coated with a thin layer of V3D3 copolymer and PTFE, namely P(V3D3-co-TFE)/TiO_2 hybrid structures, after 427 days from 1st measurements, to different gases and VOCs (hydrogen, methane, carbon dioxide, ethanol, acetone, acetone, 2-propanol, n-butanol, ammonia) at different operating temperatures for relative humidity of (a) ~10% RH and (b) ~45% RH; Figure S2. Dynamic response of TiO_2 sensing structure coated with a thin layer of V3D3 copolymer and PTFE, namely P(V3D3-co-TFE)/TiO_2 hybrid structures, after 427 days from 1st measurements, to hydrogen vapor at different operating temperatures and different relative humidity: (a) 300 °C, ~10% RH; (b) 300 °C, ~45% RH; (c) 350 °C, ~10% RH and (d) 350 °C, ~45% RH; Figure S3. Limit-of-detection graph for H_2-sensing hybrid sensor, namely P(V3D3-co-TFE)/TiO_2 hybrid structures, at different concentrations, for different relative humidity and operating temperatures: (1) 300 °C, ~10% RH; (2) 300 °C, ~45% RH; (3) 350 °C, ~10% RH and (4) 350 °C, ~45% RH.

Author Contributions: Conceptualization, O.L., F.F., and S.S.; methodology, O.L., M.B., R.A., and S.S.; software, S.S., T.A., M.B., and L.S.; validation, F.F., R.A., O.L., S.S., T.A., T.S., L.Z., T.P., and B.V.; formal analysis, M.B., B.V., and L.S.; investigation, M.B., L.S., B.V., L.Z., O.L., and T.P.; resources, F.F., R.A., T.P., and B.V.; data curation, O.L., T.A., and T.S.; writing—original draft preparation, M.B., S.S., L.S., and O.L.; writing—review and editing, O.L, R.A., T.P., B.V., T.A., T.S., and L.Z.; visualization, L.S., M.B., and S.S.; supervision T.S., O.L., F.F., and T.A.; project administration, O.L. and F.F.; funding acquisition, F.F., R.A., and T.P. All authors have read and agreed to the published version of the manuscript.

Funding: We would also like to acknowledge the Government of the Republic of Moldova, Ministry of Education and Science for partial support via the State Program LIFETECH No. 020404 at the Technical University of Moldova. Mihai Brinza gratefully acknowledges Kiel University, Department of Materials Science, Chair for Multicomponent Materials, Chair of Functional Nanomaterials, Germany for internship positions in 2024-2025 and TUM for constant support. In addition, this work was partially funded by the German Research Foundation (DFG)—Project-ID 541212717 and Project-ID 286471992-SFB1261 project A2.

Institutional Review Board Statement: Not applicable.

Informed Consent Statement: Not applicable.

Data Availability Statement: The original contributions presented in this study are included in the article. Further inquiries can be directed to the corresponding authors.

Acknowledgments: O.L. gratefully acknowledges CNRS Council for support as an expert scientist at IRCP Chimie ParisTech, Paris, France. Mihai Brinza gratefully acknowledges Kiel University, Department of Materials Science, Chair for Multicomponent Materials, Chair of Functional Nanomaterials, Germany for internship positions in 2024–2025 and TUM for constant support. We would also like to acknowledge the Government of the Republic of Moldova for partial support via the State Program LIFETECH No. 020404 at the Technical University of Moldova.

Conflicts of Interest: The authors declare no conflicts of interest.

References

1. Meixner, H.; Lampe, U. Metal Oxide Sensors. *Sens. Actuators B Chem.* **1996**, *33*, 198–202. [CrossRef]
2. Lenz, J.; Edelstein, S. Magnetic Sensors and Their Applications. *IEEE Sens. J.* **2006**, *6*, 631–649. [CrossRef]
3. Loock, H.-P.; Wentzell, P.D. Detection Limits of Chemical Sensors: Applications and Misapplications. *Sens. Actuators B Chem.* **2012**, *173*, 157–163. [CrossRef]
4. Arcos, J.M.M.; Santos, D.M.F. The Hydrogen Color Spectrum: Techno-Economic Analysis of the Available Technologies for Hydrogen Production. *Gases* **2023**, *3*, 25–46. [CrossRef]
5. Okolie, J.A.; Patra, B.R.; Mukherjee, A.; Nanda, S.; Dalai, A.K.; Kozinski, J.A. Futuristic Applications of Hydrogen in Energy, Biorefining, Aerospace, Pharmaceuticals and Metallurgy. *Int. J. Hydrogen Energy* **2021**, *46*, 8885–8905. [CrossRef]
6. Falcone, P.M.; Hiete, M.; Sapio, A. Hydrogen Economy and Sustainable Development Goals: Review and Policy Insights. *Curr. Opin. Green Sustain. Chem.* **2021**, *31*, 100506. [CrossRef]
7. Barthelemy, H.; Weber, M.; Barbier, F. Hydrogen Storage: Recent Improvements and Industrial Perspectives. *Int. J. Hydrogen Energy* **2017**, *42*, 7254–7262. [CrossRef]
8. Murugesan, K.; Senthamarai, T.; Chandrashekhar, V.G.; Natte, K.; Kamer, P.C.J.; Beller, M.; Jagadeesh, R.V. Catalytic Reductive Aminations Using Molecular Hydrogen for Synthesis of Different Kinds of Amines. *Chem. Soc. Rev.* **2020**, *49*, 6273–6328. [CrossRef]
9. Ciriminna, R.; Albanese, L.; Meneguzzo, F.; Pagliaro, M. Hydrogen Peroxide: A Key Chemical for Today's Sustainable Development. *ChemSusChem* **2016**, *9*, 3374–3381. [CrossRef]
10. Iida, A.; Nosaka, N.; Yumoto, T.; Knaup, E.; Naito, H.; Nishiyama, C.; Yamakawa, Y.; Tsukahara, K.; Terado, M.; Sato, K.; et al. The Clinical Application of Hydrogen as a Medical Treatment. *Acta Med. Okayama* **2016**, *70*, 331–337. [CrossRef]
11. Zhang, J.-Y.; Liu, C.; Zhou, L.; Qu, K.; Wang, R.; Tai, M.; Lei Lei, J.W.; Wu, Q.F.; Wang, Z. A Review of Hydrogen as a New Medical Therapy. *Hepatogastroenterology* **2012**, *59*, 1026. [CrossRef] [PubMed]
12. Ostojic, S. Molecular Hydrogen in Sports Medicine: New Therapeutic Perspectives. *Int. J. Sports Med.* **2014**, *36*, 273–279. [CrossRef] [PubMed]
13. Huang, C.-S.; Kawamura, T.; Toyoda, Y.; Nakao, A. Recent Advances in Hydrogen Research as a Therapeutic Medical Gas. *Free Radic. Res.* **2010**, *44*, 971–982. [CrossRef] [PubMed]
14. Ohta, S. Recent Progress Toward Hydrogen Medicine: Potential of Molecular Hydrogen for Preventive and Therapeutic Applications. *Curr. Pharm. Des.* **2011**, *17*, 2241–2252. [CrossRef]
15. Brinza, M.; Schröder, S.; Ababii, N.; Gronenberg, M.; Strunskus, T.; Pauporté, T.; Adelung, R.; Faupel, F.; Lupan, O. Two-in-One Sensor Based on PV4D4-Coated TiO_2 Films for Food Spoilage Detection and as a Breath Marker for Several Diseases. *Biosensors* **2023**, *13*, 538. [CrossRef]
16. Ghawanmeh, A.A.; A.Tanash, S.; Al-Rawashdeh, N.A.F.; Albiss, B. The Recent Progress on Nanomaterial-Based Chemosensors for Diagnosis of Human Exhaled Breath: A Review. *J. Mater. Sci.* **2024**, *59*, 8573–8605. [CrossRef]
17. Kerlin, P.; Wong, L. Breath Hydrogen Testing in Bacterial Overgrowth of the Small Intestine. *Gastroenterology* **1988**, *95*, 982–988. [CrossRef] [PubMed]
18. De Geyter, C.; Van de Maele, K.; Hauser, B.; Vandenplas, Y. Hydrogen and Methane Breath Test in the Diagnosis of Lactose Intolerance. *Nutrients* **2021**, *13*, 3261. [CrossRef]
19. Marton, A.; Xue, X.; Szilagyi, A. Meta-analysis: The Diagnostic Accuracy of Lactose Breath Hydrogen or Lactose Tolerance Tests for Predicting the North European Lactase Polymorphism C/T-13910. *Aliment. Pharmacol. Ther.* **2012**, *35*, 429–440. [CrossRef]
20. Lupan, O.; Brinza, M.; Piehl, J.; Ababii, N.; Magariu, N.; Zimoch, L.; Strunskus, T.; Pauporté, T.; Adelung, R.; Faupel, F.; et al. Influence of Silsesquioxane-Containing Ultra-Thin Polymer Films on Metal Oxide Gas Sensor Performance for the Tunable Detection of Biomarkers. *Chemosensors* **2024**, *12*, 76. [CrossRef]
21. Lupan, O.; Brinza, M.; Schroder, S.; Ababii, N.; Strunskus, T.; Viana, B.; Pauporté, T.; Adelung, R.; Faupel, F. Sensors Based on Hybrid Materials for Environmental, Industrial and Biomedical Applications. In Proceedings of the 2024 IEEE 14th International

Conference "Nanomaterials: Applications and Properties" (NAP 2024), Riga, Latvia, 8–13 September 2024; Institute of Electrical and Electronics Engineers Inc.: Piscataway, NJ, USA, 2024. [CrossRef]

22. Lupan, O.; Postica, V.; Ababii, N.; Reimer, T.; Shree, S.; Hoppe, M.; Polonskyi, O.; Sontea, V.; Chemnitz, S.; Faupel, F.; et al. Ultra-Thin TiO_2 Films by Atomic Layer Deposition and Surface Functionalization with Au Nanodots for Sensing Applications. *Mater. Sci. Semicond. Process.* **2018**, *87*, 44–53. [CrossRef]
23. Fan, G.; Guan, J.; Yu, H.; Zhu, Q.; Han, N.; Mo, J.; Chen, Y. Highly Sensitive Formaldehyde Gas Sensor Based on SnO_2/Zn_2SnO_4 Hybrid Structures. *Build. Environ.* **2024**, *262*, 111781. [CrossRef]
24. Sakalley, S.; Saravanan, A.; Kathiravan, D.; Tang, J.-C.; Cheng, W.-C.; Chen, S.-C.; Sun, H.; Huang, B.-R. Enhanced Hydrogen Gas Sensing through the Utilization of a Hybrid Nanostructure Combining ZnO Nanotubes and HiPIMS Cu3N Thin Film. *Sens. Actuators B Chem.* **2024**, *402*, 135107. [CrossRef]
25. Nirmala Devi, K.; Vadivel, S. Room Temperature Fiber Optic Gas Sensor Technology Based Zn3(VO4)2/PMMA Hybrid Composites via Facile Hydrothermal Route. *Opt. Fiber Technol.* **2024**, *82*, 103642. [CrossRef]
26. Lupan, O.; Santos-Carballal, D.; Ababii, N.; Magariu, N.; Hansen, S.; Vahl, A.; Zimoch, L.; Hoppe, M.; Pauporté, T.; Galstyan, V.; et al. $TiO_2/Cu_2O/CuO$ Multi-Nanolayers as Sensors for H_2 and Volatile Organic Compounds: An Experimental and Theoretical Investigation. *ACS Appl. Mater. Interfaces* **2021**, *13*, 32363–32380. [CrossRef]
27. Sari, Y.; Gareso, P.L.; Armynah, B.; Tahir, D. A Review of TiO_2 Photocatalyst for Organic Degradation and Sustainable Hydrogen Energy Production. *Int. J. Hydrogen Energy* **2024**, *55*, 984–996. [CrossRef]
28. Seremak, W.; Jasiorski, M.; Baszczuk, A.; Winnicki, M. Durability Assessment of Low-Pressure Cold-Sprayed TiO_2 Photocatalytic Coatings: Photocatalytic and Mechanical Stability. *Surf. Coat. Technol.* **2025**, *497*, 131740. [CrossRef]
29. Li, M.; Zhao, N.; Wang, C.; Zou, Y.; Chen, X.; Mei, Y. Improving Thermal Stability and Electrochemical Performance of Polymer Solid Electrolyte Membranes with a Novel TiO_2@PA-APP Composite Additive. *Colloids Surf. A Physicochem. Eng. Asp.* **2025**, *705*, 135640. [CrossRef]
30. Ferroni, M.; Guidi, V.; Martinelli, G.; Faglia, G.; Nelli, P.; Sberveglieri, G. Characterization of a Nanosized TiO2 Gas Sensor. *Nanostruct. Mater.* **1996**, *7*, 709–718. [CrossRef]
31. Li, Z.; Yao, Z.; Haidry, A.A.; Plecenik, T.; Xie, L.; Sun, L.; Fatima, Q. Resistive-Type Hydrogen Gas Sensor Based on TiO2: A Review. *Int. J. Hydrogen Energy* **2018**, *43*, 21114–21132. [CrossRef]
32. Malallah Rzaij, J.; Mohsen Abass, A. Review on: TiO_2 Thin Film as a Metal Oxide Gas Sensor. *J. Chem. Rev.* **2020**, *2*, 114–121. [CrossRef]
33. Lupan, O.; Ababii, N.; Mishra, A.K.; Bodduluri, M.T.; Magariu, N.; Vahl, A.; Krüger, H.; Wagner, B.; Faupel, F.; Adelung, R.; et al. Heterostructure-Based Devices with Enhanced Humidity Stability for H_2 Gas Sensing Applications in Breath Tests and Portable Batteries. *Sens. Actuators A Phys.* **2021**, *329*, 112804. [CrossRef]
34. Wang, C.; Yin, L.; Zhang, L.; Xiang, D.; Gao, R. Metal Oxide Gas Sensors: Sensitivity and Influencing Factors. *Sensors* **2010**, *10*, 2088–2106. [CrossRef]
35. Wicker, S.; Guiltat, M.; Weimar, U.; Hémeryck, A.; Barsan, N. Ambient Humidity Influence on CO Detection with SnO_2 Gas Sensing Materials. A Combined DRIFTS/DFT Investigation. *J. Phys. Chem. C* **2017**, *121*, 25064–25073. [CrossRef]
36. Enachi, M.; Lupan, O.; Braniste, T.; Sarua, A.; Chow, L.; Mishra, Y.K.; Gedamu, D.; Adelung, R.; Tiginyanu, I. Integration of Individual TiO_2 Nanotube on the Chip: Nanodevice for Hydrogen Sensing. *Phys. Status Solidi-Rapid Res. Lett.* **2015**, *9*, 171–174. [CrossRef]
37. Mahdavi, H.; Rahbarpour, S.; Hosseini-Golgoo, S.-M.; Jamaati, H. Reducing the Destructive Effect of Ambient Humidity Variations on Gas Detection Capability of a Temperature Modulated Gas Sensor by Calcium Chloride. *Sens. Actuators B Chem.* **2021**, *331*, 129091. [CrossRef]
38. Yamazoe, N.; Shimanoe, K. Fundamentals of Semiconductor Gas Sensors. In *Semiconductor Gas Sensors*; Elsevier: Amsterdam, The Netherlands, 2020; pp. 3–38. [CrossRef]
39. Schröder, S.; Ababii, N.; Lupan, O.; Drewes, J.; Magariu, N.; Krüger, H.; Strunskus, T.; Adelung, R.; Hansen, S.; Faupel, F. Sensing Performance of $CuO/Cu_2O/ZnO$:Fe Heterostructure Coated with Thermally Stable Ultrathin Hydrophobic PV3D3 Polymer Layer for Battery Application. *Mater. Today Chem.* **2022**, *23*, 100642. [CrossRef]
40. Guo, Y.; Zhou, D.; Li, D.; Zhao, W.; Wang, Y.; Pang, L.; Shi, Z.; Zhou, T.; Sun, S.; Singh, C.; et al. Improved Energy Storage Performance of Sandwich-Structured P(VDF-HFP)-Based Nanocomposites by the Addition of Inorganic Nanoparticles. *J. Mater. Chem. C* **2023**, *11*, 6999–7009. [CrossRef]
41. Lupan, C.; Kohlmann, N.; Petersen, D.; Bodduluri, M.T.; Buzdugan, A.; Jetter, J.; Quandt, E.; Kienle, L.; Adelung, R.; Lupan, O. Hydrogen Nanosensors Based on Core/Shell ZnO/Al_2O_3 and $ZnO/ZnAl_2O_4$ Single Nanowires. *Mater. Today Nano* **2025**, *29*, 100596. [CrossRef]

42. Trukhanov, A.V.; Turchenko, V.O.; Bobrikov, I.A.; Trukhanov, S.V.; Kazakevich, I.S.; Balagurov, A.M. Crystal Structure and Magnetic Properties of the $BaFe_{12-X}Al_XO_{19}$ (X=0.1–1.2) Solid Solutions. *J. Magn. Magn. Mater.* **2015**, *393*, 253–259. [CrossRef]
43. Trukhanov, S.V.; Troyanchuk, I.O.; Pushkarev, N.V.; Szymczak, H. Magnetic Properties of Anion-Deficient $La_{1-x}Ba_xMnO_{3-x/2}$ ($0 \leq x \leq 0.30$) Manganites. *J. Exp. Theor. Phys.* **2003**, *96*, 110–117. [CrossRef]
44. Trukhanov, S.V.; Trukhanov, A.V.; Stepin, S.G.; Szymczak, H.; Botez, C.E. Effect of the Size Factor on the Magnetic Properties of Manganite $La_{0.50}Ba_{0.50}MnO_3$. *Phys. Solid State* **2008**, *50*, 886–893. [CrossRef]
45. Schröder, S.; Ababii, N.; Brînză, M.; Magariu, N.; Zimoch, L.; Bodduluri, M.T.; Strunskus, T.; Adelung, R.; Faupel, F.; Lupan, O. Tuning the Selectivity of Metal Oxide Gas Sensors with Vapor Phase Deposited Ultrathin Polymer Thin Films. *Polymers* **2023**, *15*, 524. [CrossRef] [PubMed]
46. Schröder, S.; Brinza, M.; Cretu, V.; Zimoch, L.; Gronenberg, M.; Ababii, N.; Railean, S.; Strunskus, T.; Pauporté, T.; Adelung, R.; et al. *A New Approach in Detection of Biomarker 2-Propanol with PTFE-Coated TiO2 Nanostructured Films. Sontea, V., Tiginyanu, I., Railean, S., Eds.*; Springer Nature: Cham, Switzerland, 2024; pp. 75–83. ISBN 978-3-031-42782-4. [CrossRef]
47. Baker, B.B.; Kasprzak, D.J. Thermal Degradation of Commercial Fluoropolymers in Air. *Polym. Degrad. Stab.* **1993**, *42*, 181–188. [CrossRef]
48. Conesa, J.A.; Font, R. Polytetrafluoroethylene Decomposition in Air and Nitrogen. *Polym. Eng. Sci.* **2001**, *41*, 2137–2147. [CrossRef]
49. Hirai, D.; Sawai, O.; Nunoura, T.; Hiroi, Z. Facile Synthetic Route to Transition Metal Oxyfluorides via Reactions between Metal Oxides and PTFE. *J. Fluor. Chem.* **2018**, *209*, 43–48. [CrossRef]
50. Tenhaeff, W.E.; Gleason, K.K. Initiated and Oxidative Chemical Vapor Deposition of Polymeric Thin Films: ICVD and OCVD. *Adv. Funct. Mater.* **2008**, *18*, 979–992. [CrossRef]
51. Wang, P.; Shao, Z.; Ulfa, M.; Pauporté, T. Insights into the Hole Blocking Layer Effect on the Perovskite Solar Cell Performance and Impedance Response. *J. Phys. Chem. C* **2017**, *121*, 9131–9141. [CrossRef]
52. Lupan, O.; Ababii, N.; Santos-Carballal, D.; Terasa, M.-I.M.-I.; Magariu, N.; Zappa, D.; Comini, E.; Pauporté, T.; Siebert, L.; Faupel, F.; et al. Tailoring the Selectivity of Ultralow-Power Heterojunction Gas Sensors by Noble Metal Nanoparticle Functionalization. *Nano Energy* **2021**, *88*, 106241. [CrossRef]
53. Hoppe, M.; Ababii, N.; Postica, V.; Lupan, O.; Polonskyi, O.; Schütt, F.; Kaps, S.; Sukhodub, L.F.; Sontea, V.; Strunskus, T.; et al. $(CuO-Cu_2O)/ZnO$:Al Heterojunctions for Volatile Organic Compound Detection. *Sens. Actuators B Chem.* **2018**, *255*, 1362–1375. [CrossRef]
54. Schröder, S.; Polonskyi, O.; Strunskus, T.; Faupel, F. Nanoscale Gradient Copolymer Films via Single-Step Deposition from the Vapor Phase. *Mater. Today* **2020**, *37*, 35–42. [CrossRef]
55. Rappe, A.K.; Casewit, C.J.; Colwell, K.S.; Goddard, W.A.; Skiff, W.M. UFF, a Full Periodic Table Force Field for Molecular Mechanics and Molecular Dynamics Simulations. *J. Am. Chem. Soc.* **1992**, *114*, 10024–10035. [CrossRef]
56. Lupan, O.; Postica, V.; Pauporté, T.; Viana, B.; Terasa, M.-I.; Adelung, R. Room Temperature Gas Nanosensors Based on Individual and Multiple Networked Au-Modified ZnO Nanowires. *Sens. Actuators B Chem.* **2019**, *299*, 126977. [CrossRef]
57. Park, S.Y.; Kim, Y.H.; Lee, S.Y.; Sohn, W.; Lee, J.-H.J.E.; Kim, D.H.; Shim, Y.-S.; Kwon, K.C.; Choi, K.S.; Yoo, H.J.; et al. Highly Selective and Sensitive Chemoresistive Humidity Sensors Based on RGO/MoS 2 van Der Waals Composites. *J. Mater. Chem. A* **2018**, *6*, 5016–5024. [CrossRef]
58. Lupan, O.; Ababii, N.; Mishra, A.K.; Gronenberg, O.; Vahl, A.; Schürmann, U.; Duppel, V.; Krüger, H.; Chow, L.; Kienle, L.; et al. Single $CuO/Cu_2O/Cu$ Microwire Covered by a Nanowire Network as a Gas Sensor for the Detection of Battery Hazards. *ACS Appl. Mater. Interfaces* **2020**, *12*, 42248–42263. [CrossRef]
59. Postica, V. Proprietăţile Senzoriale Ale Structurilor Hibride În Bază de Oxizi Metalici Şi a Reţelelor Lor. Ph.D. Thesis, Technical Univeristy of Moldova, Chișinău, Moldova, 2020.
60. Ding, K.; Zhang, D.; Jiang, L.; Chen, J.; Qu, R.; Shi, F.; Li, B.; He, X.; Wang, L.; Wang, H. An Ionic Liquid-Present Immersion Method for Preparing Cotton Fiber-Shaped Cu_2O Nanoparticles at Room Temperature. *J. Appl. Electrochem.* **2022**, *52*, 55–65. [CrossRef]
61. Nakate, U.; Yu, Y.-T.; Park, S. Hierarchical CuO Nanostructured Materials for Acetaldehyde Sensor Application. *Microelectron. Eng.* **2022**, *251*, 111662. [CrossRef]
62. Chen, Z.; Jiang, Z.-Y.; Deng, S.; Hou, M.; Zhang, X.; Jiang, Z.; Lai, N.-C. Morphology-Controlled Synthesis of Cu_2O Encapsulated Phase Change Materials: Photothermal Conversion and Storage Performance in Visible Light Regime. *Chem. Eng. J.* **2023**, *454*, 140089. [CrossRef]
63. Helwig, N.; Schüler, M.; Bur, C.; Schütze, A.; Sauerwald, T. Gas Mixing Apparatus for Automated Gas Sensor Characterization. *Meas. Sci. Technol.* **2014**, *25*, 055903. [CrossRef]
64. Luo, S.; Shen, Y.; Wu, Z.; Cao, M.; Gu, F.; Wang, L. Enhanced Ethanol Sensing Performance of Mesoporous Sn-Doped ZnO. *Mater. Sci. Semicond. Process.* **2016**, *41*, 535–543. [CrossRef]

65. Brunetti, B.; Desimoni, E. About Estimating the Limit of Detection by the Signal to Noise Approach. *Pharm. Anal. Acta* **2015**, *06*, 1–4. [CrossRef]
66. Cretu, V.; Postica, V.; Mishra, A.K.; Hoppe, M.; Tiginyanu, I.; Mishra, Y.K.; Chow, L.; de Leeuw, N.H.; Adelung, R.; Lupan, O. Synthesis, Characterization and DFT Studies of Zinc-Doped Copper Oxide Nanocrystals for Gas Sensing Applications. *J. Mater. Chem. A* **2016**, *4*, 6527–6539. [CrossRef]
67. Biadasz, A.; Kotkowiak, M.; Łukawski, D.; Jadwiżak, J.; Rytel, K.; Kędzierski, K. A Versatile Gas Transmission Device with Precise Humidity Control for QCM Humidity Sensor Characterizations. *Measurement* **2022**, *200*, 111674. [CrossRef]
68. O'Shaughnessy, W.S.; Gao, M.; Gleason, K.K. Initiated Chemical Vapor Deposition of Trivinyltrimethylcyclotrisiloxane for Biomaterial Coatings. *Langmuir* **2006**, *22*, 7021–7026. [CrossRef]
69. Piwowarczyk, J.; Jędrzejewski, R.; Moszyński, D.; Kwiatkowski, K.; Niemczyk, A.; Baranowska, J. XPS and FTIR Studies of Polytetrafluoroethylene Thin Films Obtained by Physical Methods. *Polymers* **2019**, *11*, 1629. [CrossRef]
70. Du, W.; Tu, J.; Qiu, M.; Zhou, S.; Luo, Y.; Ong, W.-L.; Zhao, J. Temperature-Mediated Structural Evolution of Vapor–Phase Deposited Cyclosiloxane Polymer Thin Films for Enhanced Mechanical Properties and Thermal Conductivity. *Int. J. Extrem. Manuf.* **2023**, *5*, 025101. [CrossRef]
71. Ziegler, V.; Paraginski, R.T.; Ferreira, C.D. Grain Storage Systems and Effects of Moisture, Temperature and Time on Grain Quality—A Review. *J. Stored Prod. Res.* **2021**, *91*, 101770. [CrossRef]
72. Esrafili, A.; Wagner, A.; Inamdar, S.; Acharya, A.P. Covalent Organic Frameworks for Biomedical Applications. *Adv. Healthc. Mater.* **2021**, *10*, 2002090. [CrossRef]
73. Wolkoff, P.; Azuma, K.; Carrer, P. Health, Work Performance, and Risk of Infection in Office-like Environments: The Role of Indoor Temperature, Air Humidity, and Ventilation. *Int. J. Hyg. Environ. Health* **2021**, *233*, 113709. [CrossRef]
74. Ma, Z.; Fei, T.; Zhang, T. An Overview: Sensors for Low Humidity Detection. *Sens. Actuators B Chem.* **2023**, *376*, 133039. [CrossRef]
75. Sofia Paulino Mendes, A.; Rutgersson, A.; Paulsson, M. Outdoor Environmental Effects on Cleanrooms—A Study from a Swedish Hospital Pharmacy Compounding Unit. *Eur. J. Pharm. Biopharm.* **2022**, *177*, 100–106. [CrossRef] [PubMed]
76. Sikka, M.P.; Mondal, M. A Critical Review on Cleanroom Filtration. *Res. J. Text. Appar.* **2022**, *26*, 452–467. [CrossRef]
77. Meng, H.; Shiue, A.; Wang, C.; Liu, J.; Jia, L.; Leggett, G. Particle and Bacterial Colony Emissions from Garments and Humans in Pharmaceutical Cleanrooms. *J. Build. Eng.* **2024**, *97*, 110829. [CrossRef]
78. Lobanchykova, N.M.; Vakaliuk, T.A.; Korbut, V.P.; Lobanchykov, S.M.; Krasnov, Y.B. Features of Designing Systems for the Formation of an Internal Microclimate of a High Class of Cleanliness of Operating Rooms of Medical Institutions. *IOP Conf. Ser. Earth Environ. Sci.* **2024**, *1415*, 012124. [CrossRef]
79. Dey, A. Semiconductor Metal Oxide Gas Sensors: A Review. *Mater. Sci. Eng. B* **2018**, *229*, 206–217. [CrossRef]
80. Shin, W.; Hong, S.; Jeong, Y.; Jung, G.; Park, J.; Kim, D.; Choi, K.; Shin, H.; Koo, R.-H.; Kim, J.-J.; et al. Low-Frequency Noise in Gas Sensors: A Review. *Sens. Actuators B Chem.* **2023**, *383*, 133551. [CrossRef]
81. Chaoumead, A.; Joo, B.-H.; Kwak, D.-J.; Sung, Y.-M. Structural and Electrical Properties of Sputtering Power and Gas Pressure on Ti-Dope In2O3 Transparent Conductive Films by RF Magnetron Sputtering. *Appl. Surf. Sci.* **2013**, *275*, 227–232. [CrossRef]
82. Cao, R.; Ding, H.; Kim, K.-J.; Peng, Z.; Wu, J.; Culp, J.T.; Ohodnicki, P.R.; Beckman, E.; Chen, K.P. Metal-Organic Framework Functionalized Polymer Coating for Fiber Optical Methane Sensors. *Sens. Actuators B Chem.* **2020**, *324*, 128627. [CrossRef]
83. Sankar, S.; George, A.; Ramesan, M.T. Copper Alumina @ Poly (Aniline- Co -Indole) Nanocomposites: Synthesis, Characterization, Electrical Properties and Gas Sensing Applications. *RSC Adv.* **2022**, *12*, 17637–17644. [CrossRef] [PubMed]
84. Jayakrishnan, P.; Jithin, K.; Meera, K.; Ramesan, M.T. Synthesis, Characterization, Magnetoelectric Properties and Gas Sensing Application of Poly(Anthranilic Acid- Co- Indole)/ Magnetite Nanocomposites. *J. Thermoplast. Compos. Mater.* **2023**, *36*, 2523–2542. [CrossRef]
85. Arjmand, F.; Mohamadnia, Z. Fabrication of a Light-Responsive Polymer Nanocomposite Containing Spiropyran as a Sensor for Reversible Recognition of Metal Ions. *Polym. Chem.* **2022**, *13*, 937–945. [CrossRef]
86. Sankar, S.; Ramesan, M.T. Synthesis, Characterization, Conductivity, and Gas-sensing Performance of Copolymer Nanocomposites Based on Copper Alumina and Poly(Aniline-Co-pyrrole). *Polym. Eng. Sci.* **2022**, *62*, 2402–2410. [CrossRef]
87. Li, G.; Lian, J.; Zheng, X.; Cao, J. Electrogenerated Chemiluminescence Biosensor for Glucose Based on Poly(Luminol–Aniline) Nanowires Composite Modified Electrode. *Biosens. Bioelectron.* **2010**, *26*, 643–648. [CrossRef] [PubMed]
88. Hia, I.L.; Snyder, A.D.; Turicek, J.S.; Blanc, F.; Patrick, J.F.; Therriault, D. Electrically Conductive and 3D-Printable Copolymer/MWCNT Nanocomposites for Strain Sensing. *Compos. Sci. Technol.* **2023**, *232*, 109850. [CrossRef]
89. Zhu, Z.; Liu, C.; Jiang, F.; Liu, J.; Liu, G.; Ma, X.; Liu, P.; Huang, R.; Xu, J.; Wang, L. Flexible Fiber-Shaped Hydrogen Gas Sensor via Coupling Palladium with Conductive Polymer Gel Fiber. *J. Hazard. Mater.* **2021**, *411*, 125008. [CrossRef] [PubMed]
90. Sonwane, N.D.; Maity, M.D.; Kondawar, S.B. Conducting Polyaniline/SnO2 Nanocomposite for Room Temperature Hydrogen Gas Sensing. *Mater. Today Proc.* **2019**, *15*, 447–453. [CrossRef]

91. Sharma, H.J.; Jamkar, D.V.; Kondawar, S.B. Electrospun Nanofibers of Conducting Polyaniline/Al-SnO_2 Composites for Hydrogen Sensing Applications. *Procedia Mater. Sci.* **2015**, *10*, 186–194. [CrossRef]
92. Pippara, R.K.; Chauhan, P.S.; Yadav, A.; Kishnani, V.; Gupta, A. Room Temperature Hydrogen Sensing with Polyaniline/SnO_2/Pd Nanocomposites. *Micro Nano Eng.* **2021**, *12*, 100086. [CrossRef]
93. Kumar, A.; Zhang, P.; Vincent, A.; McCormack, R.; Kalyanaraman, R.; Cho, H.J.; Seal, S. Hydrogen Selective Gas Sensor in Humid Environment Based on Polymer Coated Nanostructured-Doped Tin Oxide. *Sens. Actuators B Chem.* **2011**, *155*, 884–892. [CrossRef]
94. Liu, Y.; Tian, Y.; Liu, F.; Gu, T.; Wang, B.; He, J.; Wang, C.; Meng, X.; Sun, P.; Lu, G. Multilayer Fluorine-Free MoBT x MBene with Hydrophilic Structural-Modulating for the Fabrication of a Low-Resistance and High-Resolution Humidity Sensor. *Adv. Sci.* **2024**, *11*, 2404178. [CrossRef]

Disclaimer/Publisher's Note: The statements, opinions and data contained in all publications are solely those of the individual author(s) and contributor(s) and not of MDPI and/or the editor(s). MDPI and/or the editor(s) disclaim responsibility for any injury to people or property resulting from any ideas, methods, instructions or products referred to in the content.

chemosensors

MDPI

Article

Construction of 2D TiO_2@MoS_2 Heterojunction Nanosheets for Efficient Toluene Gas Detection

Dehui Wang, Jinwu Hu, Hui Xu *, Ding Wang and Guisheng Li

School of Materials and Chemistry, University of Shanghai for Science & Technology, Shanghai 200093, China; wangding@usst.edu.cn (D.W.); liguisheng@usst.edu.cn (G.L.)
* Correspondence: xuhui@usst.edu.cn

Abstract: Monitoring trace toluene exposure is critical for early-stage lung cancer screening via breath analysis, yet conventional chemiresistive sensors face fundamental limitations, including compromised selectivity in complex VOC matrices and humidity-induced signal drift, with prevailing p–n heterojunction architectures suffering from inherent charge recombination and environmental instability. Herein, we pioneer a 2D core–shell n–n heterojunction strategy through rational design of TiO_2@MoS_2 heterostructures, where vertically aligned MoS_2 nanosheets are epitaxially grown on 2D TiO_2 derived from graphene-templated synthesis, creating built-in electric fields at the heterojunction interface that dramatically enhance charge carrier separation efficiency. At 240 °C, the TiO_2@MoS_2 sensor exhibits a superior response (R_a/R_g = 9.8 to 10 ppm toluene), outperforming MoS_2 (R_a/R_g = 2.8). Additionally, the sensor demonstrates rapid response/recovery kinetics (9 s/16 s), a low detection limit (50 ppb), and excellent selectivity against interfering gases and moisture. The enhanced performance is attributed to unidirectional electron transfer ($TiO_2 \rightarrow MoS_2$) without hole recombination losses, methyl-specific adsorption through TiO_2 oxygen vacancy alignment, and steric exclusion of non-target VOCs via size-selective MoS_2 interlayers. This work establishes a transformative paradigm in gas sensor design by leveraging n–n heterojunction physics and 2D core–shell synergy, overcoming long-standing limitations of conventional architectures.

Keywords: heterojunction; TiO_2@MoS_2; nanosheet; synergistic effects; toluene gas sensor

1. Introduction

Toluene is a common volatile organic compound (VOC) that can damage the central nervous system, causing headaches, dizziness, and liver and kidney toxicity with long-term exposure. High-concentration inhalation can lead to acute poisoning. Toluene also serves as a potential biomarker for diagnosing diseases like lung cancer and diabetes [1]. Traditional detection methods, such as gas chromatography–mass spectrometry (GC-MS), are accurate, but require expensive equipment and laboratory settings, preventing real-time monitoring. Electrochemical sensors face cross-interference issues, and photoionization detectors (PIDs) are costly to maintain. Therefore, developing high-sensitivity, low-cost, portable semiconductor toluene gas sensors through material innovation and intelligent design is crucial [2–4]. Such sensors could enable home health monitoring, industrial safety warnings, and early disease screening, promoting advancements in precision medicine and environmental governance [5].

Two-dimensional transition metal dichalcogenides (TMDs) have emerged as promising candidates for next-generation gas sensors, due to their atomic-scale thickness, tunable electronic properties, and high surface-to-volume ratios [6]. Molybdenum disulfide (MoS_2), a graphene-like layered TMD, has garnered significant attention. In its thermodynamically stable 2H phase, the crystal structure consists of S-Mo-S trilayers stacked via van der Waals interactions, with each Mo atom in trigonal prismatic coordination [7–9]. The hydrothermal synthesis of MoS_2 enables precise control over layer number (typically one to five layers) and defect density, yielding nanostructures with specific surface areas and direct bandgaps suitable for charge transfer-based gas sensors [10,11]. Notably, 2D MoS_2 nanosheets exhibit superior gas adsorption kinetics compared to their 0D quantum dot or 1D nanobelt due to their exposed basal planes (90% surface accessibility) and abundant edge sulfur vacancies, which serve as preferential adsorption sites [12]. Additionally, in the case of ultrathin MoS_2 layers, the sulfide terminations at MoS_2 edges can achieve maximal exposure. Thus, constructing heterostructures to vertically grow such thin 2D MoS_2 represents a viable strategy [13].

Heterojunction engineering effectively enhances sensing performance by lowering operating temperatures and improving sensitivity, selectivity, and humidity resistance in nanomaterials [14,15]. Due to the band structure mismatch between two sensing materials, energy bands bend to form electron- depletion layers (EDLs) or hole accumulation layers (HALs) when the Fermi levels equilibrate [16,17]. Furthermore, a hierarchical composite structure can provide additional active adsorption sites for target gases and exhibit superior catalytic activity compared to single metal oxides. Thus, heterostructures are critical for optimizing the gas-sensing properties of metal oxides [18,19]. Hermawan et al. [20] reported a p–n $CuO@SnO_2$ heterojunction that achieved a remarkable toluene response (R_a/R_g = 540 at 75 ppm) with exceptional selectivity. The enhanced performance was attributed to the synergistic effects of the p–n heterojunction, high surface area, and porous architecture. Notably, in situ formation of metallic Cu under high toluene concentrations disrupted the p–n junction and established Ohmic contact with the n-type SnO_2, further amplifying the sensing response. Liu et al. [21] fabricated a porous Co_3O_4–Fe_3O_4 composite with p–n heterojunctions by functionalizing Fe_3O_4 with ZIF-67-derived Co_3O_4. The Co_3O_4 derived from ZIF-67 exhibited a higher specific surface area than Fe_3O_4, leading to a 20-fold improvement in toluene response at 225 °C. Additionally, the sensor demonstrated a low detection limit (0.1 ppm, response = 1.68), excellent selectivity, and long-term stability. This enhancement primarily stemmed from the p–n heterojunction, increased surface area, and optimized oxygen vacancy (O_V) concentration. Therefore, heterojunction construction provides a viable pathway for developing highly sensitive and practical gas sensors based on nanostructured metal oxides. The integration of vertically aligned ultrathin MoS_2 nanosheets with a wide-bandgap metal oxide semiconductor (MOS) core constructs a 2D core–shell n–n heterojunction, where the MOS core acts as an electron donor and the MoS_2 shell serves as an electron acceptor [14]. A built-in electric field forms at the heterointerface through type II band alignment, significantly enhancing carrier separation efficiency, while the core–shell structure achieves specific gas recognition through the synergistic interface coupling effect between MOS and MOS_2 components [6].

Here, we report a unique 2D heterostructure $TiO_2@MoS_2$ nanosheet. Two-dimensional TiO_2 nanosheets were synthesized via a graphene sacrificial template method, followed by the hydrothermal growth of fish scale-like MoS_2 nanosheets on their surfaces. By adjusting the hydrothermal reaction time, MoS_2 layers with controlled thicknesses were achieved, exhibiting an average lateral size of 193 ± 21 nm. Compared to pristine MoS_2 nanosheets, the $TiO_2@MoS_2$ heterostructure demonstrated a superior toluene response

(R_a/R_g = 9.8@10 ppm) at 240 °C, along with rapid response/recovery kinetics (9 s/16 s), excellent stability, durability, and high humidity resistance. The enhanced gas-sensing performance is attributed to the n–n heterojunction and high specific surface area. The synergistic interaction between the two components and the unique core–shell hierarchical architecture with self-assembled nanostructures collectively amplify the sensor's capabilities. This study provides a rational material design strategy, offering a promising candidate for efficient detection of toluene gas.

2. Materials and Methods

2.1. Synthesis of Sensor Materials

Synthesis of TiO_2 Nanosheets: TiO_2 precursor was synthesized using a graphene oxide (GO) sacrificial template method. Specifically, 100 mg of GO was dispersed in 200 mL of ethanol and ultrasonicated for 1 h. Subsequently, 2.5 mmol of titanium butoxide (TBOT) was added dropwise under continuous stirring, and the mixture was stirred overnight (18 h) to obtain a gray suspension. The product was thoroughly washed three times with deionized water and ethanol, followed by drying at 80 °C overnight. Finally, the dried powder was calcined in a muffle furnace at 500 °C for 2 h with a heating rate of 2 °C/min, yielding white TiO_2 nanosheets.

Synthesis of TiO_2@MoS_2 Nanocomposites: 0.1 mmol of ammonium heptamolybdate tetrahydrate $(NH_4)_2MoO_4{\cdot}4H_2O$) was dissolved in 10 mL of deionized water under vigorous stirring. A predetermined amount of the as-prepared TiO_2 nanosheets was added to the solution, and the mixture was incubated at room temperature for 12 h and dried at 80 °C for 12 h. Separately, 0.1 mmol of $(NH_4)_2MoO_4{\cdot}4H_2O$ and 3.05 mmol of thiourea (CH_4N_2S) were dissolved in 40 mL of deionized water. The $(NH_4)_2MoO_4{\cdot}4H_2O$-loaded TiO_2 nanosheets were then added to the solution, and the mixture was stirred at room temperature for 1 h. The homogeneous suspension was transferred to a 50 mL Teflon-lined stainless-steel autoclave and heated at 200 °C for 8 h. After cooling to room temperature, the black precipitate was collected, washed three times with deionized water and ethanol, and dried at 60 °C overnight to obtain TiO_2@MoS_2-8 nanosheets. By adjusting the reaction time to 4 h and 12 h under identical conditions, samples labeled TiO_2@MoS_2-4 and TiO_2@MoS_2-12 were prepared. For comparison, pure MoS_2 nanosheets were synthesized using the same procedure without the addition of TiO_2 nanosheets.

2.2. Gas-Sensing Performance Measurement

The as-prepared TiO_2 NSs, TiO_2@MoS_2 NSs, and MoS_2 NSs were mixed with deionized water to form a homogeneous paste. Each paste was uniformly coated onto a ceramic substrate equipped with a pair of gold electrodes using a fine brush (Figure S1). The operating temperature of the sensor was controlled by adjusting the heating current applied to the embedded heating electrode. The gas-sensing performance was evaluated using a static test system (Elite Technology Co., Ltd. Suzhou, China). Target gas concentrations were generated via a static gas distribution method.

The gas-sensing performance was evaluated using a custom-built static testing system, where sensor resistance was monitored continuously with a source meter under controlled conditions. Target analytes were introduced into the 20 L test chamber by injecting liquid toluene with a microsyringe and vaporizing it on a hotplate. The volume Q can be determined by:

$$Q = \frac{VCM}{22.4D\rho} \times 10^{-9} \times \frac{273 + T_R}{273 + T_B} \quad (1)$$

Here, V, C, M, d, ρ, T_R, and T_B are the test chamber volume (L), vapor concentration (ppm), molecular mass (g/mol), liquid density (kg/m^3), liquid purity (%), environmental temperature (°C), and temperature (°C) in the testing chamber, respectively.

Sensitivity (S), a critical performance metric of gas sensors, is quantified as the resistance ratio under distinct gaseous environments. For n-type semiconductors, sensitivity is defined as:

$$S = R_a/R_g\ (N - type)\ (oxidizing\ gases) \quad (2)$$

$$S = R_g/R_a\ (N - type)\ (reducing\ gases) \quad (3)$$

where R_a denotes the baseline resistance in dry air and R_g represents the steady-state resistance upon exposure to the target analyte. Conversely, p-type semiconductors exhibit inverse trends.

3. Results and Discussion

3.1. Morphology and Structure Characterization

As demonstrated in Figure 1, TiO_2 nanosheets were synthesized via a graphene oxide sacrificial template method followed by annealing at 500 °C. Subsequently, fish scale-like MoS_2 nanosheet arrays were uniformly grown on the TiO_2 surfaces through a controlled hydrothermal reaction (Figure 1). Optimizing precursor concentration and hydrothermal reaction time was crucial for MoS_2 growth. The thickness of MoS_2 nanostructures could be effectively tuned by adjusting the hydrothermal duration. The facile and reproducible synthesis protocol offers significant advantages for scalable production.

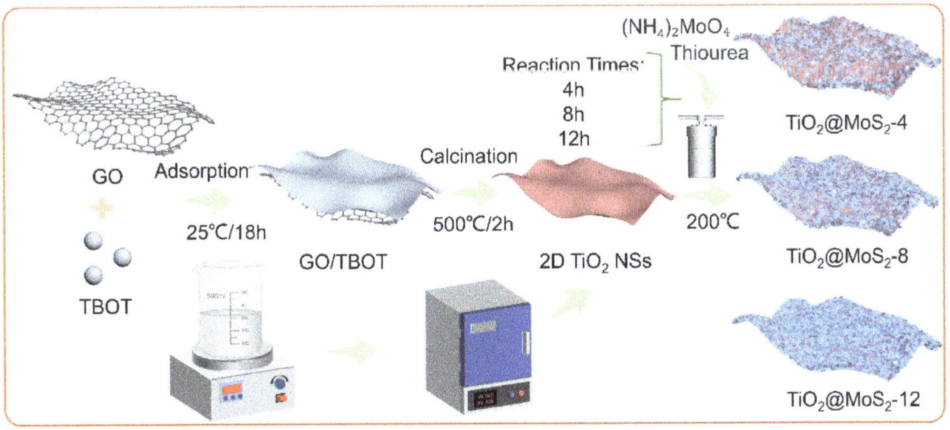

Figure 1. Schematic illustration of the synthesis process for $TiO_2@MoS_2$ nanosheets via a graphene oxide sacrificial template combined with hydrothermal reaction.

To characterize the morphology and structure of $TiO_2@MoS_2$ composites, SEM and low-magnification TEM analyses were performed. As shown in Figure 2a, ultrathin TiO_2 nanosheets (~2.1 ± 0.3 nm thickness) were successfully synthesized via the graphene oxide sacrificial template method. In contrast, pristine MoS_2 formed spherical aggregates with an average diameter of 530 ± 12 nm (Figure 2b). Figure 2c–e reveal the hierarchical heterostructure, where vertically aligned MoS_2 nanosheets (193 ± 21 nm lateral size) uniformly coat TiO_2 surfaces, forming an n–n heterojunction. The MoS_2 coverage density exhibited a positive correlation with hydrothermal duration, enabling thickness-controlled growth. TEM characterization (Figure 2f) further confirmed the 2D–2D interfacial architecture. HR-TEM analysis (Figure 2g) resolved distinct lattice spacings of 0.250 nm, corresponding to the (102) plane of MoS_2 [22–25]. The polycrystalline nature was verified by selective electron diffraction (SAED) patterns showing concentric diffraction rings (Figure 2h). Energy dispersion spectrum (EDS) elemental mapping (Figure 2i) demonstrated homogeneous spatial distribution of Ti, Mo, S, and O, conclusively confirming the formation of well-integrated TiO_2@MoS_2 heterostructures.

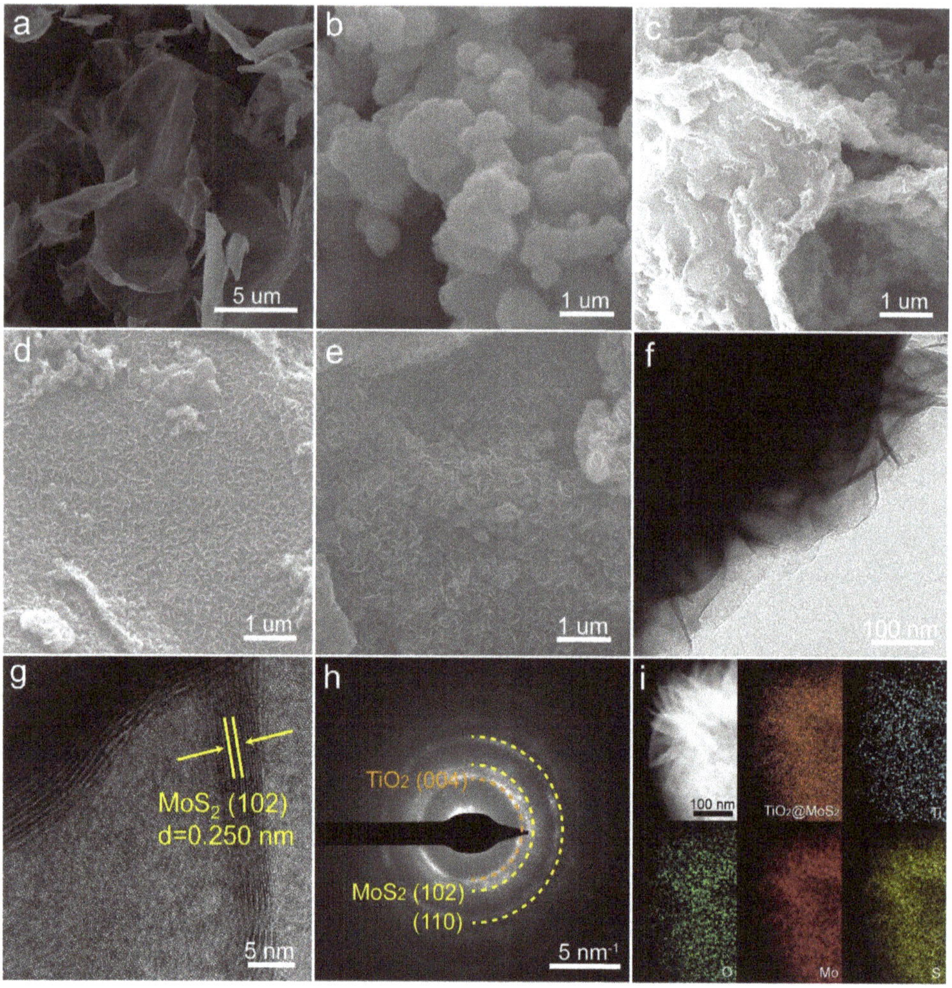

Figure 2. Morphological and structural characterization of TiO_2@MoS_2 heterostructures: SEM images of (**a**) TiO_2, (**b**) MoS_2, (**c**) TiO_2@MoS_2-4, (**d**) TiO_2@MoS_2-8, and (**e**) TiO_2@MoS_2-12; (**f–i**) TEM, HR-TEM, SAED, and EDS elemental mapping images of TiO_2@MoS_2-8, respectively.

To determine the phase composition and interfacial interactions in the TiO_2@MoS_2 heterostructure, X-ray diffraction (XRD) and Raman analyses were performed. The XRD pattern of TiO_2@MoS_2 (Figure 3a) revealed distinct diffraction peaks at 25.3° (101), 37.8° (004), and 48.0° (200), corresponding to anatase TiO_2 (JCPDS #97-002-4276), alongside peaks at 31.6° (102) and 45.5° (110), attributed to MoS_2 (JCPDS #97-064-4259). A noticeable shift in the (004) peak of TiO_2 and the (110) peak of MoS_2 was observed, indicative of lattice strain arising from the intimate interfacial coupling in the heterojunction.

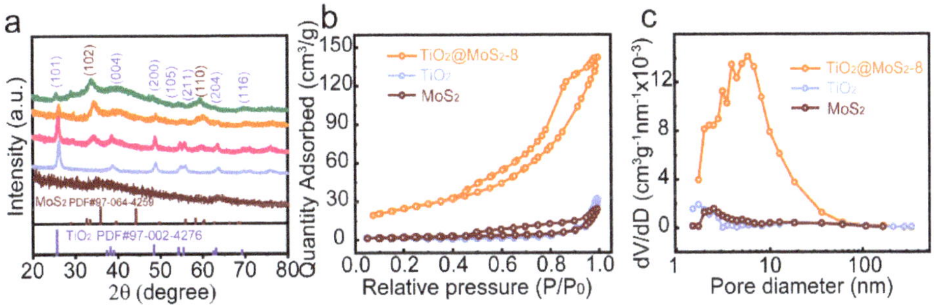

Figure 3. (**a**) XRD patterns of TiO_2, MoS_2, $TiO_2@MoS_2$-4 (pink line), $TiO_2@MoS_2$-8 (orange line), and $TiO_2@MoS_2$-12 (green line) samples; (**b**,**c**) nitrogen adsorption–desorption isotherms and BJH pore size distribution of TiO_2, MoS_2, and $TiO_2@MoS_2$-8.

To analyze the structural advantages of the $TiO_2@MoS_2$ nanocomposite, nitrogen adsorption–desorption measurements were conducted to determine its specific surface area and pore distribution using BET theory and BJH methods. As shown in Figure 3b, all three materials (TiO_2, MoS_2, $TiO_2@MoS_2$-8) exhibit type IV isotherms with H3-type hysteresis loops in the relative pressure (P/P_0) range of 0.6–1.0, characteristic of mesoporous structures. The average pore diameters determined by BJH analysis were 2.4 nm (TiO_2), 3.1 nm (MoS_2), and 8.3 nm ($TiO_2@MoS_2$-8). Notably, the $TiO_2@MoS_2$-8 composite demonstrated the highest BET surface area of 87.48 m^2/g, significantly surpassing that of pristine TiO_2 nanosheets (9.04 m^2/g) and MoS_2 nanosheets (19.68 m^2/g), confirming the heterojunction's role in enhancing porosity (Figure 3c).

X-ray photoelectron spectroscopy (XPS) was employed to investigate the surface composition and chemical states of TiO_2, MoS_2, and $TiO_2@MoS_2$-8 nanocomposites. As shown in Figure 4a, the survey spectra confirm the coexistence of Ti, Mo, S, and O in the $TiO_2@MoS_2$-8 heterostructure, consistent with EDS, XRD, and HRTEM results, with no impurity peaks except for residual carbon (C 1s). High-resolution Ti 2p spectra (Figure 4b) reveal characteristic peaks at 459.08 eV (Ti $2p_{3/2}$) and 464.87 eV (Ti $2p_{1/2}$) for $TiO_2@MoS_2$-8, with a spin–orbit splitting of 5.79 eV, typical of Ti^{4+} in TiO_2. Notably, a 0.40 eV positive binding energy shift compared to pure TiO_2 (458.68 eV and 464.47 eV) indicated electron transfer from TiO_2 to MoS_2. The Mo 3d spectrum (Figure 4c) exhibited peaks at 228.65 eV (Mo $3d_{5/2}$) and 231.87 eV (Mo $3d_{3/2}$), with a corresponding S 2s peak at 225.90 eV, while a 0.60 eV negative shift relative to pure MoS_2 further confirmed interfacial charge redistribution. Similarly, the S 2p peaks (Figure 4d) at 162.06 eV (S $2p_{3/2}$) and 163.16 eV (S $2p_{1/2}$) show a 0.15 eV shift, corroborating strong electronic interactions. Deconvolution of O 1s spectra (Figure 4e) identifies three components: lattice oxygen (O_{lat}, 529.5 ± 0.6 eV), oxygen vacancies (O_{def}, 531.6 ± 0.4 eV), and adsorbed oxygen species (O_{ads}, 532.5 ± 0.6 eV). $TiO_2@MoS_2$-8 demonstrates elevated O_{def} (22.72%) and O_{ads} (4.58%) concentrations compared to TiO_2, which aligns with its enhanced ESR signal intensity (Figure 4f) reflecting higher oxygen vacancy density. These synergistic effects of enhanced charge transfer and oxygen-mediated surface activity collectively improved toluene-sensing performance by promoting gas adsorption and redox reactions at the heterojunction interface.

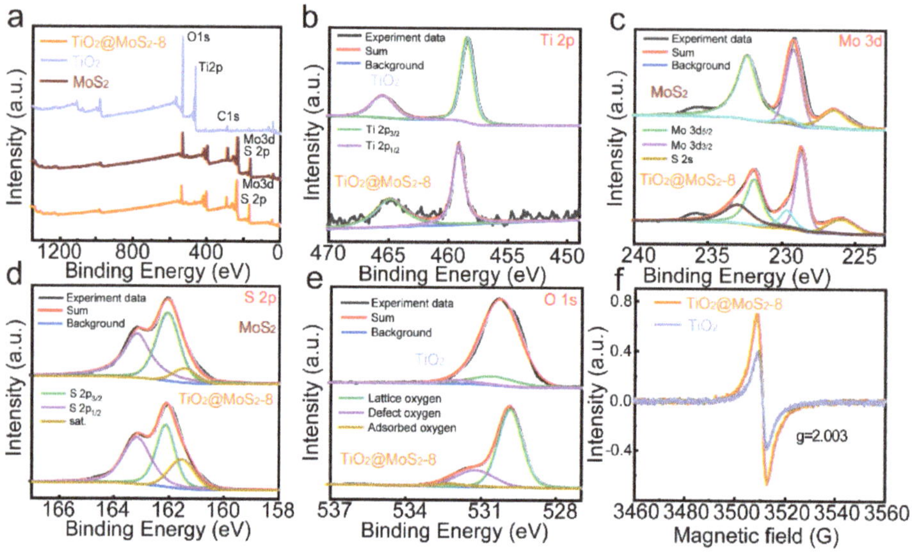

Figure 4. (**a**) X-ray photoelectron spectroscopy (XPS) survey spectra of TiO_2, MoS_2, and TiO_2@MoS_2-8; (**b**) high-resolution Ti 2p spectra of TiO_2 and TiO_2@MoS_2-8; (**c**) Mo 3d spectra of MoS_2 and TiO_2@MoS_2-8; (**d**) S 2p spectra of MoS_2 and TiO_2@MoS_2-8; (**e**) O 1s spectra of MoS_2 and TiO_2@MoS_2-8; (**f**) electron spin resonance (ESR) spectra of TiO_2 and TiO_2@MoS_2-8.

3.2. Gas-Sensitive Property

TiO_2@MoS_2 nanocomposites show great potential as chemiresistive gas sensors due to their high surface area and hierarchical 2D structures. The operating temperature affects surface resistance by altering adsorbed oxygen species and reaction kinetics. This is due to faster surface reactions at higher temperatures competing with reduced gas molecule adsorption because of fewer available active sites [26]. Temperature-dependent sensing tests were conducted from 200 to 280 °C for MoS_2, TiO_2@MoS_2-4, TiO_2@MoS_2-8, and TiO_2@MoS_2-12. The temperature-dependent response exhibited a characteristic volcano profile (Figure 5a), peaking at 240 °C. Below this temperature, inadequate thermal activation hindered oxygen ionosorption and toluene oxidation. Beyond 240 °C, accelerated desorption prevailed over surface reactions, reducing sensor response value [27]. The TiO_2@MoS_2-8 composite achieved a maximum response of 9.8@10 ppm toluene, representing a 3.5-fold enhancement over pristine MoS_2 (2.8), primarily due to oxygen vacancy-mediated toluene adsorption. Notably, TiO_2@MoS_2-8 exhibited rapid response/recovery kinetics (9 s/16 s, Figure 5b), outperforming other samples (MoS_2: 4.1 s/18 s; TiO_2@MoS_2-4: 10 s/17 s; TiO_2@MoS_2-12: 12 s/20 s), benefiting from its optimized mesoporous structure facilitating gas diffusion. Selectivity tests against xylene, benzene, triethylamine, acetone, methanol, ethanol, and ammonia (Figure 5c) revealed superior toluene selectivity. Even under simulated exhaled breath conditions containing mixed interferents (Figure 5d), the sensor maintained stable toluene detection, confirming its robustness in complex environments. These critical attributes of high sensitivity, rapid reaction kinetics, and selective discrimination collectively establish TiO_2@MoS_2 nanocomposites as promising materials for toluene detection applications.

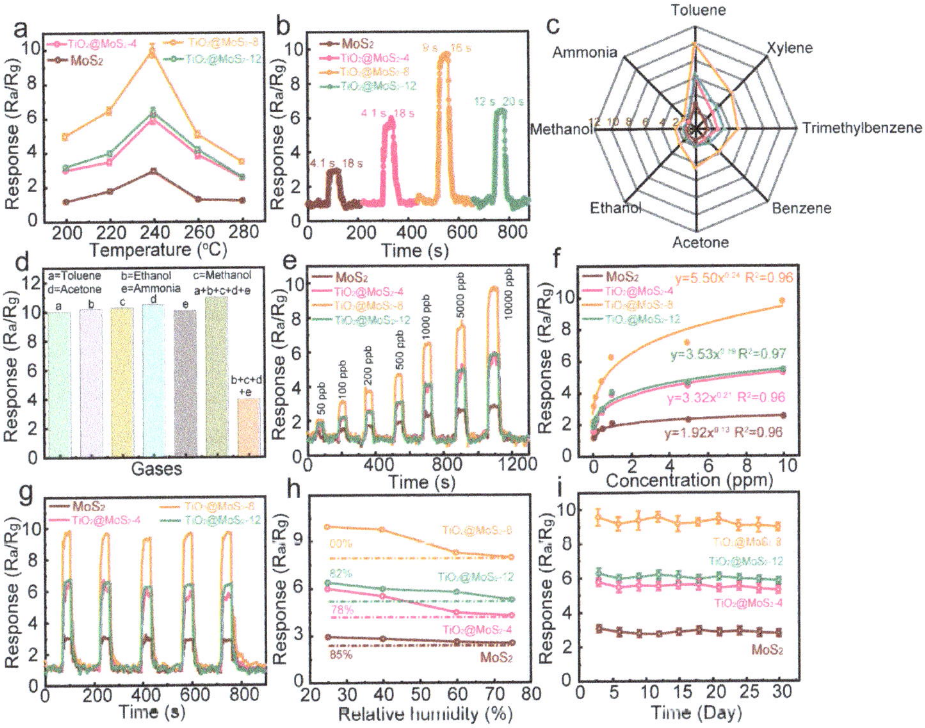

Figure 5. (**a**) Response (R_a/R_g) of MoS_2, $TiO_2@MoS_2$-4, $TiO_2@MoS_2$-8, and $TiO_2@MoS_2$-12 to 10 ppm toluene across 200–280 °C; (**b**) response/recovery time; (**c**) selectivity toward 10 ppm toluene, xylene, benzene, triethylamine, acetone, methanol, ethanol, and ammonia; (**d**) gas responses to mixed vapors at 240 °C; (**e**) dynamic response curves for varied toluene concentrations (0.05–10 ppm) at 240 °C; (**f**) relationship between sensor response and toluene concentrations (0.05–10 ppm); (**g**) reproducibility over five cycles at 10 ppm toluene (240 °C); (**h**) humidity-dependent responses (25 °C, 25–75% RH); (**i**) long-term stability (30 days, 10 ppm).

Figure 5e illustrates the dynamic response of $TiO_2@MoS_2$-based sensors to varying concentrations of toluene gas, ranging from 50 ppb to 10 ppm. The sensor response demonstrates stepwise enhancement with increasing concentration, with $TiO_2@MoS_2$ composites exhibiting superior sensitivity compared to pristine MoS_2, particularly at higher concentrations. A limit of detection (LOD) of 50 ppb was achieved for the optimal $TiO_2@MoS_2$-8 sensor. As shown in Figure 5f, all samples maintain linear responses across the 0.05–10 ppm range, validating quantitative detection capability. Repeatability tests (Figure 5g) revealed stable performance over five cycles at 10 ppm, with response variations below ±5.2%. Under humidity challenges (25%–75% RH, Figure 5h), the response attenuation remained <20%, underscoring robust moisture resistance. Long-term stability assessments (30 days exposure to 10 ppm toluene, Figure 5i) showed <8% signal drift, confirming durability for practical deployment. Table 1 shows the comparison of sensor-based toluene gas detection characteristics with different detection materials, further confirming the great potential of $TiO_2@MoS_2$-based material as a toluene sensor due to its lower detection limit and rapid response/recovery time.

Table 1. Comparison of sensor-based toluene gas detection characteristics with different detection materials.

Materials	T (°C)	Concentration (ppm)	Detection Limit (ppm)	T_{res}/T_{rec} (s)	Ref.
Pd/PdO-decorated SnO_2	285	100	0.05	50/138	[28]
V_2C MXene	25	200	0.047	14/34	[29]
SnO_2-decorated NiO	100	250	0.01	-	[30]
MoS_2-Fe_3O_4	25	20	5	58/23	[7]
ZnO@Co_3O_4	290	100	5	11.2/12.5	[16]
NiO/ZnO	100	300	0.1	2/33	[17]
TiO_2@MoS_2	240	10	0.05	9/16	This work

As shown in Figure 6a, time-resolved in situ Fourier transform infrared (in situ FTIR) spectroscopy was employed to analyze intermediate products during toluene sensing. Peaks observed in the 1450–1600 cm^{-1} range originate from C=C aromatic ring vibrations, with increasing intensity over time indicating progressive aromatic compound accumulation. Absorption bands between 1370–1450 cm^{-1} correspond to C-H bending modes of methyl groups directly bonded to the benzene ring, confirming toluene decomposition. Concurrently, diminishing toluene-specific peaks and emerging signals at 3000–3100 cm^{-1} (aromatic C-H stretching) suggest catalytic conversion pathways where toluene transforms into benzene derivatives. Resistance changes in TiO_2@MoS_2 nanocomposites arise from oxygen-mediated surface interactions: at 240 °C, chemisorbed oxygen species ($O_2{}^-$, O^-) populate oxygen vacancies, forming electron-depletion layers that reduce baseline resistance in air (Equation (4)) [31]. Upon toluene exposure, adsorbed oxygen species react with methyl groups via stepwise dehydrogenation (Equations (5)–(8)), modulating conductivity through electron transfer between the semiconductor and reaction intermediates. This mechanistic framework aligns with the observed spectral evolution and sensor response dynamics.

$$O_2(g) \rightarrow O_2(ads) \tag{4}$$

$$O_2(ads) + e^- \rightarrow O_2^-(ads) \tag{5}$$

$$O_2^-(ads) + e^- \rightarrow 2O^-(ads) \tag{6}$$

$$C_7H_8(g) + 3O^-(ads) \rightarrow C_6H_6(g) + CO_2 + H_2O(g) + 3e^- \tag{7}$$

$$C_6H_6(g) + 15O^-(ads) \rightarrow 6CO_2 + 3H_2O(g) + 15e^- \tag{8}$$

The enhanced toluene-sensing performance of the 2D stacked MoS_2@TiO_2 heterostructure compared to pristine MoS_2 and TiO_2 nanosheets arises from three synergistic mechanisms, as follows.

(I) Hierarchical architecture and surface engineering. Gas–solid interactions in chemiresistive sensors critically depend on surface-mediated oxygen adsorption and target gas diffusion. The MoS_2@TiO_2 heterostructure exhibits a superior specific surface area, providing abundant active sites for oxygen chemisorption. The vertically aligned MoS_2 nanosheets on TiO_2 substrates prevent nanosheet restacking while creating interconnected nanochannels that enhance toluene diffusion kinetics.

(II) Interface charge redistribution in n–n heterojunctions. The work function disparity between MoS_2 ($\Phi = 4.05$ eV) and TiO_2 ($\Phi = 4.77$ eV), determined by ultraviolet photoelectron spectroscopy (UPS, Figure S3), drives interfacial electron transfer from TiO_2 to MoS_2 [32,33]. This creates a built-in electric field at the heterointerface, forming an electron-depletion layer, which lowers the oxygen adsorption activation energy and facilitates hole–electron separation efficiency [34]. Band alignment analysis (Figure S2) reveals heterojunction characteristics with MoS_2 and TiO_2 bandgaps of 3.20 eV and 0.85 eV, respectively [35–37]. The conduction band offset promotes electron trapping by adsorbed oxygen species (O_2^-/O^-), amplifying resistance modulation upon toluene exposure.

(III) Influence by oxygen species. Compared with TiO_2 nanosheets, TiO_2@MoS_2 has the highest ESR intensity and the highest relative concentration of oxygen vacancy. Therefore, the TiO_2@MoS_2 hierarchical structure can provide additional O_{ads} and O_{def}, which is conducive to the reaction of the target gas with oxygen, while increasing the surface active site, which is conducive to the sensing performance of toluene [38,39].

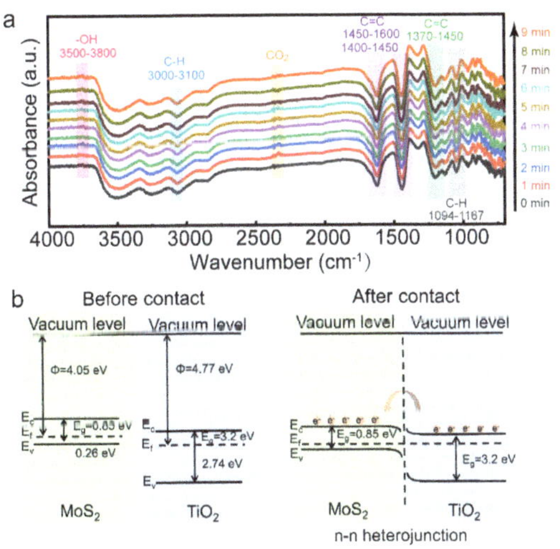

Figure 6. (**a**) In situ DRIFTS analysis of the TiO_2@MoS_2-8 nanocomposite during gas exposure; (**b**) band alignment diagrams illustrating the interfacial energy structure of TiO_2 and MoS_2 pre- and post-heterojunction formation.

4. Conclusions

In summary, TiO_2@MoS_2 core–shell heterostructures were fabricated by assembling MoS_2 nanosheet arrays onto TiO_2 nanosheets prepared via a sacrificial template method. The unique 2D hierarchical heterojunction structure resulted in a larger specific surface area and increased generation of oxygen vacancies. Compared with individual TiO_2 nanosheets and MoS_2 nanosheets, the TiO_2@MoS_2 composite demonstrated the highest response ($R_a/R_g = 9.8$@10 ppm toluene) at 240 °C. Furthermore, the hierarchical TiO_2@MoS_2 core–shell structure exhibited fast response/recovery times (9 s/16 s), a low detection limit (50 ppb), excellent selectivity, and robust stability in humid environments. The enhanced gas-sensing performance is attributed to the synergistic effects between the two components and the distinctive core–shell hierarchical architecture with self-assembled nanostructures. This work provides a rational material structure design strategy, offering a promising candidate for detecting toluene gas, thereby demonstrating significant potential value for advancing environmental or disease diagnosis detection.

Supplementary Materials: The following supporting information can be downloaded at https://www.mdpi.com/article/10.3390/chemosensors13050154/s1. Figure S1: Ceramic plate and test base; Figure S2: (a–d) UV-vis diffuse reflectance spectra (DRS) of TiO_2 and MoS_2 nanosheets with corresponding Tauc plots for bandgap determination; Figure S3: Ultraviolet photoelectron spectroscopy (UPS) spectra of pristine TiO_2 and MoS_2 recorded under He I_{α} excitation (21.22 eV), showing secondary electron cutoff (left) and valence band regions (right).

Author Contributions: D.W. (Dehui Wang): Investigation; writing original draft; conceptualization; visualization. J.H.: Investigation; conceptualization; writing original draft. H.X.: Writing, review & editing; project administration. D.W. (Ding Wang): Funding acquisition; project administration; writing, review & editing. G.L.: Funding acquisition; project administration. All authors have read and agreed to the published version of the manuscript.

Funding: The authors greatly appreciate the financial support from National Natural Science Foundation of China (62471303, 62071300, 22176127, 22301181, 22406130, 22476131), Shanghai Sailing Program (23YF1429000, 22YF1430400). We also acknowledge the USST Center for Instrumental Analysis for their support in materials characterization and High Performance Computing Center of USST for their support in the calculations.

Institutional Review Board Statement: Not applicable.

Informed Consent Statement: Not applicable.

Data Availability Statement: The raw data supporting the conclusions of this article will be made available by the authors on request.

Acknowledgments: We acknowledge the USST Center for Instrumental Analysis for their support in material characterization.

Conflicts of Interest: The authors declare no conflict of interest.

References

1. Ueda, T.; Hayashi, H.; Tsukahara, R.; Shimizu, Y.; Hyodo, T. Effects of Structure and Thickness of $Ce_{0.9}Pr_{0.1}O_2$ Electrodes of YSZ-Based Gas Sensors on VOC-Sensing Properties. *Sens. Actuators B Chem.* **2025**, *422*, 136580. [CrossRef]
2. Zhang, Y.-L.; Jia, C.-W.; Tian, R.-N.; Guan, H.-T.; Chen, G.; Dong, C.-J. Hierarchical Flower-like $NiFe_2O_4$ with Core–Shell Structure for Excellent Toluene Detection. *Rare Met.* **2021**, *40*, 1578–1587. [CrossRef]
3. Wang, T.; Liu, S.; Sun, P.; Wang, Y.; Shimanoe, K.; Lu, G. Unexpected and Enhanced Electrostatic Adsorption Capacity of Oxygen Vacancy-Rich Cobalt-Doped In_2O_3 for High-Sensitive MEMS Toluene Sensor. *Sens. Actuators B Chem.* **2021**, *342*, 129949. [CrossRef]
4. Liao, Z.; Chu, N.; Yuan, Z.; Shen, Y.; Meng, F. Highly Sensitive Toluene Sensor Based on Silver Ear-Like F-Doped Co_3O_4 Templated from Co(OH)F Guided by SiO_2 Nanospheres. *Chem. Eng. J.* **2024**, *498*, 155688. [CrossRef]
5. Zhang, H.; Hu, J.; Li, M.; Li, Z.; Yuan, Y.; Yang, X.; Guo, L. Highly Efficient Toluene Gas Sensor Based on Spinel Structured Hollow Urchin-like Core-Shell $ZnFe_2O_4$ Spheres. *Sens. Actuators B Chem.* **2021**, *349*, 130734. [CrossRef]
6. Siraj, S.; Bansal, G.; Hasita, B.; Srungaram, S.; S, S.K.; Rybicki, F.J.; Sonkusale, S.; Sahatiya, P. MXene/MoS_2 Piezotronic Acetone Gas Sensor for Management of Diabetes. *ACS Appl. Nano Mater.* **2024**, *7*, 11350–11361. [CrossRef]
7. Thayil, R.; Krishna, K.G.; Cherukulappurath, S.; Kathirvelu, V.; Parne, S.R. MoS_2 and MoS_2-Based Nanocomposites for Enhanced Toluene Sensing Response at Room Temperature. *Surf. Interfaces* **2024**, *46*, 104134. [CrossRef]
8. Tang, T.; Li, Z.; Liu, Y.Y.; Chen, Y.L.; Cheng, Y.F.; Liang, Y.; Zhuang, J.H.; Hu, X.Y.; Jannat, A.; Ou, R.; et al. Ultrathin two-dimensional titanium oxysulfide for enhanced sensitivity and stability of room temperature NO_2 sensing. *Ceram. Int.* **2025**, *51*, 3216–3223. [CrossRef]
9. Guan, X.; Zhao, L.; Zhang, P.; Liu, J.; Song, X.; Gao, L. Electrode Material of Core-Shell Hybrid MoS_2@C/CNTs with Carbon Intercalated Few-Layer MoS_2 Nanosheets. *Mater. Today Energy* **2020**, *16*, 100379. [CrossRef]
10. Gough, J.J.; McEvoy, N.; O' Brien, M.; Bell, A.P.; McCloskey, D.; Boland, J.B.; Coleman, J.N.; Duesberg, G.S.; Bradley, A.L. Dependence of Photocurrent Enhancements in Quantum Dot (QD)-Sensitized MoS_2 Devices on MoS_2 Film Properties. *Adv. Funct. Mater.* **2018**, *28*, 1706149. [CrossRef]

11. Zhang, W.; Liu, Y.; Pei, X.; Yuan, Z.; Zhang, Y.; Zhao, Z.; Hao, H.; Long, R.; Liu, N. Stretchable MoS_2 Artificial Photoreceptors for E-Skin. *Adv. Funct. Mater.* **2022**, *32*, 2107524. [CrossRef]
12. Chen, Y.L.; Li, Z.; Tang, T.; Cheng, Y.F.; Cheng, L.; Wang, X.X.; Haidry, A.A.; Jannat, A.; Ou, J.Z. Room-Temperature Optoelectronic NO_2 Sensing Using Two-Dimensional Gallium Oxyselenides. *ACS Appl. Nano Mater.* **2024**, *7*, 3229–3238. [CrossRef]
13. Wang, H.; Shao, Z.; Shi, X.; Tang, Z.; Sun, B. Rapidly Detecting the Carcinogen Acetaldehyde: Preparation and Application of a Flower-like MoS_2 Cataluminescence Sensor at Low Working Temperature. *Anal. Methods* **2023**, *15*, 5620–5629. [CrossRef] [PubMed]
14. Niu, Y.; Zeng, J.; Liu, X.; Li, J.; Wang, Q.; Li, H.; de Rooij, N.F.; Wang, Y.; Zhou, G. A Photovoltaic Self-Powered Gas Sensor Based on All-Dry Transferred MoS_2/GaSe Heterojunction for ppb-Level NO_2 Sensing at Room Temperature. *Adv. Sci.* **2021**, *8*, 2100472. [CrossRef] [PubMed]
15. Cheng, X.; Yao, Y.T.; Zheng, S.L.; Wan, Y.; Wei, C.D.; Yang, G.C.; Yuan, Y.; Tsai, H.-S.; Wang, Y.; Hao, J.Y. Te@Se Core–Shell Heterostructures with Tunable Shell Thickness for Ultra-Stable NO_2 Detection. *ACS Sens.* **2025**, *10*, 283–291. [CrossRef]
16. Shi, J.; Xiong, J.; Qiao, L.; Liu, C.; Zeng, Y. Facile MOF-on-MOF Isomeric Strategy for ZnO@Co_3O_4 Single-Shelled Hollow Cubes with High Toluene Detection Capability. *Appl. Surf. Sci.* **2023**, *609*, 155271. [CrossRef]
17. Liu, H.; Wang, Z.; Cao, G.; Pan, G.; Yang, X.; Qiu, M.; Sun, C.; Shao, J.; Li, Z.; Zhang, H. Construction of Hollow NiO/ZnO p-n Heterostructure for Ultrahigh Performance Toluene Gas Sensor. *Mater. Sci. Semicond. Process.* **2022**, *141*, 106435. [CrossRef]
18. Ma, S.; Xu, J. Nanostructured Metal Oxide Heterojunctions for Chemiresistive Gas Sensors. *J. Mater. Chem. A* **2023**, *11*, 23742–23771. [CrossRef]
19. Wang, Y.; Tan, L.; Tan, M.H.; Zhang, P.P.; Yoshiharu, Y.; Yang, G.H.; Tsubaki, N. Rationally Designing Bifunctional Catalysts as an Efficient Strategy to Boost CO_2 Hydrogenation Producing Value-Added Aromatics. *ACS Catal.* **2018**, *9*, 895–901. [CrossRef]
20. Hermawan, A.; Asakura, Y.; Inada, M.; Yin, S. A Facile Method for Preparation of Uniformly Decorated-Spherical SnO_2 by CuO Nanoparticles for Highly Responsive Toluene Detection at High Temperature. *J. Mater. Sci. Technol.* **2020**, *51*, 119–129. [CrossRef]
21. Liu, H.; Wang, Z.; Sun, C.; Shao, J.; Li, Z.; Zhang, H.; Qiu, M.; Pan, G.; Yang, X. Construction of Co_3O_4/Fe_3O_4 Heterojunctions from Metal Organic Framework Derivatives for High Performance Toluene Sensor. *Sens. Actuators B Chem.* **2023**, *375*, 132863. [CrossRef]
22. Raghu, A.V.; Karuppanan, K.K.; Nampoothiri, J.; Pullithadathil, B. Wearable, Flexible Ethanol Gas Sensor Based on TiO_2 Nanoparticles-Grafted 2D-Titanium Carbide Nanosheets. *ACS Appl. Nano Mater.* **2019**, *2*, 1152–1163. [CrossRef]
23. Cao, S.; Sui, N.; Zhang, P.; Zhou, T.; Tu, J.; Zhang, T. TiO_2 Nanostructures with Different Crystal Phases for Sensitive Acetone Gas Sensors. *J. Colloid Interf. Sci.* **2022**, *607*, 357–366. [CrossRef]
24. Zhao, Y.; Song, J.-G.; Ryu, G.H.; Ko, K.Y.; Woo, W.J.; Kim, Y.; Kim, D.; Lim, J.H.; Lee, S.; Lee, Z.; et al. Low-Temperature Synthesis of 2D MoS_2 on a Plastic Substrate for a Flexible Gas Sensor. *Nanoscale* **2018**, *10*, 9338–9345. [CrossRef]
25. Kumar, R.; Zheng, W.; Liu, X.; Zhang, J.; Kumar, M. MoS_2-Based Nanomaterials for Room-Temperature Gas Sensors. *Adv. Mater. Technol.* **2020**, *5*, 1901062. [CrossRef]
26. Verma, G.; Gokarna, A.; Kadiri, H.; Nomenyo, K.; Lerondel, G.; Gupta, A. Multiplexed Gas Sensor: Fabrication Strategies, Recent Progress, and Challenges. *ACS Sens.* **2023**, *8*, 3320–3337. [CrossRef]
27. Korotcenkov, G.; Cho, B.K. Metal Oxide Composites in Conductometric Gas Sensors: Achievements and Challenges. *Sens. Actuators B Chem.* **2017**, *244*, 182–210. [CrossRef]
28. Gong, C.; Chen, M.; Song, F.; Yin, P.; Zhao, X.; You, X.; Fu, H.; Yu, S.; Liu, X.; Zhang, K.; et al. A Highly Sensitive Toluene Gas Sensor Based on Pd/PdO Decorated SnO_2 Prepared by Electrospinning. *ACS Appl. Electron. Mater.* **2024**, *6*, 6036–6048. [CrossRef]
29. Karmakar, S.; Deepak, M.S.; Nanda, O.P.; Sett, A.; Maity, P.C.; Karmakar, G.; Sha, R.; Badhulika, S.; Bhattacharyya, T.K. 2D V_2C MXene Based Flexible Gas Sensor for Highly Selective and Sensitive Toluene Detection at Room Temperature. *ACS Appl. Electron. Mater.* **2024**, *6*, 3717–3725. [CrossRef]
30. Gao, H.; Zhao, L.; Wang, L.; Sun, P.; Lu, H.; Liu, F.; Chuai, X.; Lu, G. Ultrasensitive and Low Detection Limit of Toluene Gas Sensor Based on SnO_2-Decorated NiO Nanostructure. *Sens. Actuators B Chem.* **2018**, *255*, 3505–3515. [CrossRef]
31. Jung, G.; Shin, W.; Hong, S.; Jeong, Y.; Park, J.; Kim, D.; Bae, J.-H.; Park, B.-G.; Lee, J.-H. Comparison of the Characteristics of Semiconductor Gas Sensors with Different Transducers Fabricated on the Same Substrate. *Sens. Actuators B Chem.* **2021**, *335*, 129661. [CrossRef]
32. Hu, J.W.; Wang, F.; Yu, J.J.; Hong, Z.J.; Zhang, W.H.; Li, H.-J.; Lai, Z.C.; Wang, D.; Deng, Y.H.; Li, G.S. MOF-derived porous Co_3O_4 nanosheets array assembled on SnO_2 nanofibers for humidity-resistant high efficiency acetone detection. *Chin. Chem. Lett.* **2025**, 110863. [CrossRef]
33. Lashkov, A.V.; Fedorov, F.S.; Vasilkov, M.Y.; Kochetkov, A.V.; Belyaev, I.V.; Plugin, I.A.; Varezhnikov, A.S.; Filipenko, A.N.; Romanov, S.A.; Nasibulin, A.G.; et al. The Ti Wire Functionalized with Inherent TiO_2 Nanotubes by Anodization as One-Electrode Gas Sensor: A Proof-of-Concept Study. *Sens. Actuators B Chem.* **2020**, *306*, 127615. [CrossRef]

34. Shokri, A.; Salami, N. Gas Sensor Based on MoS_2 Monolayer. *Sens. Actuators B Chem.* **2016**, *236*, 378–385. [CrossRef]
35. Hou, J.-L.; Lin, Y.-T.; Hsueh, T.-J. Room-Temperature TiO_2 Gas Sensor with Conical-TSV Using a Thermal Oxidation Process. *ACS Appl. Electron. Mater.* **2023**, *5*, 6542–6548. [CrossRef]
36. Liu, H.; Shen, W.; Chen, X. A Room Temperature Operated Ammonia Gas Sensor Based on Ag-Decorated TiO_2 Quantum Dot Clusters. *RSC Adv.* **2019**, *9*, 24519–24526. [CrossRef]
37. Hu, J.W.; Zou, Y.D.; Deng, Y.; Li, H.-J.; Xu, H.; Wang, D.; Wu, L.M.; Deng, Y.H.; Li, G.S. Recent Advances in Non-Ionic Surfactant Templated Synthesis of Porous Metal Oxide Semiconductors for Gas Sensing Applications. *Prog. Mater. Sci.* **2025**, *150*, 101409. [CrossRef]
38. Hou, L.; Duan, J.; Xiong, F.; Carraro, C.; Shi, T.; Maboudian, R.; Long, H. Low Power Gas Sensors: From Structure to Application. *ACS Sens.* **2024**, *9*, 6327–6357. [CrossRef]
39. Doan, T.H.P.; Ta, Q.T.H.; Sreedhar, A.; Hang, N.T.; Yang, W.; Noh, J.-S. Highly Deformable Fabric Gas Sensors Integrating Multidimensional Functional Nanostructures. *ACS Sens.* **2020**, *5*, 2255–2262. [CrossRef]

Disclaimer/Publisher's Note: The statements, opinions and data contained in all publications are solely those of the individual author(s) and contributor(s) and not of MDPI and/or the editor(s). MDPI and/or the editor(s) disclaim responsibility for any injury to people or property resulting from any ideas, methods, instructions or products referred to in the content.

chemosensors

MDPI

Article

A Minimal Electronic Nose Based on Graphene Functionalized with Metalated Pyrazinoporphyrazines and Phthalocyanines for Ammonia, Benzene, and Hydrogen Sulfide Discrimination

Sonia Freddi [1,2,*], Luca Vaghi [3], Andrea Penoni [4], Luca Scapinello [4] and Luigi Sangaletti [1]

1 Surface Science and Spectroscopy Lab at I-Lamp, Department of Mathematics and Physics, Università Cattolica del Sacro Cuore, Via della Garzetta 48, 25133 Brescia, Italy
2 Institute of Photonics and Nanotechnologies-Consiglio Nazionale delle Ricerche (IFN-CNR), LNESS Laboratory, Via Anzani 42, 22100 Como, Italy
3 Dipartimento di Scienza dei Materiali, Università degli Studi di Milano—Bicocca, Via R. Cozzi 55, 20125 Milano, Italy
4 Dipartimento di Scienza e Alta Tecnologia, Università degli Studi dell'Insubria, Via Valleggio 9, 22100 Como, Italy
* Correspondence: sonia.freddi@unicatt.it or sonia.freddi@cnr.it

Abstract: The development of electronic noses is, nowadays, essential for several applications, including breath analysis and industrial security. Ammonia, benzene, and hydrogen sulfide are particularly important due to their environmental and health impacts. Here, graphene-based sensors, functionalized with unconventional in-house synthesized zinc and copper octyl-pyrazinoporphyrazines and commercially available zinc phthalocyanine, have been prepared. Enhanced solubility given by the octyl chains allowed us to exploit drop-casting as a straightforward functionalization technique. The sensors demonstrated excellent performance for detecting ammonia, benzene, and hydrogen sulfide as a single sensor, with a competitive detection limit and a high sensitivity compared to the state of the art. In particular, functionalization enabled the detection of hydrogen sulfide, for which no response is observed with bare graphene, and lowered the detection limit for all the gases compared to bare graphene. Additionally, the prepared sensors have been assembled into an e-nose that shows promising potentiality to be used for both industrial and medical applications thanks to its excellent discrimination capability of single gases and mixtures.

Keywords: graphene; electronic nose; ammonia; benzene; hydrogen sulfide; porphyrazine

1. Introduction

The development of advanced gas sensing technologies has become essential for many applications, including healthcare and medicine [1–3], environmental monitoring [3,4], food and beverage quality tracking [5,6], and security in industries [7,8]. As a consequence, researchers have been driven by market demands to develop sensors and arrays that should be highly sensitive, cost-effective, reliable, and stable, for the detection and discrimination of several volatile organic compounds (VOCs) and toxic gases.

Among these VOCs, ammonia (NH_3), benzene (C_6H_6), and hydrogen sulfide (H_2S) could represent critical targets due to their significant environmental and health impacts. These three gases need to be monitored mainly in different industrial contexts: from petroleum refining, coal gasification, and syngas production to biomass pyrolysis and natural gas processing [9–11].

C_6H_6 is known to be a carcinogen and to cause headaches, dizziness, and confusion [12,13]; inhalation of H_2S can cause nausea, pulmonary edema, and seizures [14,15], while NH_3 can provoke damage to the respiratory system, from irritation of respiratory tracks to asphyxiation, and to the neural system, from dizziness and confusion up to loss of consciousness [16,17].

The occupational exposure limit for an 8 h average is set at 1 ppm for benzene, 10 ppm for hydrogen sulfide, and 25 ppm for ammonia in the USA [18], while in Europe the recommended limits are even lower: 0.2 ppm, 5 ppm, and 20 ppm, respectively [19], and individual European Union member states may implement more stringent national limits based on their regulation.

NH_3, C_6H_6, and H_2S monitoring is also important in breath analysis, with them being biomarkers of liver and kidney failure as well as chronic obstructive pulmonary disease [20,21], lung cancer [22], and asthma [23]. The concentration of these gases in the exhaled breath is usually in the sub-ppm or ppm range [20,24]. It is, therefore, clear how for ensuring workers' safety as well as for breath analysis application the developed sensors should be able to detect very low analytes concentration, and, additionally, the effective detection of these gases should be achieved even in complex mixtures. In this context, the use of an electronic nose (e-nose), i.e., an array that comprises several sensors, presents a significant advantage over the use of a single sensor. Indeed, exploiting an array of sensors with different sensitivities and selectivities, coupled with advanced pattern recognition algorithms, can allow the discrimination of single analytes as well as effectively analyze complex gas mixtures. This capability is crucial for addressing the challenges posed by the simultaneous presence of ammonia, benzene, and hydrogen sulfide in workplace environments or diagnostic applications.

Currently, the most exploited materials to develop e-noses are metal oxide semiconductors (MOx) [25], but, besides the advantages, such as fast response and recovery times, portability, and low detection limit, they still present some drawbacks like poor selectivity, performance drift, and high operating temperature.

On the contrary, graphene presents high sensitivity to the surface adsorption of gas molecules, and thanks to its outstanding electrical properties, including high in-plane conductivity, low electrical intrinsic noise, and excellent stability at room temperature [26], it can be considered a valid alternative to MOx for the development of sensors and e-noses. Indeed, more recently, graphene has started to be exploited for the development of e-noses in chemiresistive [27,28] or field effect transistor [29,30] configurations.

The functionalization of graphene with specific chemical compounds can further tailor its sensitivity and selectivity, enabling the precise detection of diverse analytes [31,32]. Graphene could be functionalized in several ways, including both covalent and non-covalent methods [33], and different materials could be selected for the functionalization: from single atoms (mostly boron or nitrogen) [34] to nanoparticles [35] or organic molecules [36].

Here, commercially available zinc phthalocyanine (ZnPc) and zinc and copper dibenzoazepinopyrazinoporphyrazines (ZnPR and CuPR) functionalized at azepine nitrogen with octyl chains to increase solubility (Figure 1) are exploited for the functionalization of graphene via drop-casting. These types of molecules are extensive planar π-conjugated systems, and thus it is expected that a face-on stacking with graphene is mainly driven by non-covalent π-π interactions, which should favor a charge transfer during gas exposures [37,38].

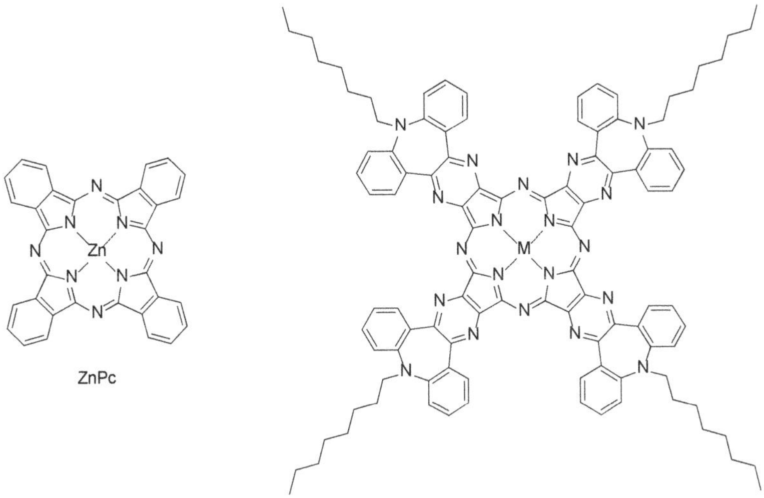

Figure 1. Molecular structures of the zinc phthalocyanine (**left side**) and metalated pyrazinoporphyrazines (**right side**) used in this study.

Very few works explore the sensing mechanism of graphene-based systems coupled with phthalocyanine beyond the mere sensor response observation. These studies mainly focus on graphene derivatives, such as reduced graphene oxide (rGO) or graphene oxide (GO) [39–41]. For instance, Guo et al. [39] report the sensing mechanism for ammonia interaction with p-type rGO-CoPc sample. The sensing mechanism of this hybrid system has been considered by starting from the initial situation of phthalocyanine in air, which interacts with oxygen, creating a $M\text{-}Pc^{+}\text{-}O_2^{-}$ complex. Ammonia can then interact with the adsorbed oxygen of the system, releasing electrons that in turn can move from the CoPc compounds to the rGO layer, since the electron transfer energy barrier between CoPc and rGO is low. Finally, the holes in the p-type rGO are recombined with the electrons from ammonia, leading to a decrease in conductivity. Although displaying remarkable sensing performances, rGO and GO present drawbacks in terms of sample reproducibility and recovery after exposures, and in this perspective, Gr represents a more reliable layer. Nevertheless, the functionalization of pristine graphene layers with Pc has not been largely investigated.

In particular, the e-nose here presented is composed of four graphene-based sensors: one graphene monolayer (Gr_bare), a graphene layer functionalized with zinc phthalocyanine (Gr_ZnPc), and two Gr layers functionalized with zinc and copper pyrazinoporphyrazines (Gr_ZnPR and Gr_CuPR). The synthesized octyl-pyrazinoporphyrazines display a higher solubility compared to commercially available phthalocyanine, allowing for a simple and cost-effective functionalization via drop-casting.

The sensors are tested towards NH_3, C_6H_6, and H_2S, drawing the calibration curves and investigating sensing parameters such as detection limit, sensitivity, selectivity, response, and recovery time. Following this first phase, the developed e-nose is tested to probe the capability of simultaneously discriminating NH_3, C_6H_6, and H_2S and their mixtures with high precision. It is worth mentioning that despite their relevance for environmental, safety, and medical applications, these gases are not commonly tested together, therefore, this work represents a first. By leveraging material engineering and functionalization strategies, we aim, on one hand, to enhance the sensitivity, selectivity, and stability of the sensors in a chemiresistor configuration, and on the other, to enhance the discrimination capability of the array towards the tested gases and their mixtures. Finally, the chemiresistive configuration of the e-nose developed allows for a rapid and simple prototype [2,42–44], avoiding complex electronics typical of other configurations.

This work represents a significant step forward in gas-sensing technology, offering a promising approach for real-world applications.

2. Materials and Methods

2.1. Synthesis of ZnPR and CuPR

The procedures for the synthesis of ZnPR and CuPR are reported in the Supplementary Materials.

2.2. Sample Preparation

A total of 4 graphene monolayers on 10×10 mm^2 Si_3N_4/Si substrates (Graphenea, Gipuzkoa, Spain) were used to prepare the array. One of the samples was used as received (Gr_bare) and three samples were functionalized. Two layers were functionalized via drop-casting exploiting the synthesized ZnPR and CuPR, while a layer with ZnPc (Sigma Aldrich, Merck, Milano, Italy). An additional solution was prepared with CuPc, but the molecules were not properly dissolving; therefore, we did not proceed with the drop-casting on graphene. In detail, the porphyrazines and phthalocyanine powders were dissolved in EtOH at a concentration of 0.1 mM and sonicated for 30 min, before 15 droplets of the solution were drop-casted on the graphene layer. After each drop, the samples were left drying in the air; upon solvent evaporation, a new drop was added. The choice of the solvent and concentration was made on the basis of ref. [45] for the ZnPc; consequently, ZnPR and CuPR were dissolved in the same condition.

The array was then composed of Gr_bare, Gr_ZnPc, Gr_ZnPR, and Gr_CuPR.

2.3. Characterization

To characterize the samples, Raman spectroscopy (Ranishaw, Pianezza, Italy) and atomic force microscopy (AFM) (Park, Hamburg, Germany) were exploited.

Raman spectroscopy was performed with a Ranishaw micro Raman, equipped with a 633 nm laser source. A 100× objective, 1800 line/mm grating, and 5 mW laser power were used to perform the measurements.

AFM images were acquired with a Park NX 10 system in tapping mode in air–solid interface with a tip working at 300 kHz. The Gwyddion software (version: 2.54) was used to process the images.

2.4. Gas Sensing Measurements

Electrical contacts were made on the 4 samples, drawing two 8 mm × 1 mm strips of silver paint at the opposite side of each layer. Then, the prepared layers were mounted on a homemade designed platform, able to collect simultaneously the signal in a chemiresistor configuration from all the samples. This platform ensured that the sensors worked in the same environmental condition and allowed for a proper comparison of the results. Two commercial sensors were mounted on the platform: a relative humidity (RH) sensor (humidity sensor HIH-4000 series—Honeywell Sensing; Charlotte, NC, USA) and a temperature sensor (Thermistor NTC PCB 5K—Murata; Kyoto, Japan). Figure 2a shows a schematic representation of the sensor array.

Gas exposure was performed at room temperature, with humidity values ranging from 10% to 60%, in a sealed chamber with a volume of 0.5 L, by introducing a mixture of synthetic air with a selected concentration of the gas molecule under investigation. Cylinders of ammonia, benzene, hydrogen sulfide, and synthetic air were used to feed the chamber through mass-flow controllers (MFCs). All the cylinders were certified by S.I.A.D. Spa (Bergamo, Italy), and loaded with 47 ppm, 1 ppm, and 10 ppm for ammonia, benzene, and hydrogen sulfide, respectively. An additional cylinder loaded with synthetic air alone was used to dilute the overall concentration of the analytes in the chamber, as well as to purge the chamber before and after each exposure. Gas from all the cylinders was premixed before entering the chamber.

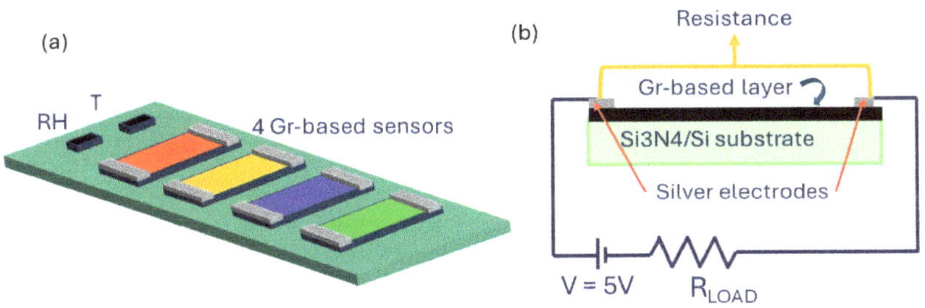

Figure 2. (**a**) Schematic representation of the homemade platform comprising the 4 developed graphene-based sensors and the two commercial sensors for RH and T detection. (**b**) Schematic representation of the chemiresistors readout scheme of each sensor.

All the sensors on the platform worked in a chemiresistor configuration and the electronic circuit that drives each sensor was composed of the sensor, a load resistance in series with the sensors, and an applied voltage of 5 V. The signal, i.e., the resistance value, was collected across the two silver electrodes (Figure 2b). The sensor response towards the gases is defined as $\Delta R/R_0$, where $\Delta R = R - R_0$, and R_0 is the baseline resistance, collected before the gas exposure.

Calibration curves were drawn plotting the sensor responses vs. concentrations for all the tested analytes. Response and recovery times were evaluated considering the time required to reach 80% of response or recovery, respectively.

2.5. Statistical Analysis

The sensor responses collected from the array were used to perform a multivariate statistical analysis, i.e., principal component analysis (PCA), implemented into the R software (1.2.5001 version). PCA is an unsupervised statistical technique aimed to reduce the dimensionality of a problem while maximizing its variance.

The aim is achieved by redistributing the total response across a series of orthogonal principal components (PCs), where PC1 captures the highest variance, PC2 captures the second highest, and so forth. The output of this analysis is a 2D or 3D graph, where a clear discrimination of the analytes is observed.

In this work, the sensor responses used to perform PCA were pre-treated only with the column mean-centering. A set of 51 exposures to the 3 analytes and their mixtures was considered for the analysis.

3. Results and Discussion

In the past, some of us reported the synthesis of this novel centrosymmetric porphyrazine system (Figure 1), arising from the formal substitution of four benzo subunits in a Pc by a 9*H*-dibenzo[*b*,*f*]pyrazino[2,3-*d*]azepine moiety, wherein the peripheral azepine nitrogen atoms were functionalized with hexadecanoyl chains to successfully address the scarce solubility in organic solvents typical of these systems [46]. However, the presence of the carbonyl in such chains that come out of the plane of the molecule could very likely interfere with the π-π interaction with graphene. Consequently, we opted for the *N*-functionalization with a linear octyl chain following the same synthetic pathway. Despite being the alkylation step challenging, we successfully prepared enough quantity of ZnPR and CuPR for this study (see Supplementary Materials for details).

Ethanolic solutions of ZnPc, ZnPR, and CuPR were easily prepared and used to functionalize the graphene monolayers supported on silicon nitride via the drop-casting method. To have a direct comparison of the new ZnPR and CuPR with commercially available metallo-phtalocyanines, we aimed to prepare an ethanolic solution of CuPc; however, the EtOH solubility of CuPc was very bad as expected [45], as well as in the other solvents, making it impossible to use with our drop-casting method. Conversely, CuPR, despite being more extensive than CuPc, was easily dissolved in EtOH thanks to the peripheral octyl chains.

Characteristic Raman spectra collected on the samples are reported in Figure 3a. All the spectra were collected in a 1200–1700 cm^{-1} range since the Gr_ZnPc and Gr_ZnPR samples showed a strong luminescence at higher wavenumber [47], which made it difficult to recognize peaks related to graphene or to the organic molecules.

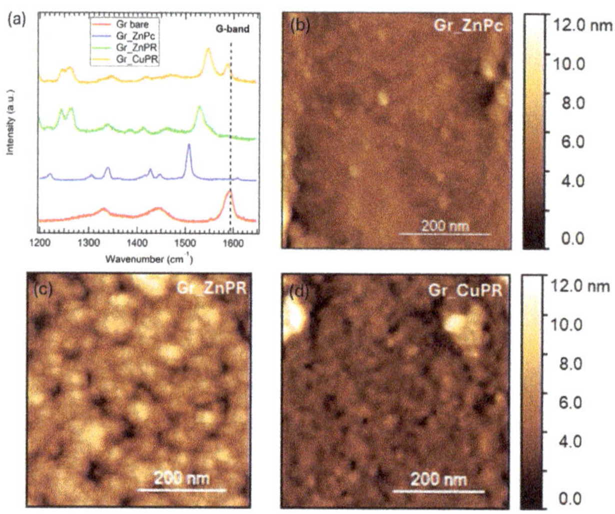

Figure 3. (**a**) Representative Raman spectra of a bare graphene layer on Si/Si3N4 (red) and with ZnPc (blue), ZnPR (green), and CuPR (yellow). AFM images were acquired on (**b**) Gr_ZnPc, (**c**) Gr_ZnPR, and (**d**) Gr_CuPR.

The bare sample (red curve) shows the G-band signal around 1585 cm^{-1}, ascribed to the in-plain stretching of sp^2 carbon-carbon atoms [48]. A band around 1335 cm^{-1} is also present; to properly assign it, we need to consider two contributions at this wavenumber: the silicon nitride substrate band [27] and the graphene D-band. The D-band is related to the presence of defects, and, in a low defective graphene monolayer, it is expected to have very low intensity. Therefore, we hypothesize that the silicon nitride broad band dominates the area around 1335 cm^{-1} in the Gr_bare spectrum.

Considering the spectra collected on the functionalized samples, the graphene G-band is strongly present in the Gr_CuPR sample (yellow curve), while it is slightly detected in the ZnPc and ZnPR layers. This is an indication that although the coating procedure was the same, the resulting CuPR layer is thinner than the ZnPR and ZnPc layers.

Finally, the Raman peaks related to the ZnPc can be found at 1223 cm^{-1}, 1305 cm^{-1}, 1340 cm^{-1}, 1430 cm^{-1}, and 1450 cm^{-1} [47], whereas the peaks related to PRs are located at 1245 cm^{-1}, 1265 cm^{-1}, 1341 cm^{-1}, and 1413 cm^{-1} [49,50] and the characteristic peak related to the metal center in the porphyrazines at 1550 cm^{-1} for CuPR [51,52] and at 1530 cm^{-1} for ZnPR [53].

Morphology was investigated through AFM, and the representative images of 500 nm × 500 nm size of the functionalized surfaces are reported in Figure 3b–d. A quite uniform distribution of the Pc and PR molecules was detected in all three cases.

After the characterization, exposures to ammonia, benzene, and hydrogen sulfide were performed.

As a consequence of functionalization, the baseline resistance of the pristine graphene layer lowered (Table S1), indicating that an effective charge transfer occurs between the Gr layer and the ZnPc, ZnPR, and CuPR layers.

As an example of dynamical responses, exposures to four different ammonia concentrations are presented in Figure 4. Firstly, all the sensors increased their resistance, denoting a p-type response, in agreement with what is expected from graphene layers. A quick response of about 30 s is observed, as well as a recovery time between about 90 s in the case of low analyte concentrations and 700 s for high ones. For the highest concentration tested, the fastest recovery is achieved for Gr_ZnPc, which shows a 100% recovery in a time where the other sensors recover 80%. These results are in line with, or even better than those reported in the literature [44,54–57].

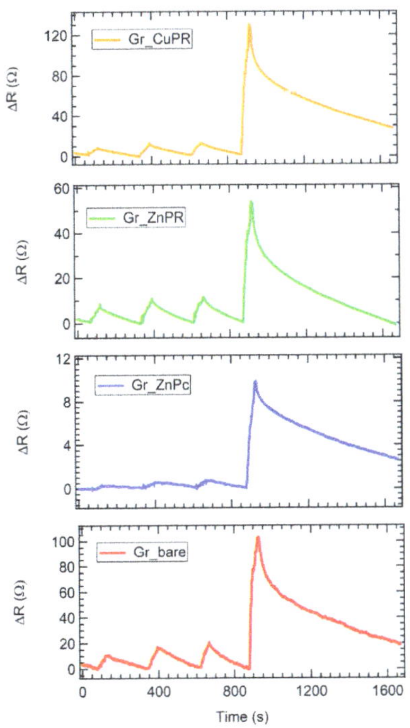

Figure 4. Resistance variation as a function of the time for all the sensors exposed to 2.6 ppm, 2.9 ppm, 3.2 ppm, and 13.6 ppm of ammonia.

Similar dynamic curves were obtained for exposures to benzene and hydrogen sulfide and were used to draw the calibration curves (i.e., response vs. concentration) for all the sensors (Figure 5). Response and recovery time were found in line with the ones obtained for ammonia exposures, ranging from 30 s for response up to about 700 s for the recovery in the case of the highest tested concentrations. All the sensors responded to the three analytes, except for Gr_bare exposed to hydrogen sulfide, for which a response was not detected. All the curves can be fitted with a Freundlich isotherm, as usually occurs for nanostructured carbon-based sensors [36,58].

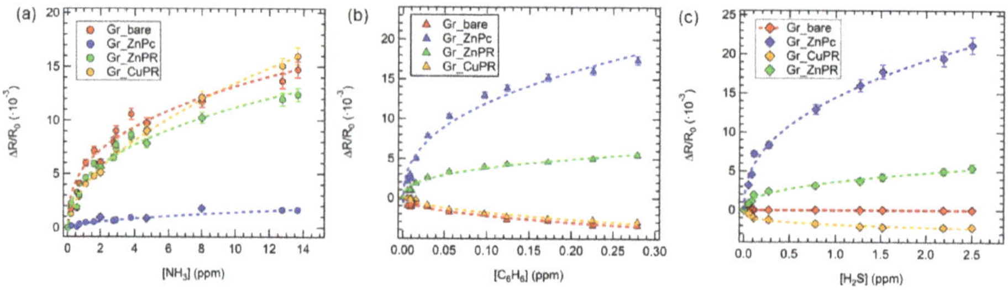

Figure 5. Calibration curves for (**a**) ammonia, (**b**) hydrogen sulfide, and (**c**) benzene for all the sensors in the array. Squares data in the ammonia calibration curves comes from exposures performed in air, in a close environment of applications. Error bars are estimated on the basis of 3 exposures to the same gas concentration.

Considering ammonia, the best-performing sensors were Gr_bare and Gr_CuPR, while in the case of benzene and hydrogen sulfide, Gr_ZnPc was clearly outperforming the others.

As regards the sensing mechanisms, ammonia has been investigated more in the literature compared to hydrogen sulfide and benzene; therefore, it is easier to rationalize the behavior of the developed sensors towards this gas.

As already mentioned, all the sensors increased the resistance upon ammonia exposure, indicating an overall p-type nature. In the case of Gr_bare, this is consistent with an electron injection from ammonia that reduces the density of holes, which are the majority carrier in graphene [59].

A recent work by Perilli et al. [60] investigated the sensing mechanism of a NiPc-graphene sensor towards ammonia, theoretically suggesting that the Ni orbital plays a key role in mediating the charge transfer between ammonia molecules and the Gr layer through a hybridization mechanism involving the frontier molecular orbitals of the analyte. CuPR, ZnPc, and ZnPR are, therefore, supposed to act as mediators in the charge transfer between ammonia and graphene [38].

Poliakov et al. [61] reported on the functionalization of reduced graphene oxide (rGO) with ZnPc and observed a lower response towards ammonia for the rGO-ZnPc sensor compared to the pristine rGO, ascribing it to a reduction in the defect sites due to functionalization, which act as active sites for the chemisorption of ammonia through hydrogen bonding. Moreover, they concluded that the higher the ZnPc coverage, the lower the response. This is in agreement with the lower ammonia response for Gr_ZnPc compared to Gr_bare that we observed, and also with a better response of Gr_CuPR compared to Gr_ZnPc and Gr_ZnPR. Indeed, as suggested by the Raman measurements, the CuPR layer was thinner than the ZnPc and ZnPR, resulting in a lower coverage.

Regarding hydrogen sulfide and benzene, little evidence can be found in the literature for the sensing mechanism of the former, while no works disclose the interaction of the latter with graphene. Zhang et al. [62] theoretically demonstrated that no charge transfer occurs from H_2S to pristine graphene layers; thus, functionalization is required. This is in agreement with the observed no response in the case of Gr_bare. We suggest that the porphyrazine layers act as charge transfer mediators and probably also as gas molecule concentrators; however, theoretical calculation will be required to properly hypothesize the sensing mechanism, going beyond the scope of the present work.

Detection limit, i.e., the lowest concentration that the sensors can detect, and sensitivity, defined as a percentage of sensor response normalized over the tested analytes concentrations, were evaluated and compared to the literature (Table 1). The benchmarking was performed considering only chemiresistive sensors based on graphene. In the case of ammonia, the works of the last 10 years have been considered, while for benzene and hydrogen sulfide, which are far less investigated, all the published works have been considered. It is worth mentioning that in the case of H_2S, quite a few theoretical works on graphene-based H_2S sensors can be found [52–65], but only a couple of works report experimental results.

Table 1. Benchmarking for dynamical tested range, sensitivity S range ($S = (|\Delta R/R_0| \times 100)/[gas]$), and detection limit (dl) of the prepared samples against the values for other graphene-based chemiresistors reported in the literature [27,54–57,60,66–78]. Of note, only articles reporting gas concentration and sensor response/sensitivity have been taken into account for this benchmarking. In the case of ammonia, only works of the last 10 years are reported.

Sensor Type	Gas: Dynamic Tested Range (ppm)	Sensitivity Range (%ppm^{-1})	Detection Limit (ppb)	Reference
Gr_bare	NH_3: 0.2–14	7.75–0.11	170	Present work
Gr_ZnPc	NH_3: 0.2–14	0.75–0.012	50	Present work
Gr_ZnPR	NH_3: 0.2–14	1.124–0.09	11	Present work
Gr_CuPR	NH_3: 0.2–14	0.62–0.12	5	Present work
Gr_NiPc	NH_3: 0.04–4.64	8.6–0.5	3.3	[60]
Gr_CuO NPs	NH_3: 100	0.83	41	[66]
MXene deposited Gr	NH_3: 0.5–100	4–2.5	56	[56]
NiPc-Gr	NH_3: 5–10	0.47–0.19	-	[67]
Gr_CoPt	NH_3: 2.2–36.0	0.8–0.24	0.1	[68]
Gr_NiPc	NH_3: 0.5–13.6	0.64–0.11	50	[27]
Nitric acid treated Gr	NH_3: 20–100	0.7–0.4	27	[69]
S-ZnO@Gr	NH_3: 5–15	1.9–1.1	-	[70]
Gr_NBD	NH_3: 0.05–8.4	10.78–0.45	-	[71]
Gr-TCN	NH_3: 0.85–22.5	6.3–1.5	4.2	[54]
B-doped Gr	NH_3: 16–256	0.12–0.09	-	[72]
CVD graphene	NH_3: 100–800	0.05–0.01	-	[55]
Laser written Gr	NH_3: 75–400	0.04–0.075	-	[73]
B-doped Gr	NH_3: 1–20	0.04–0.042	59.9	[74]
TiO_2@PPy–GN	NH_3: 10–200	2.4–1.3	1000	[57]
graphene–PEDOT:PSS	NH_3: 25–1000	0.2–0.019	10,000	[75]
Gr_ZnPc	H_2S: 0.05–2.5	6.41–0.85	12.6	Present work
Gr_ZnPR	H_2S: 0.05–2.5	0.86–0.21	23.0	Present work

Table 1. *Cont.*

Sensor Type	Gas: Dynamic Tested Range (ppm)	Sensitivity Range (%ppm^{-1})	Detection Limit (ppb)	Reference
Gr_CuPR	H_2S: 0.05–2.5	1.57–0.09	29.3	Present work
Gr and AgNPs	H_2S: 0.5–50	8–2.6	100	[76]
CuO_2 NCs on Gr	H_2S: 0.05–0.1	15–35	5	[77]
Gr_bare	C_6H_6: 0.005–0.3	22.10–1.22	5	Present work
Gr_ZnPc	C_6H_6: 0.005–0.3	44.00–6.22	2.7	Present work
Gr_ZnPR	C_6H_6: 0.005–0.3	17.8–1.97	2.8	Present work
Gr_CuPR	C_6H_6: 0.005–0.3	1.11–0.93	2.2	Present work
Polyaniline–Gr nanoplatelets	C_6H_6: 1000–22,000	0.1–0.8	-	[78]

In detail, the detection limit was evaluated through the formula $3[gas]/((R - R_0)/\sigma)$ [36,79], where σ is the fluctuation of the electrical signal. The results for the functionalized Gr layers are among the best so far reported for all the analytes. Also, the sensitivity is among the best achieved so far, especially in the case of benzene. It is worth noting that not many works investigated the detection of benzene and hydrogen sulfide, and additionally, in the case of benzene, the present work clearly pushed down the detection range of the analyte, getting closer to a real application. The repeatability of the response towards the same analyte concentration is demonstrated, as shown in Supplementary Materials Figure S1.

Finally, the stability and reproducibility of the sensor response have been proven for ammonia exposures. The results, reported in Figure 6, demonstrate that the response towards different ammonia concentrations is stable for up to 4 months from the sensor preparation.

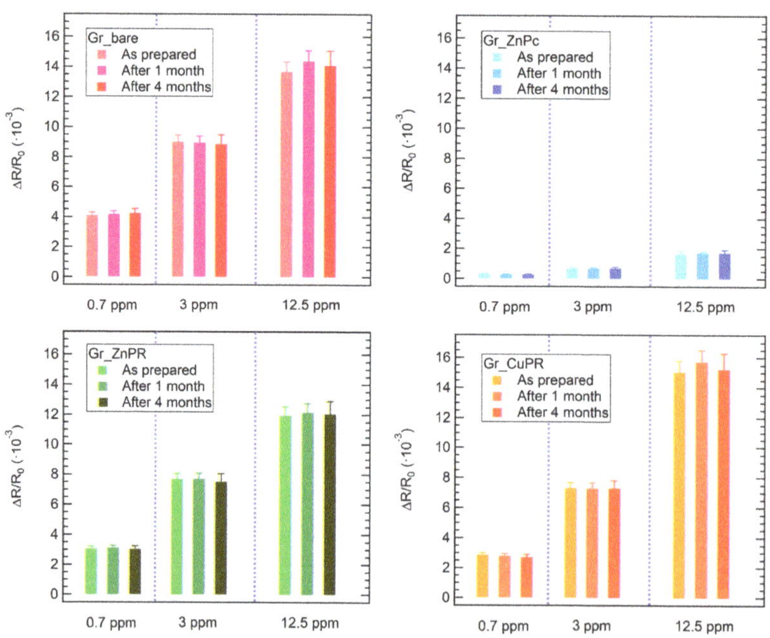

Figure 6. Stability of the sensors' response towards 3 different ammonia concentrations up to 4 months from the sensors' preparation.

When exposed to the same concentration of analytes (270 ppb), significant response variations are observed: in general, the highest responses were detected in the case of benzene exposure for all the sensors. Considering Figures 5 and 7, while for ammonia all the sensors respond with a positive resistance variation, it is interesting to observe that for H_2S a negative response was obtained for Gr_CuPR, while for benzene Gr_bare and Gr_CuPR showed a negative response. The developed sensors are clearly not selective and the variation in sensing responses to different analytes presents a significant starting point for data analysis and classification using machine learning algorithms. Indeed, an easy approach to overcome selectivity issues for gas sensors is to consider an electronic nose and proceed with statistical analysis to achieve discrimination among the tested analytes. In this regard, to build a matrix to feed the PCA algorithm, sensor responses towards ammonia, benzene, and hydrogen sulfide, as reported in the calibration curves, were used. Additionally, exposures to mixtures were performed: ammonia + hydrogen sulfide, ammonia + benzene, and benzene + hydrogen sulfide.

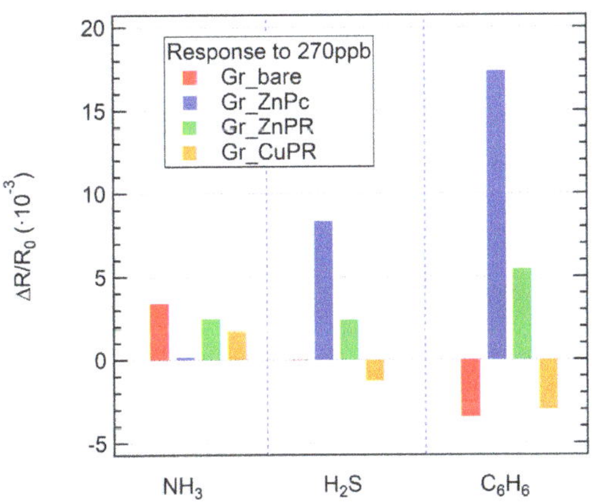

Figure 7. Sensor response to the same concentration of analytes (270 ppb) for all the sensors.

The results are reported in Figure 8. The space defined by PC1 and PC2 allows for clear discrimination of all the tested analytes, for both single gases (NH_3, C_6H_6, and H_2S) and their binary mixtures, as there is no overlap among the clusters corresponding to different gases. Additionally, similar to what has been already reported for arrays based on carbon nanomaterials [20], a concentration gradient is evident within each cluster: the arrows in Figure 8 indicate the trend of an increase in concentration for each cluster of a single gas.

This clustering indicates that the developed e-nose produces distinct and reproducible response profiles for each gas or mixture, demonstrating strong selectivity. For example, NH_3 (green circles) and NH_3 + C_6H_6 (red stars) occupy well-separated regions in the PCA space, confirming the sensor array's ability to differentiate between individual gases and mixtures.

Additionally, the spatial separation between C_6H_6 (blue triangles), H_2S (purple diamonds), and their mixture (e.g., C_6H_6 + H_2S, yellow diamonds) suggests that the sensor responses contain rich, analyte-specific features that can be effectively captured through PCA. Importantly, the positions of the binary gas mixtures tend to lie between the clusters of the corresponding individual gases, reflecting the composite nature of their chemical signatures. This intermediate clustering behavior indicates that the sensor array not only discriminates between single analytes but also captures additive or synergistic effects present in gas mixtures. This aspect is particularly relevant in view of a possible application, where the accurate identification of complex or coexisting gases is essential, including in environmental monitoring, industrial leak detection, or medical diagnostics (e.g., breath analysis), where gases rarely need to be monitored individually. The ability to resolve mixed analytes enhances the practical utility of the developed array, supporting robust classification in dynamic and chemically rich environments.

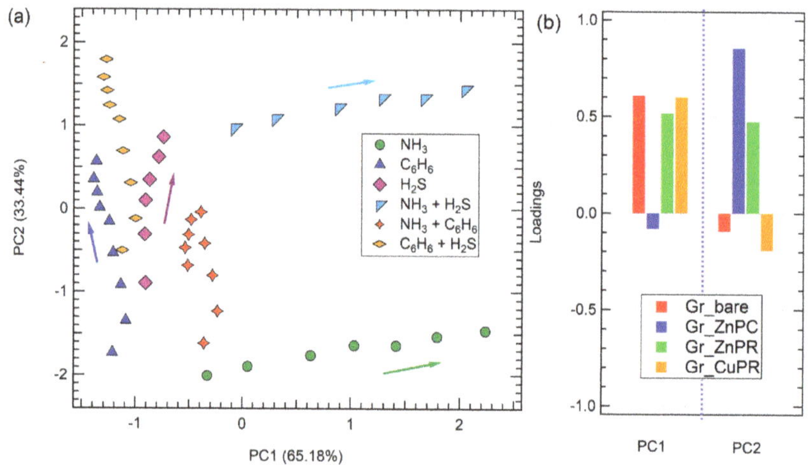

Figure 8. (**a**) PCA of the sensor array responses to exposure to hydrogen sulfide, benzene, ammonia, and their mixtures. (**b**) Loadings for PC1 and PC2. Arrow in panel (**a**) indicates the concentration trend.

Notably, while the developed sensors are not particularly selective, their combined response in the PCA enables the array to fully discriminate the contribution of each single gas and also mixtures. The overall variance explaining the data in the PCA space is 98.62%.

Finally, PCA loading plots offer valuable insights into the significance of individual sensors in contributing to the overall discrimination capability of the array (Figure 8b).

Specifically, discrimination along the PC1, meaning discrimination between on the one hand ammonia and benzene; hydrogen sulfide and benzene and their combination, and on the other ammonia + hydrogen sulfide and benzene, hydrogen sulfide and benzene and their combination, and ammonia + benzene, is achieved quite equally thanks to all sensors except Gr_ZnPc, while Gr_ZnPc was mainly responsible for discrimination along PC2, meaning discrimination between ammonia and ammonia + hydrogen sulfide.

This suggests that different functionalizations respond preferentially to different gases, enabling the dimensionality reduction to preserve chemically relevant variance.

As anticipated, achieving effective discrimination does not require all the sensors to exhibit high responses to every gas. Instead, variability in responses across the sensing layers is a critical factor.

These findings highlight the potential of the proposed sensor system for practical applications such as gas classification, electronic nose systems, and complex environment monitoring. The clear separation of analytes in PCA space supports robust pattern recognition, making the e-nose suitable for deployment in applications requiring rapid and reliable gas discrimination.

4. Conclusions

In this work, we presented the development and test of an electronic nose based on 4 sensors: bare graphene, and three graphene layers functionalized via drop-casting with unconventional metallated *N*-octyldibenzoazepinopyrazinoporphyrazines and commercial zinc phthalocyanine.

After characterization through Raman spectroscopy and atomic force microscopy (AFM), the developed sensors have been exposed to ammonia (NH_3), benzene (C_6H_6), and hydrogen sulfide (H_2S) in a chemiresistor configuration. Calibration curves have been drawn and a detection limit evaluated. The developed sensors show competitive sensing parameters compared to the literature: low limit of detection, high sensitivity, stability up to 4 months after the sample preparation, and a tested concentration range compatible with breath analysis and industrial applications. Finally, the discrimination capability of the e-nose has been assessed for single analytes and for mixtures of the investigated gases. It is worthing to mention that the selected three gases are not commonly tested together, despite their relevance, and therefore, this work represents a first in the literature. By leveraging the unique interaction mechanisms between the functionalized graphene surfaces and the analyte molecules, highlighting the synthesis and characterization of the sensing materials, their integration into the e-nose platform, and the performance evaluation in detecting and discriminating NH_3, C_6H_6, H_2S, and their mixtures, this e-nose represents a promising approach for real-time, selective, and sensitive gas detection in several applications.

Supplementary Materials: The following supporting information can be downloaded at https://www.mdpi.com/article/10.3390/chemosensors13050165/s1. Procedures for the synthesis of ZnPR and CuPR; Scheme S1: synthesis of ZnPR and CuPR. Figure S1: repeatability of ammonia exposure; Table S1: baseline resistances.

Author Contributions: Conceptualization, S.F. and L.S. (Luigi Sangaletti); methodology, S.F. and L.V.; software, S.F.; validation, S.F., L.V., A.P. and L.S. (Luca Scapinello); formal analysis, S.F., investigation, S.F., L.V. and A.P.; resources, L.S. (Luigi Sangaletti); data curation, S.F.; writing—original draft preparation, S.F.; writing—review and editing, S.F., L.V. and L.S. (Luigi Sangaletti); visualization, S.F.; supervision, L.S. (Luigi Sangaletti); funding acquisition, L.S. (Luigi Sangaletti), S.F. All authors have read and agreed to the published version of the manuscript.

Funding: S.F. acknowledges Fondazione Cariplo for financial support, within the grant BREATH-SENSE 2023-1506—'Giovani Ricercatori 2023' call.

Institutional Review Board Statement: Not applicable.

Informed Consent Statement: Not applicable.

Data Availability Statement: Data are available from the authors.

Acknowledgments: L.V. gratefully acknowledges Giovanni Palmisano for the fruitful discussion. S.F. and L.S. (Luigi Sangaletti) acknowledge Antonio Papagni for the preliminary discussion on the system functionalization.

Conflicts of Interest: The authors declare no conflicts of interest.

References

1. Bag, A.; Lee, N.E. Recent advancements in development of wearable gas sensors. *Adv. Mater. Technol.* **2021**, *6*, 2000883. [CrossRef]
2. Freddi, S.; Sangaletti, L. Trends in the Development of Electronic Noses Based on Carbon Nanotubes Chemiresistors for Breathomics. *Nanomaterials* **2022**, *12*, 2992. [CrossRef] [PubMed]
3. Chen, X.; Leishman, M.; Bagnall, D.; Nasiri, N. Nanostructured gas sensors: From air quality and environmental monitoring to healthcare and medical applications. *Nanomaterials* **2021**, *11*, 1927. [CrossRef] [PubMed]

4. Seesaard, T.; Kamjornkittikoon, K.; Wongchoosuk, C. A comprehensive review on advancements in sensors for air pollution applications. *Sci. Total Environ.* **2024**, *951*, 175696. [CrossRef]
5. Weston, M.; Geng, S.; Chandrawati, R. Food sensors: Challenges and opportunities. *Adv. Mater. Technol.* **2021**, *6*, 2001242. [CrossRef]
6. Kumar, A.; Castro, M.; Feller, J.F. Review on sensor array-based analytical technologies for quality control of food and beverages. *Sensors* **2023**, *23*, 4017. [CrossRef]
7. Reis, T.; Moura, P.C.; Gonçalves, D.; Ribeiro, P.A.; Vassilenko, V.; Fino, M.H.; Raposo, M. Ammonia Detection by Electronic Noses for a Safer Work Environment. *Sensors* **2024**, *24*, 3152. [CrossRef]
8. Tan, L.; Feng, Z.; Zheng, H.; Yao, Z.; Weng, X.; Wang, F.; Chang, Z. Development trend of electronic nose technology in closed cabins gas detection: A review. *Appl. Sci.* **2022**, *12*, 9326. [CrossRef]
9. Groysman, A.; Groysman, A. Process units in oil refineries and petrochemical plants. In *Corrosion Problems and Solutions in Oil Refining and Petrochemical Industry*; Springer: Berlin/Heidelberg, Germany, 2017; pp. 1–7.
10. Torres, W.; Pansare, S.S.; Goodwin Jr, J.G. Hot gas removal of tars, ammonia, and hydrogen sulfide from biomass gasification gas. *Catal. Rev.* **2007**, *49*, 407–456. [CrossRef]
11. Marcantonio, V.; Bocci, E.; Ouweltjes, J.P.; Del Zotto, L.; Monarca, D. Evaluation of sorbents for high temperature removal of tars, hydrogen sulphide, hydrogen chloride and ammonia from biomass-derived syngas by using Aspen Plus. *Int. J. Hydrogen Energy* **2020**, *45*, 6651–6662. [CrossRef]
12. Cronkite, E.P.; Drew, R.T.; Inoue, T.; Hirabayashi, Y.; Bullis, J.E. Hematotoxicity and carcinogenicity of inhaled benzene. *Environ. Health Perspect.* **1989**, *82*, 97–108. [CrossRef] [PubMed]
13. Duarte-Davidson, R.; Courage, C.; Rushton, L.; Levy, L. Benzene in the environment: An assessment of the potential risks to the health of the population. *Occup. Environ. Med.* **2001**, *58*, 2–13. [CrossRef] [PubMed]
14. Van Aalst, J.A.; Isakov, R.; Polk, J.D.; Van Antwerp, A.D.; Yang, M.; Fratianne, R.B. Hydrogen sulfide inhalation injury. *J. Burn Care Rehabil.* **2020**, *21*, 248–253. [CrossRef]
15. Guidotti, T.L. Hydrogen sulfide intoxication. In *Handbook of Clinical Neurology*; Elsevier: Amsterdam, The Netherlands, 2015; Volume 131, pp. 111–133.
16. Leduc, D.; Gris, P.; Lheureux, P.; Gevenois, P.A.; De Vuyst, P.; Yernault, J.C. Acute and long term respiratory damage following inhalation of ammonia. *Thorax* **1992**, *47*, 755–757. [CrossRef]
17. Vadysinghe, A.N.; Attygalle, U.; Ekanayake, E.M.K.B.; Dharmasena, E.G.I.A. Ammonia exposure: A review of six cases. *Am. J. Forensic Med. Pathol.* **2021**, *42*, 373–378. [CrossRef]
18. Regulations (Standards—29 CFR). Available online: https://www.osha.gov/laws-regs/regulations/standardnumber/1910 (accessed on 16 January 2025).
19. Recommendations of the SCOEL. Available online: https://echa.europa.eu/recommendations-of-the-scoel (accessed on 16 January 2025).
20. Freddi, S.; Emelianov, A.V.; Bobrinetskiy, I.I.; Drera, G.; Pagliara, S.; Kopylova, D.S.; Chiesa, M.; Santini, G.; Mores, N.; Moscato, U.; et al. Development of a sensing array for human breath analysis based on swcnt layers functionalized with semiconductor organic molecules. *Adv. Healthc. Mater.* **2020**, *9*, 2000377. [CrossRef]
21. Popa, C.; Dutu, D.C.; Cernat, R.; Matei, C.; Bratu, A.M.; Banita, S.; Dumitras, D.C. Ethylene and ammonia traces measurements from the patients' breath with renal failure via LPAS method. *Appl. Phys. B* **2011**, *105*, 669. [CrossRef]
22. Chatterjee, S.; Castro, M.; Feller, J.F. An e-nose made of carbon nanotube based quantum resistive sensors for the detection of eighteen polar/nonpolar VOC biomarkers of lung cancer. *J. Mater. Chem. B* **2013**, *1*, 4563. [CrossRef]
23. Suzuki, Y.; Saito, J.; Uematsu, M.; Fukuhara, A.; Sato, S.; Togawa, R.; Sato, Y.; Misa, K.; Nikaido, T.; Fukuhara, N.; et al. Clinical application of hydrogen sulfide (H2S) as a marker of asthma management. *Eur. Respir. J.* **2016**, *48*, PA3379.
24. Pham, Y.L.; Beauchamp, J. Breath biomarkers in diagnostic applications. *Molecules* **2021**, *26*, 5514. [CrossRef]
25. Li, Y.; Wang, Z.; Zhao, T.; Li, H.; Jiang, J.; Ye, J. Electronic nose for the detection and discrimination of volatile organic compounds: Application, challenges, and perspectives. *TrAC Trends Anal. Chem.* **2024**, *180*, 117958. [CrossRef]
26. Varghese, S.S.; Lonkar, S.; Singh, K.K.; Swaminathan, S.; Abdala, A. Recent advances in graphene based gas sensors. *Sens. Actuators B* **2015**, *218*, 160–183. [CrossRef]
27. Freddi, S.; Marzuoli, C.; Pagliara, S.; Drera, G.; Sangaletti, L. Targeting biomarkers in the gas phase through a chemoresistive electronic nose based on graphene functionalized with metal phthalocyanines. *RSC Adv.* **2023**, *13*, 251–263. [CrossRef] [PubMed]
28. Zanotti, M.; Freddi, S.; Sangaletti, L. Physical Virtualization of a GFET for a Versatile, High-Throughput, and Highly Discriminating Detection of Target Gas Molecules at Room Temperature. *Adv. Mater. Technol.* **2025**, *10*, 2400985. [CrossRef]

29. Capman, N.S.; Zhen, X.V.; Nelson, J.T.; Chaganti, V.S.K.; Finc, R.C.; Lyden, M.J.; Williams, T.L.; Freking, M.; Sherwood, G.J.; Bühlmann, P.; et al. Machine learning-based rapid detection of volatile organic compounds in a graphene electronic nose. *ACS Nano* **2022**, *16*, 19567–19583. [CrossRef]
30. Alzate-Carvajal, N.; Park, J.; Bargaoui, I.; Rautela, R.; Comeau, Z.J.; Scarfe, L.; Menard, J.-M.; Darling, S.B.; Luican-Mayer, A. Arrays of functionalized graphene chemiresistors for selective sensing of volatile organic compounds. *ACS Appl. Electron. Mater.* **2023**, *5*, 1514–1520. [CrossRef]
31. Chen, Z.; Wang, J.; Wang, Y. Strategies for the performance enhancement of graphene-based gas sensors: A review. *Talanta* **2021**, *235*, 122745. [CrossRef]
32. Recum, P.; Hirsch, T. Graphene-based chemiresistive gas sensors. *Nanoscale Adv.* **2024**, *6*, 11–31. [CrossRef]
33. Georgakilas, V.; Otyepka, M.; Bourlinos, A.B.; Chandra, V.; Kim, N.; Kemp, K.C.; Hobza, P.; Zboril, R.; Kim, K.S. Functionalization of Graphene: Covalent and Non-Covalent Approaches, Derivatives and Applications. *Chem. Rev.* **2012**, *112*, 6156–6214. [CrossRef]
34. Wang, H.; Maiyalagan, T.; Wang, X. Review on Recent Progress in Nitrogen-Doped Graphene: Synthesis, Characterization, and Its Potential Applications. *ACS Catal.* **2012**, *2*, 781–794. [CrossRef]
35. Ghanbari, R.; Safaiee, R.; Sheikhi, M.H.; Golshan, M.M.; Horastani, Z.K. Graphene decorated with silver nanoparticles as a low-temperature methane gas sensor. *ACS Appl. Mater. Interfaces* **2019**, *11*, 21795–21806. [CrossRef] [PubMed]
36. Freddi, S.; Rodriguez Gonzalez, M.C.; Casotto, A.; Sangaletti, L.; Feyter, S. Machine learning-aided NO_2 discrimination with an array of graphene chemiresistors covalently functionalized by diazonium chemistry. *Chem. A Eur. J.* **2023**, *29*, e202302154. [CrossRef] [PubMed]
37. Calmeiro, J.M.; Tomé, J.P.; Lourenço, L.M. Supramolecular graphene–phthalocyanine assemblies for technological breakthroughs. *J. Mater. Chem. C* **2020**, *8*, 8344–8361. [CrossRef]
38. Casotto, A.; Drera, G.; Perilli, D.; Freddi, S.; Pagliara, S.; Zanotti, M.; Schio, L.; Verdini, A.; Floreano, L.; Di Valentin, C.; et al. π-Orbital mediated charge transfer channels in a monolayer Gr–NiPc heterointerface unveiled by soft X-ray electron spectroscopies and DFT calculations. *Nanoscale* **2022**, *14*, 13166–13177. [CrossRef]
39. Guo, Z.; Wang, B.; Wang, X.; Li, Y.; Gai, S.; Wu, Y.; Cheng, X.-L. A high-sensitive room temperature gas sensor based on cobalt phthalocyanines and reduced graphene oxide nanohybrids for the ppb-levels of ammonia detection. *RSC Adv.* **2019**, *9*, 37518. [CrossRef]
40. Wang, B.; Wang, X.; Guo, Z.; Gai, S.; Li, Y.; Wu, Y. A highly sensitive ppb-level H_2S gas sensor based on fluorophenoxy-substituted phthalocyanine cobalt/rGO hybrids at room temperature. *RSC Adv.* **2021**, *11*, 5993–6001. [CrossRef]
41. Yabaş, E.; Biçer, E.; Altındal, A. Novel reduced graphene oxide/zinc phthalocyanine and reduced graphene oxide/cobalt phthalocyanine hybrids as high sensitivity room temperature volatile organic compound gas sensors. *J. Mol. Struct.* **2023**, *1271*, 134076. [CrossRef]
42. Shooshtari, M.; Salehi, A. An electronic nose based on carbon nanotube-titanium dioxide hybrid nanostructures for detection and discrimination of volatile organic compounds. *Sens. Acts B Chem.* **2022**, *357*, 131418. [CrossRef]
43. Verma, G.; Gupta, A. Next-Generation Chemiresistive Wearable Breath Sensors for Non-Invasive Healthcare Monitoring: Advances in Composite and Hybrid Materials. *Small* **2025**, *21*, 2411495. [CrossRef]
44. Tang, X.; Debliquy, M.; Lahem, D.; Yan, Y.; Raskin, J.P. A review on functionalized graphene sensors for detection of ammonia. *Sensors* **2021**, *21*, 1443. [CrossRef]
45. Ghani, F.; Kristen, J.; Riegler, H. Solubility Properties of Unsubstituted Metal Phthalocyanines in Different Types of Solvents. *J. Chem. Eng. Data* **2012**, *57*, 439–449. [CrossRef]
46. Parravicini, M.; Vaghi, L.; Cravotto, G.; Masciocchi, N.; Maspero, A.; Palmisano, G.; Penoni, A. A novel porphyrazine ligand tailored to homogeneous metal catalyzed transformations. *Arkivoc* **2014**, *2014*, 72–85. [CrossRef]
47. Murray, C.; Dozova, N.; McCaffrey, J.G.; FitzGerald, S.; Shafizadeh, N.; Crepin, C. Infra-red and Raman spectroscopy of free-base and zinc phthalocyanines isolated in matrices. *Phys. Chem. Chem. Phys.* **2010**, *12*, 10406–10422. [CrossRef]
48. Ferrari, A.C.; Basko, D.M. Raman spectroscopy as a versatile tool for studying the properties of graphene. *Nat. Nanotechnol.* **2013**, *8*, 235–246. [CrossRef]
49. Liu, Z.; Zhang, X.; Zhang, Y.; Jiang, J. The molecular, electronic structures and IR and Raman spectra of metal-free, N, N′-dideuterio, and magnesium tetra-2,3-pyrazino-porphyrazines: Density functional calculations. *Vib. Spectrosc.* **2017**, *43*, 447–459. [CrossRef]
50. Zhabanov, Y.A.; Eroshin, A.V.; Ryzhov, I.V.; Kuzmin, I.A.; Finogenov, D.N.; Stuzhin, P.A. Molecular structure, thermodynamic and spectral characteristics of metal-free and nickel complex of tetrakis (1, 2, 5-thiadiazolo) porphyrazine. *Molecules* **2021**, *26*, 2945. [CrossRef]
51. Li, D.; Peng, Z.; Deng, L.; Shen, Y.; Zhou, Y. Theoretical studies on molecular structure and vibrational spectra of copper phthalocyanine. *Vib. Spectrosc.* **2005**, *39*, 191–199. [CrossRef]

52. Gorski, A.; Gawinkowski, S.; Starukhin, A.; Gladkov, L.; Chizhova, N.; Mamardashvili, N.; Scheblykin, I.; Waluk, J. Resonance Raman and FTIR spectra of Mg-porphyrazines. *J. Mol. Struct.* **2014**, *1058*, 197–204. [CrossRef]
53. Angeloni, S.; Bauer, E.M.; Ercolani, C.; Popkova, I.A.; Stuzhin, P.A. Tetrakis (selenodiazole) porphyrazines. 2: Metal complexes with Mn (II), Co (II), Ni (II), and Zn (II). *J. Porphyr. Phthalocyanines* **2001**, *5*, 881–888. [CrossRef]
54. Freddi, S.; Perilli, D.; Vaghi, L.; Monti, M.; Papagni, A.; Di Valentin, C.; Sangaletti, L. Pushing down the limit of NH_3 detection of graphene-based chemiresistive sensors through functionalization by thermally activated tetrazoles dimerization. *ACS Nano* **2022**, *16*, 10456–10469. [CrossRef]
55. Liang, T.; Liu, R.; Lei, C.; Wang, K.; Li, Z.; Li, Y. Preparation and test of NH_3 gas sensor based on single-layer graphene film. *Micromachines* **2020**, *11*, 965. [CrossRef] [PubMed]
56. Li, Q.; Xu, M.; Jiang, C.; Song, S.; Li, T.; Sun, M.; Chen, W.; Peng, H. Highly sensitive graphene-based ammonia sensor enhanced by electrophoretic deposition of MXene. *Carbon* **2023**, *202*, 561–570. [CrossRef]
57. Xiang, C.; Jiang, D.; Zou, Y.; Chu, H.; Qiu, S.; Zhang, H.; Xu, F.; Sun, L.; Zheng, L. Ammonia sensor based on polypyrrole–graphene nanocomposite decorated with titania nanoparticles. *Ceram. Int.* **2015**, *41*, 6432–6438. [CrossRef]
58. Singh, S.; Deb, J.; Kumar, S.; Sarkar, U.; Sharma, S. Selective N, N-dimethylformamide vapor sensing using $MoSe_2$/multiwalled carbon nanotube composites at room temperature. *ACS Appl. Nano Mater.* **2022**, *5*, 3913–3924. [CrossRef]
59. Schedin, F.; Geim, A.K.; Morozov, S.V.; Hill, E.W.; Blake, P.; Katsnelson, M.I.; Novoselov, K.S. Detection of individual gas molecules adsorbed on graphene. *Nat. Mater.* **2007**, *6*, 652–655. [CrossRef]
60. Perilli, D.; Freddi, S.; Zanotti, M.; Drera, G.; Casotto, A.; Pagliara, S.; Schio, L.; Sangaletti, L.; Di Valentin, C. Design of highly responsive chemiresistor-based sensors by interfacing NiPc with graphene. *Commun. Mater.* **2024**, *5*, 254. [CrossRef]
61. Polyakov, M.S.; Basova, T.V.; Göksel, M.; Şenocak, A.; Demirbaş, E.; Durmuş, M.; Kadem, B.; Hassan, A. Effect of covalent and non-covalent linking of zinc (II) phthalocyanine functionalised carbon nanomaterials on the sensor response to ammonia. *Synth. Met.* **2017**, *227*, 78–86. [CrossRef]
62. Zhang, Y.H.; Han, L.F.; Xiao, Y.H.; Jia, D.Z.; Guo, Z.H.; Li, F. Understanding dopant and defect effect on H_2S sensing performances of graphene: A first-principles study. *Comput. Mater. Sci.* **2013**, *69*, 222–228. [CrossRef]
63. Reshak, A.H.; Auluck, S. Adsorbing H_2S onto a single graphene sheet: A possible gas sensor. *J. Appl. Phys.* **2014**, *116*, 103702. [CrossRef]
64. Khodadadi, Z. Evaluation of H_2S sensing characteristics of metals–doped graphene and metals-decorated graphene: Insights from DFT study. *Phys. E Low-Dimens. Syst. Nanostruct.* **2018**, *99*, 261–268. [CrossRef]
65. Peng, X.; Liu, D.; Zhao, F.; Tang, C. Gas sensing properties of Mg-doped graphene for H_2S, SO_2, SOF_2, and SO_2F_2 based on DFT. *Int. J. Quantum Chem.* **2022**, *122*, e26989. [CrossRef]
66. Tsymbalenko, O.; Lee, S.; Lee, Y.M.; Nam, Y.S.; Kim, B.C.; Kim, J.Y.; Lee, K.B. High-sensitivity NH_3 gas sensor using pristine graphene doped with CuO nanoparticles. *Microchim. Acta* **2023**, *190*, 134. [CrossRef] [PubMed]
67. Freddi, S. Enhancing the response to ammonia and nitrogen dioxide of bare graphene through NiPc functionalization. *Il Nuovo C.* **2023**, *100*, 46.
68. Freddi, S.; Vergari, M.; Pagliara, S.; Sangaletti, L. A Chemiresistor sensor array based on graphene nanostructures: From the detection of ammonia and possible interfering VOCs to chemometric analysis. *Sensors* **2023**, *23*, 882. [CrossRef]
69. Li, Q.; Chen, W.; Liu, W.; Sun, M.; Xu, M.; Peng, H.; Wu, H.; Song, S.; Li, T.; Tang, S. Highly sensitive graphene ammonia sensor enhanced by concentrated nitric acid treatment. *Appl. Surf. Sci.* **2022**, *586*, 152689. [CrossRef]
70. Jagannathan, M.; Dhinasekaran, D.; Rajendran, A.R.; Subramaniam, B. Selective room temperature ammonia gas sensor using nanostructured ZnO/CuO graphene on paper substrate. *Sens. Actuators B Chem.* **2022**, *350*, 130833. [CrossRef]
71. Freddi, S.; Rodriguez Gonzalez, M.C.; Carro, P.; Sangaletti, L.; De Feyter, S. Chemical defect-driven response on graphene-based chemiresistors for sub-ppm ammonia detection. *Angew. Chem. Int. Ed.* **2022**, *61*, e202200115. [CrossRef]
72. Srivastava, S.; Jain, S.K.; Gupta, G.; Senguttuvan, T.D.; Gupta, B.K. Boron-doped few-layer graphene nanosheet gas sensor for enhanced ammonia sensing at room temperature. *RSC Adv.* **2020**, *10*, 1007–1014. [CrossRef]
73. Wu, D.; Peng, Q.; Wu, S.; Wang, G.; Deng, L.; Tai, H.; Wang, L.; Yang, Y.; Dong, L.; Zhao, Y.; et al. A simple graphene NH_3 gas sensor via laser direct writing. *Sensors* **2018**, *18*, 4405. [CrossRef]
74. Lv, R.; Chen, G.; Li, Q.; McCreary, A.; Botello-Méndez, A.; Morozov, S.V.; Liang, L.; Declerck, X.; Perea-López, N.; Cullen, D.A.; et al. Ultrasensitive gas detection of large-area boron-doped graphene. *Proc. Natl. Acad. Sci. USA* **2015**, *112*, 14527–14532. [CrossRef]
75. Seekaew, Y.; Lokavee, S.; Phokharatkul, D.; Wisitsoraat, A.; Kerdcharoen, T.; Wongchoosuk, C. Low-cost and flexible printed graphene–PEDOT: PSS gas sensor for ammonia detection. *Org. Electron.* **2014**, *15*, 2971–2981. [CrossRef]

76. Ovsianytskyi, O.; Nam, Y.S.; Tsymbalenko, O.; Lan, P.T.; Moon, M.W.; Lee, K.B. Highly sensitive chemiresistive H_2S gas sensor based on graphene decorated with Ag nanoparticles and charged impurities. *Sens. Actuators B Chem.* **2018**, *257*, 278–285. [CrossRef]
77. Zhou, L.; Shen, F.; Tian, X.; Wang, D.; Zhang, T.; Chen, W. Stable Cu_2O nanocrystals grown on functionalized graphene sheets and room temperature H_2S gas sensing with ultrahigh sensitivity. *Nanoscale* **2013**, *5*, 1564–1569. [CrossRef] [PubMed]
78. Al-Hartomy, O.A.; Khasim, S.; Roy, A.; Pasha, A. Highly conductive polyaniline/graphene nano-platelet composite sensor towards detection of toluene and benzene gases. *Appl. Phys. A* **2019**, *125*, 12. [CrossRef]
79. Kumar, D.; Chaturvedi, P.; Saho, P.; Jha, P.; Chouksey, A.; Lal, M.; Rawat, J.S.B.S.; Tandon, R.P.; Chaudhury, P.K. Effect of single wall carbon nanotube networks on gas sensor response and detection limit. *Sens. Actuators B* **2017**, *240*, 1134–1140. [CrossRef]

Disclaimer/Publisher's Note: The statements, opinions and data contained in all publications are solely those of the individual author(s) and contributor(s) and not of MDPI and/or the editor(s). MDPI and/or the editor(s) disclaim responsibility for any injury to people or property resulting from any ideas, methods, instructions or products referred to in the content.

chemosensors

MDPI

Article

The Use of Low-Cost Gas Sensors for Air Quality Monitoring with Smartphone Technology: A Preliminary Study

Domenico Suriano [1,*], Francis Olawale Abulude [2] and Michele Penza [1]

[1] ENEA—Italian National Agency for New Technologies, Energy and Sustainable Economic Development—Brindisi Research Center, Strada Statale 7, Appia, Km 706, I-72100 Brindisi, Italy
[2] Environmental and Sustainable Research Group, Science and Education Development Institute, Akure 340001, Nigeria
* Correspondence: domenico.suriano@enea.it; Tel.: +39-0831201490

Abstract: In the past decades, both low-cost gas sensors for air quality monitoring and smartphone devices have experienced a remarkable spread in the worldwide market. Smartphone devices have become a unique tool in everyday life, whilst the use of low-cost gas sensors in air quality monitors has allowed for a better understanding of the personal exposure to air pollutants. The traditional technologies for measuring air pollutant concentrations, even though they provide accurate data, cannot assure the necessary spatio-temporal resolution for assessing personal exposure to the various air pollutants. In this respect, one of the most promising solutions appears to be the use of smartphones together with the low-cost miniaturized gas sensors, because it allows for the monitoring of the air quality characterizing the different environments frequented in everyday life by leveraging the capability to perform mobile measurements. In this research, a handheld air quality monitor based on low-cost gas sensors capable of connecting to smartphone devices via Bluetooth link has been designed and implemented to explore the different ways of its use for assessing the personal exposure to air pollutants. For this purpose, two experiments were carried out: the first one was indoor monitoring of CO and NO_2 concentrations performed in an apartment occupied by four individuals and the second one was mobile monitoring of CO and NO_2 performed in a car cabin. During the indoor measurements, the maximum value for the CO concentrations was equal to 12.3 ppm, whilst the maximum value for NO_2 concentrations was equal to 64 ppb. As concerns the mobile measurements, the maximum concentration of CO was equal to 8.3 ppm, whilst the maximum concentration of NO_2 was equal to 38 ppb. This preliminary study has shown that this system can be potentially used in all those situations where the use of traditional chemical analyzers for measuring gas concentrations in everyday life environments is hardly feasible, but also has highlighted some limits concerning the performance of such systems.

Keywords: gas sensors; environmental monitoring; air quality monitoring; electronic noses; air quality sensors; electrochemical sensors; smartphones; personal exposure

1. Introduction

According to the WHO [1], air pollution is one of the biggest concerns for human health seeing as over 90% of people live in regions where the safe limits are systematically exceeded. Poor air quality affects both outdoor and indoor environments, as proven by several institutions [1–4]. For this reason, air quality monitoring is of the utmost importance because it can provide a tool for governments and regulatory bodies to effectively assess the levels of pollution to which the population is exposed.

The instruments traditionally used by the dedicated institutions to measure air pollutant concentrations are bulky, expensive, and requires relevant infrastructure and resources for their operation [5–8]. As a consequence, in most cases, few monitoring fixed stations are available on a territory, and therefore, it is hardly feasible to gather air pollutant concentration maps with a good spatio-temporal resolution. Moreover, it can happen that no monitoring station is available to cover whole regions, as is the case in many emerging countries [9].

To address this issue, remarkable attention has been given recently to low-cost miniaturized gas sensors (LCSs) and monitors (LCMs) based on their functioning [10–12]. Some of them have achieved a significant level of maturity and are already available on the worldwide market (e.g., resistive sensors, electrochemical cells, non-dispersive infrared radiation absorption sensors, and photo-ionization sensors) [13–15], whilst, in other cases, their development can be considered still in a preliminary study phase [16,17]. In both cases, such devices are considered an appealing option due to their high grade of miniaturization level, affordability, low power consumption, and ease of use [18,19].

Despite all these positive aspects, the use of LCSs and LCMs in air quality monitoring is still a matter of investigation for research groups due to some limitations affecting the technology that make them less accurate when compared to the traditional instruments [20–22]. The performance of LCSs, and consequently, of LCMs, is influenced by both meteorological and environmental conditions [23–25], and also by the type of technology implementing them. In particular, the presence of interfering gases [26–28], the temperature, and ambient humidity changes can represent a disturbing factor which affects the LCS measurements to different degrees [29–31], depending not only on the type of sensor, but also considering different copies of the same sensor manufactured in the same batch. To mitigate the adverse influence of these variables, the scientific community has explored various countermeasures, mostly consisting of the elaboration of sensor output signals performed by suitable software through which various algorithms are implemented, aiming at correcting the LCS or LCM measurements [32–34].

Even though it is commonly recognized that the performance of LCMs cannot still be compared with the traditional and professional instruments usually used by governmental agencies [35], it appears clear that in some circumstances they are the only viable way to carry out air pollutant concentration measurements in real time. As a matter of fact, households or common citizens cannot afford bulky and expensive instruments to monitor their homes or apartments, or even to assess their personal exposure to air pollutants during their everyday life in vehicle cabins, public buildings, or on public transport. If we recognize that air pollutant concentrations can significantly vary in urban microenvironments or streets very close each other [36], or that indoor air quality can be remarkably worse than the outdoor one [37] in some cases, it also clearly appears that personal exposure assessment to air pollutants cannot be accomplished by sparse, fixed monitoring stations. Therefore, the use of LCMs appears to be the only way to obtain indications about personal exposure to air contaminants, especially if we consider that the most effective way to accomplish this task is to perform mobile measurements.

To carry out such measurements, it is preferable that the tool or device to use has specific features, such as handheld sizes, battery-power operated, capability to operate both in indoor and in outdoor environments, and capability to track or store the measures. LCMs already available on the market do not always present all these characteristics; moreover, most of times, the manufacturers of such devices do not provide information about their accuracy or data about their calibration, therefore, the reliability of their measurements can be questionable.

This preliminary study is aimed at exploring the capabilities of a handheld monitor based on electrochemical gas sensors and designed to operate together with a smartphone device. Using a smartphone in conjunction with handheld monitors provides more flexibility for air quality monitoring in all the situations in which research-grade and accurate instruments, such as gas chemical analyzers, cannot be deployed or are hard to arrange and use due to their bulkiness. For example, more handheld monitors can be used to set up an indoor air quality monitoring network composed of several nodes, each of them represented by a handheld monitor deployed in a room, whilst a smartphone can act as a hub which collects the measurements of the various nodes through Bluetooth connections. Combining the use of such monitors with smartphones enables a very easy and quick way to set up local monitoring networks for the air quality monitoring of indoor environments composed of multiple rooms or spaces without using third services, such as, cloud services or wi-fi local networks. Considering that almost all citizens own a smartphone, this feature can provide immediate access to the air quality data of public spaces, such as, airports, hospitals, or public offices, for any user. The same structure can also be used for performing mobile measurements, such as the air quality monitoring in a car cabin during a trip.

In this preliminary study, two experiments were performed to test the capabilities of a handheld monitor, called "DSnasus", designed and implemented in our laboratories to be used with a smartphone. In the first experiment, two monitoring units were deployed in two rooms of an apartment to form a little monitoring network with the smartphone, as earlier briefly explained, to measure carbon monoxide and nitrogen dioxide concentrations. In the second experiment, one of the two monitors was used in a car cabin during a trip to perform mobile measurements of carbon monoxide and nitrogen dioxide concentrations.

2. Materials and Methods

In this section, we are going to describe the structure and functions of the devices under evaluation, the methods used to calibrate the sensors, and the setting up of the two experiments.

2.1. The Devices and the Sensors Used for the Experiments

Two units of the designed handheld monitor were implemented for the experiments. Moreover, a smartphone, featuring an Android operating system, was used to complete the necessary hardware (see Figure 1).

Figure 1. On the left of the figure we can find two monitors, DSnasus, implemented in our laboratory, while on the right there is the smartphone device used for the experiments showing the App running. The sizes of the DSnasus monitors are 17 cm × 8.5 cm.

On each of the two monitoring units, two electrochemical gas sensor types were mounted. The first one was the CO-B4 by Alphasense, dedicated to CO concentration measurements, and the second one was the NO2-B43F, dedicated to NO_2 concentration measurements. These sensors are designed for measuring ambient concentrations of gas, as stated by the manufacturer. In Table 1, a summary of their technical specifications provided by Alphasense is reported to give an idea of their capabilities.

Table 1. Main technical specifications of CO-B4 and NO2-B43F sensors.

Sensor	Sensitivity	Response Time	Zero Current	Range
CO-B4	+420/+650 nA/ppm	<30 s	+30/−250 nA	1000 ppm
NO2-B43F	−200/−650 nA/ppm	<80 s	−80/+80 nA	20 ppm

Both sensors are enclosed in identical cases and are the same size. They are amperometric sensors featuring four electrodes, as can be seen in Figure 2. Their working principle is exposed in the Baron and Saffell article [38]. Basically, they have four electrodes; the counter and the reference ones are necessary for the correct operation of the sensors, whilst the auxiliary and the working electrodes provide the sensor outputs, represented by weak electric currents. The counter, reference, and working electrodes are separated by the wetting filters. On the working electrode the reduction or oxidation of the gas molecules takes place (depending on the gas species). It is coated with a catalyst to facilitate the reaction with them. This reaction generates electronic charges which are balanced by a second reaction at the counter electrode, forming with the first one a redox pair of chemical reactions. When the sensor is exposed to clean air, the counter electrode is at the same potential as the working one, while in the presence of the target gas their potential changes. In this way, a potential difference between the working and counter electrodes is generated, and consequently, an electric current at the working electrode starts to flow, providing the output of the sensor. The auxiliary electrode is not exposed to the target gas, but it reflects the fluctuations of the sensor baseline. The electrochemical sensors are sensitive to the temperature and humidity changes, which affect both the sensor sensitivity and their zero current. Another challenge posed by the use of electrochemical sensors is the so-called cross-sensitivity effect. As a matter of fact, sensors designed for measuring a specific gas, for example, nitrogen dioxide, can be influenced by the presence of other gases (named interfering gases), which modify the correct measurement depending on their concentrations.

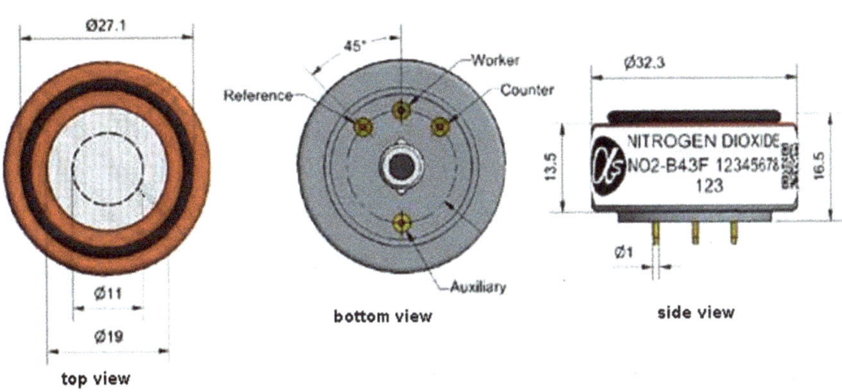

Figure 2. A representation of the NO2-B43F sensor. The CO-B4 sensor has same size and features.

Despite the aforementioned adverse aspects, electrochemical sensors present some advantages if compared with other sensor types commonly used for pollutant gas monitoring, such as the chemiresistive sensors. They feature a lower power consumption and also a lower cross-sensitivity [39], and for these reasons they were selected to implement this monitor. As stated by the manufacturer, both the sensors can measure very low gas concentrations (less than 100 ppb) [40,41] if supported by an adequate electronic board. To achieve this goal, the electronic circuitry must be able to adequately amplify the weak electric currents representing the sensor output and limit the electronic noise affecting these signals at the same time. We designed and implemented an on-purpose electronic board for the sensor use, and its characteristics are exposed in detail in a previous article [42].

The sensors used in this experiment are equipped with a membrane which helps to filter out some interfering gases, limiting the cross-sensitivity effect for some gas types. The NO2-B43F sensor has a membrane made of MnO_2/PTFE microparticles which helps to filter out the main sensor interfering gas: ozone [28,40,43]. As concerns the CO-B4 sensor, the manufacturer reports in its datasheet that it is equipped with a filter for limiting the effects of one of the main interfering gases: hydrogen sulfide [41]. Other interfering gases are hydrogen for the CO-B4 and chlorine for the NO2-B4 [40,41]. In this respect, the effects of such gases are expected to be reasonably negligible due to their expected very low concentrations in the environments explored in this study.

As concerns the design of the handheld monitor, it is based on three modules implemented by three printed circuit boards: the sensor board [42], the data elaboration/interface board, and the Bluetooth module. The first two modules were designed in our laboratory, while the Bluetooth module is the RN42XV by Microchip. The electronics featuring the sensor board are able to read the weak electric currents representing the outputs of the sensors, opportunely amplify them, and convert them into voltage signals [42]. These constitute the inputs for the data elaboration/interface board, which digitalizes them. The voltage signals reflect the gas concentrations affecting the sensors; their digitalization and the subsequent elaboration is performed by the microprocessor PIC18F4652 by Microchip, which is the core of the module. The firmware running on the microprocessor is in charge of performing both the data elaboration function and the interface function. The data elaboration consists of clearing the input signals by removing or limiting the electronic noise affecting them through a numerical filter based on the rolling average smoothing technique. Subsequently, the mathematical law for converting the sensor signals into values expressing the gas concentrations is applied. The measures of the gas concentrations composing the outputs of the elaboration so far described can be transmitted via Bluetooth to a smartphone, acting as the master device. Optionally, they can also be transmitted via the USB port available on this module. In both cases, the connection between the handheld monitor and a smartphone device, or a PC connected via USB, follows the "master-slave" scheme; in this respect, at the sampling rate selected by the user, the smartphone requests the monitor for the current gas concentrations, which will transmit them using the Bluetooth or USB channel. The data elaboration/interface board also has the necessary electronic circuitry for managing powering the monitor. It is typically powered by its internal ion-lithium battery, which is recharged using the USB socket. A standard smartphone battery charger plugged into the USB socket can be used to recharge its battery, or optionally, the battery recharge will be active whenever the monitor is connected to a PC USB port.

As mentioned earlier, the connection of the DSnasus monitor with the smartphone or PC follows the master/slave architecture. Therefore, the specific App running on the smartphone (which is the device acting as master) sends to the monitor (which is the device acting as slave) the specific command at the sampling rate selected by the user to retrieve the current gas measures. These measures are shown to the user and written in a file stored on the local memory of the smartphone to form the dataset. Obviously, the file containing the measure records can be subsequently downloaded from the smartphone for further analysis. As concerns the selection of the sampling rate, it can be carried out by the smartphone user through the App earlier mentioned, designed also to perform all the necessary tasks, comprising the searching of DSnasus monitors operating in the vicinity and the Bluetooth connection management. This App, designed in our laboratory, was developed for Android operating systems. The main advantage provided by the use of smartphone devices with handheld monitors connected via Bluetooth consists of the possibility of easily setting up multiple star-shaped local networks, as depicted in Figure 3, without the need for a third support, such as, cloud services or similar. The maximum number of connections, and therefore of DSnasus monitors, that the system can manage is 255. The use of local networks composed of several monitors can be helpful for the air quality monitoring of indoor environments, such apartments or public offices, formed of several rooms or spaces which can be separated by doors or structures that create micro-environments not necessarily connected to each other. For this preliminary study, we prepared just two monitors to form a simple network for performing a preliminary test of the idea at the base of the system, as described in this document.

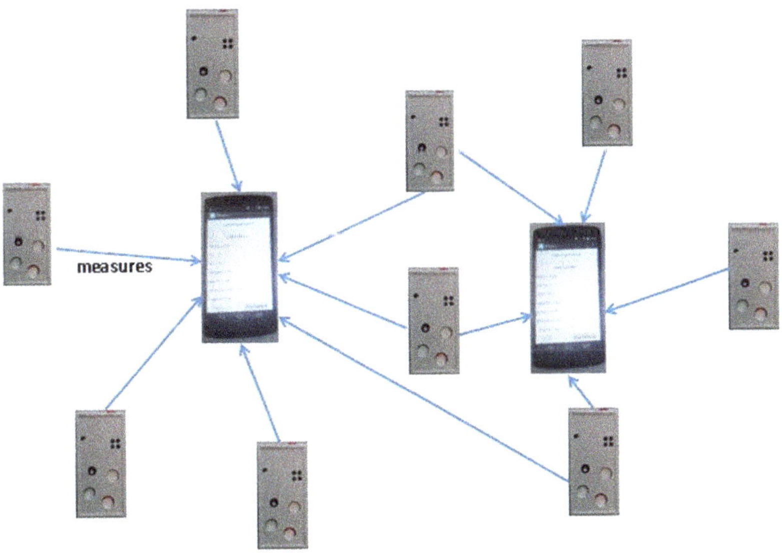

Figure 3. A system composed of smartphone devices and more DSnasus's can be used to form multiple star-shaped networks based on Bluetooth connections, where the smartphone is the center, or hub, to which the measures converge, while the DSnasus monitors are the branches providing the measures.

2.2. The Calibration of the Sensors

An important step of the experiment was the calibration of the sensors. It consisted of finding the mathematical law which allowed us to convert the voltage signals into gas concentrations. To perform this task, laboratory tests were performed by using a test chamber in which the two monitors were introduced. In this chamber, known gas concentrations were injected using a mass flow controller and the real gas concentration inside it was monitored by regulatory-grade instruments which acted as reference for the measurements.

By having the sensors a highly linear response, we decided to use the multivariate regression technique to convert the voltages into gas concentrations. Thus, the conversion law is as follows:

$$GC = aV_W + bV_A + c \quad (1)$$

where GC is the gas concentration expressed in ppm, in the case of CO measurements, or in ppb, in the case of NO_2 measurements; V_w and V_A are the voltage signal of the working and auxiliary electrode, respectively; and a, b, and c are the coefficient to find for calibrating the sensors. The method used to find the a, b, and c coefficients was the ordinary least square (OLS) algorithm, available in many software tools or libraries for scientific purposes. In this study, the software Origin 7.0 by Originlab was used for calculating the coefficients. The quality of the calibrations carried out by the laboratory tests was evaluated by using the coefficient of determination (R^2) and the mean absolute error (MAE) computed by comparing the reference data with the ones produced by the calibration law of each sensor.

2.3. The Indoor Monitoring

For the indoor monitoring, two units of the DSnasus monitor were used in an apartment occupied by four individuals. This monitoring was carried out during their everyday activities. The first monitor was placed in the kitchen of the apartment, while the second monitor was deployed in the living room (see Figures 4 and 5). In this way, it was possible to test a little local network composed of the two monitors. The two rooms were separated by a sliding door kept almost always open. The decision to deploy the monitors in this area of the apartment was taken because a previous study [28] has shown that this is the place where one is likely to measure gas emissions due to the presence of various sources, such as, methane gas burners, cigarette smoke, or candles. The monitors were fixed to the wall of the rooms, roughly two meters high. The living room had a door giving access to a balcony, same also for the kitchen. During the experiment, the two doors were partially opened for some minutes to ensure a minimum of air exchange, but avoiding significantly modifying the temperature and the humidity inside the apartment. The thermostat of the apartment heating system was kept at 20 °C to ensure steady thermal conditions as close as possible to the conditions registered during the calibration process. As concerns the humidity values, a dehumidifier placed in the living room was set to a relative humidity equal to 50% to reproduce the environmental conditions met during the calibration process. The unit placed in the kitchen had the sensors denoted as NO_2(1) and CO(1), while the unit placed in the living room had the sensors denoted as NO_2(2) and CO(2). The distance between the two units was 7.2 m.

Before deploying the units in the apartment, the firmware of the microprocessor inside the monitors was updated by inserting the conversion law with the values of the coefficients found after the calibration process. The possibility to update the conversion law or change its coefficients or parameters is an important aspect of this monitoring system. It is important to underline that monitors or systems based on low-cost gas sensors having the same feature are not always available on the worldwide market currently.

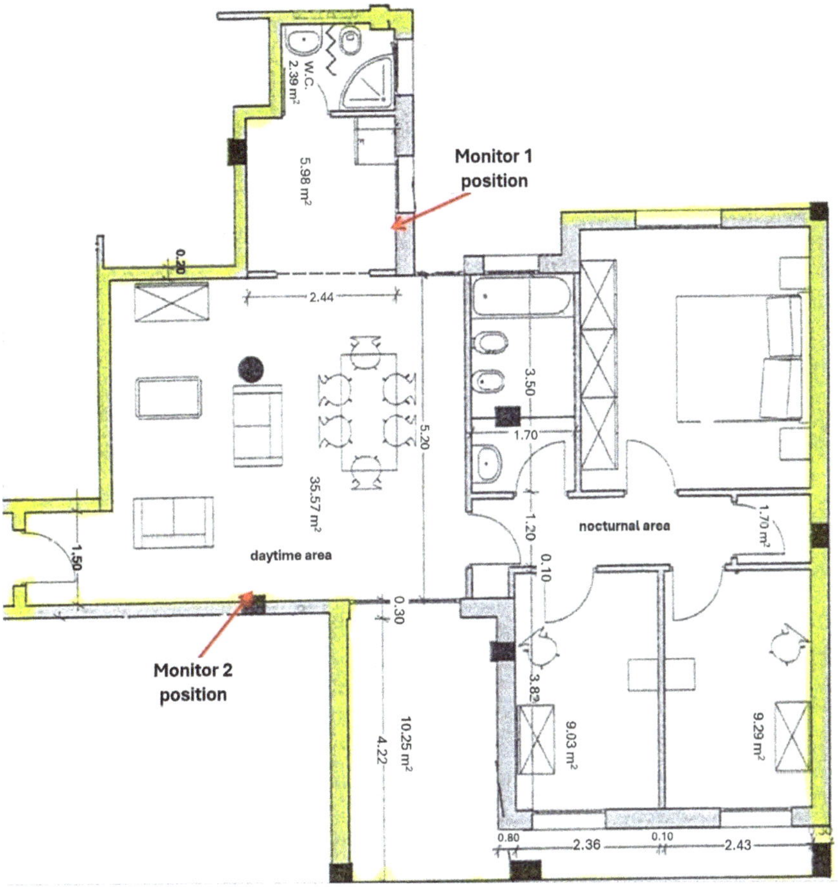

Figure 4. The apartment map, its size, and the position of the two monitors forming the monitoring network. The height of the apartment walls is 2.75 m.

(**a**) (**b**)

Figure 5. The position of the monitor placed in the kitchen (monitor 1) (**a**), and the monitor placed in the living room (monitor 2) (**b**), as reported also in Figure 4.

2.4. The Mobile Monitoring

The unit used for monitoring the concentration of CO and NO_2 in the car cabin was placed on the car dashboard using an ordinary smartphone holder available in any store selling car accessories (see Figure 6). The air conditioning system of the car was set to 20 °C, whilst the car windows were kept closed during the duration of the experiment and the forced ventilation was set to the minimum. These precautions were aimed at preserving the car microclimate as close as possible to the conditions characterizing the calibration operations. The route run by the car was formed of extra-urban and urban roads to ensure a certain amount of variability of environments and gas concentrations. For this preliminary study, we carried out the experiment by making a trip in a diesel fueled car on which the particulate filter was absent.

Figure 6. The monitor used for mobile measurements arranged on the car dashboard by using a standard smartphone holder.

3. Results

The data concerning calibration of the sensors, along with the data related to the indoor and mobile measurements, are provided in the Supplementary Materials section. Hereinafter, the results related to the calibration, indoor, and mobile measurements are exposed in three distinct subsections for an easier reading.

3.1. The Results of the Sensor Calibration

The calibration of the sensors was performed using a dataset formed of records containing both the sensor voltage output signals and the measures of the reference instrumentations. The sampling rate was set to one minute, whilst the concentration of CO and NO_2 injected into the test chamber were selected to cover the typical range measurable in the indoor or outdoor environments (see Figures 7–10). The quality of the calibration can be evaluated by inspecting Table 2, where the coefficient of determination and the MAE are exposed for each sensor. In Table 3, we can see the values of the coefficients of the conversion law found for each sensor. The figures here are exposed and the values of R^2 greater than 0.9 found for each sensor show a good quality of the calibration. As a matter of fact, values of R^2 very near to 1 denote a good capacity of the monitors to follow the variations in the real gas concentrations. The good quality of the calibration, and therefore of the conversion law, is confirmed by the MAE value of each sensor as they are in the range of 0.001 ppm to 0.0009 ppm for the CO sensors, and 0.053 ppb to 1.415 ppb for the NO_2 sensors. As concerns the temperature and the relative humidity values, we observed very limited variations during the calibration of each sensor, as expected, ranging from 18.6 °C to 21.2 °C for the temperature and from 39.5% to 53.7% for the relative humidity. As can be found on the sensor manufacturer website, the variations registered for both these environmental parameters do not significantly affect the response of the sensors due to the fact that their range is limited enough.

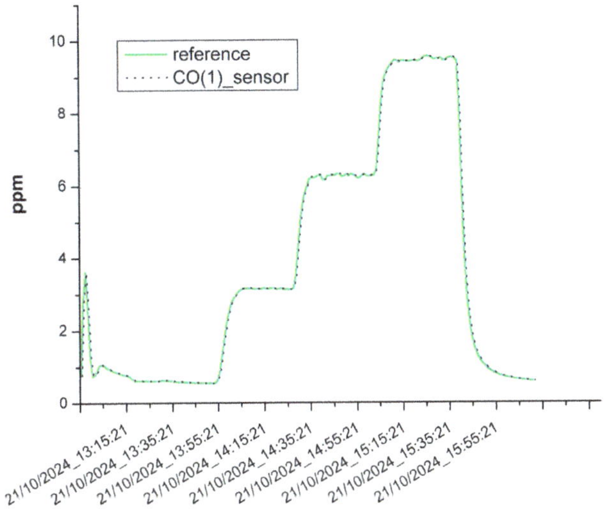

Figure 7. The calibration of the first CO-B4 sensor, herein denoted as CO(1), and the CO concentrations read by the reference instruments.

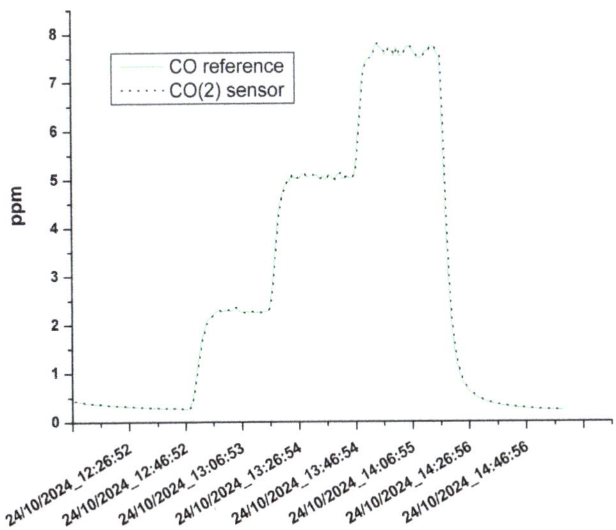

Figure 8. The calibration of the second CO-B4 sensor, herein denoted as CO(2), and the CO concentrations read by the reference instruments.

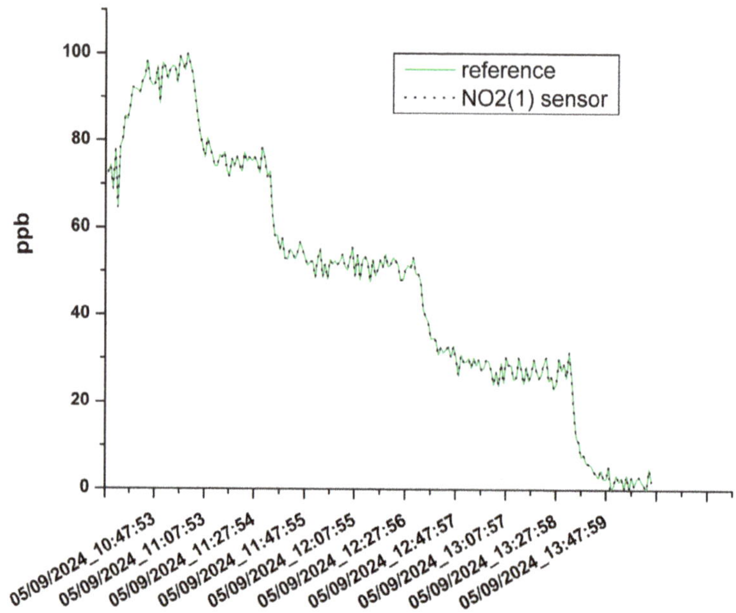

Figure 9. The calibration of the first NO2-B43F sensor, herein denoted as $NO_2(1)$, and the NO_2 concentrations read by the reference instruments.

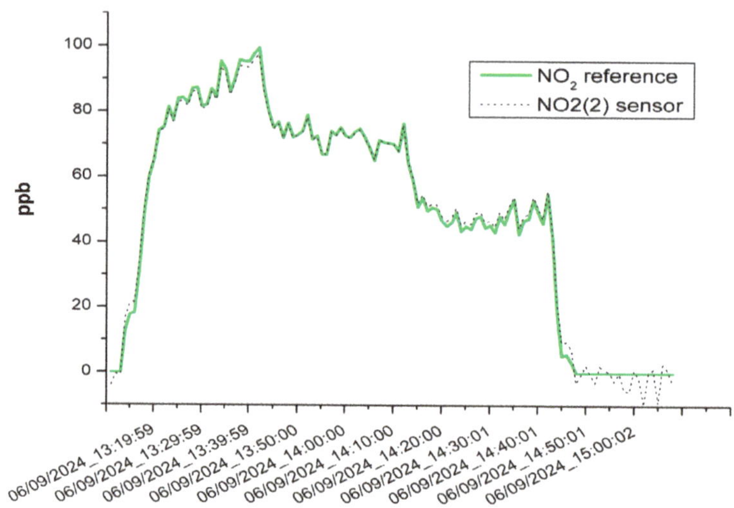

Figure 10. The calibration of the second NO2-B43F sensor, herein denoted as $NO_2(2)$, and the NO_2 concentrations read by the reference instruments.

Table 2. Coefficient of determination (R^2) and MAE found for each sensor after calibration.

Sensor	R^2	MAE
CO(1)	0.999	0.001 ppm
CO(2)	0.999	0.0009 ppm
$NO_2(1)$	0.999	0.053 ppb
$NO_2(2)$	0.995	1.415 ppb

Table 3. Coefficients of the conversion law found for each sensor.

Sensor	a	b	c
CO(1)	−3150 ppm/V	3.127 ppm/V	6.243 ppm
CO(2)	−2.762 ppm/V	2.705 ppm/V	5.239 ppm
NO_2(1)	104.933 ppb/V	−103.909 ppb/V	93.010 ppb
NO_2(2)	98.294 ppb/V	−97.607 ppb/V	91.989 ppb

3.2. The Results of the Indoor Monitoring

The indoor experiment started on the 23rd of November 2024 and finished on the 25th of November 2024. The sampling rate of both the monitors was set to one minute, producing a dataset composed of 2930 records. The temperature and the relative humidity did not vary significantly during the experiment, ranging from 16.7 °C to 21.2 °C, and from 38.8% to 54.2% for the humidity. Moreover, the mean and median values for the temperature were, respectively, equal to 19.5 °C and 19.9 °C, while the mean and median values for the relative humidity were, respectively, equal to 45.2% and 45.8%. The values of both these environmental variables are very close to the ones experienced by the sensors during the calibration process, thus, we can consider that the environmental conditions in the test chamber and in the apartment are the same. The time series of the gas measurements are shown in Figures 11 and 12.

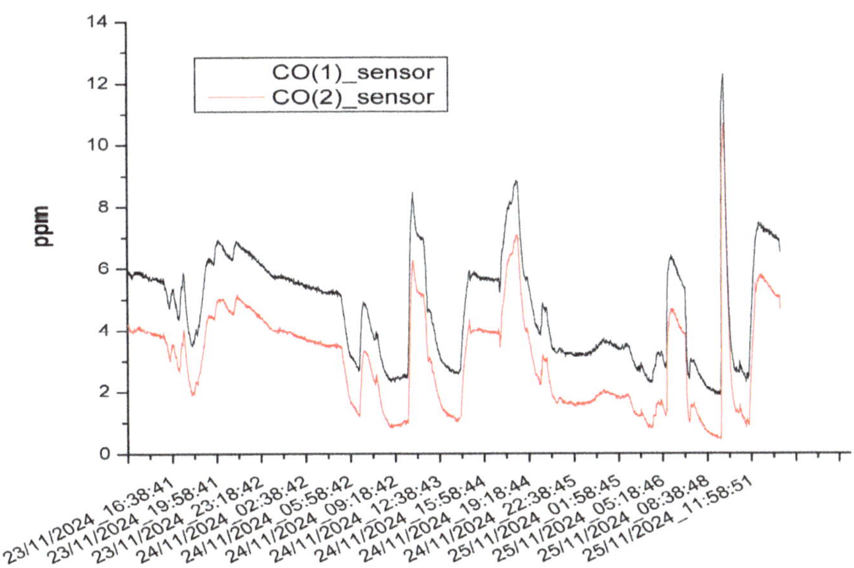

Figure 11. The CO concentrations measured by the CO(1) sensor placed in the apartment kitchen and the ones measured by the CO(2) sensor placed in the living room.

The maximum concentration of CO was equal to 12.3 ppm, while the maximum concentration of NO_2 was equal to 64 ppb. Both these values were registered by the monitor placed in the kitchen, which are named CO(1) and NO_2(1).

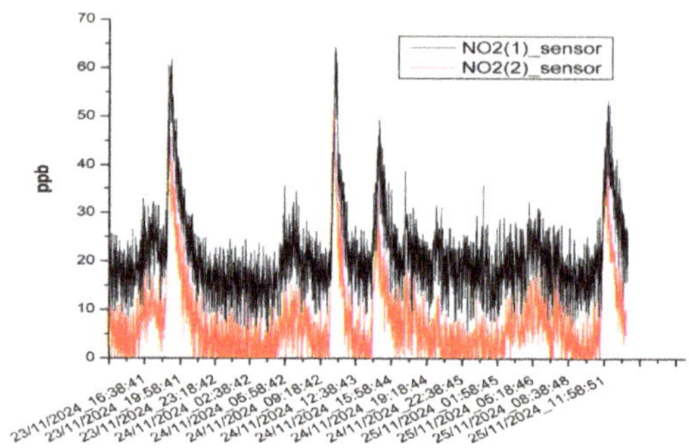

Figure 12. The NO_2 concentrations measured by the NO_2(1) sensor placed in the apartment kitchen and the ones measured by the NO_2(2) sensor placed in the living room.

3.3. The Results of the Mobile Measurements

The mobile measurements were carried out by making a trip in a diesel fueled car during rush hour on urban and extra-urban roads. In this test, the monitor placed in the kitchen during the indoor experiment was used. The mobile measurements were performed on the 18 November 2024 and, by being mobile monitoring, the sampling rate of the monitor was set to 4 s, which is the fastest sampling rate possible for this device. Therefore, the resulting dataset was composed of 5486 measurements. The maximum value for the CO concentrations was equal to 8.3 ppm, while the maximum value for NO_2 concentrations was equal to 38 ppb. As regards to the temperature and relative humidity ranges, they were, respectively, from 11.5 °C to 21 °C and from 22.3% to 59.8%. The ranges of temperature and humidity are wider than the ones found during the apartment test, but if we consider the mean and median values for the temperature (respectively, equal to 20.1 °C and 20.2 °C), it can be seen that they are comparable with the values registered during the calibration phase. As concerns the humidity, we registered a continuous decrease in its values, which resulted in a mean value equal to 39% and a median value equal to 38.7%. The time series of these measurements are shown in Figures 13 and 14, while Figure 15 highlights a moment in which we registered a spike in the NO_2 and CO concentrations due to a prolonged stop while we were waiting in a queue caused by heavy traffic. The trends of temperature and relative humidity measured in the car cabin are exposed, respectively, in Figures 16 and 17.

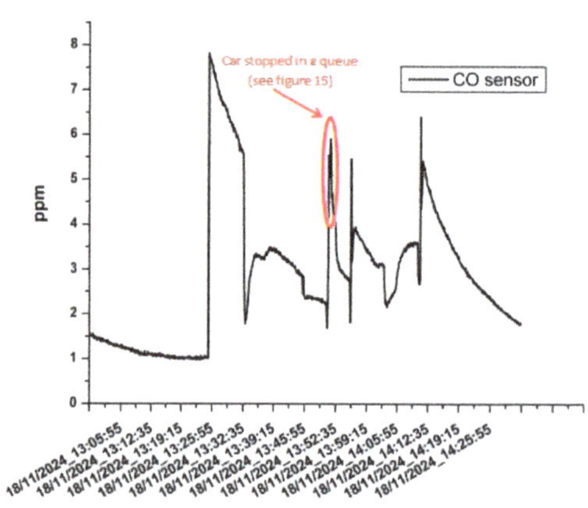

Figure 13. The time series of the CO concentrations measured in the mobile measurement test. In this figure is shown the spike registered during a stop in a queue caused by heavy traffic caught in Figure 15.

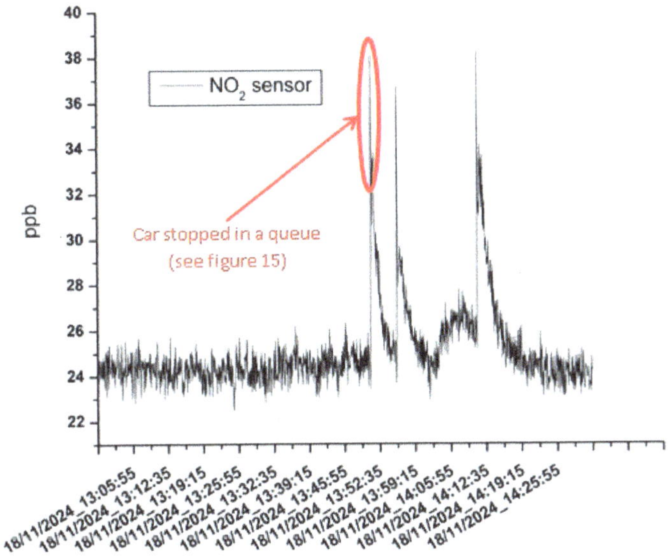

Figure 14. The time series of the NO_2 concentrations measured in the mobile measurement test. In this figure is shown the spike registered during a stop in a queue caused by heavy traffic caught in Figure 15.

Figure 15. This figure shows the car stopping in a queue caused by heavy traffic, in which a remarkable peak of CO and NO_2 concentrations was registered by the monitor arranged on the car dashboard.

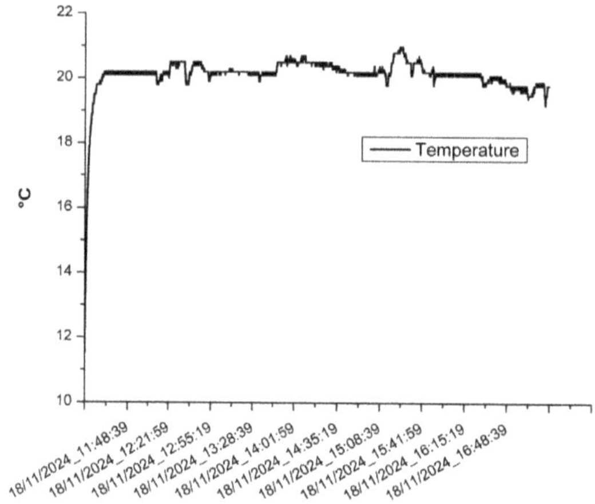

Figure 16. The time series of the temperature measured in the car cabin.

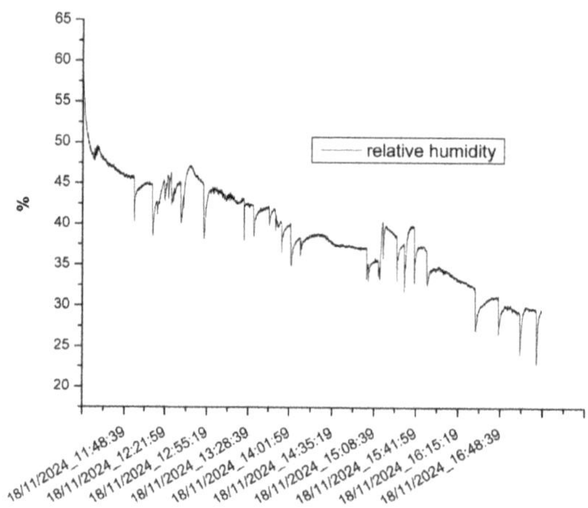

Figure 17. The time series of the relative humidity measured in the car cabin.

4. Discussion

The calibration of the sensors for air quality monitoring is an important step for a good quality of data to be produced by the monitors. The scientific community has explored different calibration techniques that can be summarized in two big general categories: laboratory calibrations and on-field calibrations [5,8]. Laboratory calibrations are carried out in a protected environment, such as a laboratory, where the device to calibrate is placed into a test chamber in which all the variables affecting the sensor response are under control and also the interfering gases can be excluded. On the contrary, on-field calibrations are performed in the place where the sensor will be deployed in co-location with reference instrumentations [6]. In this second case, temperature, relative humidity, and interfering gases cannot be controlled, but their effects are taken into account in the conversion law which is to be applied to the sensor under calibration. Both of the two approaches have advantages and disadvantages, but in general, calibration in a laboratory provides better performance indicator values in terms of R^2 [5,8,12].

In the review proposed by Kang et al. [12], it has been found that, in the case of CO and NO_2 sensors, laboratory evaluations produced R^2 values ranging between 0.99 and 1.

The values found during the calibration of the sensors used for this work, ranging from 0.995 to 0.999, are in line with the study carried out by Kang.

Other researchers have performed laboratory calibration of electrochemical sensors for their work. Zimmerman et al. [27] used nine CO-B4 sensors and fourteen NO2-B43F sensors to implement several devices to monitor CO and NO_2. For the laboratory calibrations they used the multivariate linear regression technique to find the conversion law. After the calibration process, they found an average R^2 equal to 0.98 for both of the sensor types, and an average value for MAE equal to 132 ppb in the case of the CO-B4 sensor and 35 ppb in the case of the NO2-B43F sensor. In this study, similar values were found for both of the performance indicators.

Castell et al. [8] evaluated in their study 24 units of the same monitor model (the v3.5 type) produced by AQmesh for measuring CO, NO_2, NO, and O_3. These monitors were based on the CO-B4 sensor for CO measurements and on the NO2B42 sensor for NO_2 measurements, a model very similar to the NO2-B43F considered in our study. Two of the twenty-four units were calibrated in the laboratory and the linear regression technique was used for this purpose. After the calibration process, they found R^2 values equal to 0.99 for both sensor models, which is comparable with the values found after the calibration performed for our study.

Castell et al., along with many other researcher groups, highlighted that meteorological conditions influence the response of electrochemical sensors. More specifically, the relative humidity, and more heavily the temperature, can play a significant role in determining the final sensor outputs. For this reason, it is important to take into account their variations in the conversion law, or alternatively, to deploy them in environments where temperature and humidity have limited variations.

In our preliminary study we did not include these variables in the conversion law, but we designed the tests by reproducing the laboratory conditions. During the laboratory calibrations the temperature ranged from 18.6 °C to 21.2 °C while the relative humidity ranged from 39.5% to 53.7%. By examining the documents and the datasheets published by the manufacturer of the sensors, these ranges do not have a remarkable influence on the sensor response due to the fact that they are limited enough. As desired, these values were comparable with the ones registered during the indoor experiment (16.7 °C to 21.2 °C for the temperature and 38.8% to 54.2% for the humidity).

As concerns the mobile measurements performed in the car cabin, the ranges of temperature and humidity are both wider (from 11.5 °C to 21 °C for the temperature and from 22.3% to 59.8% for the humidity) than the ones measured during the laboratory calibration. The reason for this result can be explained by considering that, before starting the experiment, the car used for this test was parked in an underground carpark, where environmental conditions were different from the ones planned for the experiment. The measurements started when the car was in the carpark, and the car conditioning system was set to 20 °C as soon as the car was started. As you can see in Figure 16, the temperature values very quickly reached their planned level, and subsequently they did not move out of the range measured during the calibration for the whole duration of the experiment. All these elements explain the trend of the temperature exposed in Figure 16. By considering that the temperature was out of the range measured during the calibration for a very short time, we can judge the measures carried out through this test to be reliable. Similar considerations can be drawn in the case of the relative humidity, even though its trend showed that the calibration range was reached more gradually (see Figure 17).

In the indoor experiment, the concentrations of CO and NO_2 gases were measured during the everyday activities which typically take place in an apartment: cooking, cleaning, or smoking. The results highlight that the peaks of gas emissions (see Figures 11 and 12) were registered mainly during smoking and cleaning. While it was expected that cigarette smoke can emit CO or NO_2 gases, to a certain extent, it was unexpected that peaks of concentrations of CO or NO_2 can be measured also when some household cleaning products are used for cleaning activities. It is very likely, or better we can surely consider, that those products did not emit CO or NO_2, but rather some interfering gases or VOCs (volatile organic compounds), even though we could not identify which type of them could be. Another important element of this experiment is given by the similar trends registered by both the monitors, in addition to the fact that the monitor placed in the kitchen measured concentrations higher than the one placed in the living room. This is explained by the position of pollutant sources, which were all located in the kitchen, and also by considering that the sliding door between the living room and the kitchen was always kept open. The maximum value of NO_2 concentration was 64ppb, which is way lower than the WHO limit set to 106 ppb (1 h average) in the 2023 air quality guidelines [44]. As regarding the CO concentrations, the same considerations can be made as the maximum concentration measured in the apartment was equal to 12.3 ppm, which is lower than the WHO limit equal to 26 ppm.

As concerns the mobile measurements carried out in the car cabin, we noted that the peaks of CO and NO_2 occurred when the car was stopped in traffic very close to other vehicles, forming the queue as shown in Figure 15. The rapid increase in gas concentrations and their slower decrease shown in Figures 13 and 14 are explained by the car ventilation being set to the minimum in conjunction with the car windows being kept closed. It was necessary to keep them closed to avoid excessive variations in temperature and humidity in the car cabin. During this experiment the WHO limits were not exceeded as the maximum concentration for NO_2 was equal to 38 ppb, whilst we recorded 8.3 ppm for the CO.

5. Conclusions

A new device called DSNasus for monitoring personal exposure to air pollutant gases has been designed and implemented. For its operation, a smartphone device and electrochemical gas sensors by Alphasense are needed. In this preliminary work, its functionalities have been tested by performing two experiments. The first one was aimed at forming a little indoor network composed of two monitors measuring CO and NO_2 concentrations and a smartphone acting as a hub for the measure records. In the second experiment, mobile measurements were performed in a car cabin during a trip through urban and extra-urban roads. Considering the results of both the experiments and considering that in some environments interfering gases or volatile compounds can be still an issue, we can conclude that this device can be used as a personal air quality indicator tool rather than a tool for assessing the personal exposure to specific pollutant gases.

Supplementary Materials: The following supporting information can be downloaded at: https://www.mdpi.com/article/10.3390/chemosensors13050189/s1, Data files concerning the sensor calibration and the measures carried out during the tests.

Author Contributions: Conceptualization, D.S.; methodology, D.S.; software, D.S.; validation, D.S.; formal analysis, D.S.; investigation, D.S.; resources, M.P.; data curation, D.S.; writing—original draft preparation, D.S.; writing—review, F.O.A. and M.P.; visualization, D.S.; supervision, D.S. All authors have read and agreed to the published version of the manuscript.

Funding: This research received no external funding.

Institutional Review Board Statement: Not applicable.

Informed Consent Statement: Not applicable.

Data Availability Statement: Data are provided in the Supplementary Materials section.

Conflicts of Interest: The authors declare no conflicts of interest.

References

1. Gumy, S.; Prüss-Üstün, A. *Ambient Air Pollution: A Global Assessment of Exposure and Burden of Disease*; WHO: Geneva, Switzerland, 2016.
2. European Commission. Indoor Air Pollution: New EU Research Reveals Higher Risks than Previously Thought. 2017. Available online: https://ec.europa.eu/commission/presscorner/detail/en/IP_03_1278 (accessed on 10 December 2024).
3. WHO. How Air Pollution Is Destroying Our Health. 2018. Available online: https://www.who.int/news-room/spotlight/how-air-pollution-is-destroying-our-health (accessed on 10 December 2024).
4. EPA. The Inside Story: A Guide to Indoor Air Quality. 2017. Available online: https://www.epa.gov/indoor-air-quality-iaq/inside-story-guide-indoor-air-quality (accessed on 10 December 2024).
5. Karagulian, F.; Barbiere, M.; Kotsev, A.; Spinelle, L.; Gerboles, M.; Lagler, F.; Redon, N.; Crunaire, S.; Borowiak, A. Review of the Performance of Low-Cost Sensors for Air Quality Monitoring. *Atmosphere* **2019**, *10*, 506. [CrossRef]
6. Spinelle, L.; Gerboles, M.; Villani, M.G.; Aleixandre, M.; Bonavitacola, F. Field calibration of a cluster of low-cost commercially available sensors for air quality monitoring. Part B: NO, CO and CO_2. *Sens. Actuators B Chem.* **2017**, *238*, 706–715. [CrossRef]
7. Mead, M.I.; Popoola, O.A.M.; Stewart, G.B.; Landshoff, P.; Calleja, M.; Hayes, M.; Baldovi, J.J.; McLeod, M.W.; Hodgson, T.F.; Dicks, J.; et al. The use of electrochemical sensors for monitoring urban air quality in low-cost, high-density networks. *Atmos. Environ.* **2013**, *70*, 186–203. [CrossRef]
8. Castell, N.; Dauge, F.R.; Schneider, P.; Vogt, M.; Lerner, U.; Fishbain, B.; Broday, D.; Bartonova, A. Can commercial low-cost sensor platforms contribute to air quality monitoring and exposure estimates? *Environ. Int.* **2017**, *99*, 293–302. [CrossRef]
9. McKercher, G.R.; Salmond, J.A.; Vanos, J.K. Characteristics and applications of small, portable gaseous air pollution monitors. *Environ. Pollut.* **2016**, *223*, 102–110. [CrossRef] [PubMed]
10. Kumar, P.; Morawska, L.; Martani, C.; Biskos, G.; Neophytou, M.; Di Sabatino, S.; Bell, M.; Norford, L.; Britter, R. The rise of low-cost sensing for managing air pollution in cities. *Environ. Int.* **2015**, *75*, 199–205. [CrossRef]
11. Snyder, E.G.; Watkins, T.H.; Solomon, P.A.; Thoma, E.D.; Williams, R.W.; Hagler, G.S.; Shelow, D.; Hindin, D.A.; Kilaru, V.J.; Preuss, P.W. The Changing Paradigm of Air Pollution Monitoring. *Environ. Sci. Technol.* **2013**, *47*, 11369–11377. [CrossRef]
12. Kang, Y.; Aye, L.; Ngo, T.D.; Zhou, J. Performance evaluation of low-cost air quality sensors: A review. *Sci. Total Environ.* **2022**, *818*, 151769. [CrossRef] [PubMed]
13. Suriano, D. SentinAir system software: A flexible tool for data acquisition from heterogeneous sensors and devices. *SoftwareX* **2020**, *12*, 100589. [CrossRef]
14. Tryner, J.; Phillips, M.; Quinn, C.; Neymark, G.; Wilson, A.; Jathar, S.H.; Carter, E.; Volckens, J. Design and testing of a low-cost sensor and sampling platform for indoor air quality. *Build. Environ.* **2021**, *206*, 108398. [CrossRef]
15. Jo, J.; Jo, B.; Kim, J.; Kim, S.; Han, W. Development of an IoT-Based Indoor Air Quality Monitoring Platform. *J. Sens.* **2020**, *2020*, 8749764. [CrossRef]
16. Mandal, D.; Banerjee, S. Surface Acoustic Wave (SAW) Sensors: Physics, Materials, and Applications. *Sensors* **2022**, *22*, 820. [CrossRef]
17. Han, T.; Nag, A.; Mukhopadhyay, S.C.; Xu, Y. Carbon nanotubes and its gas-sensing applications: A review. *Sens. Actuators A Phys.* **2019**, *291*, 107–143. [CrossRef]
18. Sayahi, T.; Garff, A.; Quah, T.; Lê, K.; Becnel, T.; Powell, K.M.; Gaillardon, P.E.; Butterfield, A.E.; Kelly, K.E. Long term calibration models to estimate ozone concentrations with a metal oxide sensor. *Environ. Pollut.* **2020**, *267*, 1–10. [CrossRef] [PubMed]
19. Velasco, A.; Ferrero, R.; Gandino, F.; Montrucchio, B.; Rebaudengo, M. A mobile and low-cost system for environmental monitoring: A case study. *Sensors* **2016**, *16*, 710. [CrossRef]
20. Spinelle, L.; Gerboles, M.; Villani, M.G.; Aleixandre, M.; Bonavitacola, F. Field calibration of a cluster of low-cost available sensors for air quality monitoring. Part A: Ozone and nitrogen dioxide. *Sens. Actuators B Chem.* **2015**, *215*, 249–257. [CrossRef]
21. Mijling, B.; Jiang, Q.; De Jonge, D.; Bocconi, S. Field calibration of electrochemical NO2 sensors in a citizen science context. *Atmos. Meas. Tech.* **2018**, *11*, 1297–1312. [CrossRef]
22. Hagan, D.H.; Isaacman-VanWertz, G.; Franklin, J.P.; Wallace, L.M.; Kocar, B.D.; Heald, C.L.; Kroll, J.H. Calibration and assessment of electrochemical air quality sensors by co-location with regulatory-grade instruments. *Atmos. Meas. Tech.* **2018**, *11*, 315–328. [CrossRef]
23. Demanega, I.; Mujan, I.; Singer, B.C.; Anđelković, A.S.; Babich, F.; Licina, D. Performance assessment of low-cost environmental monitors and single sensors under variable indoor air quality and thermal conditions. *Build. Environ.* **2021**, *187*, 107415. [CrossRef]

24. Wei, P.; Sun, L.; Anand, A.; Zhang, Q.; Huixin, Z.; Deng, Z.; Wang, Y.; Ning, Z. Development and evaluation of a robust temperature sensitive algorithm for long term NO_2 gas sensor network data correction. *Atmos. Environ.* **2020**, *230*, 1–10. [CrossRef]
25. Wei, P.; Ning, Z.; Ye, S.; Sun, L.; Yang, F.; Wong, K.C.; Westerdahl, D.; Louie, P.K.K. Impact Analysis of Temperature and Humidity Conditions on Electrochemical Sensor Response in Ambient Air Quality Monitoring. *Sensors* **2018**, *18*, 59. [CrossRef] [PubMed]
26. Smith, K.R.; Edwards, P.M.; Ivatt, P.D.; Lee, J.D.; Squires, F.; Dai, C.; Peltier, R.E.; Evans, M.J.; Sun, Y.; Lewis, A.C. An improved low–power measurement of ambient NO_2 and O_3 combining electrochemical sensor clusters and machine learning. *Atmos. Meas. Tech.* **2019**, *12*, 1325–1336. [CrossRef]
27. Zimmerman, N.; Presto, A.A.; Kumar, S.P.N.; Gu, J.; Hauryliuk, A.; Robinson, E.S.; Robinson, A.L.; Subramanian, R. A machine learning calibration model using random forests to improve sensor performance for lower-cost air quality monitoring. *Atmos. Meas. Tech.* **2018**, *11*, 291–313. [CrossRef]
28. Hossain, M.; Saffell, J.; Baron, R. Differentiating NO2 and O3 at low cost air quality amperometric gas sensors. *ACS Sens.* **2016**, *1*, 1291–1294. [CrossRef]
29. Bigi, A.; Mueller, M.; Grange, S.K.; Ghermandi, G.; Hueglin, C. Performance of NO, NO2 low cost sensors and three calibration approaches within a real world application. *Atmos. Meas. Tech.* **2018**, *11*, 3717–3735. [CrossRef]
30. Suriano, D.; Penza, M. Assessment of the Performance of a Low-Cost Air Quality Monitor in an Indoor Environment through Different Calibration Models. *Atmosphere* **2022**, *13*, 567. [CrossRef]
31. Cordero, J.M.; Borge, R.; Narros, A. Using statistical methods to carry out in field calibrations of low cost air quality sensors. Sens. *Actuators B Chem.* **2018**, *267*, 245–254. [CrossRef]
32. Trizio, L.; Brattoli, M.; De Gennaro, G.; Suriano, D.; Rossi, R.; Alvisi, M.; Cassano, G.; Pfister, V.; Penza, M. Application of artificial neural networks to a gas sensor-array database for environmental monitoring. In *Sensors and Microsystems*; Springer: Boston, MA, USA, 2012; pp. 139–144.
33. Cross, E.S.; Williams, L.R.; Lewis, D.K.; Magoon, G.R.; Onasch, T.B.; Kaminsky, M.L.; Worsnop, D.R.; Jayne, J.T. Use of electrochemical sensors for measurement of air pollution: Correcting interference response and validating measurements. *Atmos. Meas. Tech.* **2017**, *10*, 3575–3588. [CrossRef]
34. Liang, Y.; Wu, C.; Jiang, S.; Li, Y.J.; Wu, D.; Li, M.; Cheng, P.; Yang, W.; Cheng, C.; Li, L.; et al. Field comparison of electrochemical gas sensor data correction algorithms for ambient air measurements. *Sens. Actuators B Chem.* **2021**, *327*, 1–13. [CrossRef]
35. Han, P.; Mei, H.; Liu, D.; Zeng, N.; Tang, X.; Wang, Y.; Pan, Y. Calibrations of Low-Cost Air Pollution Monitoring Sensors for CO, NO_2, O_3, and SO_2. *Sensors* **2021**, *21*, 256. [CrossRef]
36. Dai, Y.; Mazzeo, A.; Zhong, J.; Cai, X.; Mele, B.; Toscano, D.; Murena, F.; MacKenzie, A.R. Modelling of Deep Street Canyon Air Pollution Chemistry and Transport: A Wintertime Naples Case Study. *Atmosphere* **2023**, *14*, 1385. [CrossRef]
37. Mendoza, D.L.; Benney, T.M.; Boll, S. Long-term analysis of the relationships between indoor and outdoor fine particulate pollution: A case study using research grade sensors. *Sci. Total Environ.* **2021**, *776*, 145778. [CrossRef] [PubMed]
38. Baron, R.; Saffell, J. Amperometric gas sensors as a low cost emerging technology platform for air quality monitoring applications: A review. *ACS Sens.* **2017**, *2*, 1553–1566. [CrossRef] [PubMed]
39. Najafi, P.; Ghaemi, A. Chemiresistor gas sensors: Design, Challenges, and Strategies: A comprehensive review. *Chem. Eng. J.* **2024**, *498*, 154999. [CrossRef]
40. NO2-B43F Datasheet. Available online: https://ametekcdn.azureedge.net/mediafiles/project/oneweb/oneweb/alphasense/products/datasheets/alphasense_no2-b43f_datasheet_en_4.pdf (accessed on 27 March 2025).
41. CO2-B4 Datasheet. Available online: https://ametekcdn.azureedge.net/mediafiles/project/oneweb/oneweb/alphasense/products/datasheets/alphasense_co-b4_datasheet_en_2.pdf (accessed on 27 March 2025).
42. Suriano, D. Design and Development of an Electronic Board for Supporting the Operation of Electrochemical Gas Sensors. *Hardware* **2024**, *2*, 173–189. [CrossRef]
43. Alphasense Ozone Sensors. Available online: https://www.alphasense.com/products/view-by-target-gas/ozone-sensors-o3 (accessed on 27 March 2025).
44. WHO. World Health Organization (WHO) Air Quality Guidelines (AQGs) and Estimated Reference Levels (RLs). Available online: https://www.eea.europa.eu/publications/status-of-air-quality-in-Europe-2022/europes-air-quality-status-2022/world-health-organization-who-air (accessed on 27 March 2025).

Disclaimer/Publisher's Note: The statements, opinions and data contained in all publications are solely those of the individual author(s) and contributor(s) and not of MDPI and/or the editor(s). MDPI and/or the editor(s) disclaim responsibility for any injury to people or property resulting from any ideas, methods, instructions or products referred to in the content.

chemosensors

MDPI

Article

Enhanced Triethylamine-Sensing Characteristics of $SnS_2/LaFeO_3$ Composite

Hong Wu [1], Xiaobing Wang [2], Yuxiang Chen [2] and Xiaofeng Wang [1,*]

[1] School of General Education, Dalian University of Technology, 2 Dagong Road, Liaodongwan New District, Panjin 124221, China; whong@mail.dlut.edu.cn

[2] School of Chemical Engineering, Ocean and Life Sciences, Dalian University of Technology, 2 Dagong Road, Liaodongwan New District, Panjin 124221, China; xicewang@gmail.com (X.W.); reche@mail.dlut.edu.cn (Y.C.)

* Correspondence: wangxf@dlut.edu.cn

Abstract: Triethylamine (TEA), a volatile organic compound (VOC), has important applications in industrial production. However, TEA has an irritating odor and potential toxicity, making it necessary to develop sensitive TEA gas sensors with high efficiency. This study focused on preparing $LaFeO_3$ nanoparticles modified by SnS_2 nanosheets ($SnS_2/LaFeO_3$ composite) using a hydrothermal method together with sol–gel technique. According to the comparison results of the gas-sensing performance between pure $LaFeO_3$ and $SnS_2/LaFeO_3$ composite with varying composition ratios, 5% $SnS_2/LaFeO_3$ sensor had a sensitivity for TEA that was 3.2 times higher than pure $LaFeO_3$ sensor. The optimized sensor operates at 140 °C and demonstrates strong stability, selectivity, and long-term durability. Detailed analyses revealed that the SnS_2 nanosheets enhanced oxygen vacancy (O_V) content and carrier mobility through heterojunction formation with $LaFeO_3$. This study provides insights into improving gas-sensing performance via p-n heterostructure design and proposes a novel $LaFeO_3$-based material for TEA detection.

Keywords: $LaFeO_3$ nanoparticles; SnS_2 nanosheets; p-n heterojunction; gas sensor

1. Introduction

The release and leakage of volatile organic compounds (VOCs) have severely affected human survival and safety over the last couple of decades [1–3]. Triethylamine (TEA) is a hazardous VOC primarily serving as a preservative, catalyst, and solvent in the chemical industry [4,5]. As a toxic gas, it is colorless, is transparent, and has an irritating odor. Being exposed to a TEA environment for a long term possibly induces many health issues of headaches, nausea, and skin and respiratory irritation, and in severe cases, death [6,7]. In food safety, TEA is the primary characteristic malodorous gas produced during the spoilage of aquatic products, such as fish. Its concentration directly correlates with fish freshness, making it a critical biomarker for assessing aquatic product quality and enabling food safety monitoring [8,9]. The Occupational Safety and Health Administration has confirmed an allowable exposure concentration for triethanolamine in air not to exceed 10 ppm [10,11]. Although conventional gas detection techniques are effective in accurately identifying toxic gases, they have several limitations. Therefore, efforts need to be made to develop a gas sensor with a relatively simple operation and good performance for industrial and daily life applications.

Metal oxide semiconductor (MOS)-based nanostructured materials have emerged as innovative solutions to overcome these challenges, finding extensive applications across multiple technological domains with notable implementation in catalytic processes and chemical detection systems [12–14]. MOS gas sensors are characterized by simple operation and low synthesis cost, and they are capable of detecting low concentrations of toxic gases [15,16]. Perovskite-type ABO_3 compounds, characterized by their ternary oxide configuration comprising A- and B-site metal cations with distinct ionic radii, exhibit structurally robust crystalline frameworks that demonstrate considerable potential in gas sensor applications [17–20]. Recent advancements in $LaFeO_3$-based sensing materials have demonstrated varied performance characteristics under different synthetic conditions [21]. Wang et al. fabricated hollow $LaFeO_3$ microspheres with a yolk-shell structure that exhibited a gas response value of 25.5 to 100 ppm ethanol at 225 °C [22]. Li et al. fabricated Au-$LaFeO_3$ nanocomposites with a simple wet-chemical method. The optimally proportioned Au-$LaFeO_3$ sensor had a response value of 44 to 100 ppm ethanol at 200 °C, representing a 27-fold enhancement compared to the pristine $LaFeO_3$ sensor, with a recovery time reduced to 8 s [23]. Xiao et al. successfully synthesized uniform porous $LaFeO_3$ microspheres via a hydrothermal method. According to gas-sensing tests, the porous $LaFeO_3$ microspheres had a response value of 29 to 100 ppm acetone at 260 °C [24].

Two-dimensional SnS_2 with a nanosheet morphology exhibits a narrow band gap, excellent adsorption kinetics, and low gas adsorption energy, making it well applied in the gas-sensing field [25]. The SnS_2 sensor developed by Yan et al. demonstrates a sensitivity of 13,000% toward 9 ppm NO_2, a strong selectivity, and a low detection limit for NO_2 [26]. Dong et al. adopted a hydrothermal treatment process for synthesizing α-MoO_3@SnS_2 nanosheets, which exhibited a response value of 114.9 to 100 ppm TEA at 175 °C. The foam-like structure, with its excellent permeability, facilitates enhanced gas-sensing performance [27]. Xu et al. applied a hydrothermal method for synthesizing SnS_2/ZnS flower-like microspheres presenting a high specific surface area (SSA), achieving a response of ~11 toward 50 ppm TEA at 180 °C [28].

By reasonably building p-n heterojunctions in MOS nanomaterials to modulate the electron and hole concentration, gas-sensing performance can be effectively enhanced [29]. Shuai et al. successfully synthesized NiO/$BiVO_4$ p-n heterojunction microspheres via hydrothermal treatment. The 10% NiO-$BiVO_4$ sensor presented a gas response of 183.11 to 30 ppm TEA versus the pure $BiVO_4$ sensor [30]. Wu et al. synthesized In_2O_3-decorated Mn_2O_3 nanorod-based multidimensional nanostructures via a hydrothermal method, which exhibited an enhanced SSA. The In_2O_3/Mn_2O_3 composite with an In:Mn ratio of 1:0.12 demonstrated a response value of 44.3 toward 100 ppm TEA at 180 °C [31].

The study successfully synthesized SnS_2/$LaFeO_3$ composite by combining the hydrothermal method with high-temperature annealing treatment. In the SnS_2/$LaFeO_3$ composite material, the combination of p-type $LaFeO_3$ and n-type SnS_2 forms a p-n heterojunction. This creates a built-in electric field at the interface, enabling more efficient carrier separation and transport. The introduction of SnS_2 increases the specific surface area of the composite material, providing more reactive sites for the target gas and thereby enhancing gas-sensing performance. The 5% SnS_2/$LaFeO_3$ sensor demonstrates a response of 124.5 to 100 ppm TEA, along with excellent selectivity and long-term stability. Compared to similar sensors, the performance of our sensor significantly outperforms comparable counterparts. For example, Hao et al. fabricated a RuO_2/$LaFeO_3$ gas sensor exhibiting a response value of 60.4 towards 100 ppm TEA at 260 °C [32]. Ma et al. fabricated a WO_3-$SnWO_4$ nanorod-based gas sensor showing a response value of 80.2 towards 100 ppm TEA at 130 °C [33]. Zhao et al. fabricated a gas sensor based on WO_3-modified $ErVO_4$ nanoparticles. At its optimal operating temperature of 200 °C, the $ErVO_4$/WO_3 composite exhibited a response of 15.79 towards 100 ppm TEA [34]. The $LaFeO_3/ZnO$ composite prepared by Ni et al. exhibited a response value of 50 to 100 ppm TEA at 200 °C [35].

2. Experiment

2.1. Preparation of $LaFeO_3$ Nanoparticles and SnS_2 Nanosheets

The preparation of $LaFeO_3$ nanoparticles was achieved by using the citric acid sol–gel method [36] by first dissolving $Fe(NO_3)_3{\cdot}9H_2O$ and $La(NO_3)_3{\cdot}6H_2O$ in 50 mL of deionized water (DW) to form an orange solution A and then dissolving citric acid (210.0 mg, 1 mmol) in 30 mL of DW to obtain a colorless transparent solution B. The two solutions were mixed under magnetic agitation, followed by 5 h of continuous stirring at 80 °C until obtaining a reddish-brown gel. After 12 h of drying treatment, the gel underwent 2 h of calcination at 350 °C and 4 h of calcination at 700 °C in succession to synthesize the $LaFeO_3$.

A one-step hydrothermal method was applied for synthesizing SnS_2 nanosheets [37]. Firstly, a colorless transparent solution was obtained by completely dissolving $SnCl_4{\cdot}5H_2O$ (17.5 mg, 0.05 mmol) and thiourea (22.8 mg, 0.3 mmol) in 50 mL of DW. The hydrothermal reaction proceeded for 12 h at 180 °C, followed by the collection of the generated yellow product.

2.2. Synthesis of $SnS_2/LaFeO_3$ Composite

$LaFeO_3$ powder (100 mg) was dissolved in a 40 mL ethylene glycol (EG) solution using 30 min of ultrasonication. SnS_2 powder with varying masses ultrasonically dispersed in 20 mL of EG solution was added dropwise into the $LaFeO_3$ solution before half an hour of stirring. The mixture underwent 12 h of hydrothermal heating at 160 °C to obtain a powder sample. The powder sample was sequentially rinsed using DW and ethanol and dried again. The $SnS_2/LaFeO_3$ composite with different contents (5 wt%, 10 wt%, and 20 wt%) could be made. The resulting materials were designated as 5% $SnS_2/LaFeO_3$, 10% $SnS_2/LaFeO_3$, and 20% $SnS_2/LaFeO_3$, respectively.

2.3. Control Experiment: In Situ Synthesis of $SnS_2/LaFeO_3$ Composite

$LaFeO_3$ powder (100 mg) was dissolved in a 40 mL ethylene glycol (EG) solution using 30 min of ultrasonication. Afterwards, specific quantities of $SnCl_4{\cdot}5H_2O$ (0.025, 0.05, and 0.075 mmol) and thiourea (0.15, 0.3, and 0.45 mmol) were completely dissolved in 20 mL of ethylene glycol and then added dropwise to the $LaFeO_3$ solution under magnetic stirring. The mixture underwent 12 h of hydrothermal heating at 160 °C to obtain a powder sample. The powder sample was sequentially rinsed using DW and ethanol and dried again. The obtained $SnS_2/LaFeO_3$ composite with different molar ratios are designated as $SnS_2/LaFeO_3$-1, $SnS_2/LaFeO_3$-2, and $SnS_2/LaFeO_3$-3.

2.4. Characterizations of Gas-Sensitive Materials

X-ray diffraction (XRD) was employed to analyze the composition, crystal type, and grain size of the samples. The model of the diffractometer used in this study was a SHIMADZU XRD-7000S (SHIMADZU, Kyoto, Japan). The scanning 2θ angle range during testing was 5–80°, with a scanning speed of 5°/min. Field-emission scanning electron microscopy (SEM) was used to observe the microscopic morphology of the synthetic material precursors and target samples. The model of the SEM used in this study was an FEI Nova NanoSEM450 (FEI, Hillsboro, OR, USA). Prior to testing, high-purity ethanol, clean silicon wafers, conductive adhesive, and other materials were prepared, and the accelerating voltage was set to 18 kV. To test the specific surface area, pore volume, and pore size distribution of the samples, a fully automated physical adsorption instrument produced by Beijing JWGB Technology Co., Ltd. (Beijing, China), model BK100C, was used to analyze the experimental materials. Before testing, cleaned BET tubes and a certain amount of liquid nitrogen were prepared. To determine the chemical composition, elemental valence states, and their percentages on the surface of the target samples, X-ray photoelectron spectroscopy (XPS) was performed using an X-ray photoelectron spectrometer manufactured and designed by Thermo Fisher Scientific (Waltham, MA, USA), model ESCALABTM 250Xi.

2.5. Fabrication of the Gas Sensors

Gas sensors were constructed and measured following a previous description [38]. Initially, a specific amount of pure $LaFeO_3$, pure SnS_2, and $SnS_2/LaFeO_3$ composite was dispersed in DW. This suspension was then used to prepare a slurry with the appropriate viscosity for coating onto the ceramic tube for fabricating the sensors (Figure 1). A CGS-8 intelligent gas-sensitive analysis system (Beijing Elite Technology, Beijing, China) was employed for testing the sensor's sensitivity (R_a/R_g or R_g/R_a). R_a and R_g denote the resistance values of the sensor in air and the target gas, respectively. The response and recovery time denote the duration necessary for the sensor to achieve 90% of the total resistance variation in the target gas and air.

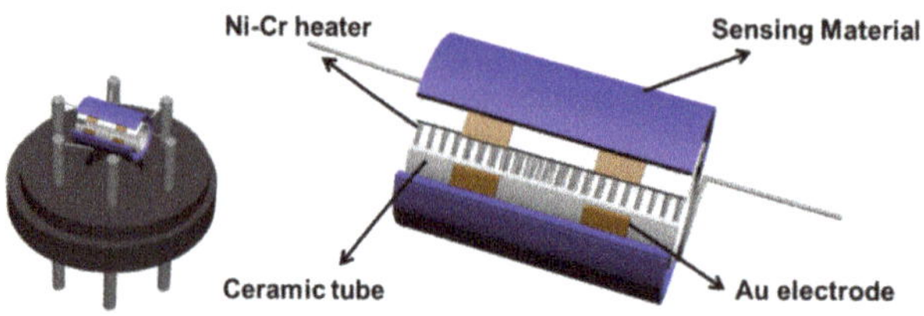

Figure 1. The structure of the gas-sensing device.

3. Results and Discussion

3.1. Structural and Morphological Characteristics

Figure 2 presents the microscopic morphology and elemental distribution of pure SnS_2, pure $LaFeO_3$, and the 5% $SnS_2/LaFeO_3$ composite. As shown in Figure 2a, $LaFeO_3$ primarily consists of aggregated nanoparticle structures, while SnS_2 in Figure 2b exhibits a well-defined hexagonal morphology. Figure 2c,d reveal that SnS_2 nanosheets in the 5% $SnS_2/LaFeO_3$ composite are randomly and densely anchored onto the $LaFeO_3$ nanoparticle framework, with no significant alterations in their original dimensions or morphology after composite formation. Elemental mapping images (Figure 2e–j) demonstrate a homogeneous spatial distribution of Sn, S, La, Fe, and O components, further confirming the successful synthesis of the $SnS_2/LaFeO_3$ composite.

According to Figure 3, the $SnS_2/LaFeO_3$ composite with varying content ratios has diffraction peaks corresponding to the pure $LaFeO_3$ crystal phase (JCPDS Card No. 75-0541) [39]. With rising SnS_2 content, the intensities of the diffraction peaks at 15.02° and 41.87° in the $SnS_2/LaFeO_3$ composite gradually become more pronounced, matching those of pure SnS_2 crystals (JCPDS Card No. 83-1705) [40]. The $SnS_2/LaFeO_3$ composite is successfully synthesized, as evidenced by the impurity peaks not being detected in the patterns.

The results from Figure 4 and Table 1 show that specific surface areas (SSAs) of $LaFeO_3$ and 5% $SnS_2/LaFeO_3$ are 10.165 m^2/g and 12.433 m^2/g, respectively. Incorporating SnS_2 elevates the material's SSA. A larger SSA offers more reactive sites, contributing to improved gas-sensing performance. Although the sample of 10% $SnS_2/LaFeO_3$ has the largest SSA, there are still other key factors that affect the gas-sensing performance, such as the sensors' baseline resistances in air, which is discussed later.

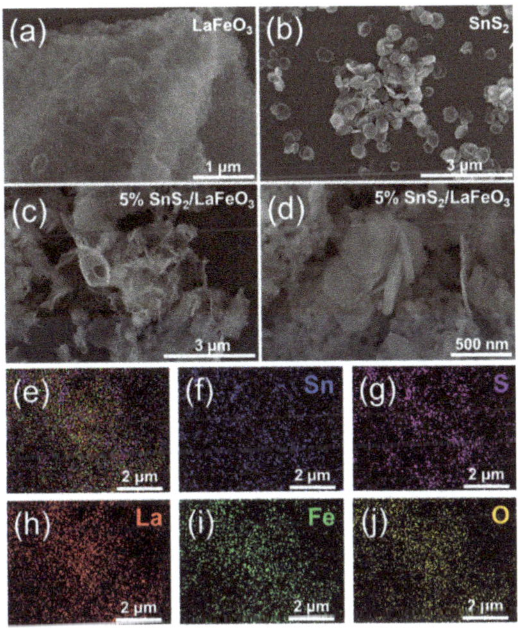

Figure 2. SEM images of (**a**) $LaFeO_3$ nanoparticles, (**b**) SnS_2 nanosheets, and (**c**,**d**) 5% $SnS_2/LaFeO_3$ composite. EDS spectrum of (**e**–**j**) for the 5% $SnS_2/LaFeO_3$ composite.

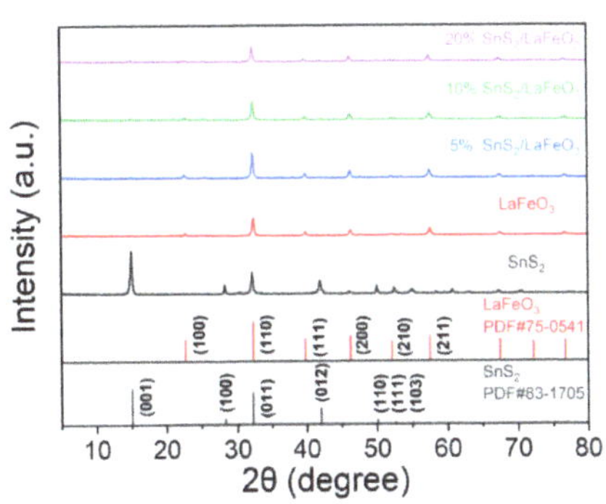

Figure 3. XRD patterns of pure $LaFeO_3$, pure SnS_2, and $SnS_2/LaFeO_3$ composite.

Table 1. Samples' textural parameters from BET analysis.

Sample	Surface Area ($m^2\ g^{-1}$)	Pore Volume ($cm^3\ g^{-1}$)	Average Pore Size (nm)
$LaFeO_3$	10.165	0.075	29.434
5% $SnS_2/LaFeO_3$	12.433	0.086	27.582
10% $SnS_2/LaFeO_3$	12.768	0.078	24.544
20% $SnS_2/LaFeO_3$	11.295	0.077	25.339

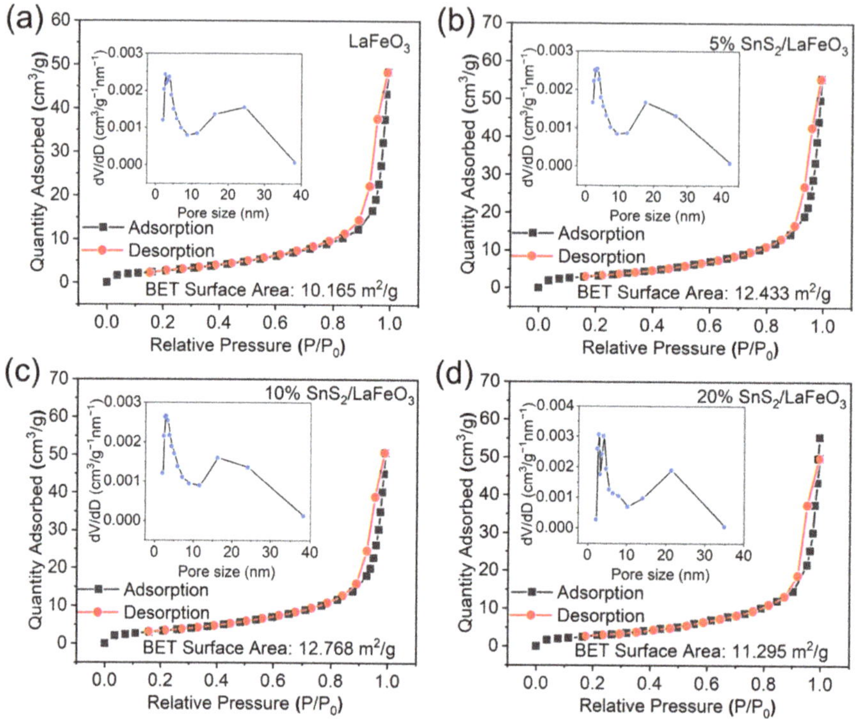

Figure 4. N_2 adsorption/desorption curves: (**a**) $LaFeO_3$; (**b–d**) 5%, 10%, and 20% $SnS_2/LaFeO_3$.

The XPS analysis results revealed the chemical states of various elements in pure SnS_2, pure $LaFeO_3$, and $SnS_2/LaFeO_3$ composite with different ratios. As shown in Figure 5a, the doublet peaks at 487.2 and 495.7 eV are assigned to the Sn^{4+} $3d_{5/2}$ and Sn^{4+} $3d_{3/2}$ orbitals of SnS_2 [41]. Compared with pure SnS_2, the Sn^{4+} $3d_{5/2}$ and Sn^{4+} $3d_{3/2}$ peaks in the $SnS_2/LaFeO_3$ composite exhibited a 0.9 eV shift to lower binding energies. Similarly, the doublet peaks at 162.2 and 163.4 eV in Figure 5b are assigned to the S^{2-} $2p_{3/2}$ and S^{2-} $2p_{1/2}$ orbitals of SnS_2, respectively [41]. These S^{2-} 2p peaks in the $SnS_2/LaFeO_3$ composite also showed a 0.9 eV negative shift relative to those in pure SnS_2. The observed binding energy shifts of the elements demonstrate a strong electronic interaction between SnS_2 and $LaFeO_3$, ultimately promoting $SnS_2/LaFeO_3$ heterojunctions to be formed.

In Figure 5c, the peaks located at 833.7, 838.2, 850.6, and 855.0 eV are assigned to the La^{3+} $3d_{5/2}$ and La^{3+} $3d_{3/2}$ orbitals in pure $LaFeO_3$, accompanied by three satellite peaks. Compared to pure $LaFeO_3$, the La 3d peaks in the $SnS_2/LaFeO_3$ composite exhibit a shift toward higher binding energy by approximately 0.1–0.2 eV [42]. In Figure 5d, the peaks at 710.0 and 723.7 eV target the Fe^{3+} $2p_{3/2}$ and $2p_{1/2}$ orbitals, while those at 712.2 and 725.5 eV target the Fe^{4+} $2p_{3/2}$ and $2p_{1/2}$ orbitals [43,44]. Binding energy presents a slight shift resulting from the heterostructure effect at the interface of the $SnS_2/LaFeO_3$ composite.

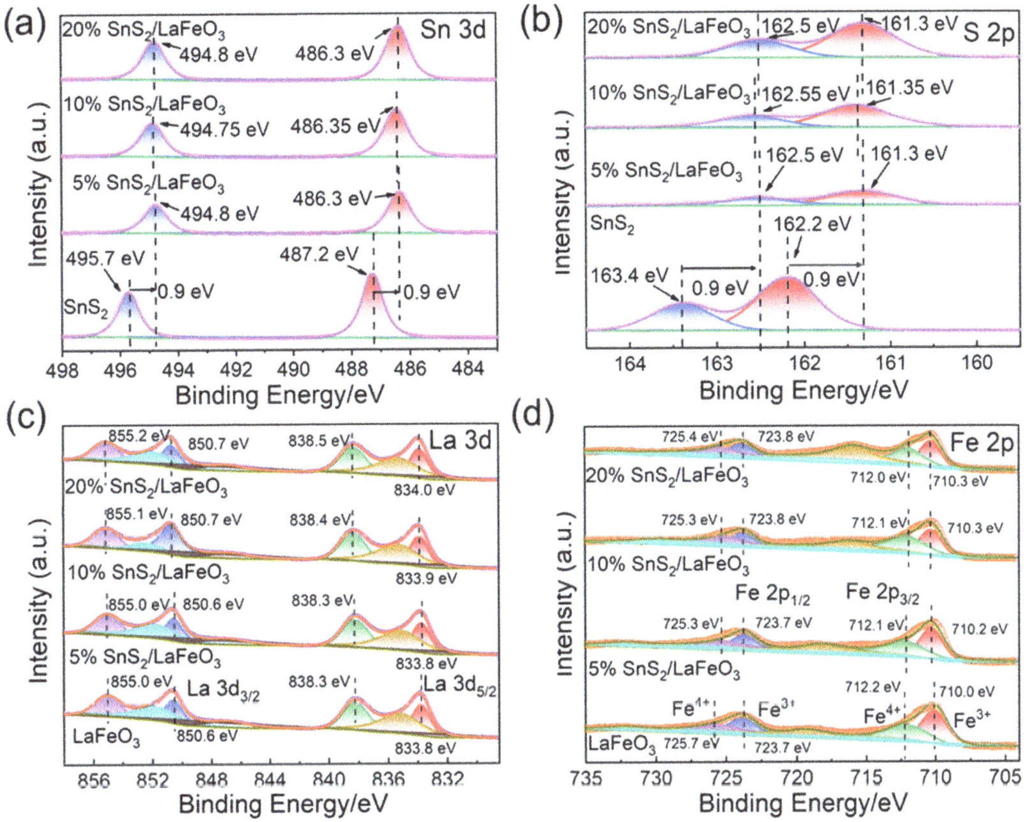

Figure 5. The high-resolution XPS spectra of Sn 3d (**a**), S 2p (**b**), La 3d (**c**), and Fe 2p (**d**) for all samples.

In Figure 6, the O 1s binding energy peak is deconvoluted into three peaks located at 529.3, 531.25, and 532.8 eV, targeting lattice oxygen (O_L), oxygen vacancies (O_V), and chemisorbed oxygen (O_C), respectively [45,46], with relevant peak area ratios displayed in Table 2. The 5% $SnS_2/LaFeO_3$ composite exhibit a higher O_V content than pure $LaFeO_3$. Oxygen vacancies (O_V) are active sites for surface oxygen adsorption, capable of accelerating surface redox reactions [47]. Generally, a higher O_V content more remarkably strengthens gas-sensing performance.

Table 2. XPS O1s peak area ratios of varying oxygen species.

Sample	O_L (%)	O_V (%)	O_C (%)
$LaFeO_3$	66.17	26.67	7.16
5% $SnS_2/LaFeO_3$	65.17	31.33	3.50
10% $SnS_2/LaFeO_3$	65.32	28.84	5.84
20% $SnS_2/LaFeO_3$	65.75	27.74	6.51

According to the analyses from XRD, SEM, EDS, and XPS, the synthesis of the $SnS_2/LaFeO_3$ composite was successfully accomplished.

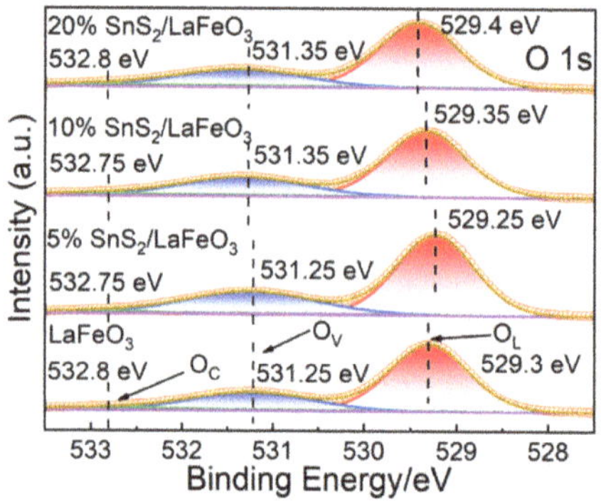

Figure 6. The high-resolution XPS spectra of O 1s for all samples.

3.2. Gas-Sensing Properties of the Nanomaterials

The optimal operating temperature is a critical indicator in gas sensor testing. According to Figure 7a,b, these sensors exhibit two distinct optimal operating temperatures and demonstrate significantly divergent response behaviors to TEA gas. The pure SnS_2-based sensor achieves optimal performance at 80 °C, with a response value of 14.2 to 100 ppm TEA. Comparatively, $LaFeO_3$-based sensors reach their optimal working temperature at 140 °C. Notably, the 5% $SnS_2/LaFeO_3$ sensor delivers a response value of 124.5 to 100 ppm TEA, representing an approximately three-fold enhancement compared to the pure $LaFeO_3$ sensor. According to Figure 7c,d, as temperature increases, the semiconductor's thermal excitation effect dominates, inducing a monotonic decline in baseline resistance with rising temperature. With the increased addition of SnS_2, the baseline resistance of the $SnS_2/LaFeO_3$ composite starts to increase. Meanwhile, the difference in R_g/R_a gradually decreases, thus reducing the sensitivity.

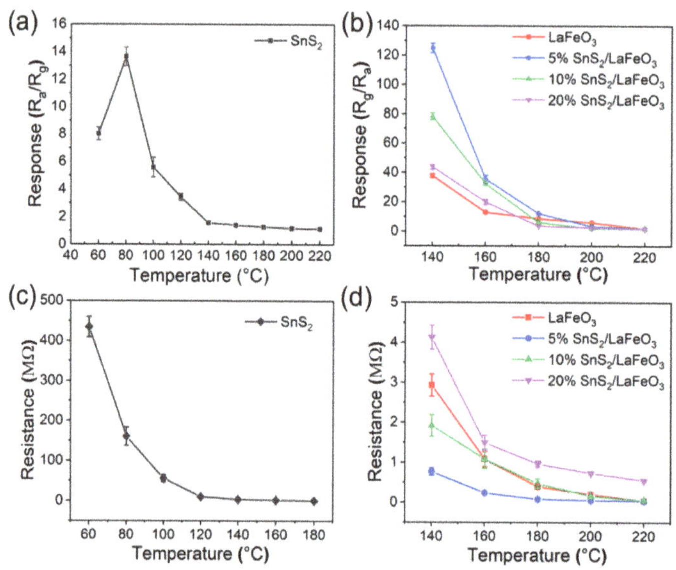

Figure 7. (**a**,**b**) Responses of pure SnS_2, pure $LaFeO_3$, and $SnS_2/LaFeO_3$ composite to 100 ppm TEA at 60–220 °C; (**c**,**d**) the sensors' baseline resistances in air at different temperatures.

Figure 8a–c shows the gas sensors' time-varying response–recovery curves to 100 ppm TEA concentration. In terms of response time, the 5% $SnS_2/LaFeO_3$ sensor does not show a significant response advantage, and with regard to the recovery time, the pure $LaFeO_3$ and pure SnS_2 sensor have a recovery time of 165 and 175 s, respectively, while the 5% $SnS_2/LaFeO_3$ sensor has a recovery time elevated to 64 s, which is significantly faster than that of the two single material sensors. Figure 8d illustrates the resistance response variation curves of different sensors to 100 ppm TEA at the optimal working temperature.

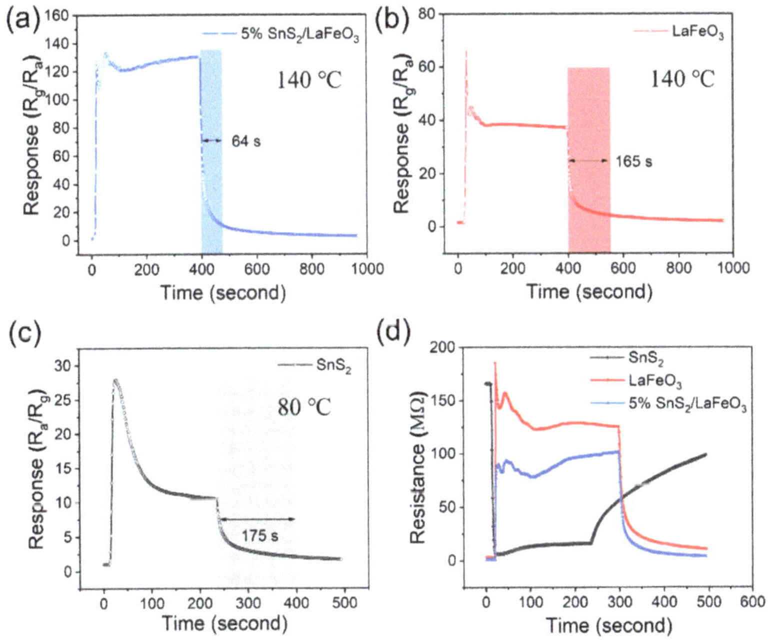

Figure 8. (**a–c**) Time-varying response curves of 3 sensors to 100 ppm TEA. (**d**) Resistance response variation curves of 3 sensors at the optimal operating temperature.

Selectivity is also a fundamental criterion for gas sensors. The sensors were used to test the selectivity at 140 °C to 100 ppm of acetone, ethanol, n-propanol, isopropanol, methanol, xylene, acetic acid, hexane, formaldehyde, ammonia, and other gases (Figure 9a). Compared to other tested gases (such as acetone with a C=O bond energy of 728 kJ/mol and methanol with an O-H bond energy of 458.8 kJ/mol), the 5% $SnS_2/LaFeO_3$ sensor exhibits significantly enhanced response characteristics toward TEA. This superiority primarily stems from the lower bond energy (305 kJ/mol) of the C-N bond in TEA molecules. The disparity in bond energy facilitates easier surface interfacial reactions between TEA and the gas-sensitive material, thereby enhancing the gas selectivity.

These sensors present excellent stability and reproducibility (Figure 9b). Notably, the 5% $SnS_2/LaFeO_3$ gas sensor exhibits more prominent response and recovery characteristics at 1 ppm TEA, achieving a response value of 10.2 (Figure 9c). This result indicates that the 5% $SnS_2/LaFeO_3$ gas sensor possesses a lower detection limit. In addition, in Figure 9d, a linear fitting was conducted for all sensors in 1–100 ppm. The 5% $SnS_2/LaFeO_3$ sensor exhibited the fitting equation of $y = 10.838 + 1.109x$ (R^2 = 0.998), demonstrating a strong linear relationship.

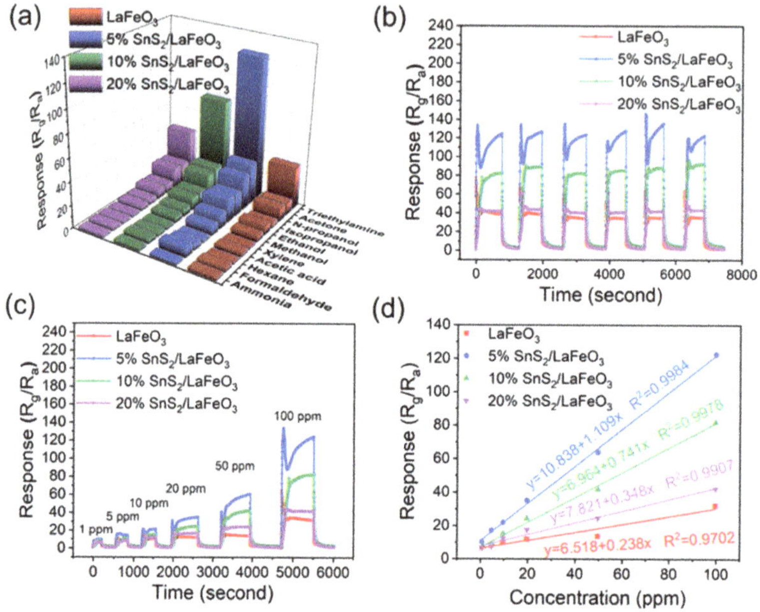

Figure 9. (**a**) Selectivity of 4 sensors to 100 ppm of varying VOC gases at 140 °C. (**b**,**c**) Cyclic stability and dynamic response curve of sensors. (**d**) Fit curve of the 5% $SnS_2/LaFeO_3$ sensor responses.

Relative humidity (RH) also critically affects the sensor performance. Figure 10a illustrates the humidity resistance tests of four gas sensors toward 100 ppm TEA. With the RH rising from 20% to 80%, the 5% $SnS_2/LaFeO_3$ sensor has a response declining from 123.4 to 9.6. This decline arises due to water molecules in the air competing with the target gas for active sites on the material surface, impeding the formation of adsorbed oxygen species. Consequently, the baseline resistance increases, leading to reduced gas-sensing performance [48]. Figure 10b shows a continuous 30-day gas sensitivity test of the 5% $SnS_2/LaFeO_3$ gas sensor, which maintains a stable response value to 100 ppm TEA at 140 °C, providing excellent long-term stability. Additionally, anti-interference tests were conducted using mixed gases containing 50 ppm TEA and 50 ppm interfering gases (formaldehyde, acetic acid, methanol, ammonia, and hexane), as shown in Figure 10c. The 5% $SnS_2/LaFeO_3$ sensor displays minimal response variations to the mixed gases, confirming its superior anti-interference capability.

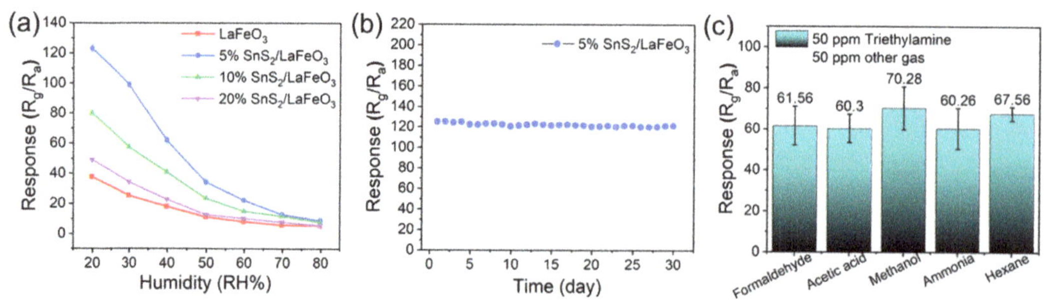

Figure 10. (**a**) Humidity resistance tests and (**b**) long-term stability of four gas sensors; (**c**) the responses of the 5% $SnS_2/LaFeO_3$ gas sensor in the different gas mixtures.

To demonstrate that the hydrothermal method of combining SnS_2 and $LaFeO_3$ materials is simple and that the synthesized $SnS_2/LaFeO_3$ composites have excellent properties, we conducted a control experiment of the in situ synthetic method. Figure 11a presents a control experiment on the sensing properties of in situ-grown SnS_2 on $LaFeO_3$. The $SnS_2/LaFeO_3$-2 composite shows a response of 26.2 to 100 ppm TEA at 140 °C and 40% RH. Compared to pure $LaFeO_3$, there is no significant improvement in gas-sensing performance. Figure 11b shows the resistance variation curves of the three sensors. As the SnS_2 loading content increases, the baseline resistance demonstrates an upward trend (Figure 11c).

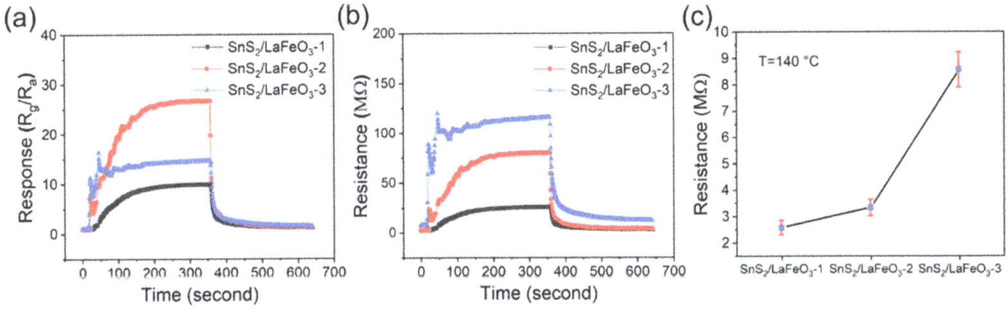

Figure 11. (**a**,**b**) Time-varying response and resistance curves of the in situ $SnS_2/LaFeO_3$-1, -2, and -3 sensors to 100 ppm TEA at 140 °C and 40% RH; (**c**) baseline resistance values of the three sensors at 140 °C.

Figure 12a presents the selectivity tests of the three sensors based on the in situ synthesized samples toward eight different gases. The $SnS_2/LaFeO_3$-2 sensor exhibits relatively high responses to gases such as acetone, n-propanol, and iso-propanol, but shows no significant selective advantage for the target analyte. Figure 12b displays the cyclic stability test results, where the response values remain essentially unchanged over multiple cycles. Notably, the sensing performance improved during testing due to a decrease in ambient humidity to approximately 35% RH. Figure 12c,d show the concentration-dependent response curves of the three sensors to TEA from 1 ppm to 100 ppm. The $SnS_2/LaFeO_3$-2 composite demonstrates superior gas response values, with its concentration–response relationship following the linear fitting equation (y = 4.254 + 0.214x, R^2 = 0.962).

3.3. Gas-Sensing Mechanism

During the high-temperature annealing treatment of the $LaFeO_3$ sol–gel precursor in air, cation vacancies (La and Fe) generated under high-temperature conditions induce hole formation, triggering a p-type $LaFeO_3$ semiconductor. The equations below interpret the corresponding defect reaction [49,50]:

$$V_{La} \rightarrow V_{La}^{'''} + 3h^{\cdot} \tag{1}$$

$$V_{Fe} \rightarrow V_{Fe}^{'''} + 3h^{\cdot} \tag{2}$$

It is allowed to explain the sensing mechanism by examining the reversible conductivity variation of the sensing layer attributed to the process through which the sensing material interacts with adsorbed gas molecules [51]. The equations below interpret the overall steps:

$$O_{2(gas)} \rightarrow O_{2(ads)} \tag{3}$$

$$O_{2(ads)} + e^- \rightarrow O^-_{2(ads)} (T < 100\ ^\circ\text{C}) \tag{4}$$

$$O^-_{2(ads)} + e^- \rightarrow 2O^-_{(ads)} (100\ ^\circ\text{C} < T < 300\ ^\circ\text{C}) \tag{5}$$

$$(C_2H_5)_3N_{(gas)} + 39O^-_{(ads)} \leftrightarrow N_{2(gas)} + 12CO_{2(gas)} + 15H_2O_{(gas)} + 39e^- \tag{6}$$

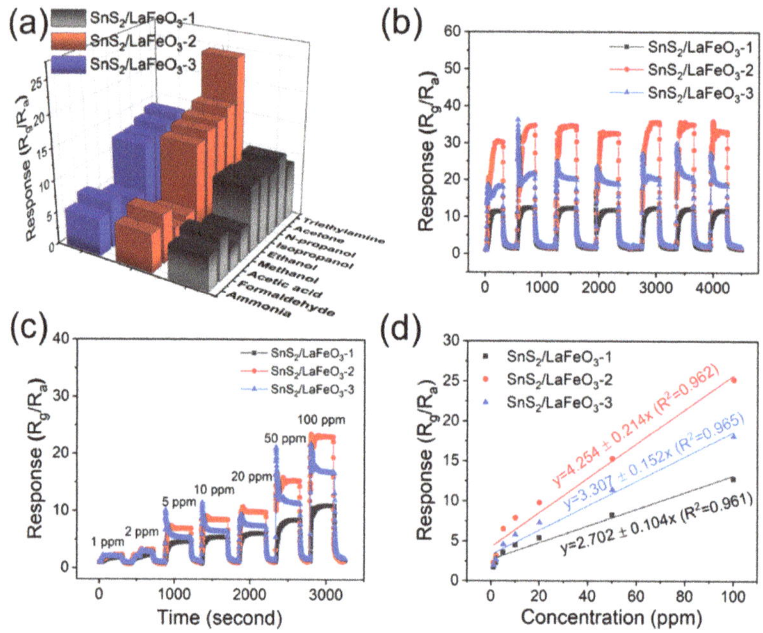

Figure 12. (**a**) Selectivity tests of the three sensors toward different gases; (**b**) cyclic stability test; (**c**,**d**) response curves to increasing TEA concentrations (1–100 ppm) and corresponding fitting curves for the three sensors.

Upon the exposure of the $LaFeO_3$ to atmospheric air, oxygen molecules (O_2) adsorbed onto the surface are converted to $O^-_{2(ads)}$ and $O^-_{(ads)}$ through capturing electrons from the conduction band, forming a hole accumulation layer (HAL) on the $LaFeO_3$ surface and reducing the electronic resistance of $LaFeO_3$. Upon the exposure to TEA gas, the reaction between the TEA molecules and the adsorbed oxygen species present on the $LaFeO_3$ surface promoted the release of electrons back into the materials, increasing the resistance.

As shown in Figure 13, $LaFeO_3$ is a p-type semiconductor [52], and SnS_2 is an n-type semiconductor [53]. Affected by their different work functions, the electrons of $LaFeO_3$ can flow to SnS_2 and the holes of SnS_2 flow to $LaFeO_3$ until the Fermi level achieves equilibrium [54,55]. This charge transfer can be verified by XPS characterization (Figure 5).

This further increases the hole concentration in $LaFeO_3$ and thickens the HAL in an air atmosphere, thereby reducing the electronic resistance R_a and providing the potential for a lager change between R_a and R_g. As shown in Figure 7b, a small amount (5%) of SnS_2-composition significantly reduces the resistance value of R_a. Moreover, the p-n heterojunction improves the charge transport during the reaction process on the surface, enhancing the sensing response. The 5% $SnS_2/LaFeO_3$ composite exhibits the highest content of O_V and O_C. This promotes the participation of more target gas molecules in surface redox reactions. Furthermore, BET surface area measurements indicate that the introduction of SnS_2 effectively increases the specific surface area of the material. This provides more active sites and accelerates the redox reaction process, thus enhancing the gas-sensing performance.

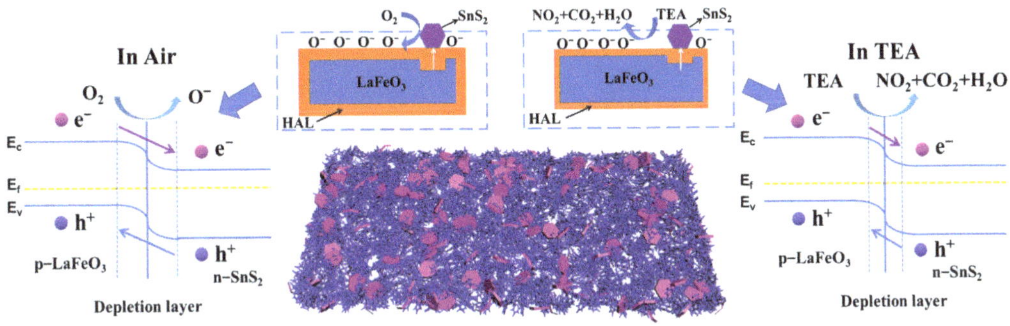

Figure 13. Schematic diagram of the sensing mechanism of the $SnS_2/LaFeO_3$ sensor.

4. Conclusions

In conclusion, the $SnS_2/LaFeO_3$ composites were successfully synthesized by a high-temperature annealing treatment and hydrothermal method. The 5% $SnS_2/LaFeO_3$ sensor demonstrates a response of 124.5 to 100 ppm TEA at 140 °C, 3.2 times higher than the pure $LaFeO_3$ sensor. Furthermore, for the $SnS_2/LaFeO_3$ sensor, the sensor response exhibits a linear relevance to the gas concentration, anti-interference capability, and long-term stability. Furthermore, the formed heterojunction elevates the O_V content and expands the HAL on the surface, thereby improving the TEA-sensing property. The findings in the study will assist researchers in designing and synthesizing MOS gas sensors for TEA detection.

Author Contributions: Conceptualization, X.W. (Xiaofeng Wang); methodology, H.W. and X.W. (Xiaofeng Wang); data curation, H.W.; validation, X.W. (Xiaobing Wang) and Y.C.; investigation, X.W. (Xiaobing Wang) and Y.C.; writing—original draft preparation, H.W. and X.W. (Xiaofeng Wang); writing—review and editing, X.W. (Xiaofeng Wang); supervision, X.W. (Xiaofeng Wang); funding acquisition, X.W. (Xiaofeng Wang). All authors have read and agreed to the published version of the manuscript.

Funding: This research was funded by the Natural Science Foundation of Liaoning Province of China (No. 2023-MSBA-012) and the Fundamental Research Funds for the Central Universities of China (DUT24MS012).

Institutional Review Board Statement: Not applicable.

Informed Consent Statement: Not applicable.

Data Availability Statement: The data that support the findings of this study are available from the corresponding authors upon request.

Conflicts of Interest: The authors declare no conflicts of interest.

References

1. Banga, I.; Paul, A.; Poudyal, D.C.; Muthukumar, S.; Prasad, S. Recent Advances in Gas Detection Methodologies with a Special Focus on Environmental Sensing and Health Monitoring Applications—A Critical Review. *ACS Sens.* **2023**, *8*, 3307–3319. [CrossRef] [PubMed]

2. Bilge, S.; Dogan-Topal, B.; Yücel, A.; Sınağ, A.; Ozkan, S.A. Recent advances in flower-like nanomaterials: Synthesis, characterization, and advantages in gas sensing applications. *TrAC Trends Anal. Chem.* **2022**, *153*, 116638. [CrossRef]
3. Bulemo, P.M.; Kim, D.-H.; Shin, H.; Cho, H.-J.; Koo, W.-T.; Choi, S.-J.; Park, C.; Ahn, J.; Güntner, A.T.; Penner, R.M.; et al. Selectivity in Chemiresistive Gas Sensors: Strategies and Challenges. *Chem. Rev.* **2025**, *125*, 4111–4183. [CrossRef] [PubMed]
4. Li, Q.; Zeng, W.; Li, Y. Metal oxide gas sensors for detecting NO_2 in industrial exhaust gas: Recent developments. *Sens. Actuators B Chem.* **2022**, *359*, 131579. [CrossRef]
5. Zhu, M.; Yang, T.; Zhai, C.; Du, L.; Zhang, J.; Zhang, M. Fast triethylamine gas sensing response properties of Ho-doped SnO_2 nanoparticles. *J. Alloys Compd.* **2020**, *817*, 152724. [CrossRef]
6. Wang, H.; Luo, Y.; Li, K.; Liu, B.; Gao, L.; Duan, G. Porous α-Fe_2O_3 gas sensor with instantaneous attenuated response toward triethylamine and its reaction kinetics. *Chem. Eng. J.* **2022**, *427*, 131631. [CrossRef]
7. Gui, Y.; Tian, K.; Liu, J.; Yang, L.; Zhang, H.; Wang, Y. Superior triethylamine detection at room temperature by {-112} faceted WO_3 gas sensor. *J. Hazard. Mater.* **2019**, *380*, 120876. [CrossRef]
8. Pereira, P.F.M.; de Sousa Picciani, P.H.; Calado, V.; Tonon, R.V. Electrical gas sensors for meat freshness assessment and quality monitoring: A review. *Trends Food Sci. Technol.* **2021**, *118*, 36–44. [CrossRef]
9. Wu, K.; Debliquy, M.; Zhang, C. Metal-oxide-semiconductor resistive gas sensors for fish freshness detection. *Compr. Rev. Food Sci. Food Saf.* **2023**, *22*, 913–945. [CrossRef]
10. Zeng, J.; Rong, Q.; Xiao, B.; Yu, R.; Zi, B.; Kuang, X.; Deng, X.; Ma, Y.; Zhang, J.; Wu, J.; et al. Single-atom silver loaded on tungsten oxide with oxygen vacancies for high performance triethylamine gas sensors. *J. Mater. Chem. A* **2021**, *9*, 8704–8710. [CrossRef]
11. Yang, J.; Han, W.; Jiang, B.; Wang, X.; Sun, Y.; Wang, W.; Lou, R.; Ci, H.; Zhang, H.; Lu, G. Electrospinning Derived $NiO/NiFe_2O_4$ Fiber-in-Tube Composite for Fast Triethylamine Detection under Different Humidity. *ACS Sens.* **2022**, *7*, 995–1007. [CrossRef] [PubMed]
12. Khan, I.; Sun, N.; Wang, Y.; Li, Z.; Qu, Y.; Jing, L. Synthesis of SnO_2/yolk-shell $LaFeO_3$ nanocomposites as efficient visible-light photocatalysts for 2,4-dichlorophenol degradation. *Mater. Res. Bull.* **2020**, *127*, 110857. [CrossRef]
13. Kou, X.; Meng, F.; Chen, K.; Wang, T.; Sun, P.; Liu, F.; Yan, X.; Sun, Y.; Liu, F.; Shimanoe, K.; et al. High-performance acetone gas sensor based on Ru-doped SnO_2 nanofibers. *Sens. Actuators B Chem.* **2020**, *320*, 128292. [CrossRef]
14. Lee, J.-H. Gas sensors using hierarchical and hollow oxide nanostructures: Overview. *Sens. Actuators B Chem.* **2009**, *140*, 319–336. [CrossRef]
15. Lee, J.; Jung, Y.; Sung, S.-H.; Lee, G.; Kim, J.; Seong, J.; Shim, Y.-S.; Jun, S.C.; Jeon, S. High-performance gas sensor array for indoor air quality monitoring: The role of Au nanoparticles on WO_3, SnO_2, and NiO-based gas sensors. *J. Mater. Chem. A* **2021**, *9*, 1159–1167. [CrossRef]
16. Zhang, S.; Song, P.; Zhang, J.; Yan, H.; Li, J.; Yang, Z.; Wang, Q. Highly sensitive detection of acetone using mesoporous In_2O_3 nanospheres decorated with Au nanoparticles. *Sens. Actuators B Chem.* **2017**, *242*, 983–993. [CrossRef]
17. Hao, P.; Qu, G.-M.; Song, P.; Yang, Z.-X.; Wang, Q. Synthesis of Ba-doped porous $LaFeO_3$ microspheres with perovskite structure for rapid detection of ethanol gas. *Rare Met.* **2021**, *40*, 1651–1661. [CrossRef]
18. Nakhostin Panahi, P.; Rasoulifard, M.H.; Babaei, S. Photocatalytic activity of cation (Mn) and anion (N) substitution in $LaCoO_3$ nanoperovskite under visible light. *Rare Met.* **2020**, *39*, 139–146. [CrossRef]
19. Voznyy, O. Realizing ultra-pure red emission with Sn-based lead-free perovskites. *Rare Met.* **2020**, *39*, 330–331. [CrossRef]
20. Ochoa-Muñoz, Y.H.; Mejía de Gutiérrez, R.; Rodríguez-Páez, J.E.; Gràcia, I.; Vallejos, S. Gas Sensors Based on Porous Ceramic Bodies of $MSnO_3$ Perovskites (M = Ba, Ca, Zn): Formation and Sensing Properties towards Ethanol, Acetone, and Toluene Vapours. *Molecules* **2022**, *27*, 2889. [CrossRef]
21. Xiao, C.; Zhang, X.; Ma, Z.; Yang, K.; Gao, X.; Wang, H.; Jia, L. Formaldehyde gas sensor with 1 ppb detection limit based on In-doped $LaFeO_3$ porous structure. *Sens. Actuators B Chem.* **2022**, *371*, 132558. [CrossRef]
22. Wang, B.; Yu, Q.; Zhang, S.; Wang, T.; Sun, P.; Chuai, X.; Lu, G. Gas sensing with yolk-shell $LaFeO_3$ microspheres prepared by facile hydrothermal synthesis. *Sens. Actuators B Chem.* **2018**, *258*, 1215–1222. [CrossRef]
23. Li, F.; Wang, S.; Wu, Z.; Xiong, X.; Li, J.; Zhou, J.; Gao, X. Excellent ethanol sensor based on $LaFeO_3$ modified with gold nanoparticles. *J. Mater. Sci. Mater. Electron.* **2021**, *32*, 27587–27595. [CrossRef]
24. Xiao, H.; Xue, C.; Song, P.; Li, J.; Wang, Q. Preparation of porous $LaFeO_3$ microspheres and their gas-sensing property. *Appl. Surf. Sci.* **2015**, *337*, 65–71. [CrossRef]
25. Liu, W.-D.; Xiong, Y.; Shen, A.; Wang, X.-Z.; Chang, X.; Lu, W.-B.; Tian, J. SnS_2 nanosheets decorated SnO_2 hollow multishelled nanostructures for enhanced sensing of triethylamine gas. *Rare Met.* **2024**, *43*, 2339–2348. [CrossRef]
26. Yan, W.; Lv, C.; Zhang, D.; Chen, Y.; Zhang, L.; Ó Coileáin, C.; Wang, Z.; Jiang, Z.; Hung, K.-M.; Chang, C.-R.; et al. Enhanced NO_2 Sensitivity in Schottky-Contacted n-Type SnS_2 Gas Sensors. *ACS Appl. Mater. Interfaces* **2020**, *12*, 26746–26754. [CrossRef]

27. Dong, X.; Han, Q.; Kang, Y.; Li, H.; Huang, X.; Fang, Z.; Yuan, H.; Elzatahry, A.A.; Chi, Z.; Wu, G.; et al. Rational construction and triethylamine sensing performance of foam shaped α-MoO_3@SnS_2 nanosheets. *Chin. Chem. Lett.* **2022**, *33*, 567–572. [CrossRef]
28. Xu, X.; Ma, S.; Xu, X.; Pei, S.; Han, T.; Liu, W. Transformation synthesis of heterostructured SnS_2/ZnS microspheres for ultrafast triethylamine detection. *J. Alloys Compd.* **2021**, *868*, 159286. [CrossRef]
29. Gao, X.; Li, F.; Wang, R.; Zhang, T. A formaldehyde sensor: Significant role of p-n heterojunction in gas-sensitive core-shell nanofibers. *Sens. Actuators B Chem.* **2018**, *258*, 1230–1241. [CrossRef]
30. Shuai, Y.; Peng, R.; He, Y.; Liu, X.; Wang, X.; Guo, W. NiO/$BiVO_4$ p-n heterojunction microspheres for conductometric triethylamine gas sensors. *Sens. Actuators B Chem.* **2023**, *384*, 133625. [CrossRef]
31. Wu, R.; Liu, T.; Chen, X.; Yin, X.-T. Shape-controlled multi-dimensional In_2O_3/Mn_2O_3 p-n heterojunction for triethylamine detection. *J. Alloys Compd.* **2023**, *960*, 170527. [CrossRef]
32. Hao, P.; Song, P.; Yang, Z.; Wang, Q. Synthesis of novel RuO_2/$LaFeO_3$ porous microspheres its gas sensing performances towards triethylamine. *J. Alloys Compd.* **2019**, *806*, 960–967. [CrossRef]
33. Ma, N.; Ma, S.; Guo, J.; Fan, G.; Ni, P.; Wang, Y.; Zhu, J.; Wang, H. Enhanced triethylamine sensing with novel WO_3 composite $SnWO_4$ nanorod-based gas sensor. *Surf. Interfaces* **2025**, *58*, 105870. [CrossRef]
34. Zhao, Y.; Lu, S.; Lv, M.; Liu, J.; Chen, Y.; Wang, Y.; Zhou, H.; Che, X.; Kang, B.; Zhang, Q. An ultra-sensitive triethylamine sensor based on WO_3 modified $ErVO_4$ nanoparticles. *Vacuum* **2025**, *239*, 114440. [CrossRef]
35. Ni, P.; Ma, S.; Cai, Y.; Guo, J.; Fan, G.; Ma, N.; Wang, H.; Wang, Y.; Zhu, J. Triethylamine sensing behavior of p-$LaFeO_3$/n-ZnO composite synthesized by one-step hydrothermal method. *Sens. Actuators B Chem.* **2025**, *442*, 138051. [CrossRef]
36. Dhahri, R.; Benamara, M.; Nassar, K.I.; Elkenany, E.B.; Al-Syadi, A.M. Zinc oxide-based sensor prepared by modified sol–gel route for detection of low concentrations of ethanol, methanol, acetone, and formaldehyde. *Semicond. Sci. Technol.* **2024**, *39*, 115021. [CrossRef]
37. Bharathi, P.; Harish, S.; Shimomura, M.; Krishna Mohan, M.; Archana, J.; Navaneethan, M. Controlled growth and fabrication of edge enriched SnS_2 nanostructures for room temperature NO_2 gas sensor applications. *Mater. Lett.* **2023**, *335*, 133691. [CrossRef]
38. Liang, H.; Liu, N.; Lin, J.; Meng, Y.; Ni, J.; Gao, B.; Tan, Z.; Song, X.-Z.; Wang, X. Narrow-Band $HoFeO_3$ Nanostructure-Based Gas Sensor for n-Propanol Detection. *ACS Appl. Nano Mater.* **2024**, *7*, 7573–7581. [CrossRef]
39. Shingange, K.; Swart, H.; Mhlongo, G.H. Ultrafast Detection of Low Acetone Concentration Displayed by Au-Loaded $LaFeO_3$ Nanobelts owing to Synergetic Effects of Porous 1D Morphology and Catalytic Activity of Au Nanoparticles. *ACS Omega* **2019**, *4*, 19018–19029. [CrossRef]
40. Wei, H.; Hou, C.; Zhang, Y.; Nan, Z. Scalable low temperature in air solid phase synthesis of porous flower-like hierarchical nanostructure SnS_2 with superior performance in the adsorption and photocatalytic reduction of aqueous Cr(VI). *Sep. Purif. Technol.* **2017**, *189*, 153–161. [CrossRef]
41. Liu, Y.; Geng, P.; Wang, J.; Yang, Z.; Lu, H.; Hai, J.; Lu, Z.; Fan, D.; Li, M. In-situ ion-exchange synthesis Ag_2S modified SnS_2 nanosheets toward highly photocurrent response and photocatalytic activity. *J. Colloid Interface Sci.* **2018**, *512*, 784–791. [CrossRef] [PubMed]
42. Shao, J.; Sun, C.; Liu, H.; He, P.; Liu, Q.; Sun, J.; Li, J.; Pan, G.; Yang, X. Insight into Au functionalization on core-shell $LaFeO_3$ spheres for high-response and selectivity n-butanol gas sensors with DFT study. *Sens. Actuators B Chem.* **2023**, *382*, 133506. [CrossRef]
43. Haron, W.; Wisitsoraat, A.; Wongnawa, S. Nanostructured perovskite oxides—$LaMO_3$ (M=Al, Co, Fe) prepared by co-precipitation method and their ethanol-sensing characteristics. *Ceram. Int.* **2017**, *43*, 5032–5040. [CrossRef]
44. Liu, X.; Hu, J.; Cheng, B.; Qin, H.; Zhao, M.; Yang, C. First-principles study of O_2 adsorption on the $LaFeO_3$ surface. *Sens. Actuators B Chem.* **2009**, *139*, 520–526. [CrossRef]
45. Wang, S.; Xiong, H.; Tang, Y.; Zhang, W.; Zhang, Y.; Liu, Q.; Zhang, J. High sensitivity and selectivity triethylamine gas sensor based on ZnO–$SmFeO_3$ molecular imprinted polymers. *Mater. Res. Bull.* **2023**, *161*, 112147. [CrossRef]
46. Zhang, H.; Liu, L.; Huang, C.; Liang, S.; Jiang, G. Enhanced acetone gas sensing performance of ZnO polyhedrons decorated with $LaFeO_3$ nanoparticles. *Mater. Res. Express* **2023**, *10*, 095902. [CrossRef]
47. Cao, E.; Zhang, Y.; Zhang, Y.; Hao, W.; Sun, B.; Sun, L. Acetone sensing characteristics of TiO_2-$LaFeO_3$ nanocomposites. *Mater. Lett.* **2023**, *351*, 135056. [CrossRef]
48. Wu, J.; Wu, Z.; Ding, H.; Wei, Y.; Huang, W.; Yang, X.; Li, Z.; Qiu, L.; Wang, X. Three-Dimensional Graphene Hydrogel Decorated with SnO_2 for High-Performance NO_2 Sensing with Enhanced Immunity to Humidity. *ACS Appl. Mater. Interfaces* **2020**, *12*, 2634–2643. [CrossRef]
49. Wærnhus, I.; Vullum, P.E.; Holmestad, R.; Grande, T.; Wiik, K. Electronic properties of polycrystalline $LaFeO_3$. Part I: Experimental results and the qualitative role of Schottky defects. *Solid State Ion.* **2005**, *176*, 2783–2790. [CrossRef]

50. Wærnhus, I.; Grande, T.; Wiik, K. Electronic properties of polycrystalline $LaFeO_3$. Part II: Defect modelling including Schottky defects. *Solid State Ion.* **2005**, *176*, 2609–2616. [CrossRef]
51. Baek, J.W.; Kim, Y.H.; Ahn, J.; Kim, D.-H.; Shin, H.; Ko, J.; Park, S.; Park, C.; Shin, E.; Jang, J.-S.; et al. Galvanic replacement reaction in perovskite oxide for superior chemiresistors. *J. Mater. Chem. A* **2022**, *10*, 23282–23293. [CrossRef]
52. Ma, Z.; Yang, K.; Xiao, C.; Jia, L. C-doped $LaFeO_3$ Porous Nanostructures for Highly Selective Detection of Formaldehyde. *Sens. Actuators B Chem.* **2021**, *347*, 130550. [CrossRef]
53. Yang, Z.; Su, C.; Wang, S.; Han, Y.; Chen, X.; Xu, S.; Zhou, Z.; Hu, N.; Su, Y.; Zeng, M. Highly sensitive NO_2 gas sensors based on hexagonal SnS_2 nanoplates operating at room temperature. *Nanotechnology* **2020**, *31*, 075501. [CrossRef] [PubMed]
54. Kumarage, G.W.C.; Jayawardena, S.; Zappa, D.; Shimomura, M.; Comini, E. Unlocking superior acetone sensitivity with Co_3O_4/ZnO nanowire innovations. *Sens. Actuators B Chem.* **2025**, *431*, 137450. [CrossRef]
55. Vikraman, H.K.; George, J.; Ghuge, R.S.; Painappallil Reji, R.; Jayaraman, S.V.; Kawazoe, Y.; Sivalingam, Y.; Mangalampalli, S.R.N.K. Highly Selective, Room-Temperature Triethylamine Sensor Using Humidity-Resistant Novel TiZn Alloy Nanoparticles-Decorated MoS_2 Nanosheets. *Small* **2025**, *21*, 2408500. [CrossRef]

Disclaimer/Publisher's Note: The statements, opinions and data contained in all publications are solely those of the individual author(s) and contributor(s) and not of MDPI and/or the editor(s). MDPI and/or the editor(s) disclaim responsibility for any injury to people or property resulting from any ideas, methods, instructions or products referred to in the content.

chemosensors

MDPI

Article

Modeling the Frequency–Amplitude Characteristics of a Tunable SAW Oscillator

Ionut Nicolae and Cristian Viespe *

National Institute for Laser Plasma and Radiation Physics, Atomistilor 409, 077125 Magurele, Romania; ionut.nicolae@inflpr.ro

* Correspondence: cristian.viespe@inflpr.ro

Abstract: The resonant frequency of an SAW oscillator can be modulated by varying the signal amplitude, due to non-linear acoustic interactions within the chemoselective layer. In this study, we developed an explicit model to describe the amplitude–frequency behavior of a tunable SAW oscillator. A polymeric layer of variable thickness was deposited in a circular area (radius 1.1 mm) at the center of the piezoactive surface. Increasing the oscillator loop attenuation resulted in a continuous increase in the resonant frequency by up to 1.8 MHz. The layer was modeled as a succession of non-interacting sub-layers of varying thicknesses. As a result, the function model consists of a superposition of terms, each corresponding to a layer region of distinct length and thickness. The maximum difference between the experimental data and function model (also known as residual of the fit) was below 1% (13.02 kHz) of the resonant frequency variation, thus supporting the validity of our approach. While our model proved successful, the results suggest that some interactions are unaccounted for, as evidenced by the periodicity of the residuals of fit and unrealistically large variation in acoustic wave velocity.

Keywords: analyte discrimination; non-linear response; gas sensors; non-gravimetric effects; SAW sensors

1. Introduction

The continuous expansion of human civilization has led to a growing demand for sensing devices that are accurate, affordable, and high-performing—a trend that shows no signs of slowing down. This demand pertains not only to quantity but also to quality. As an example, a complex air chemical composition assessment is needed for the food industry as a means to evaluate the state of perishable food items. In addition, due to the continuous industrial development, environmental pollution has become ubiquitous, posing a problem even in indoor domestic spaces [1]. This is because a plethora of complex chemical compounds is slowly released into the air by furniture and domestic appliances, and many of them are a health hazard even in low concentrations [2]. Air toxins such as formaldehyde, acetaldehyde, 1,4-dichlorobenzene, chloroform, carbon tetrachloride, benzene, and naphthalene have been identified as pollutants of potential concern because of several risk-based analyses [3,4]. Multiple disorders associated with VOC indoor air pollution were amply reported, such as irritation of the eyes and nose, respiratory dysfunctions, sick-building syndrome, and leukemia, especially in children [5–13]. As a result, the specialized literature abounds in reports on new developments in this field, most reporting advances in the field of materials used for chemoselective layers. An important aspect limiting the

SAW (surface acoustic wave) sensors' success is the complex interactions governing their functionality, some of which are related to device-layer acoustic interactions. Still, it is due to this complexity that SAW devices are a natural option for complex environmental and material assessment. Sustained efforts have been made to harness the inherent complexity of these acoustic devices. The design and implementation of multi-mode/multi-frequency SAW devices have also been tested for the integration of the functionality of an entire sensor array into one single device. This approach consists of designing an SAW device capable of generating different modes simultaneously [14], multiple-order resonance [15], and both multi-mode and multi-frequency responses [16]. In the same vein, an interesting approach was to probe wavefront scattering by using multiple receiving IDTs [17]. Device-layer acoustic interaction is the basis of our approach to expanding the acoustic wave sensor's capability. Resonant frequency variation due to non-linear reactions in a chemoselective layer within an SAW loop oscillator has the potential to greatly expand the application area of SAW devices. This technique works best in combination with soft polymeric chemoselective layers. However, proper calibration of the device is necessary in order to extract valid sensing information. Until now, this has proved a difficult task, due to the technique's sensitivity to the slightest differences in layer morphology. Therefore, the analytical information contained in the SAW sensor's amplitude–frequency characteristic must be deconvoluted from the effects of the morphological features of the layer. Developing an explicit functional model for the layer reaction would allow for the quantification of its physical properties and therefore account for their changes. Such a model could then be used to calibrate SAW sensors in spite of the differences in the morphology of their chemoselective layers, and it could serve as a foundation for an analytical protocol based on a single SAW sensor. This perspective becomes even more promising in light of recent insights [18] into the physics of SAW propagation, which suggest the potential for implementing hypersensitive SAW sensors using this technique. In our previous article [19], the effect of layer morphology and position on the amplitude–frequency characteristics of a tunable SAW oscillator was qualitatively evaluated. One important result was that the frequency trend appeared to be almost linearly dependent on attenuation. This behavior of the resonant frequency rendered our previous model [20] unusable for the characteristics tested in [19], presumably due to its limited generality. In [20], the layer covers nearly the entire piezoactive area; in the second case, over 90% of it remains exposed. For this type of layer, we propose a new model that accounts for the general case in which the layer partially covers the SAW device and successfully fits the experimental data.

2. Materials and Methods

The tunable oscillator consists of an SAW device with a central frequency of 70 MHz connected in a closed-loop configuration with a DHPVA-200 FEMTO amplifier (Messtechnik GmbH, Berlin, Germany). The SAW sensor consists of an ST-X cut quartz substrate with interdigital transducers (IDTs) made of 200 nm thick gold electrodes deposited on a 10 nm chromium adhesion layer, patterned using standard photolithography. A double/double finger design was employed, featuring 50 electrode pairs with a periodicity of 11 μm, resulting in a resonant frequency of approximately 69 MHz. A potentiometer included in the oscillator's feedback loop allowed resonant frequency control, via non-linear acoustic reaction of the PEI (Polyethylenimine) layer deposited on the SAW device. The SAW was placed in an RF-isolated, closed chamber filled with synthetic air with zero humidity. On the center of the SAW device, a roughly circular, 2.2 mm diameter PEI layer was deposited. This was created by drop casting a 0.2 μL 1% by weight PEI solution in ethanol. Visual inspection revealed that the PEI film resulting after solvent evaporation had random variations in thickness, with the thickest point being located close to the center of the layer. Signal attenuation in the feedback loop influences the effective amplitude at the SAW surface, thereby modulating the non-linear response of the polymer layer. Changing the resonant frequency of the SAW oscillator was possible using a rotary potentiometer inserted into the feedback loop. A computer-controlled, high-precision, motorized rotary stage was used to actuate the potentiometer, thus providing fast, precise, and fine-grained frequency control. For maintaining stable experimental conditions, the SAW device was placed on a 87 Watt Peltier element which maintained a constant temperature of 25 ± 0.1 °C. Automated, real-time temperature adjustment was made by a TEC-1089 Meerstetter Peltier controller. A schematic of the setup is presented in Figure 1.

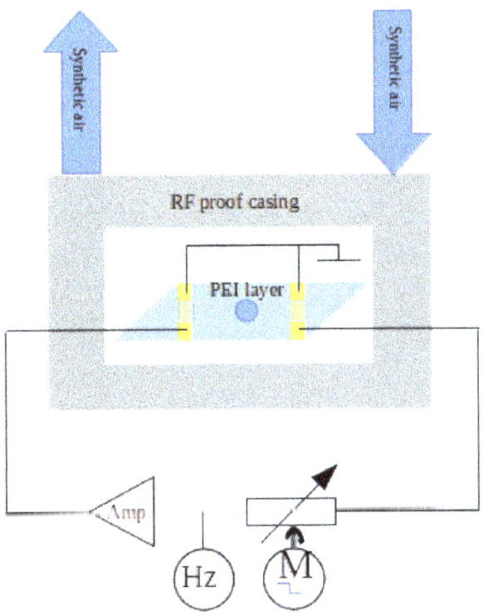

Figure 1. Experimental setup for the tunable SAW oscillator.

A CNT-90-type frequency counter (Spectracom Corp, Rochester, NY, USA) was used to acquire the resonant frequency values. Custom-made software automated the entire measurement process, consisting of successive potentiometer actuations and frequency data acquisition.

3. Modeling

The most prominent methods for addressing the problem under study are parametric non-linear regression and the finite element method (FEM). Although FEM is a powerful and widely used simulation tool, it presents significant challenges in this context. At the operating frequency (∼60 MHz), FEM requires an extremely fine mesh to accurately resolve wave propagation across the millimeter-scale geometry of the device, resulting in prohibitively high computational demands. Furthermore, the non-linear viscoelastic behavior of the polymer layer is difficult to parameterize due to the lack of precise higher-order material constants. To address a more general case than that of uniform coverage, the polymeric layer in our configuration was intentionally deposited in a localized shape, covering only a portion of the piezoactive SAW region. Additionally, due to the nature of the deposition technique, the layer's shape was irregular in both contour and thickness. This kind of partial and asymmetric coverage further complicates any FEM-based simulation, as it would require detailed spatial characterization and fine spatial meshing—conditions that are rarely practical in experimental setups. By contrast, the parametric analytical model we developed accounts for such interactions in a simplified yet physically meaningful way. It enables efficient evaluation without the need for fine spatial meshing or exhaustive material data. Moreover, the model captures the non-linear behavior of the polymer layer through fitted parameters, making it well suited for non-linear regression with experimental data.

In an SAW oscillator, the central resonant frequency is the result of interdigital transducers' spatial frequency and acoustic wave propagation velocity. Besides that, the resonant frequency must fulfill the standing wave condition, i.e., the acoustic and electric path length all across the oscillators loop amounts to an integer number of wavelengths. Depositing a layer on the SAW surface increases the inertia of the material that undergoes acoustic oscillations. With a constant imparted oscillatory energy, the conservation law dictates a reduction in surface wave propagation velocity and wavelength. As a result, the resonant frequency of the oscillator will shift so that it regains the standing wave condition, adjusting for the wavelength reduction due to a decrease in velocity. Also, in an acoustically thick layer, the top surface shear oscillation lags behind the oscillation at the layer–SAW interface. In other words, the shear oscillation reflected by the top surface of the layer exhibits a significant phase difference Φ with respect to the shear waves of the piezoelectric surface. The interference between the two shear waves results in the wavefront at the SAW surface being delayed or advanced, resulting in either a decrease or an increase in wave propagation velocity and a subsequent resonant frequency adjustment. The two effects briefly outlined above are called gravimetric and acoustic effects, and they are both accounted for by the following equation [21]:

$$\frac{\Delta v}{v} = -Im\Big(\sum_{j=x,y,z}\Big(\frac{c_j\beta_j M_j}{\omega} tanh(i\beta_j h)\Big)\Big), \tag{1}$$

where $\beta_j = \omega((\rho - E_j/v_0^2)/M_j)^{\frac{1}{2}}$ and $E_x = G, E_y = 0, E_z = 4G(3K+G)/(3K+4G)$. In the above equations, v is the surface wave propagation velocity, *Im()* represents the imaginary part of the expression, index j gives the oscillation's polarization, i is the imaginary unit, c_j is the SAW–film coupling parameter [21], *tanh()* is the hyperbolic tangent function, ω is the wave pulsation $2\pi f$ with f being the oscillation frequency of the SAW, v_0 is the surface acoustic wave velocity of an uncoated SAW, ρ is the layer density, and $M_x = M_z = G, My = K$. In this model, phase difference $\Phi_j = Re(i\beta_j h)$, where *Re()* represents the real part of the expression in parenthesis.

We will approximate the layer as a succession of m adjacent sub-layers of different lengths l_k (k = 1…m), with constant thicknesses h_k (see Figure 2). Under our experimental conditions, the number of sub-layers yielding the best fit was eleven (m = 11). In the general case, the layer does not cover the entire piezoactive area, so a length $L - \sum l_k$ of the free SAW surface will be included, where L is the distance between the two IDTs. Accordingly, we approximate the total time T needed for the wavefront to cross the space between the two IDTs as the sum of the times required to traverse all the sub-layers, plus the time needed to cross the uncovered portion of the SAW surface. The total wavefront propagation time is therefore given by

$$T = \sum \frac{l_k}{v_k} + \frac{L - \sum l_k}{v_0}. \tag{2}$$

Considering the standing wave condition, the total propagation time is equal to an integer number of wave periods, i.e., $T = nT_0$, where n is an integer and T_0 is the wave period. By substituting wave period T_0 with resonant frequency f in Equation (2), we obtain

$$f = \frac{n}{\sum \frac{l_k}{v_k} + \frac{L-\sum l_k}{v_0}}. \quad (3)$$

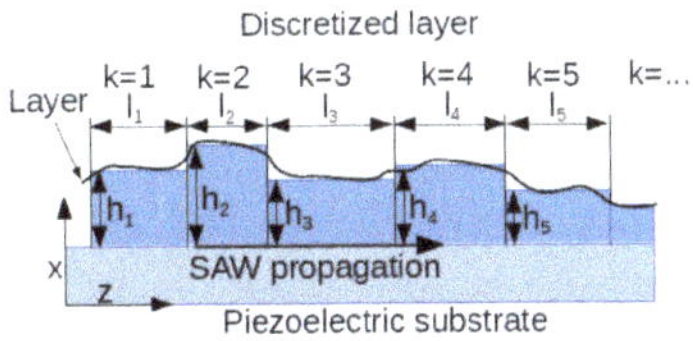

Figure 2. PEI layer decomposition in sub-layers.

For the chemoselective layer, we preferred a polymer, for its pronounced non-linear response. Shear thinning is specific to polymeric materials, and consists of viscosity reduction proportional to the shear rate. In our setup, the SAW oscillator resonant frequency shift due to amplitude variation is the effect of the loss coefficient G″ dependence on the oscillation amplitude, i.e., shear thinning [22,23]. For the purposes of this study, both the argument and the scaling factor of the hyperbolic tangent function in Equation (1) are approximated using first-degree complex polynomials in attenuation x. For each oscillation polarization, an independent set of parameters is provided. To alleviate the departure from the non-linearity of the original formula and to account for the differences between them, distinct parameters will be used for the argument and the scaling factor of the hyperbolic tangent function. In addition to varying the purely complex part, we will introduce parameters B_j and BB_j to account for non-linear effects in the purely real part of the expressions, thereby enhancing the model's flexibility. Thus, the multiplicative factor of the tangent function will be represented by a variable complex quantity, with a linear dependence on the signal attenuation x, $P1_j(x) = A_j + xB_j + iC_j + ixD_j$. This quantity is polarization-mode-specific (indexed by j) in Equation (1), resulting in three sets of parameters. As mentioned, we opted for an independent expression $P2_j(x) = AA_j + xBB_j + iCC_j + ixDD_j$ for the argument of the hyperbolic tangent, also with distinct parameters for each oscillation mode j. Due to the energy amount dissipated along the way, the oscillatory amplitude varies across sub-layer locations. This is due to the attenuation of acoustic energy during wave propagation, which leads to a spatially varying amplitude profile. Thus, a parameter S_k will be provided to scale the attenuation x accordingly, with its value constrained within the [0, 1] interval. Although this choice may seem counterintuitive—since attenuation increases with propagation and one might expect a scaling factor greater than 1—it is important to note that part of the attenuation effect is implicitly included in other multiplicative terms such as B_j, BB_j, D_j and DD_j. Therefore, the [0, 1] range for S_k is effectively equivalent to a higher-than-unity range. We adopted this formulation to enhance computational efficiency. Considering all of the above, the velocity v_k, corresponding to a portion of the sub-layer of thickness h_k, can be written as follows:

$$v_k = \sum_j v_0(1 - Im(P1_j(xS_k)tanh(P2_j(xS_k))). \quad (4)$$

The final form of the model used for data fitting is obtained by substituting Equation (4) into Equation (3) and carrying out the summation over eleven sub-layers, indexed from $k = 1$ to 11.

The goodness of fit will be quantified based on the coefficient of determination (R2) and the maximum of absolute fit residual. The coefficient of determination gives a measure of how close the response function is to experimental results. It is given by subtracting the ratio between the residual sum of squares and the total sum of function response squares from unity. It takes values between 1 and 0, and it reaches unity in the ideal case of perfect fit, i.e., the residual sum of squares is null. Measurement units were normalized to centimeters and seconds by constants v_0 and L, with values of 1.4×10^4 cm/s and 1 cm, respectively. Constraints were imposed on sub-layer lengths and thicknesses, such that $l_k > 0$, $h_k > 0$, and $\sum l_k < 0.22$ cm. Additionally, the number of wavelentghs n, corresponding to resonant standing wave mode, was constrained to the interval [150, 200]. All other parameters were left unconstrained.

4. Results

The amplitude–frequency dependence of the SAW oscillator was probed for 245 different attenuation values. Its profile closely resembles those presented in (20) and consists of an increasing resonant frequency from 59.5 MHz to well above 60 MHz. It is characterized by the presence of two sigmoids and a significant number of curvatures and sharp, small-scale variations. Also, an overall slight curvature of the frequency variation can be noted.

We used the NonlinearModelFit function in Wolfram Mathematica to perform the parameter estimation. The model, as defined by Equations (3) and (4), was provided as input along with the corresponding experimental resonance frequencies. The fitting procedure was based on a least-squares minimization of the differences between experimental and theoretical resonance frequencies, returning both the optimal parameters and the goodness-of-fit metrics. Fitting attempts were made for different numbers of component sub-layers. By successive testing, the optimal number of component sub-layers was established to eleven (k = 1 to 11), the optimization criteria being the maximization of the coefficient of determination (R2). In Figure 3, the fitting model and residuals are presented side by side.

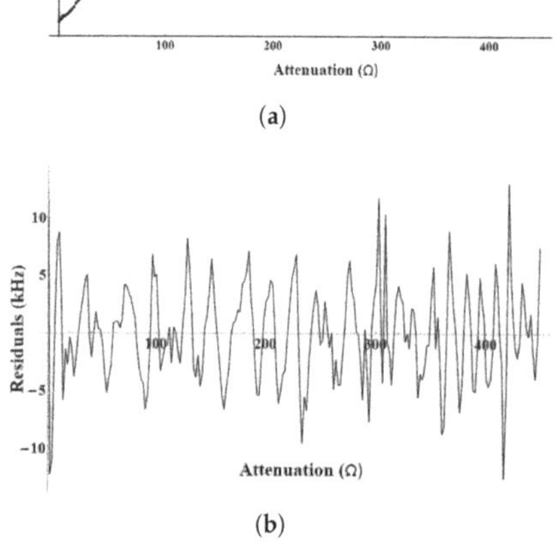

(a)

(b)

Figure 3. (**a**) Model fit and experimental data. (**b**) Fit residuals.

The graphical representation of the model fit versus experimental data in Figure 3 demonstrates the excellent agreement between the model and the measurements. This is also confirmed by the coefficient of determination (R^2), which has a value of more than 0.99997, only marginally less than unity. This value signifies exceptional accuracy of the proposed model. While the large number of parameters certainly played a role, it is still a strong indication that the theoretical reasoning behind the model appears well founded. Additionally, Figure 4 shows the residual histogram. While the maximum residual was 13.020 kHz, the vast majority of residuals were below 7 kHz, and more than 50% of them fell below 5 kHz.

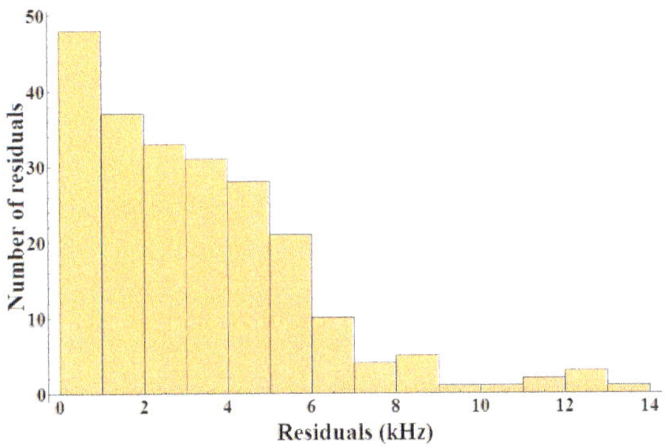

Figure 4. Histogram of residuals.

As shown in Figure 3, the variation in fit residuals exhibits a discernible oscillatory pattern. A clear peak–valley alternation is observable, which contrasts with the expected random distribution originating from stochastic processes and suggests a systematic origin for the model–data discrepancies. One possible explanation could be unaccounted wave interference between adjacent sub-layers of differing thicknesses. This hypothesis is strongly supported by the periodic nature of the residuals.

In Table 1, oscillation mode (j)-related and sub-layer (k)-related parameters are given. The number of wavelengths, n, is 150.52. The parameters associated with the elasticity module (B and BB) are small, two and three orders of magnitude lower, respectively. Although this may be expected, they are non-zero and contribute partially to amplitude–frequency characteristics. Moreover, it is not necessary to tie its origin to the SAW device, since it could be caused by complex amplification and measurement electronics. This hypothesis can be readily tested, as a detection experiment involving analytes should leave their values unchanged. We emphasize that parameters h_k do not represent sub-layer thicknesses, as they are part of the multiplicative factor within the argument of the hyperbolic tangent function. While, in principle, they could offer a relative measure of sub-layer thicknesses, this is also not the case, due to the implicit approximation of the layer as a PEI strip transverse to the wave propagation direction. It remains to be experimentally tested if their values are valid indicators that could be used in a detection/identification protocol. The lengths l_k, on the other hand, have their units fixed due to the presence of the constants L and v_0 in Equation (2), though they are still presumably affected by the approximation of the layer as a transverse strip. Nevertheless, their behavior may provide valuable insights into the model's limitations. Additionally, it is possible that, even without direct physical significance, they might still function as valid indicators in a detection protocol.

Table 1. Parameter values. Top: Oscillation mode parameters. Bottom: Sub-layer component parameters.

j	0	1	2
A_j	0.271991	0.39837	0.383148
B_j	0.000184	0.000191	0.000151
C_j	0.385068	0.339205	0.368484
D_j	0.0642	0.030183	0.052371
AA_j	0.421417	0.443985	0.294993
BB_j	0.003614	−0.00331	0.000769
CC_j	0.075363	0.657828	0.659584
DD_j	0.143377	0.067499	0.176909
k	$\mathbf{S}_k$	$\mathbf{h}_k$	$\mathbf{l}_k$ **(mm)**
0	0.186418	0.546559	0.016482
1	0.200251	0.601545	0.032309
2	0.204112	0.548499	0.03191
3	0.207271	0.618024	0.025398
4	0.220155	0.61322	0.030375
5	0.249576	0.575603	0.021884
6	0.27225	0.645901	0.012917
7	0.282554	0.563273	0.016547
8	0.287698	0.531453	0.012966
9	0.288238	0.517732	0.01033
10	0.28883	0.572192	0.012563

5. Discussion

Separate plotting of the relative velocity variation brings unexpected high values to attention, such as the one seen in Figure 5. For comparison purposes, the fitting procedure was also performed using a different model (not included here), featuring a smaller number of components ($k = 1$ to 6) and independent parameter sets (A to DD) for each sub-layer, totaling 61 parameters. Although the fitting criteria were acceptable—with a maximum residual of 17.8 kHz—the resulting relative velocity variation was even higher, reaching several hundred. This strongly suggests that the elevated relative velocity values result from the lower number of components. In such a case, high-amplitude/high-frequency components become necessary to reproduce the fine details of the amplitude–frequency profile. Furthermore, the unrealistic propagation velocity excursions might stem from the unaccounted non-linearity in both the hyperbolic tangent argument and its multiplying factor.

While the magnitude of the velocity variation might suggest that the oscillatory modes in the model correspond to the x, y, and z polarizations of the acoustic waves, the presence of the PEI layer complicates this interpretation. Z-polarized modes correspond to compression waves, and the large value of the bulk modulus E typically suppresses such resonances. Therefore, it is plausible that the observed resonances are artifacts arising from layer discretization and the limited number of sub-layer components. This interpretation is further reinforced by the presence of multiple-layer resonances.

Although the finite element method (FEM) is a powerful general-purpose simulation tool, it is not well suited for this specific problem. At the operating frequency (60 MHz), FEM requires an extremely fine mesh to resolve wave behavior over the millimeter-scale geometry, resulting in prohibitively large computational loads. Additionally, the non-linear viscoelastic behavior of the polymer layer is difficult to parameterize for FEM because accurate higher-order material constants are unavailable. By contrast, our analytical model captures the non-linearity through fitted parameters and can be evaluated quickly, without the burden of meshing or detailed material data.

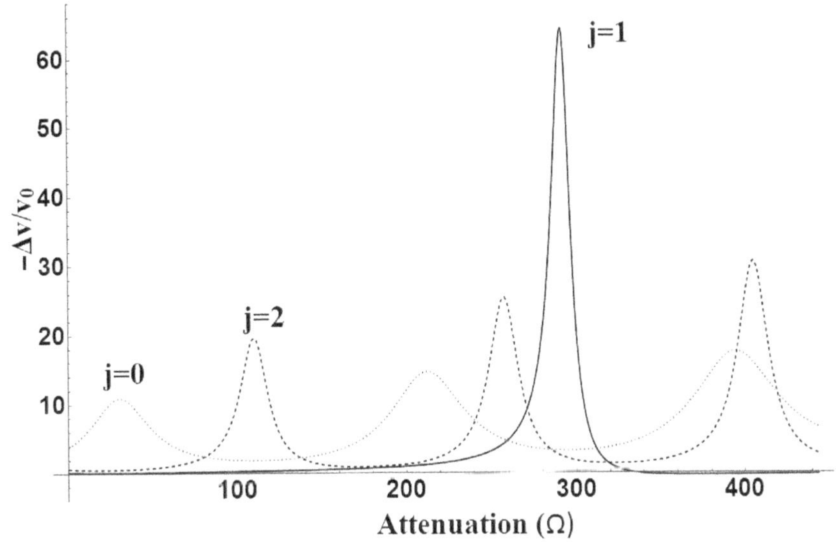

Figure 5. Relative velocity variation for the sub-layer k = 4.

In a conventional configuration—where the chemoselective layer covers the entire piezoactive surface—higher-order resonances are highly unlikely due to the strong energy dissipation in viscoelastic polymer films [21]. However, when the layer covers only a small portion of the surface, the dissipated energy is proportionally reduced. This leads to a lower effective loss modulus and, in turn, allows for higher propagation velocity excursions and potentially multiple-layer resonances. At this stage, the precise origin of the modeled relative velocity behavior remains unclear. Another anomaly is the relatively low number of wave cycles, around 150. By comparison, the number of wavelengths corresponding to the uncovered length of the SAW surface at a 60 MHz resonant frequency is 152. This mismatch can be attributed to the anomalously high wave velocity—an effect clearly shown in Figure 4. Again, this supports the hypothesis that the model's limited number of sub-layers fails to fully capture the actual physical profile of the deposited film. In reality, the PEI layer is not composed of discrete sub-layers but rather exhibits a continuous thickness gradient. A more accurate modeling strategy would involve integrating the propagation velocity across a layer with a continuously varying thickness. Provided that such an approach is mathematically and computationally viable, it represents the intended next step in our model development. On the other hand, significant velocity variations may indeed be required to account for the observed resonance excursion. To better understand this, one should examine Equation (3), where the denominator consists of the total time required for the wavefront to pass through both the covered and uncovered regions of the SAW device. For fixed values of l_k, the time needed to cross the uncovered portion is constant—approximately $0.8/(1.4 \times 10^5)$ cm/s. Since the resonant frequency is inversely proportional to this total propagation time, the magnitude of the velocity variation is amplified by the ratio between the lengths of the covered and uncovered regions. A smaller covered area (as is the case here) necessitates a larger apparent velocity variation to achieve the same total frequency shift. Thus, the high velocity variation seen in the results might be a necessary consequence of the small surface area of the PEI layer. Nevertheless, even if such variation is mathematically consistent, it is not physically plausible that it arises solely from the interactions modeled by Equation (1). This strongly suggests the presence of additional significant interactions. One such factor may be the acoustic lensing effect caused by the circular geometry of the polymer layer. In this configuration, the layer focuses the incident SAW wavefront, resulting in overlapping waves with differing phases. The resulting interference introduces additional velocity modulation that is not explicitly modeled in Equation (1) and is thus erroneously attributed to the modeled interactions, leading to artificially high velocity values.

6. Conclusions

While the model successfully captured the complex amplitude–frequency characteristics, several key questions remain open. Firstly, the issue of the unexpectedly large variations in propagation velocity is one of them. One of the primary goals of this study was to develop a model with a high degree of generality, which led to a relatively complex experimental and modeling scenario. Future investigations should aim to simplify the experimental setup and reduce the number of model parameters, thereby increasing the robustness and interpretability of the results. It is worth noting that the use of the corresponding expressions derived in Equation (1) yielded no meaningful results in this particular context. From a modeling standpoint, the implementation of a continuous variation in layer thickness is likely to be the most beneficial next step. When combined with polymeric layers that fully cover the piezoactive region, this approach would enable a more accurate estimation of acoustic wave velocity variations. Moreover, by fully covering the piezoactive area, the modeling process becomes less dependent on the exact geometry of the deposited layer. Nevertheless, this approach may still prove useful in representing transversal thickness gradients, though its effectiveness may be limited by the fact that the SAW wavefront expands laterally as it approaches the receiving interdigital transducer (IDT). As a result, only a narrow region of the layer directly influences wave propagation. Finally, an important concern remains: the potential utility of this model in the design and implementation of detection or identification protocols for multiple analytes. Targeted experiments will be developed and carried out to address this question and assess the model's practical applicability in real-world sensing scenarios.

Author Contributions: Conceptualization, I.N.; methodology, I.N.; formal analysis, I.N.; investigation, I.N.; resources, C.V.; writing—original draft preparation, I.N.; writing—review and editing, I.N. and C.V. All authors have read and agreed to the published version of the manuscript.

Funding: This research was supported (or financed) by the Romanian Ministry of Research, Innovation and Digitalization under the Romanian National Nucleus Program LAPLAS VII, with contract no. 30N/2023 and POCIDIF nr. 390008/27.11.2024.

Institutional Review Board Statement: Not applicable.

Informed Consent Statement: Not applicable

Data Availability Statement: The data presented in this study are available on request from the corresponding author due to legal reasons.

Conflicts of Interest: The authors declare no conflicts of interest.

References

1. Missia, D.A.; Demetriou, E.; Michael, N.; Tolis, E.I.; Bartzis, J.G. Indoor exposure from building materials: A field study. *Atmos. Environ.* **2010**, *44*, 4388–4395. [CrossRef]

2. Air Pollution: Current and Future Challenges. EPA-United States Environmental Protection Agency. Available online: https://www.epa.gov/clean-air-act-overview/air-pollution-current-and-future-challenges (accessed on 23 August 2024).
3. Loh, M.M.; Levy, J.I.; Spengler, J.D.; Houseman, E.A.; Bennett, D.H. Ranking Cancer Risks of Organic Hazardous Air Pollutants in the United States. *Environ. Health Perspect.* **2007**, *115*, 1160–1168. [CrossRef] [PubMed]
4. Jia, C.; D'Souza, J.; Batterman, S. Distributions of personal VOC exposures: A population-based analysis. *Environ. Int.* **2008**, *34*, 922–931. [CrossRef] [PubMed]
5. Norbäck, D.; Björnsson, E.; Janson, C.; Widström, J.; Boman, G. Asthmatic symptoms and volatile organic compounds, formaldehyde, and carbon dioxide in dwellings. *Ocupational Environ. Med.* **1995**, *52*, 388–395. [CrossRef] [PubMed]
6. Daisey, J.M.; Angell, W.J.; Apte, M.G. Indoor air quality, ventilation and health symptoms in schools: An analysis of existing information. *Indoor Air* **2003**, *13*, 53–64. [CrossRef] [PubMed]
7. Rumchev, K.; Spickett, J.; Bulsara, M.; Phillips, M.; Stick, S. Association of domestic exposure to volatile organic compounds with asthma in young children. *Thorax* **2004**, *59*, 746–751. [CrossRef] [PubMed]
8. Vanos, J.K.; Hebbern, C.; Cakma, S. Risk assessment for cardiovascular and respiratory mortality due to air pollution and synoptic meteorology in 10 Canadian cities. *Environ. Pollut.* **2013**, *185*, 322–332. [CrossRef] [PubMed]
9. Gao, Y.; Zhang, Y.; Kamijima, M.; Sakai, K.; Khalequzzaman, M.; Nakajima, T.; Shi, R.; Wang, X.; Chen, D.; Ji, X.; et al. Quantitative assessments of indoor air pollution and the risk of childhood acute leukemia in Shangha. *Environ. Pollut.* **2014** *187*, 81–89. [CrossRef]
10. Gonzalez-Flesca, N.; Bates, M.S.; Delmas, V.; Cocheo, V. Benzene exposure assessment at indoor, outdoor and personal levels. *Environ. Monit. Asses.* **2000**, *65*, 59–67. [CrossRef]
11. Salthammer, T.; Mentese, S.; Marutzky, R. Formaldehyde in the indoor environment. *Chem. Rev.* **2010**, *110*, 2536–2572. [CrossRef] [PubMed]
12. Nielsen, G.D.; Wolkoff, P. Cancer effects of formaldehyde: A proposal for an indoor air guideline value. *Arch. Toxicol.* **2010**, *84*, 423–446. [CrossRef] [PubMed]
13. Panagopoulos, I.K.; Karayannis, A.N.; Kassomenos, P.; Aravossis, K. A CFD simulation study of VOC and formaldehyde indoor air pollution dispersion in an apartment as part of an indoor pollution management plan. *Aerosol Air Qual. Res.* **2011**, *11*, 758–762. [CrossRef]
14. Lamanna, L.; Rizzi, F.; Bhethanabotla, V.R.; De Vittorio, M. GHz AlN-based multiple mode SAW temperature sensor fabricated on PEN substrate. *Sens. Act. A* **2020**, *315*, 112268. [CrossRef]
15. Schmalz, J.; Kittmann, A.; Durdaut, P.; Spetzler, B.; Faupel, F.; Höft, M.; Quandt, E.; Gerken, M. Multi-Mode Love Wave SAW Magnetic-Field Sensors. *Sensors* **2020**, *20*, 3421. [CrossRef] [PubMed]
16. Seidel, W.; Hesjedal, T. Multimode and multifrequency gigahertz surface acoustic wave sensors. *Appl. Phys. Lett.* **2004**, *84*, 1407–1409. [CrossRef]
17. Pamwani, L.; Habib, A.; Melandsø, F.; Ahluwalia, B.S.; Shelke, A. Single-Input and Multiple-Output Surface Acoustic Wave Sensing for Damage Quantification in Piezoelectric Sensors. *Sensors* **2017**, *18*, 67. [CrossRef] [PubMed]
18. Hamdaoui, M. Uncertainty Propagation and Global Sensitivity Analysis of a Surface Acoustic Wave Gas Sensor Using Finite Elements and Sparse Polynomial Chaos Expansions. *Vibration* **2023**, *6*, 610–624. [CrossRef]
19. Nicolae, I.; Bojan, M.; Viespe, C. Acoustic Effects of Uneven Polymeric Layers on Tunable SAW Oscillator. *Processes* **2024**, *12*, 1217. [CrossRef]
20. Nicolae, I.; Viespe, C.; Miu, D.; Marcu, A. Analyte discrimination by SAW sensor variable loop amplification probing. *Sens. Actuators B Chem.* **2022**, *358*, 131493. [CrossRef]
21. Martin, S.J.; Frye, G.C.; Senturia, S.D. Dynamics and response of polymer-coated surface acoustic wave devices: Effect of viscoelastic properties and film resonance. *Anal. Chem.* **1994**, *66*, 2201–2219. [CrossRef]
22. Shaw, M.T.; MacKnight, W.J. *Introduction to Polymer Viscoelasticity*, 3rd ed.; John Wiley & Sons: Hoboken, NJ, USA, 2005; pp. 25–27.
23. Ferry, J.D. *Viscoelastic Properties of Polymers*, 3rd ed.; John Wiley & Sons: New York, NY, USA, 1980; pp. 51–52.

Disclaimer/Publisher's Note: The statements, opinions and data contained in all publications are solely those of the individual author(s) and contributor(s) and not of MDPI and/or the editor(s). MDPI and/or the editor(s) disclaim responsibility for any injury to people or property resulting from any ideas, methods, instructions or products referred to in the content.

chemosensors

Article

Synthesis, Characterization, and Evaluation of Photocatalytic and Gas Sensing Properties of $ZnSb_2O_6$ Pellets

Jacob Morales-Bautista [1], Héctor Guillén-Bonilla [1,*], Lucia Ivonne Juárez-Amador [1], Alex Guillén-Bonilla [2], Verónica-María Rodríguez-Betancourtt [3], Jorge Alberto Ramírez-Ortega [3], José Trinidad Guillén-Bonilla [4] and María de la Luz Olvera-Amador [5]

[1] Departamento de Ingeniería de Proyectos, Centro Universitario de Ciencias Exactas e Ingenierías, Universidad de Guadalajara, M. García Barragán 1421, Guadalajara 44410, Jalisco, Mexico; jacob.mbautista@academicos.udg.mx (J.M.-B.); lucia.juarez@academicos.udg.mx (L.I.J.-A.)

[2] Departamento de Ciencias Computacionales e Ingenierías, Centro Universitario de los Valles, Universidad de Guadalajara, Carretera Guadalajara-Ameca Km 45.5, Ameca 46600, Jalisco, Mexico; alex.guillen@academicos.udg.mx

[3] Departamento de Física, Centro Universitario de Ciencias Exactas e Ingenierías, Universidad de Guadalajara, M. García Barragán 1421, Guadalajara 44410, Jalisco, Mexico; veronica.rbetancourtt@academicos.udg.mx (V.-M.R.-B.); jorge.rortega@academicos.udg.mx (J.A.R.-O.)

[4] Departamento de Electro-Fotónica, Centro Universitario de Ciencias Exactas e Ingenierías, Universidad de Guadalajara, M. García Barragán 1421, Guadalajara 44410, Jalisco, Mexico; trinidad.guillen@academicos.udg.mx

[5] Departamento de Ingeniería Eléctrica-Seccion de Electronica del Estado Solido, Centro de Investigación y de Estudios Avanzados del Instituto Politécnico Nacional, Av Instituto Politécnico Nacional 2508, San Pedro Zacatenco, Gustavo A. Madero, Mexico City 07360, Mexico; molvera@cinvestav.mx

* Correspondence: hector.guillen1775@academicos.udg.mx; Tel.: +52-33-1378-5900 (ext. 27689)

Abstract: This work reports a low-cost, microwave-assisted wet chemistry synthesis of zinc antimonate ($ZnSb_2O_6$) powders with a trirutile structure, yielding highly homogeneous, nanometric particles. X-ray diffraction (XRD) confirmed the formation of the trirutile phase with lattice parameters of a = 4.664 Å and c = 9.263 Å, and an estimated crystallite size of 42 nm. UV–vis spectroscopy revealed a bandgap of 3.35 eV. Scanning electron microscopy (SEM) showed that ethylenediamine, as a chelating agent, formed porous microstructures of microrods and cuboids, ideal for enhanced gas adsorption. Brunauer–Emmett–Teller (BET) analysis revealed a specific surface area of 6 m^2/g and a total pore volume of 0.0831 cm^3/g, indicating a predominantly mesoporous structure. The gas sensing properties of $ZnSb_2O_6$ pellets were evaluated in CO and C_3H_8 atmospheres at 100, 200, and 300 °C. The material exhibited high sensitivity at 300 °C, where the maximum responses were 5.86 for CO at 300 ppm and 1.04 for C_3H_8 at 500 ppm. The enhanced sensitivity at elevated temperatures was corroborated by a corresponding decrease in electrical resistivity. Furthermore, the material demonstrated effective photocatalytic activity, achieving up to 60% degradation of methylene blue and 50% of malachite green after 300 min of UV irradiation, with the process following first-order reaction kinetics. These results highlight that $ZnSb_2O_6$ synthesized by this method is a promising bifunctional material for gas sensing and photocatalytic applications.

Keywords: microwave radiation; $ZnSb_2O_6$ pellets; gas sensing; photocatalytic performance

 https://doi.org/10.3390/chemosensors13090329

1. Introduction

$ZnSb_2O_6$, also known as ordoñezite, is a mineral first discovered by G. Switzer and W. F. Foshag at the Santín mine, Guanajuato, Mexico [1,2]. This compound has been synthesized and reported as an n-type semiconductor oxide due to its electron (e^-) charge carrier mobility [3,4]. $ZnSb_2O_6$ fits the general formula AB_2O_6 for trirutile-type crystal structures [5], where A can be substituted with divalent ions such as Mg^{2+}, Ca^{2+}, Mn^{2+}, Co^{2+}, Cu^{2+}, and Zn^{2+}, while B can be occupied by pentavalent cations like Sb^{5+} or Ta^{5+} [6]. Specifically for $ZnSb_2O_6$, site A is occupied by the divalent zinc ion (Zn^{2+}) and site B by the pentavalent antimony cation (Sb^{5+}) [7,8]. $ZnSb_2O_6$ belongs to the antimonate family characterized by trirutile crystal structures [9]. It is isomorphous with $MgSb_2O_6$ (Byströmite) [1], as both belong to the P4/mnm space group with cell parameters of a = 4.68 Å and c = 9.21 Å [9]. According to references [10,11], the crystal structure of $ZnSb_2O_6$ is composed of SbO_6 octahedra that share edges along the c-axis. These, in turn, share corners with other SbO_6 octahedra, and both Zn and Sb atoms are octahedrally coordinated by six oxygen atoms [11,12]. This electronic arrangement gives rise to interesting physical and chemical properties that depend on the oxide's synthesis process [12,13].

Various wet chemistry processes, which are simple and easy to implement, have been employed to prepare $ZnSb_2O_6$. Roper et al. [1] prepared the oxide using the Bystrom method, obtaining the parameters mentioned above. However, other authors have synthesized the compound through alternative routes such as the colloidal method [13], solid-state reaction (ceramic method) [14], sol–gel [15], microwave-assisted solution method [16], hydrothermal method [17], precipitation method [18], and a simple sonochemical process [11]. Despite this variety, many of these routes require high calcination temperatures, often exceeding 800 °C, which increase energy consumption and can lead to undesirable particle sintering, thereby reducing the effective surface area for sensing and photocatalytic applications. According to such reports, those preparation routes enable the production of particle sizes smaller than 100 nm and various morphologies, including nanoparticles, nanorods, nanowires, nanospheres, and nanodiscs [7,19–21], among others.

Overall, chemical methods have enabled the synthesis of nanostructures with specific morphological characteristics at relatively low temperatures (below 1000 °C). These methods limit excessive particle growth by increasing the specific surface area of the material [7,19]. This results in better particle dispersion and greater uniformity in size distribution, which are fundamental aspects for improving the material's efficiency in applications such as chemical gas sensors and photocatalysis [19–22].

These microstructural features significantly impact the properties of $ZnSb_2O_6$, including its electrical, thermal, optical, and carrier density properties [4,11,23]. As a result, $ZnSb_2O_6$ emerges as a promising candidate in photocatalysis, since the limited literature suggests that its unique properties could be effectively exploited for such applications [10,11]. Due to this, the oxide has also been applied in anti-static agents for plastics, in lithium-ion batteries (LIBs), in supercapacitors (SCs), and in solar cells [4,7,8,11,23].

On the other hand, $ZnSb_2O_6$ has also been reported as a potential gas sensor. For instance, in reference [15], thick films made from nanostructured $ZnSb_2O_6$ powders were successfully tested in CO, O_2, and CO_2 atmospheres. In that study, a good dynamic response is reported by applying frequencies from 0.1 to 100 kHz at 400 °C. In contrast, in reference [16], pellets were fabricated and subjected to static atmospheres of CO and C_3H_8. Good thermal stability and high sensitivity (6.66 for CO and 1.2 for C_3H_8) were reported at a concentration of 300 ppm and a temperature of 250 °C. Similarly, in reference [24], thick films of $ZnSb_2O_6$ were prepared via the dip-coating method, yielding good stability and selectivity in H_2S concentrations; the favorable response was attributed to the porosity of the films. Considering these findings, the good dynamic response, high sensitivity, and thermal stability of $ZnSb_2O_6$ in various atmospheres are attributed to the microstructure obtained during the preparation process.

To overcome the challenge of high calcination temperatures, this work reports a microwave-assisted wet chemistry method that successfully yields the pure trirutile phase of $ZnSb_2O_6$ at a significantly lower calcination temperature of 600 °C. This process is not only cost-effective but also utilizes ethylenediamine to control the growth of unique porous microstructures, which are ideal for enhancing functional performance. The powders generated from the synthesis were microstructurally characterized. Subsequently, pellets were prepared and subjected to varying concentrations of carbon monoxide and propane at temperatures of 100, 200, and 300 °C. As expected, the compound showed a high sensitivity, capacity, and thermal stability as the concentration of the test gases and the operating temperature increased.

As a relatively new and less-explored material, $ZnSb_2O_6$ holds significant potential for advancing gas sensing and photocatalysis technologies. Studying and optimizing its properties through efficient synthesis can lead to more sensitive, stable, and sustainable devices, thereby expanding the possibilities for environmental applications and material innovation.

2. Experimental

2.1. Synthesis of $ZnSb_2O_6$ Particles

$ZnSb_2O_6$ powders were synthesized using an easy and cost-effective microwave radiation-assisted method. The synthesis involved the use of reagent-grade chemicals, including zinc nitrate hexahydrate ($Zn(NO_3)_2 \bullet 6H_2O$), antimony trichloride ($SbCl_3$), ethylenediamine ($C_2H_8N_2$), and ethanol (C_2H_6O). Initially, three separate solutions were simultaneously prepared. These included a 2 M solution of antimony trichloride, a 1 M solution of zinc nitrate hexahydrate, and a 7.5 M solution of ethylenediamine, each dissolved in 5 mL of ethanol. These solutions were stirred for 10 min. Subsequently, the zinc nitrate solution was added dropwise to the ethylenediamine solution, followed by the addition of antimony trichloride solution to the mixture. The resulting solution was kept under vigorous magnetic stirring at room temperature for 24 h to ensure homogeneity. The ethanol solvent was then removed via microwave irradiation at 1 min intervals, with 3 min cooling periods between each cycle, over a total irradiation time of 20 min, until a dried paste was obtained. This paste was further dried in air for eight hours at 200 °C. Finally, the dried material was calcined in air at 600 °C for five hours to achieve the trirutile phase of $ZnSb_2O_6$.

2.2. Physical Characterization

The crystal structure of the calcined $ZnSb_2O_6$ powders was characterized by X-ray diffraction (XRD) using a Panalytical Empyrean diffractometer with Cu Kα radiation (λ = 1.541 Å). The diffraction patterns were recorded over a 2θ range of 10° to 90°, with a scan step size of 0.02° and a time of 30 s per step.

The optical bandgap of the $ZnSb_2O_6$ powders was determined by diffuse reflectance spectroscopy. Spectra were acquired in the wavelength range of 250 to 800 nm using a Jasco V-670 UV–vis spectrophotometer equipped with an integrating sphere. The Kubelka–Munk transformation was applied to estimate the bandgap energy.

The main Raman spectroscopy vibrational modes of the calcined powders were analyzed using a Thermo Scientific DXR Raman microscope with a 633 nm excitation source. The spectra were recorded from 200 to 800 cm^{-1} with an exposure time of 60 s. Additionally, the microstructure of the calcined $ZnSb_2O_6$ powders was examined by scanning electron microscopy (SEM) using a Jeol model JSM-6390LV microscope, operated at an accelerating voltage of 20 kV. The BET analysis was carried out using the ASAP 2020 Micromeritics instrument.

2.3. Gas Sensing Measurements

For the gas detection experiments, 0.39 g of $ZnSb_2O_6$ powders was first pressed into pellets using an Ital-Mexicana manual hydraulic press. The pellets were prepared by applying a pressure of 5 tons for 20 min to ensure mechanical integrity. The resulting pellets had a diameter of 9 mm and a thickness of 2 mm.

For electrical characterization, two ohmic contacts were manually applied (and verified using a curve tracer) to the pellet surfaces using colloidal silver paint (Alfa Aesar, 99%). The pellets were placed on a heater located inside a vacuum chamber. A low vacuum of 10^{-3} Torr was set within the chamber, and the concentration of the test gases was controlled by gradually increasing the chamber pressure with the respective gases, monitored via a Leybold TM20 Vacuum Controller.

Gas sensing measurements were conducted using C_3H_8 (1, 5, 50, 100, 200, 300, 400, and 500 ppm) and CO (1, 5, 50, 100, 200, and 300 ppm) at three operating temperatures (100, 200, and 300 °C). The gases used for the measurements were separate mixtures of 1000 ppm of CO and C_3H_8 in N_2. The final gas concentrations in the chamber were calculated based on the partial pressures using the ideal gas law. The variation in electrical resistance of the pellets was recorded using a Keithley digital multimeter (model 2001). A schematic diagram of the sensing measurement setup is illustrated in Figure 1.

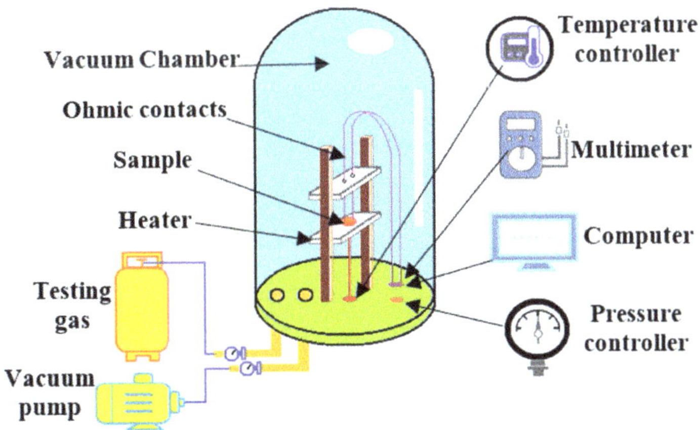

Figure 1. Diagram of the gas sensing equipment used to monitor the electrical resistance change in the sensing materials.

2.4. Photocatalytic Activity Test

The photocatalytic activity of the $ZnSb_2O_6$ powders was evaluated through the degradation of three organic dyes: malachite green, methylene blue, and methyl orange. Initially, 0.3 g of the $ZnSb_2O_6$ powders was pressed into pellets with a diameter of 9 mm and a thickness of 1 mm using an Ital-Mexicana manual hydraulic press.

Subsequently, each pellet was immersed in a quartz cell containing 3.5 mL of an aqueous dye solution at a concentration of 1×10^{-5} M. To establish adsorption–desorption equilibrium, the suspensions were kept in the dark for 30 min. Following this period, the samples were transferred to an annular photoreactor equipped with a 15 W UV-254 nm lamp, with the samples positioned 5 cm from the light source. The samples were then irradiated for 300 min. The decolorization of each dye solution was monitored at 60 min intervals by measuring the absorbance spectra using a JASCO V-670 UV–vis spectrophotometer.

3. Results

3.1. XRD Analysis

The X-ray diffraction pattern of the crystalline phase of $ZnSb_2O_6$ powders calcined at 600 °C is shown in Figure 2. Based on PDF 96-900-4698, peaks associated with the crystalline phase of the oxide were located at points 2θ = 19.35°, 21.31°, 27.10°, 33.37°, 35.04°, 38.50°, 40.11°, 43.23°, 44.38°, 48.12°, 52.97°, 55.72, 60.03°, 66.82°, 73.55°, 80.55°, 82.79°, 86.38°, and 88.96°. A tetragonal phase structure was identified, with cell parameters a = 4.6640 and c = 9.2630, and a space group of $P4_2/mnm$. These parameters are characteristic of a trirutile structure. In the diffractogram, the crystalline phase of the $ZnSb_2O_6$ is observed without the presence of any secondary phase. In addition, the diffractogram shows broad peaks, indicating a very small particle size (in the nanometer range), while the peak intensity indicates a high purity of the material. This result can be attributed to the synthesis method and calcination temperature employed.

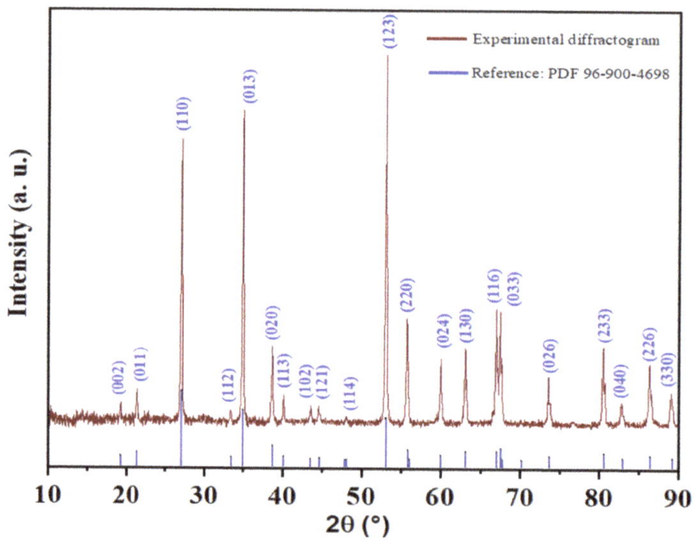

Figure 2. X-ray diffraction (XRD) pattern of $ZnSb_2O_6$ powders calcined at 600 °C.

Our results were consistent with those reported by other research groups, which synthesized the same compound by different processes [8,11,13].

To calculate the crystallite size of $ZnSb_2O_6$ (Figure 2), Scherrer's equation was used as follows [25]:

$$t = \frac{0.9\lambda}{\beta \cos\theta} \quad (1)$$

where λ is the wavelength of the radiation (Cu = 1.5406 Å), β is the peak full width at half maximum (FWHM), and θ is the Bragg angle. To determine the crystallite size, all the peaks in the diffractogram of Figure 2 were used. The estimated crystallite size was approximately 42 nm.

The crystallite's nanoscale size is mainly attributed to the synthesis method. The size is crucial for gas sensor applications, as reduced particle sizes increase the surface area, thereby enhancing chemical reactions between the test gas and the oxide surface [25,26]. Our results, obtained with a calcination temperature of 600 °C and a muffle residence time of only five hours, compare well with those reported for ZnSb2O6 synthesized by the coprecipitation (or wet chemistry) method [8,11,13]. For example, Junming Li et al. [8] obtained the crystalline phase of $ZnSb_2O_6$ by a facile coprecipitation method and post-annealing at 800 °C. Michel et al. [13] reported the crystalline phase of $ZnSb_2O_6$, ZnO, and SbO_2 at 500 °C, employing the colloidal method. According to reference [11], the material was obtained by heating it in a furnace at 900 °C for six hours in air. Therefore, our method favored the production of phase powders at low temperatures and with crystallite sizes of the nanometer order (42 nm).

3.2. UV Analysis

Figure 3 shows the diffuse reflectance UV–vis spectrum of $ZnSb_2O_6$ along with Tauc's plot which was used to estimate the bandgap of the oxide. The diffuse reflectance spectrum is commonly employed to study the optical properties of materials in powder form. In this case, wavelengths ranging from 200 to 800 nm were employed for the analysis.

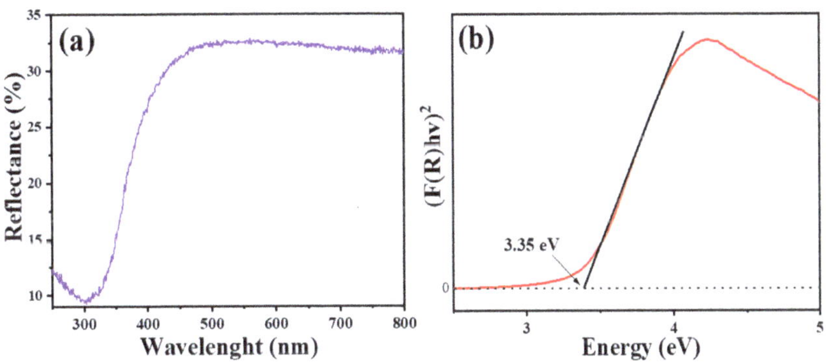

Figure 3. (**a**) Reflectance spectrum of $ZnSb_2O_6$ obtained at 600 °C, (**b**) graph to determine its bandgap value.

To determine the bandgap, the Kubelka–Munk function was applied to convert diffuse reflectance data into an absorption coefficient (α), allowing the use of Tauc's method as follows:

$$F(R) = \frac{(1-R)^{\frac{1}{n}}}{2R} \tag{2}$$

where R is the reflectance and n is a factor that depends on the type of electronic transition, equal to 1/2 for direct transitions and 2 for indirect transitions. By plotting $(F(R)hv)^2$ vs. energy (E) we estimated the bandgap [27]: 3.35 eV. Bandgap values for gas sensors typically range from 1.0 to 3.5 eV [11,17].

3.3. Raman Analysis

The Raman spectrum of the oxide is depicted in Figure 4. In general, previous studies have reported that the $ZnSb_2O_6$ crystallizes in a trirutile-type structure with space group P42/mnm (D_{14}^{4h}). The Raman spectrum exhibits vibrations in the energy range of 200–800 cm^{-1}, indicating the single-phasic nature of the products [10,16]. The bands appearing in the range of 500–800 cm^{-1} are characteristic of vibrational modes associated with the Sb_2O_{10} bonds [16]. In the range 600–800 cm^{-1}, the stretching modes of the Sb–Op bonds predominate [10]. The deformation modes of the Sb–Op bonds, coupled to the vibrations of the Zn–O bonds, are dominant in the range 400–600 cm^{-1} [10,16]. This analysis supports the results obtained by XRD.

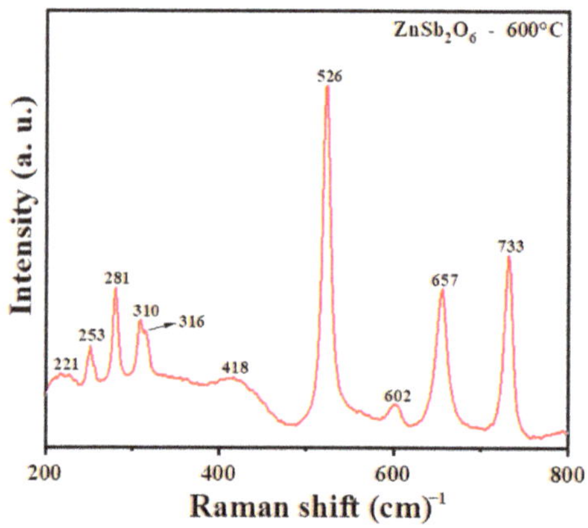

Figure 4. Raman spectrum of $ZnSb_2O_6$ with a trirutile-type structure calcined at 600 °C.

3.4. SEM Analysis

Figure 5 shows scanning electron microscopy (SEM) images of $ZnSb_2O_6$ powders calcined at 600 °C. Figure 5a,b reveal the growth of well defined microstructures, with a rod-like agglomerate morphology, having an approximate diameter of 3.7 µm. The histogram in Figure 5e was generated using ImageJ 1.52a software. It is based on 200 measurements taken from images in Figure 5a,b. This histogram depicts the length size distribution of the rods within the sample, whose average length was 57 µm. Additionally, porosity and surface area measurements obtained by using BET equipment indicate that the material possesses a mesoporous structure [28], with a total pore volume of 0.0831 cm^3/g, predominantly consisting of pores in the range of 10 to 40 nm, and a specific surface area of 6 m^2/g. The porous nature resulting from the arrangement of the microrods (Figure 5a,b) suggests the oxide's potential applications as a gas sensor, as porosity enhances gas diffusion over a larger active surface area, thereby improving sensing performance [29]. Conversely, Figure 5c,d show the formation of microstructures in the form of compact cuboids, averaging approximately 75 µm on each side. It is essential to clarify that all these microstructures correspond to the same sample, and the variations in morphologies are due to the specific synthesis conditions, which highlight the importance of the experimental parameters. The morphology of these structures, as shown in the zoomed-in image in Figure 5d and supported by the BET measurements, indicates a porosity that increases the surface-to-volume ratio. This characteristic facilitates gas diffusion and adsorption, which are essential for sensing and photocatalytic applications.

According to the literature, ethylenediamine plays a key role in the formation of microstructures such as microrods and microcuboids, primarily due to its function as a coordination agent [30,31]. Its ability to bind to metal ions, like Zn^{2+} in our system, through its two nitrogen atoms, allows it to chelate these ions effectively. This chelating action is crucial because it enables ethylenediamine to regulate the metal ion concentration within the reaction, thereby exerting significant control over the nucleation and growth processes that determine the final microstructure morphology [31].

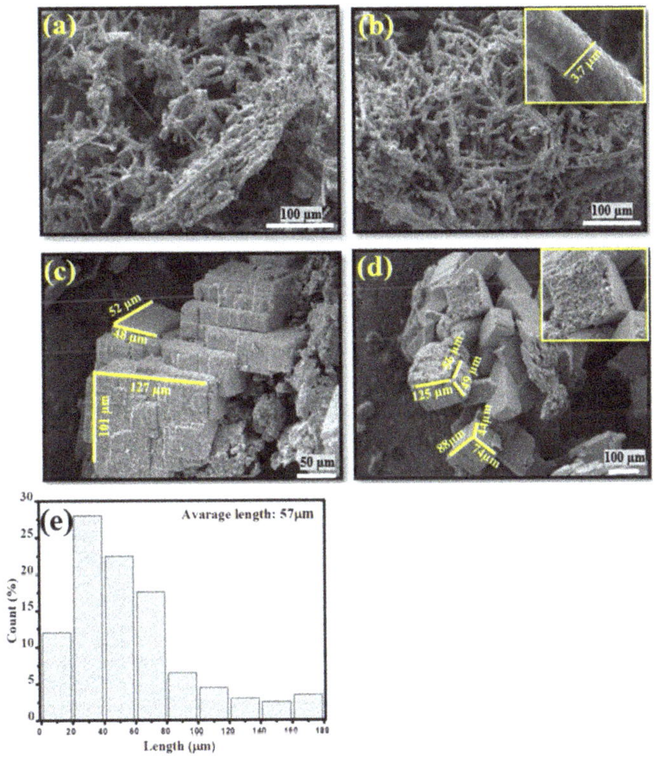

Figure 5. (**a**,**b**) SEM micrographs showing microrods, (**c**,**d**) microcuboids, and (**e**) length distribution of the microrods.

The formation of microrods analogous to those observed in this study (Figure 5a,b) has been documented in the literature. For example, Liu et al. [30] reported the growth of rod-shaped microstructures achieved by varying the concentration of ethylenediamine. Deng et al. [32] observed the formation of microstructures with rectangular morphologies, like those obtained in this work (see Figure 5c,d), attributing these specific morphologies to the use of ethylenediamine. They proposed that ethylenediamine influences the nucleation and growth mechanisms during synthesis, thereby determining the resulting microstructure. These findings suggest that the concentration of ethylenediamine plays a crucial role in controlling the morphological characteristics of the synthesized materials [33].

The surface area of a material is a fundamental parameter in gas sensing and photocatalytic applications. A higher surface area provides more active sites for molecule adsorption, thereby facilitating greater interaction between the target analyte and the sensor material. As shown, microrods and microcubes, morphologies that can enhance the active surface area, were obtained in this work. According to the literature, such cuboidal microstructures can improve sensor response. For example, in [34], the authors demonstrated that cubic-shaped Fe_2O_3 structures enhance the detection of organic vapors. In [35], an increased sensing response was reported for Co_3O_4 cuboid structures in ethanol atmospheres. Additionally, in [36], ZnO microrods showed a high response to CO, ethanol, and acetaldehyde. These findings highlight the significance of microstructural morphology, including microrods and microcubes, in enhancing the sensing performance of materials.

3.5. Gas Sensing Tests

To investigate the potential application of zinc antimonate powders calcined at 600 °C as sensors for CO and C_3H_8 atmospheres, measurements were conducted using the two-point probe method with direct current (DC) signals. For these tests, the concentrations used were 1, 50, 100, 200, and 300 ppm for CO, and 1, 5, 50, 100, 200, 300, 400, and 500 ppm for the C_3H_8. The experiments were performed at temperatures of 100, 200, and 300 °C for both gases. The sensing tests in static CO and C_3H_8 atmospheres involved four basic steps. In the first step, the pellets fabricated with powders calcined at 600 °C were placed inside the chamber, which was equipped with two electrodes (a two-point probe), positioned on the ohmic contacts on the pellets' surface. As a second step, the chamber was evacuated to a high vacuum for the electrical tests. The third step involved conducting the experiments in air at temperatures of 100, 200, and 300 °C. At each temperature, the pellets were allowed to stabilize for 5 min, after which the different concentrations of the test gases were alternately injected. Upon contact between the pellet surface and the test gases, changes in electrical resistance were recorded at each applied temperature. Finally, in the fourth step, the response was calculated as a function of gas concentration and operating temperature using the following equation [37–39]:

$$\text{Response} = \frac{R_O}{R_{gas}} \tag{3}$$

where R_O is the chamber's reference electrical resistance in vacuum and R_{gas} is the electrical resistance in the presence of the test gas, with the experiments conducted in static CO and C_3H_8 atmospheres at 100, 200, and 300 °C. The results are shown in Figures 6 and 7.

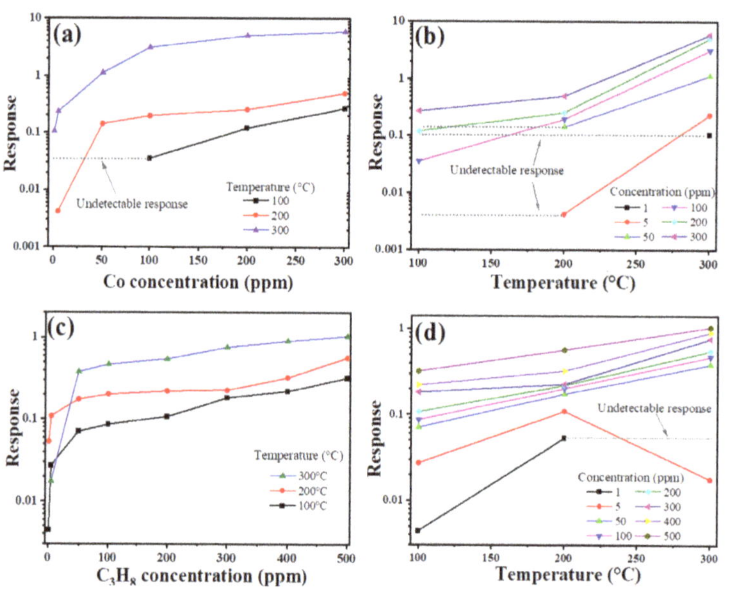

Figure 6. (**a**) Response of the zinc antimonate ($ZnSb_2O_6$) pellets as a function of CO concentration, (**b**) response in CO as a function of temperature, (**c**) response as a function of C_3H_8 concentration, (**d**) response in C_3H_8 as a function of temperature.

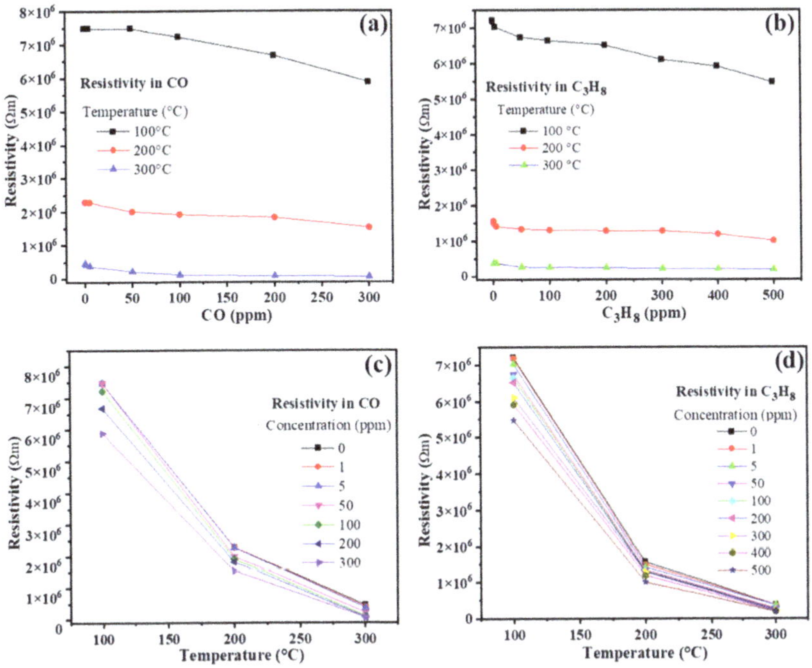

Figure 7. Electrical resistivity variation of zinc antimonate ($ZnSb_2O_6$) pellets as a function of (**a**) CO concentration at various temperatures, (**b**) C_3H_8 concentration at various temperatures, (**c**) temperature for various CO concentrations, (**d**) temperature for various C_3H_8 concentrations.

According to Figure 6, measurements at 100 °C showed that the pellets had a maximum response of 0.27 in 300 ppm of CO (Figure 6a,b), while a maximum response of 0.32 was recorded in 500 ppm of C_3H_8 (Figure 6c,d). We observed that increasing the temperature to 200 °C significantly raised the compound's maximum response to 0.5 for 300 ppm of CO (Figure 6a). Meanwhile, in C_3H_8 atmospheres, a maximum response of 0.56 was obtained (Figure 6c,d). The increasing response trend of $ZnSb_2O_6$ was corroborated when the operating temperature was raised to 300 °C for both test gases (Table 1). It is essential to mention that the undetectable responses shown in Figure 6 were due to the resistance changes being negligible; therefore, the multimeter used could not detect them.

Table 1. Electrical characterization results of $ZnSb_2O_6$ pellets.

CO (300 ppm)				C_3H_8 (500 ppm)			
Temp. (°C)	**Resistance (MΩ)**	**Resistivity (Ω·m)**	**Sensor Response**	**Temp. (°C)**	**Resistance (MΩ)**	**Resistivity (Ω·m)**	**Sensor Response**
23	250	7.95×10^6	0	23	250	7.95×10^6	0
100	185	5.88×10^6	0.27	100	172	5.47×10^6	0.32
200	48.1	1.52×10^6	0.5	200	31.4	0.99×10^6	0.56
300	2.1	0.06×10^6	5.86	300	5.7	0.18×10^6	1.04

We verified that as the operating temperature and test gas concentration increased, the pellets' ability to detect both test gases improved substantially, with the best performance observed at 300 °C. The operating temperature plays a crucial role in promoting the chemical reaction between the test gases and the oxygen species (O^-) present on the pellet's surface, leading to an increase in the gas sensing response. The enhanced response of semiconductor oxides at elevated temperatures is widely reported in the literature [40,41]. Therefore, the high response of the oxide is attributed to the oxygen adsorption and desorption processes (O_2^-, O^-, or O^{2-}) on the pellets' surface, enhanced by the rising operating temperature [29,42]. The results are summarized in Table 1.

The maximum responses obtained at 300 °C were 5.86 for 300 ppm of CO and 1.04 for 500 ppm of C_3H_8. As previously discussed, the increase in response is attributed to the enhanced oxygen adsorption and desorption process at higher temperatures [43–45]. The elevated temperature promotes a strong interaction between the test gas molecules and highly reactive oxygen species (O^- and O^{2-} [46–48]) on the pellets' surface. It has been documented that these reactive oxygen species cause the oxidation of test gases, contributing to an increase in sensitivity [19,49,50], which aligns with our results (Figure 6).

In general, at 300 °C, a physicochemical equilibrium exists between the adsorption of CO and the previously adsorbed oxygen species. It has been reported that at temperatures below 300 °C, the sensor's response decreases significantly because the operating temperature is insufficient to promote redox reactions between reducing gases like CO and the oxygen on the sensor's surface [40,42,44]. But, at temperatures above 300 °C, the response could also decrease [38,45]. This phenomenon is primarily attributed to the following: (1) a reduction in available catalytic sites due to the desorption of oxygen species from the sensor's surface, (2) possible sensor instability at elevated temperatures, (3) a decreased interaction time between molecules due to the greater desorption of the target gas (CO) [29,43,45]. The excellent response to CO (5.86 at 300 ppm) at 300 °C is attributed to a higher efficiency in the reaction kinetics between the reducing gas and the oxygen species (O^- and O^{2-}) adsorbed onto the material's surface [44,45]. The equilibrium established between these gases favors greater charge transfer and, consequently, a larger change in the sensor's response. Furthermore, the mobility of charge carriers in the material is notably increased, which contributes significantly to an improvement in the electrical signal due to the gas–solid interactions that take place [45].

On the other hand, a crucial parameter for the application of a semiconductor oxide as a gas sensor is the variation in electrical resistivity (ρ). We calculated the resistivity as a function of the gas concentration and the operating temperature using the following expression [51]:

$$\rho = \frac{RA}{t} \tag{4}$$

where R is the pellets' electrical resistance in the test gases, A is the pellets' cross-sectional area, and t is their thickness (=2 mm; diameter = 9 mm). Figure 7 shows the variation in resistivity as a function of CO and C_3H_8 concentrations and operating temperature.

As depicted in Figure 7, and as expected, at 100 °C, the material exhibits high electrical resistivity, regardless of the test gas concentration (Figure 7a,b). Upon increasing the temperature to 200 °C, a decrease in resistivity was recorded for both gases. At 100 and 200 °C, the resistivity remains relatively linear. It shows little dependence on the concentration of either gas, which is attributed to the temperature having the dominant effect on the resistivity variation [48]. This was corroborated by increasing the temperature to 300 °C, at which the most significant decrease in resistivity was observed (Figure 7c,d).

The observed trends are attributed mainly to the fact that with an increasing temperature, the mobility of charge carriers (electrons) is enhanced due to the available higher energy at the pellet's surface, which contributes to an increase in the material's conductivity as its resistivity decreases [33,42–44,47].

The temperature-enhanced oxygen adsorption, which subsequently reacts with the test gases on the pellets' surface, causes a decrease in resistivity and, consequently, an increase in the material's response. These trends are commonly reported for ternary semiconductor oxides used as gas sensors, particularly for trirutile-type antimonate oxides [19,37,48,52,53]. The values for resistivity are listed in Table 1.

3.6. Reaction Mechanism

The mechanism explaining the response of zinc antimonate in CO and C_3H_8 atmospheres is based on the chemical reaction on the pellets' surface between the test gases and the oxygen species (O_2^-, O^-, or O^{2-}) present at temperatures above 150 °C. For an n-type semiconductor like $ZnSb_2O_6$, an electron transfer is induced on the material's surface in CO and C_3H_8 atmospheres, driven by the oxygen adsorption–desorption phenomenon [42]. The literature reports that this transfer of charge carriers (electrons) is reflected in changes in electrical resistivity (or electrical response), which depends on both the reaction between the adsorbed oxygen and the gases [19,37,52], and the movement of free electrons on the pellets' surface. Therefore, at 300 °C in high vacuum, oxygen species adsorb onto the material due to the temperature, trapping electrons from the conduction band [19,43,44,49], a process that ultimately results in an increased response from the $ZnSb_2O_6$.

Chemical reactions that occur during the chemisorption of CO and C_3H_8 at 300 °C on the surface of a semiconductor have been widely studied by various authors [39–41,46,54]. In our CO experiments, when the gas was injected into the measurement system, it immediately reacted with the oxygen on the surface of the pellets, producing CO_2 and releasing electrons. According to the literature, a possible chemical reaction for the chemisorption of CO at 300 °C on the $ZnSb_2O_6$ pellets is as follows [43,44]:

$$\begin{aligned} 2CO + O_2^-(ads) &\rightarrow 2CO_2 + e^- \\ CO + O^-(ads) &\rightarrow CO_2 + e^- \\ CO + O^{2-}(ads) &\rightarrow CO_2 + 2e^- \end{aligned} \quad (5)$$

where the suffix (ads) represents the adsorbed species [43,44]. These reactions release thermal energy that promotes greater charge carrier displacement, thus increasing the compound's response. A similar reaction may occur during the chemisorption of C_3H_8 at 300 °C. According to the literature, the reaction of adsorbed C_3H_8 molecules with highly reactive oxygen species (O^- and O^{2-} above 150 °C, [33,43,46–48,55]) is the most probable cause for the variation in electrical resistance and increased response to C_3H_8 at moderate temperatures (300 °C in our case). As the operating temperature increased, the highly reactive oxygen ions exhibited an enhanced solid–gas interaction [46], which favored an increase in the sensor's response.

According to references [29,44,45], the adsorption of oxygen ions that react with the test gas leads to faster oxidation at 300 °C. The oxidation reaction that may occur is reported in references [19,37,56] and is as follows:

$$C_3H_8 + 10O^- \rightarrow 3CO_2 + 4H_2O + 10e^- \quad (6)$$

which indicates that when C_3H_8 molecules are adsorbed on the pellets' surface, they dissociate before reacting with the ionosorbed oxygen [46], producing CO_2, H_2O, and free electrons [56], which favors an increase in the pellets' response, as observed in Figures 6 and 7. Such results are commonly reported in the literature for binary (ZnO and SnO) [39,40,57] or ternary oxides with perovskite ($LaCoO_3$, $LaFeO_3$) [58], spinel ($ZnAl_2O_4$, $AlCo_2O_4$) [46,59], and trirutile ($MgSb_2O_6$, $ZnSb_2O_6$) [19,37,47] structures.

Beyond its promising capabilities in gas detection, the bifunctional nature of the synthesized $ZnSb_2O_6$ was further explored. The material's calculated bandgap of 3.35 eV indicates activation by UV light, while its high surface area morphology is ideal for facilitating surface reactions. Therefore, to fully characterize its potential for environmental remediation applications, the material's photocatalytic performance in degrading organic dyes was systematically evaluated.

3.7. Photocatalytic Tests

The photocatalytic performance of $ZnSb_2O_6$ pellets was evaluated using photodecolorization tests with aqueous solutions of dyes as model pollutants: malachite green (MG), methylene blue (MB), and methyl orange (MO). To determine the concentration of the dyes in the samples, calibration curves were constructed by measuring the absorbance at specific wavelengths corresponding to the maximum absorption of each dye (664 nm for MB, 617 nm for MG, and 465 nm for MO).

The calibration curves were generated using serial dilutions of standard solutions, covering a relevant concentration range for each dye. The resulting correlation coefficients (R) demonstrated excellent linearity, with values of 0.9933 for malachite green (MG), 0.9974 for methylene blue (MB), and 0.9947 for methyl orange (MO), indicating strong linear relationships between absorbance and concentration. Figure 8 presents the calibration curves for each dye, along with their respective linear equations, which were used to accurately quantify the dye concentrations in the samples.

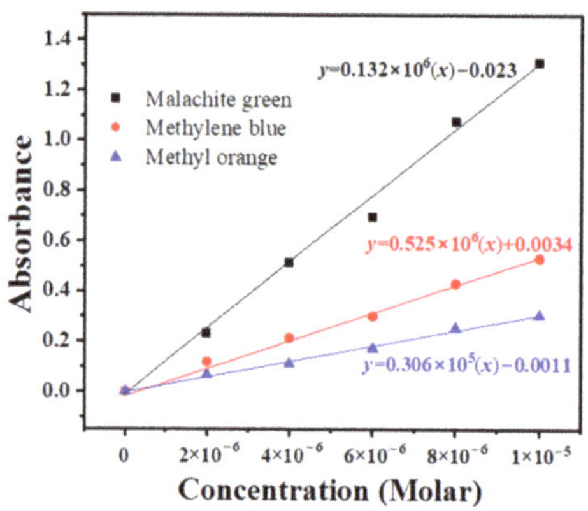

Figure 8. Calibration curve for aqueous solutions of malachite green (MG), methylene blue (MB), and methyl orange (MO).

As shown in Figure 9a–c, after a 30 min dark adsorption period, the $ZnSb_2O_6$ pellets adsorbed approximately 1% of MB, 9% of MG, and about 11% of MO. These values reflect the adsorption effects, which help distinguish the photocatalytic contribution observed in the shaded area of Figure 9. When exposed to UV light irradiation, the concentration of MG decreased significantly, indicating effective photocatalytic degradation. Specifically, up to 50% of MG was decolorized after 300 min of irradiation, and approximately 60% of MB was degraded, while methyl orange showed no notable degradation throughout the experiment. These results demonstrate the promising photocatalytic potential of $ZnSb_2O_6$, similar to other antimoniates that have been investigated in photocatalytic applications [9–12].

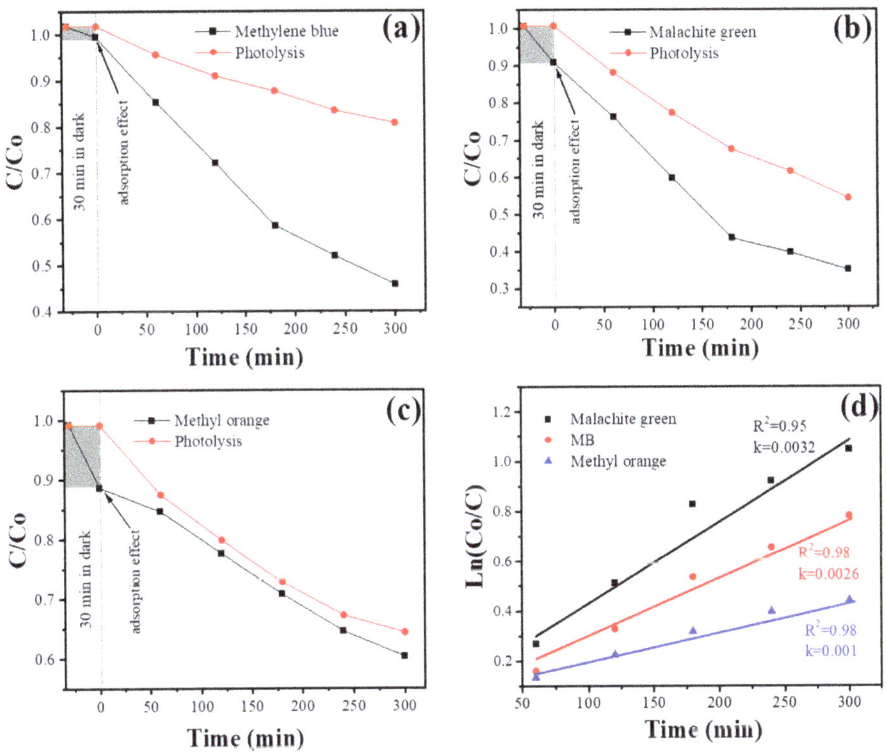

Figure 9. Photocatalytic discoloration of (**a**) methylene blue, (**b**) malachite green, (**c**) methyl orange, and (**d**) kinetic model of discoloration.

Table 2 presents a literature review of the photocatalytic performance of $ZnSb_2O_6$, comparing our results with those in the literature. As observed, our results demonstrate competitive activity, highlighting the potential and versatility of $ZnSb_2O_6$ for photocatalytic applications. Table 2 also indicates that the use of multiple light sources, particularly multiple UV sources, correlates with higher degradation efficiencies. Moreover, the incorporation of dopants into $ZnSb_2O_6$ consistently leads to enhanced photocatalytic performance, likely due to improved charge separation and increased carrier mobility. Therefore, it is essential to emphasize the critical roles of light source energy and material modification in optimizing photocatalytic efficiency. Consequently, further investigation into these parameters will be undertaken in the future to maximize the performance of our results.

Table 2. Literature review of the photocatalytic performance of $ZnSb_2O_6$.

Light Source	Analyte	Doping	Degradation Percentage (%)	Degradation Time (min)	Reference
6 UV lamps	Rhodamine b	-	~95	180	[10]
6 UV lamps	Rhodamine b	N	~97	80	
4 UV lamps	Rhodamine b	-	~95	95	[11]
4 UV lamps	Rhodamine b	Tb/Eu	~95	95	
Tungsten lamp	Methyl violet	-	~40	180	[17]
Tungsten lamp	Methyl violet	Carbon	~92	180	

Table 2. *Cont.*

Light Source	Analyte	Doping	Degradation Percentage (%)	Degradation Time (min)	Reference
Xe lamp	Rhodamine b	-	~10	120	[60]
Xe lamp	Rhodamine b	N	~70	120	
Full arc Xe lamp	Rhodamine b	-	~10	120	
Full arc Xe lamp	Rhodamine b	N	~95	120	
4 UV lamps	Methyl orange	-	~80	180	[61]
4 UV lamps	Rhodamine b	-	~99	60	
UV lamp	Methylene blue	-	~60	300	This work
UV lamp	Malachite green	-	~50	300	

Materials with photocatalytic properties possess specific key properties and parameters that enable them to catalyze chemical reactions under the action of light. These parameters determine their efficiency in processes such as the photodegradation of pollutants. Among the main parameters are bandgap, light absorption capacity, and surface area [62,63]. According to the literature, antimoniates with a trirutile crystalline structure have demonstrated good light absorption capacity [64,65], and their bandgap values typically range between 2 and 3.5 eV, making them suitable for photocatalytic applications [64]. In our case, we obtained a bandgap value of 3.35 eV. Additionally, the relative ease of controlling the morphology and, consequently, the surface area of these materials, makes them promising candidates for photocatalytic applications [9–12]. The observed microrod and microcuboid morphologies (Figure 5) suggest a high surface area and porosity, which is supported by BET measurements. This increased surface area plays a key role in the photocatalytic activity of the synthesized $ZnSb_2O_6$.

To describe the kinetics of the photo-discoloration process and estimate the kinetic parameter k_{ap}, the following first-order kinetic model was applied [61,66]:

$$\ln\frac{C_0}{C} = k_{ap}t \tag{7}$$

where k_{ap} is the apparent rate constant, C_0 denotes the initial dye concentration, and C represents the dye concentration at any irradiation time t. Figure 9d shows that the photocatalytic reaction of $ZnSb_2O_6$ pellets fits well with a first-order model. The k_{ap} for MB, MG, and MO were 0.0026, 0.0032, and 0.001, respectively.

The estimated bandgap for the synthesized $ZnSb_2O_6$ powders was 3.35 eV, indicating that the material requires radiation with wavelengths equal to or less than 372 nm to initiate photocatalytic reactions. Consequently, the synthesized $ZnSb_2O_6$ powders are activated by UV light. This activation energy promotes the formation of electron–hole pairs, which are primarily responsible for generating highly reactive oxidizing species such as hydroxyl radicals ($^{\bullet}OH$) and superoxide anions ($O_2^{\bullet-}$). These radicals play a crucial role in the degradation of pollutants [67].

The photocatalytic reactions involve a series of processes initiated by the absorption of UV radiation in $ZnSb_2O_6$, leading to the generation of electron–hole pairs (see Equation (8)). The holes (h^+) oxidize water, producing $^{\bullet}OH$ radicals (Equation (9)), while the electrons (e^-) reduce dissolved oxygen, forming $O_2^{\bullet-}$ radicals (Equation (10)). These reactive species participate in subsequent reactions: $O_2^{\bullet-}$ radicals can react with H^+ to generate more radicals (Equation (11)), facilitating chain reactions that promote the mineralization of organic contaminants. Additionally, $HO_2^{\bullet}$, and $O_2^{\bullet-}$ radicals react to form hydrogen peroxide H_2O_2 (Equation (12)), which can be reduced by electrons to produce more $^{\bullet}OH$ radicals (Equation (13)). Furthermore, OH^- radicals can interact with holes (h^+)

to form more $^{\bullet}OH$ radicals (Equation (14)). As mentioned, collectively, these processes lead to the degradation and mineralization of organic pollutants [9,66,68,69], such as the dyes studied in this work.

$$ZnSb_2O_6 + hv \rightarrow e^- + h^+ \quad (8)$$

$$h^+ + H_2O \rightarrow {}^{\bullet}OH + H^+ \quad (9)$$

$$O_2 + e^- \rightarrow O_2^{\bullet-} \quad (10)$$

$$O_2^{\bullet-} + H^+ \rightarrow HO_2^{\bullet} \quad (11)$$

$$HO_2^{\bullet} + O_2^{\bullet-} \rightarrow H_2O_2 + O_2 \quad (12)$$

$$H_2O_2 + e^- \rightarrow {}^{\bullet}OH + OH^- \quad (13)$$

$$OH^- + h^+ \rightarrow {}^{\bullet}OH \quad (14)$$

4. Discussion and Limitations

While this study successfully demonstrates the synthesis of $ZnSb_2O_6$ and its promising bifunctional properties, it is essential to acknowledge certain limitations that provide avenues for future research.

A key parameter for any practical gas sensor is its selectivity—the ability to respond to a target gas in the presence of other interfering gases. In this study, while the sensor demonstrated a high sensitivity to CO and C_3H_8, its response to other common reducing gases (e.g., H_2, ethanol) or oxidizing gases (e.g., NO_2) was not evaluated. Future work should focus on comprehensive selectivity tests to determine the practical applicability of $ZnSb_2O_6$. Furthermore, long-term stability, response, and recovery times are critical for real-world performance. The current study provides a snapshot of the material's sensitivity, but further investigations are needed to assess its stability over extended periods and multiple sensing cycles. Quantifying the response and recovery times would also provide a more complete picture of the sensor's dynamic performance.

On the other hand, in the photocatalytic evaluation, the material effectively degraded the cationic dyes malachite green and methylene blue. However, it showed no notable degradation of the anionic dye, methyl orange. This suggests that surface charge interactions may influence the photocatalytic activity. It is plausible that the surface of $ZnSb_2O_6$ under the experimental conditions is negatively charged, leading to the electrostatic repulsion of the anionic methyl orange molecules. This hypothesis warrants further investigation through zeta potential measurements and testing with other anionic pollutants. Additionally, to better contextualize the photocatalytic efficiency, a direct comparison against a standard benchmark material, such as TiO_2 P25, under identical experimental conditions would be highly valuable. Additionally, the material can be further optimized through the strategies explored in recent electrocatalytic studies [70], which have demonstrated that enhancing oxidation reaction efficiency by controlling the interaction between metal centers and modifying the electronic structure can significantly enhance the catalytic mechanism in water oxidation and, consequently, improve photocatalysis. This would enable a more objective evaluation of the material's performance regarding the current state-of-the-art techniques.

5. Conclusions

This study successfully demonstrated the synthesis of $ZnSb_2O_6$ pellets using a microwave-assisted wet chemistry method, resulting in microstructures with high homogeneity and a low production cost. Structural characterization confirmed the formation of a tetragonal trirutile phase of the $ZnSb_2O_6$ semiconductor. UV–vis spectroscopy revealed a bandgap of 3.35 eV, indicating that the material is activated by UV light. Morphological analysis revealed porous microstructures in the form of microrods and cuboids. Additionally, porosity and surface area measurements obtained using BET equipment indicated that the material possesses a mesoporous structure, with a total pore volume of 0.0831 cm^3/g, predominantly consisting of pores in the range of 10 to 40 nm, and a specific surface area of 6 m^2/g. These characteristics contribute significantly to the material's effectiveness in gas sensing and photocatalytic activity. The gas sensing tests indicated that $ZnSb_2O_6$ pellets exhibit a high sensitivity to CO and C_3H_8, with maximum responses of 5.86 and 1.04 at 300 °C. The response increased with both temperature and gas concentration, attributable to the enhanced adsorption and desorption processes of surface oxygen species at elevated temperatures. The observed decrease in electrical resistivity with rising temperature further corroborates the effective material charge transfer mechanisms during gas detection. Photocatalytic evaluations demonstrated that $ZnSb_2O_6$ degrades organic dyes such as malachite green and methylene blue under UV irradiation, following first-order kinetics. The bandgap and surface properties facilitate the generation of reactive oxygen species, which supports its photocatalytic activity. Overall, the synthesized $ZnSb_2O_6$ exhibits promising properties for dual applications in toxic gas sensing and photocatalysis, combining low-cost synthesis and functional performance. These findings support the further development and optimization of $ZnSb_2O_6$-based devices for environmental monitoring and the degradation of water pollutants.

Author Contributions: Conceptualization, J.M.-B., H.G.-B., L.I.J.-A., A.G.-B., V.-M.R.-B., J.A.R.-O., J.T.G.-B., and M.d.l.L.O.-A.; methodology, H.G.-B., V.-M.R.-B., and J.A.R.-O.; software, A.G.-B., and J.T.G.-B.; validation, H.G.-B., A.G.-B., V.-M.R.-B., and J.T.G.-B.; formal analysis, L.I.J.-A., A.G.-B., J.M.-B., M.d.l.L.O.-A., J.T.G.-B., and J.A.R.-O.; investigation, H.G.-B., J.M.-B., J.A.R.-O., V.-M.R.-B., L.I.J.-A., and A.G.-B.; resources, H.G.-B., A.G.-B., V.-M.R.-B., and J.T.G.-B.; data curation, H.G.-B., A.G.-B., and J.T.G.-B.; writing—original draft preparation, J.M.-B., J.A.R.-O., M.d.l.L.O.-A., and H.G.-B.; writing—review and editing, H.G.-B., J.M.-B., V.-M.R.-B., A.G.-B., L.I.J.-A., J.T.G.-B., and J.A.R.-O.; visualization, H.G.-B., J.M.-B., V.-M.R.-B., A.G.-B., L.I.J.-A., J.T.G.-B.; supervision, H.G.-B., A.G.-B., V.-M.R.-B., and J.T.G.-B.; project administration, H.G.-B., A.G.-B., V.-M.R.-B., and J.T.G.-B.; funding acquisition, H.G.-B., and V.-M.R.-B. All authors have read and agreed to the published version of the manuscript.

Funding: This research received no external funding.

Acknowledgments: The authors thank Mexico's Secretary of Science, Humanities, Technology and Innovation (Secihti), and the University of Guadalajara for the support granted. Jacob Morales-Bautista thanks Secihti for the scholarships received. Likewise, we thank Darío Pozas Zepeda, Víctor-Manuel Soto–García, René Morán Salazar, Sergio Gómez Salazar, and Miguel–Ángel Luna-Arias for their technical assistance. This research was carried out following the line of research "Nanostructured Semiconductor Oxides" of the academic group UDG–CA–895 "Nanostructured Semiconductors" of CUCEI, University of Guadalajara.

Conflicts of Interest: The authors declare no conflicts of interest.

References

1. Roper, A.; Leverett, P.; Murphy, T.; Williams, P. Stabilities of byströmite, $MgSb_2O_6$, ordoñezite, $ZnSb_2O_6$ and rosiaite, $PbSb_2O_6$, and their possible roles in limiting antimony mobility in the supergene zone. *Mineral. Mag.* **2015**, *79*, 537–544. [CrossRef]
2. Webmineral. Ordonezite. Available online: https://www.webmineral.com/data/Ordonezite.shtml (accessed on 21 May 2025).
3. Jackson, A.J.; Parrett, B.J.; Willis, J.; Ganose, A.M.; Leung, W.W.; Liu, Y.; Williamson, B.A.D.; Kim, T.K.; Hoesch, M.; Veiga, L.S.I.; et al. Computational Prediction and Experimental Realization of Earth-Abundant Transparent Conducting Oxide Ga-Doped $ZnSb_2O_6$. *ACS Energy Lett.* **2022**, *7*, 3807–3816. [CrossRef]
4. Dąbrowska, G. Reactivity of $ZnSb_1O_6$ with $ZnTa_2O_6$ and some properties of new limited $ZnSb_{2-x}Ta_xO_6$ solid solution with tri-rutile structure. *Thermochim. Acta* **2015**, *614*, 62–67. [CrossRef]
5. Hussoyn, E.; Repelin, Y.; Brussrt, H.; Cerez, A. Spectres de vibration et calcul du champ de force des antimoniates et des tantalates de structure trirutile. *Spectrochim. Acta* **1979**, *35A*, 1177–1187. [CrossRef]
6. Haeuseler, H. Infrared and Raman spectra and normal coordinate calculations on trirutile-type compounds. *Spectrochim. Acta Part A* **1981**, *37*, 487–495. [CrossRef]
7. Balasubramaniam, M.; Balakumar, S. Nanostructuring of a one-dimensional zinc antimonate electrode material through a precipitation strategy for use in supercapacitors. *New J. Chem.* **2018**, *42*, 6613–6616. [CrossRef]
8. Li, J.; Du, K.; Lai, Y.; Chen, Y.; Zhang, Z. $ZnSb_2O_6$: An advanced anode material for Li-ion batteries. *J. Mater. Chem. A* **2017**, *5*, 10843–10848. [CrossRef]
9. Nagarajan, A.; Naraginti, S. Facile synthesis of N-$MgSb_2O_6$ trirutile antimonate and its enhanced photocatalytic performance. *Int. J. Environ. Anal. Chem.* **2020**, *102*, 7938–7952. [CrossRef]
10. Nagarajan, A.; Vijayaraghavan, R. Enhanced photocatalytic activity of nanocrystalline N-doped $ZnSb_2O_6$: Role of N doping, cation ordering, particle size and crystallinity. *RSC Adv.* **2014**, *4*, 65223–65231.
11. Dutta, D.P.; Ballal, A.; Singh, A.; Fulekar, M.H.; Tyagi, A.K. Multifunctionality of rare earth doped nano $ZnSb_2O_6$, $CdSb_2O_6$ and $BaSb_2O_6$: Photocatalytic properties and white light emission. *Dalton Trans.* **2013**, *42*, 16887–16897. [CrossRef] [PubMed]
12. Nagarajan, A. Band gap engineering and photocatalytic activity of new trirutile structure $Zn_{1-x}Mg_xSb_2O_6$ ($0 < x < 1$) solid solution. *Indian J. Chem.* **2021**, *60A*, 220–227.
13. Michel, C.R.; Lopez Contreras, N.L.; Lopez-Alvarez, M.A.; Martinez-Preciado, A.H. Gas selectivity of nanostructured $ZnSb_2O_6$ synthesized by a colloidal method. *Sens. Actuators B Chem.* **2012**, *171–172*, 686–690. [CrossRef]
14. Mizoguchi, H.; Woodward, P.M. Electronic structure studies of main group oxides possessing edge-sharing octahedra: Implications for the design of transparent conducting oxides. *Chem. Mater.* **2004**, *16*, 5233–5248. [CrossRef]
15. Jiang, L.; Dong, C.; Lai, Y.; Chen, Y.; Zhang, Z.; Lv, S.; Wang, J.; Pan, S.; Wang, C.; Sun, P.; et al. Mixed potential type methanol gas sensor based on $Gd_2Zr_2O_7$ solid state electrolyte and $ZnSb_2O_6$ sensing electrode. *Sens. Actuators B Chem.* **2023**, *397*, 134630. [CrossRef]
16. Guillen-Bonilla, H.; Rodríguez-Betancourtt, V.-M.; Guillén-Bonilla, J.-T.; Reyes-Gómez, J.; Gildo-Ortiz, L.; Flores-Martínez, M.; Olvera-Amador, M.L. CO and C_3H_8 Sensitivity Behavior of Zinc Antimonate Prepared by a Microwave-Assisted Solution Method. *J. Nanomater.* **2015**, *2015*, 979543. [CrossRef]
17. Manasa, S.; Gundeboina, R.; Perala, V.; Kammara, V.; Vithal, M. Carbon nanospheres supported visible-light-driven $ZnSb_2O_6$: Synthesis, characterization and photocatalytic degradation studies. *SN Appl. Sci.* **2019**, *1*, 1046.
18. Balasubramaniam, M.; Balakumar, S. Contribution of Tin in Electrochemical Properties of Zinc Antimonate Nanostructures: An Electrode Material for Supercapacitors. *AIP Conf. Proc.* **2018**, *1942*, 140013. [CrossRef]
19. Singh, S.; Singh, A.; Singh, A.; Tandon, P. A stable and highly sensitive room-temperature liquefied petroleum gas sensor based on nanocubes/cuboids of zinc antimonate. *RSC Adv.* **2020**, *10*, 20349–20357. [CrossRef] [PubMed]
20. Ding, W.; Wang, X.; Yang, C.; Wang, P.; Tian, W.; Zhao, K.; Zhang, K. Machine Learning-Motivated Trace Triethylamine Identification by Bismuth Vanadate/Tungsten Oxide Heterostructures. *J. Colloid Interface Sci.* **2025**, *682*, 1140–1150. [CrossRef]
21. Teymourian, H.; Salimi, A.; Hallah, R. An Electrochemical Sensor Based on Gold Nanoparticle Decorated Carbon Nanotubes for Simultaneous Determination of Hydrazine and Hydroquinone. *Int. J. Electrochem. Sci.* **2019**, *14*, 6267–6275.
22. Liu, K.; Xu, Y.; Tian, X.; Zhou, W.; Chen, H.; Zhang, X.; Li, Y.; Wang, M.; Yang, L.; Liu, F. Stacking Growth of Ionically Conductive MOF on Biofabrics Enables Reliable NH_3 Sensor for Hepatic Encephalopathy Diagnosis. *NPJ Flex. Electron.* **2025**, *9*, 67. [CrossRef]
23. Balasubramaniam, M.; Balakumar, S. A review on multifunctional attributes of zinc antimonate nanostructures towards energy and environmental applications. *Chem. Pap.* **2020**, *74*, 55–75. [CrossRef]
24. Tamaki, J.; Yamada, Y.; Yamamoto, Y.; Matsuoka, M.; Ota, I. Sensing properties to dilute hydrogen sulfide of $ZnSb_2O_6$ thick-film prepared by dip-coating method. *Sens. Actuators B Chem.* **2000**, *66*, 70–73. [CrossRef]
25. Biswas, A.; Bayer, I.S.; Biris, A.S.; Wang, T.; Dervishi, E.; Faupel, F. Advances in top–down and bottom–up surface nanofabrication: Techniques, applications & future prospects. *Adv. Colloid Interface Sci.* **2012**, *170*, 2–27. [CrossRef]

26. Dhand, C.; Dwivedi, N.; Loh, X.J.; Jie Ying, A.N.; Verma, N.K.; Beuerman, R.W.; Lakshminarayanan, R.; Ramakrishna, S. Methods and strategies for the synthesis of diverse nanoparticles and their applications: A comprehensive overview. *RSC Adv.* **2015**, *5*, 105003–105037. [CrossRef]
27. Anbuvannan, M.; Ramesh, M.; Viruthagiri, G.; Shanmugam, N.; Kannadasan, N. Synthesis, characterization and photocatalytic activity of ZnO nanoparticles prepared by biological method. *Spectrochim. Acta Part A Mol. Biomol. Spectrosc.* **2015**, *143*, 304–308. [CrossRef] [PubMed]
28. Sing, S.W. Characterization of Porous Solids: An Introductory Survey. *Stud. Surf. Sci. Catal.* **1991**, *62*, 1–9.
29. Lin, T.; Lv, X.; Li, S.; Wang, Q. The Morphologies of the Semiconductor Oxides and Their Gas-Sensing Properties. *Sensors* **2017**, *17*, 2779. [CrossRef]
30. Qi, X.; Zhu, X.; Wu, J.; Wu, Q.; Li, X.; Gu, M. Controlled synthesis of $BiVO_4$ with multiple morphologies via an ethylenediamine-assisted hydrothermal method. *Mater. Res. Bull.* **2014**, *59*, 435–441. [CrossRef]
31. Mo, M.-S.; Wang, D.; Du, X.; Ma, J.; Qian, X.; Chen, D.; Qian, Y. Engineering of nanotips in ZnO submicrorods and patterned arrays. *Cryst. Growth Des.* **2009**, *9*, 2. [CrossRef]
32. Deng, Z.-X.; Wang, C.; Sun, X.-M.; Li, Y.-D. Structure-directing coordination template effect of ethylenediamine in formations of ZnS and ZnSe nanocrystallites via solvothermal route. *Inorg. Chem.* **2002**, *41*, 869−873. [CrossRef]
33. Gao, X.; Li, X.; Yu, W. Flowerlike ZnO nanostructures via hexamethylenetetramine-assisted thermolysis of zinc-ethylenediamine complex. *J. Phys. Chem. B* **2005**, *109*, 1155–1161. [CrossRef]
34. Shen, X.; Liu, Q.; Ji, Z.; Zhu, G.; Zhou, H.; Chen, K. Controlled synthesis and gas sensing properties of porous Fe_2O_3/NiO hierarchical nanostructures. *CrystEngComm* **2015**, *17*, 5522–5529. [CrossRef]
35. Jamil, S.; Jing, X.; Wang, J.; Li, S.; Liu, J.; Zhang, M. The synthesis of porous Co_3O_4 micro cuboid structures by solvothermal approach and investigation of its gas sensing properties and catalytic activity. *Mater. Res. Bull.* **2013**, *48*, 4513–4520. [CrossRef]
36. Prabhakar Rai, H.; Song, H.-M.; Kim, Y.-S.; Song, M.-K.; Oh, P.-R.; Yoon, J.-M.; Yu, Y.-T. The synthesis and characterization of layered double hydroxides with enhanced thermal stability. *Mater. Lett.* **2012**, *68*, 90–93.
37. Juárez-Amador, L.I.; Guillén-Bonilla, H.; Guillén-Bonilla, A.; Guillén-Bonilla, J.T.; Morales-Bautista, J.; Casillas-Zamora, A.; Rodríguez-Betancourtt, V.M.; Olvera-Amador, M.L. Photocatalytic and sensing properties in propane atmospheres of $MgSb_2O_6$ nanoparticles synthesized by a chemical method. *J. Mater. Sci. Mater. Electron.* **2024**, *35*, 1857. [CrossRef]
38. Carbajal-Franco, G.; Tiburcio-Silver, A.; Domínguez, J.M.; Sánchez-Júarez, A. Thin film tin oxide-based propane gas sensors. *Thin Solid Films* **2000**, *373*, 141–144. [CrossRef]
39. Gómez-Pozos, H.; Karthik, T.V.; Olvera, M.d.l.L.; Barrientos, A.G.; Pérez Cortés, O.; Vega-Pérez, J.; Maldonado, A.; Pérez-Hernández, R.; Rodríguez-Lugo, V. ZnO thin films as propane sensors: Band structure models to explicate the dependence between the structural and morphological properties on gas sensitivity. *J. Phys. Chem. Solids* **2017**, *106*, 16–28. [CrossRef]
40. Karthik, T.V.K.; Olvera-Amador, M.L.; Maldonado, A.; Hernandez, A.G.; Gómez-Pozos, H. Propane gas-sensing properties of pure and Pd-doped tin oxide nanostructures. *J. Mater. Sci. Mater. Electron.* **2023**, *34*, 228. [CrossRef]
41. Singh, S.; Singh, A.; Singh, A.; Rathore, S.; Yadav, B.C.; Tandon, P. Nanostructured cobalt antimonate: A fast responsive and highly stable sensing material for liquefied petroleum gas detection at room temperature. *RSC Adv.* **2020**, *10*, 33770–33781. [CrossRef]
42. Meena, D.; Jain, M.; Bhatnagar, M.C. Resistive gas sensors based on nanostructured ternary metal oxide: A review. *J. Mater. Sci.* **2024**, *59*, 12177–12218. [CrossRef]
43. Hsiao, C.-C.; Luo, L.-S. A Rapid Process for Fabricating Gas Sensors. *Sensors* **2014**, *14*, 12219–12232. [CrossRef]
44. Baraton, M.-I.; Merhari, L. Electrical behavior of semiconducting nanopowders *versus* environment. *Adv. Mater. Sci.* **2003**, *4*, 15–24.
45. Gao, X.; Zhang, T. An overview: Facet-dependent metal oxide semiconductor gas sensors. *Sens. Actuators B Chem.* **2018**, *277*, 604–633. [CrossRef]
46. Gildo-Ortiz, L.; Rodríguez-Betancourtt, V.-M.; Ramírez Ortega, J.A.; Blanco-Alonso, O. An Alternative Approach for the Synthesis of Zinc Aluminate Nanoparticles for CO and Propane Sensing Applications. *Chemosensors* **2023**, *11*, 105. [CrossRef]
47. Guillen Bonilla, J.T.; Jiménez Rodríguez, M.; Bonilla, H.G.; Bonilla, A.G.; Padilla, E.H.; Morales, M.E.S.; Flores Jiménez, A.B.; Gutiérrez, J.C.E. The Design, Simulation, and Construction of an O_2, C_3H_8, and CO_2 Gas Detection System Based on the Electrical Response of $MgSb_2O_6$ Oxide. *Technologies* **2025**, *13*, 79. [CrossRef]
48. Guillén-Bonilla, A.; Guillén-Bonilla, J.T.; Guillén-Bonilla, H.; Huízar-Padilla, E.; Zamora, A.C.; Olvera Amador, M.d.L.L.; Rodríguez-Betancourtt, V.M. A theoretical and experimental study on the dynamic response in propane atmospheres of sensors made from trirutile magnesium antimonate powders. *J. Mater. Sci. Mater. Electron.* **2024**, *35*, 1849. [CrossRef]
49. Yu, S.; Jia, X.; Zhang, J.; Yang, W.; Song, H. Recent advances in different materials for moisture resistance of metal oxide-based gas sensors: A review. *Chem. Eng. J.* **2025**, *505*, 159639. [CrossRef]

MDPI AG
Grosspeteranlage 5
4052 Basel
Switzerland
Tel.: +41 61 683 77 34

Chemosensors Editorial Office
E-mail: chemosensors@mdpi.com
www.mdpi.com/journal/chemosensors

Disclaimer/Publisher's Note: The title and front matter of this reprint are at the discretion of the Guest Editor. The publisher is not responsible for their content or any associated concerns. The statements, opinions and data contained in all individual articles are solely those of the individual Editor and contributors and not of MDPI. MDPI disclaims responsibility for any injury to people or property resulting from any ideas, methods, instructions or products referred to in the content.

www.ingramcontent.com/pod-product-compliance
Lightning Source LLC
Chambersburg PA
CBHW061815210926
51850CB00006B/715
9783725854738